VECTOR AND TENSOR ANALYSIS

MONOGRAPHS AND TEXTBOOKS IN PURE AND APPLIED MATHEMATICS

1. *K. Yano,* Integral Formulas in Riemannian Geometry (1970)
2. *S. Kobayashi,* Hyperbolic Manifolds and Holomorphic Mappings (1970)
3. *V. S. Vladimirov,* Equations of Mathematical Physics (A. Jeffrey, ed.; A. Littlewood, trans.) (1970)
4. *B. N. Pshenichnyi,* Necessary Conditions for an Extremum (L. Neustadt, translation ed.; K. Makowski, trans.) (1971)
5. *L. Narici et al.,* Functional Analysis and Valuation Theory (1971)
6. *S. S. Passman,* Infinite Group Rings (1971)
7. *L. Dornhoff,* Group Representation Theory. Part A: Ordinary Representation Theory. Part B: Modular Representation Theory (1971, 1972)
8. *W. Boothby and G. L. Weiss, eds.,* Symmetric Spaces (1972)
9. *Y. Matsushima,* Differentiable Manifolds (E. T. Kobayashi, trans.) (1972)
10. *L. E. Ward, Jr.,* Topology (1972)
11. *A. Babakhanian,* Cohomological Methods in Group Theory (1972)
12. *R. Gilmer,* Multiplicative Ideal Theory (1972)
13. *J. Yeh,* Stochastic Processes and the Wiener Integral (1973)
14. *J. Barros-Neto,* Introduction to the Theory of Distributions (1973)
15. *R. Larsen,* Functional Analysis (1973)
16. *K. Yano and S. Ishihara,* Tangent and Cotangent Bundles (1973)
17. *C. Procesi,* Rings with Polynomial Identities (1973)
18. *R. Hermann,* Geometry, Physics, and Systems (1973)
19. *N. R. Wallach,* Harmonic Analysis on Homogeneous Spaces (1973)
20. *J. Dieudonné,* Introduction to the Theory of Formal Groups (1973)
21. *I. Vaisman,* Cohomology and Differential Forms (1973)
22. *B.-Y. Chen,* Geometry of Submanifolds (1973)
23. *M. Marcus,* Finite Dimensional Multilinear Algebra (in two parts) (1973, 1975)
24. *R. Larsen,* Banach Algebras (1973)
25. *R. O. Kujala and A. L. Vitter, eds.,* Value Distribution Theory: Part A; Part B: Deficit and Bezout Estimates by Wilhelm Stoll (1973)
26. *K. B. Stolarsky,* Algebraic Numbers and Diophantine Approximation (1974)
27. *A. R. Magid,* The Separable Galois Theory of Commutative Rings (1974)
28. *B. R. McDonald,* Finite Rings with Identity (1974)
29. *J. Satake,* Linear Algebra (S. Koh et al., trans.) (1975)
30. *J. S. Golan,* Localization of Noncommutative Rings (1975)
31. *G. Klambauer,* Mathematical Analysis (1975)
32. *M. K. Agoston,* Algebraic Topology (1976)
33. *K. R. Goodearl,* Ring Theory (1976)
34. *L. E. Mansfield,* Linear Algebra with Geometric Applications (1976)
35. *N. J. Pullman,* Matrix Theory and Its Applications (1976)
36. *B. R. McDonald,* Geometric Algebra Over Local Rings (1976)
37. *C. W. Groetsch,* Generalized Inverses of Linear Operators (1977)
38. *J. E. Kuczkowski and J. L. Gersting,* Abstract Algebra (1977)
39. *C. O. Christenson and W. L. Voxman,* Aspects of Topology (1977)
40. *M. Nagata,* Field Theory (1977)
41. *R. L. Long,* Algebraic Number Theory (1977)
42. *W. F. Pfeffer,* Integrals and Measures (1977)
43. *R. L. Wheeden and A. Zygmund,* Measure and Integral (1977)
44. *J. H. Curtiss,* Introduction to Functions of a Complex Variable (1978)
45. *K. Hrbacek and T. Jech,* Introduction to Set Theory (1978)
46. *W. S. Massey,* Homology and Cohomology Theory (1978)
47. *M. Marcus,* Introduction to Modern Algebra (1978)
48. *E. C. Young,* Vector and Tensor Analysis (1978)
49. *S. B. Nadler, Jr.,* Hyperspaces of Sets (1978)
50. *S. K. Segal,* Topics in Group Kings (1978)
51. *A. C. M. van Rooij,* Non-Archimedean Functional Analysis (1978)
52. *L. Corwin and R. Szczarba,* Calculus in Vector Spaces (1979)

53. *C. Sadosky,* Interpolation of Operators and Singular Integrals (1979)
54. *J. Cronin,* Differential Equations (1980)
55. *C. W. Groetsch,* Elements of Applicable Functional Analysis (1980)
56. *I. Vaisman,* Foundations of Three-Dimensional Euclidean Geometry (1980)
57. *H. I. Freedan,* Deterministic Mathematical Models in Population Ecology (1980)
58. *S. B. Chae,* Lebesgue Integration (1980)
59. *C. S. Rees et al.,* Theory and Applications of Fourier Analysis (1981)
60. *L. Nachbin,* Introduction to Functional Analysis (R. M. Aron, trans.) (1981)
61. *G. Orzech and M. Orzech,* Plane Algebraic Curves (1981)
62. *R. Johnsonbaugh and W. E. Pfaffenberger,* Foundations of Mathematical Analysis (1981)
63. *W. L. Voxman and R. H. Goetschel,* Advanced Calculus (1981)
64. *L. J. Corwin and R. H. Szcarba,* Multivariable Calculus (1982)
65. *V. I. Istrăţescu,* Introduction to Linear Operator Theory (1981)
66. *R. D. Järvinen,* Finite and Infinite Dimensional Linear Spaces (1981)
67. *J. K. Beem and P. E. Ehrlich,* Global Lorentzian Geometry (1981)
68. *D. L. Armacost,* The Structure of Locally Compact Abelian Groups (1981)
69. *J. W. Brewer and M. K. Smith, eds.,* Emily Noether: A Tribute (1981)
70. *K. H. Kim,* Boolean Matrix Theory and Applications (1982)
71. *T. W. Wieting,* The Mathematical Theory of Chromatic Plane Ornaments (1982)
72. *D. B. Gauld,* Differential Topology (1982)
73. *R. L. Faber,* Foundations of Euclidean and Non-Euclidean Geometry (1983)
74. *M. Carmeli,* Statistical Theory and Random Matrices (1983)
75. *J. H. Carruth et al.,* The Theory of Topological Semigroups (1983)
76. *R. L. Faber,* Differential Geometry and Relativity Theory (1983)
77. *S. Barnett,* Polynomials and Linear Control Systems (1983)
78. *G. Karpilovsky,* Commutative Group Algebras (1983)
79. *F. Van Oystaeyen and A. Verschoren,* Relative Invariants of Rings (1983)
80. *I. Vaisman,* A First Course in Differential Geometry (1984)
81. *G. W. Swan,* Applications of Optimal Control Theory in Biomedicine (1984)
82. *T. Petrie and J. D. Randall,* Transformation Groups on Manifolds (1984)
83. *K. Goebel and S. Reich,* Uniform Convexity, Hyperbolic Geometry, and Nonexpansive Mappings (1984)
84. *T. Albu and C. Năstăsescu,* Relative Finiteness in Module Theory (1984)
85. *K. Hrbacek and T. Jech,* Introduction to Set Theory: Second Edition (1984)
86. *F. Van Oystaeyen and A. Verschoren,* Relative Invariants of Rings (1984)
87. *B. R. McDonald,* Linear Algebra Over Commutative Rings (1984)
88. *M. Namba,* Geometry of Projective Algebraic Curves (1984)
89. *G. F. Webb,* Theory of Nonlinear Age-Dependent Population Dynamics (1985)
90. *M. R. Bremner et al.,* Tables of Dominant Weight Multiplicities for Representations of Simple Lie Algebras (1985)
91. *A. E. Fekete,* Real Linear Algebra (1985)
92. *S. B. Chae,* Holomorphy and Calculus in Normed Spaces (1985)
93. *A. J. Jerri,* Introduction to Integral Equations with Applications (1985)
94. *G. Karpilovsky,* Projective Representations of Finite Groups (1985)
95. *L. Narici and E. Beckenstein,* Topological Vector Spaces (1985)
96. *J. Weeks,* The Shape of Space (1985)
97. *P. R. Gribik and K. O. Kortanek,* Extremal Methods of Operations Research (1985)
98. *J.-A. Chao and W. A. Woyczynski, eds.,* Probability Theory and Harmonic Analysis (1986)
99. *G. D. Crown et al.,* Abstract Algebra (1986)
100. *J. H. Carruth et al.,* The Theory of Topological Semigroups, Volume 2 (1986)
101. *R. S. Doran and V. A. Belfi,* Characterizations of C*-Algebras (1986)
102. *M. W. Jeter,* Mathematical Programming (1986)
103. *M. Altman,* A Unified Theory of Nonlinear Operator and Evolution Equations with Applications (1986)
104. *A. Verschoren,* Relative Invariants of Sheaves (1987)
105. *R. A. Usmani,* Applied Linear Algebra (1987)
106. *P. Blass and J. Lang,* Zariski Surfaces and Differential Equations in Characteristic p > 0 (1987)
107. *J. A. Reneke et al.,* Structured Hereditary Systems (1987)

108. *H. Busemann and B. B. Phadke,* Spaces with Distinguished Geodesics (1987)
109. *R. Harte,* Invertibility and Singularity for Bounded Linear Operators (1988)
110. *G. S. Ladde et al.,* Oscillation Theory of Differential Equations with Deviating Arguments (1987)
111. *L. Dudkin et al.,* Iterative Aggregation Theory (1987)
112. T. Okubo, *Differential Geometry* (1987)
113. *D. L. Stancl and M. L. Stancl,* Real Analysis with Point-Set Topology (1987)
114. *T. C. Gard,* Introduction to Stochastic Differential Equations (1988)
115. *S. S. Abhyankar,* Enumerative Combinatorics of Young Tableaux (1988)
116. *H. Strade and R. Farnsteiner,* Modular Lie Algebras and Their Representations (1988)
117. *J. A. Huckaba,* Commutative Rings with Zero Divisors (1988)
118. *W. D. Wallis,* Combinatorial Designs (1988)
119. *W. Więsław,* Topological Fields (1988)
120. *G. Karpilovsky,* Field Theory (1988)
121. *S. Caenepeel and F. Van Oystaeyen,* Brauer Groups and the Cohomology of Graded Rings (1989)
122. *W. Kozlowski,* Modular Function Spaces (1988)
123. *E. Lowen-Colebunders,* Function Classes of Cauchy Continuous Maps (1989)
124. *M. Pavel,* Fundamentals of Pattern Recognition (1989)
125. *V. Lakshmikantham et al.,* Stability Analysis of Nonlinear Systems (1989)
126. *R. Sivaramakrishnan,* The Classical Theory of Arithmetic Functions (1989)
127. *N. A. Watson,* Parabolic Equations on an Infinite Strip (1989)
128. *K. J. Hastings,* Introduction to the Mathematics of Operations Research (1989)
129. *B. Fine,* Algebraic Theory of the Bianchi Groups (1989)
130. *D. N. Dikranjan et al.,* Topological Groups (1989)
131. *J. C. Morgan II,* Point Set Theory (1990)
132. *P. Biler and A. Witkowski,* Problems in Mathematical Analysis (1990)
133. *H. J. Sussmann,* Nonlinear Controllability and Optimal Control (1990)
134. *J.-P. Florens et al.,* Elements of Bayesian Statistics (1990)
135. *N. Shell,* Topological Fields and Near Valuations (1990)
136. *B. F. Doolin and C. F. Martin,* Introduction to Differential Geometry for Engineers (1990)
137. *S. S. Holland, Jr.,* Applied Analysis by the Hilbert Space Method (1990)
138. *J. Okniński,* Semigroup Algebras (1990)
139. *K. Zhu,* Operator Theory in Function Spaces (1990)
140. *G. B. Price,* An Introduction to Multicomplex Spaces and Functions (1991)
141. *R. B. Darst,* Introduction to Linear Programming (1991)
142. *P. L. Sachdev,* Nonlinear Ordinary Differential Equations and Their Applications (1991)
143. *T. Husain,* Orthogonal Schauder Bases (1991)
144. *J. Foran,* Fundamentals of Real Analysis (1991)
145. *W. C. Brown,* Matrices and Vector Spaces (1991)
146. *M. M. Rao and Z. D. Ren,* Theory of Orlicz Spaces (1991)
147. *J. S. Golan and T. Head,* Modules and the Structures of Rings (1991)
148. *C. Small,* Arithmetic of Finite Fields (1991)
149. *K. Yang,* Complex Algebraic Geometry (1991)
150. *D. G. Hoffman et al.,* Coding Theory (1991)
151. *M. O. González,* Classical Complex Analysis (1992)
152. *M. O. González,* Complex Analysis (1992)
153. *L. W. Baggett,* Functional Analysis (1992)
154. *M. Sniedovich,* Dynamic Programming (1992)
155. *R. P. Agarwal,* Difference Equations and Inequalities (1992)
156. *C. Brezinski,* Biorthogonality and Its Applications to Numerical Analysis (1992)
157. *C. Swartz,* An Introduction to Functional Analysis (1992)
158. *S. B. Nadler, Jr.,* Continuum Theory (1992)
159. *M. A. Al-Gwaiz,* Theory of Distributions (1992)
160. *E. Perry,* Geometry: Axiomatic Developments with Problem Solving (1992)
161. *E. Castillo and M. R. Ruiz-Cobo,* Functional Equations and Modelling in Science and Engineering (1992)
162. *A. J. Jerri,* Integral and Discrete Transforms with Applications and Error Analysis (1992)
163. *A. Charlier et al.,* Tensors and the Clifford Algebra (1992)

164. *P. Biler and T. Nadzieja*, Problems and Examples in Differential Equations (1992)
165. *E. Hansen*, Global Optimization Using Interval Analysis (1992)
166. *S. Guerre-Delabrière*, Classical Sequences in Banach Spaces (1992)
167. *Y. C. Wong*, Introductory Theory of Topological Vector Spaces (1992)
168. *S. H. Kulkarni and B. V. Limaye*, Real Function Algebras (1992)
169. *W. C. Brown*, Matrices Over Commutative Rings (1993)
170. *J. A. Loustau and M. I. Dillon*, Linear Geometry with Computer Graphics (1993)
171. *W. V. Petryshyn*, Approximation-Solvability of Nonlinear Functional and Differential Equations (1993)
172. *E. C. Young*, Vector and Tensor Analysis: Second Edition (1993)

Additional Volumes in Preparation

VECTOR AND TENSOR ANALYSIS

Second Edition, Revised and Expanded

Eutiquio C. Young
Department of Mathematics
Florida State University
Tallahassee, Florida

CRC Press
Taylor & Francis Group
Boca Raton London New York

CRC Press is an imprint of the
Taylor & Francis Group, an **informa** business

CRC Press
Taylor & Francis Group
6000 Broken Sound Parkway NW, Suite 300
Boca Raton, FL 33487-2742

First issued in paperback 2019

ISBN-13: 978-0-8247-8789-9 (hbk)
ISBN-13: 978-0-367-40253-2 (pbk)

Library of Congress catalog number: 92-33741

Library of Congress Cataloging-in-Publication Data

Catalog record is available from the Library of Congress

Visit the Taylor & Francis Web site at
http://www.taylorandfrancis.com

and the CRC Press Web site at
http://www.crcpress.com

To My Wife

Nena

PREFACE TO THE SECOND EDITION

In this new edition we have tried to maintain the objective of the first edition, namely, to acquaint students with the fundamental concepts of vector and tensor analysis together with some of their physical applications and geometrical interpretations, and to enable students to attain some degree of proficiency in the manipulation and application of the mechanics and techniques of the subject. We have tried to retain the qualities and features of the previous edition, placing great emphasis on intuitive understanding and development of basic techniques and computational skills.

In this edition each chapter has been rewritten and certain chapters have been reorganized. For example, in Chapter 3 the section on directional derivatives of vector fields has been deleted, the section on transformation of rectangular cartesian coordinate systems, together with the invariance of the gradient, divergence and the curl has been incorporated in the discussion of tensors. In Chapter 4 the section on test for independence of path has been combined with the section on path independence. In each chapter we have expanded discussions and provided more examples and figures to demonstrate computational techniques as well as to help clarify concepts. Whenever it is helpful we have introduced subtitles in each section to alert students to discussion of new topics. Throughout the book, we have written statements of definitions and theorems in boldface letters for easy identification.

The author will appreciate receiving information about any errors or suggestions for the improvement of this book. The author also wishes to thank Miss Deirdre Griese, Production Editor, and her staff for assistance rendered in the revision of this book.

Eutiquio C. Young

PREFACE TO THE FIRST EDITION

This book is intended for an introductory course in vector and tensor analysis. In writing the book, the author's objective has been to acquaint the students with the various fundamental concepts of vector and tensor analysis together with some of their corresponding physical and geometric interpretations, as well as to enable the students to attain some degree of proficiency in the manipulation and application of the mechanics and techniques of the subject.

Throughout the book, we place great emphasis on intuitive understanding as well as geometric and physical illustrations. To help achieve this end, we have included a great number of examples drawn from the physical sciences, such as mechanics, fluid dynamics, and electromagnetic theory, although prior knowledge of these subjects is not assumed. We stress the development of basic techniques and computational skills and deliberately de-emphasize highly complex proofs. Teaching experience at this level suggests that highly technical proofs of theorems are difficult for students and serve little purpose toward understanding the significance and implications of the theorems. Thus we have presented the classical integral theorems of Green, Gauss, and Stokes only intuitively and in the simplest geometric setting. At the end of practically every section, there are exercises of varying degree of difficulty to test students' comprehension of the subject matter presented and to make the students proficient in the basic computation and techniques of the subject.

The book contains more than enough material for a one-year or two-quarter course at the junior or senior level or even at the beginning graduate level for physical sciences majors. Omitting Secs. 3.9 through 3.12, Chaps. 1 through 4 can serve as material for a one-semester course in vector analysis, or for a one-quarter course with further deletion of topics depending on the interest of the class. Preceded by Secs. 3.9 and 3.11, the material of Chaps. 5 and 6 can then be used for a second-semester or a one-quarter course in tensor analysis.

As a prerequisite for a course based on this book, the students must be familiar with the usual topics covered in a traditional elementary calculus course. Specifically, the students must know the basic rules of differentiation and integration, such as the chain rule, integration by parts, and iterated integration of multiple integrals. Although a knowledge of matrix algebra would be helpful, this is not an essential prerequisite. The book requires only the bare rudiments of this subject, and they are summarized in the text.

The author wishes to thank his colleagues Professor Steven L. Blumsack, Wolfgang Heil, David Lovelady, and Kenneth P. Yanosko for reviewing portions of the manuscript and offering valuable comments and suggestions, and Professors Chiu Yeung Chan and Christopher K. W. Tam for testing the material on tensors in their classes during the developmental stage of the book. Last but not least, the author acknowledges with gratitude the assistance rendered by the production and editorial department of the publisher.

Eutiquio C. Young

CONTENTS

VECTOR AND TENSOR ANALYSIS

1

VECTOR ALGEBRA

1.1 INTRODUCTION

In the study of physics and engineering, we encounter many important quantities which can be described by the specification of their magnitude alone in terms of some appropriate units. For example, the volume of a cube can be described by the number of cubic inches or cubic centimeters, and the temperature at a particular time of a day can be described by giving the number of degrees on a Fahrenheit scale or a Centigrade scale. Such quantities characterized by the fact that they have magnitude only are called scalar quantities. Scalar quantities are represented by real numbers, and they are also called scalars. On the other hand, there are many other physical quantities such as displacement, force, velocity, and acceleration which cannot be described by the specification of their magnitude alone. These quantities possess not only magnitude but also direction so that a complete description of any such quantity must specify these two pieces of information. Thus when a weatherman reports the wind velocity on a particular day, he specifies not only the speed of the wind (magnitude of the wind velocity), but also the direction from which the wind is blowing. Such quantities characterized by having magnitude and direction are called vector quantities.

Just as we use real numbers or scalars to represent and manipulate scalar quantities, so we use the mathematical entities called vectors to represent and manipulate vector quantities. Thus, in a sense, vectors can be thought of as generalized numbers. The study of the representation of vectors, the algebra and calculus of vectors, together with some of their various applications constitute the subject matter of vector analysis.

Scalars and vectors are hardly sufficient to treat the class of quantities that are of interest in physics, engineering, and applied mathematics. In fact, there are quantities of a more complicated structure whose description requires more than knowledge of a magnitude and a direction. For example, to describe a quantity such as stress, we need to give a force (described by a vector) and a surface on which the force acts. Such a quantity can be described and represented only by the more sophisticated mathematical entity called tensor. As we shall see later, scalars and vectors are actually special cases of tensors. This is in accord with an important characteristic of mathematics in which concepts of more general nature do incorporate as particular cases those from which the concepts originated in the first place.

For the most part in this book, we shall study vectors and tensors in the familiar two- and three-dimensional Euclidean spaces. Although in many instances the concepts and results we obtain can be immediately extended to higher dimensional spaces, we shall do so only in few cases.

In order to distinguish vectors from scalars, we will use boldfaced letters, **A**, **B**, . . , **a**, **b**, . . , to denote vectors, and lower case letters, A, B, . . , a, b, . . , to denote scalars or real numbers. Tensors will be represented by their so called components.

1.2 DEFINITION OF A VECTOR

A vector may be defined in essentially three different ways: geometrically, analytically, and axiomatically. The geometric definition of a vector makes use of the notion of a directed line segment or an arrow. A line segment determined by two points P and Q is said to be directed if one of the points, say P, is designated as the initial point and the other, Q, the terminal point. The directed line segment so obtained is then denoted by $\boldsymbol{PQ}$, and it is represented graphically by drawing an arrow from P to Q (Fig. 1.1). We denote the length of $\boldsymbol{PQ}$ by $|\boldsymbol{PQ}|$. Two directed line segments $\boldsymbol{PQ}$ and $\boldsymbol{RS}$ are said to be equal, and we write $\boldsymbol{PQ} = \boldsymbol{RS}$, if they have the same length and the same direction. Graphically, this means that the arrows representing $\boldsymbol{PQ}$ and $\boldsymbol{RS}$ are parallel, have the same length and are pointing in the same direction.

Geometrically, then, a vector is defined as the collection of all directed line segments or arrows having the same length and direction. (Such a collection is also called an equivalence class of directed

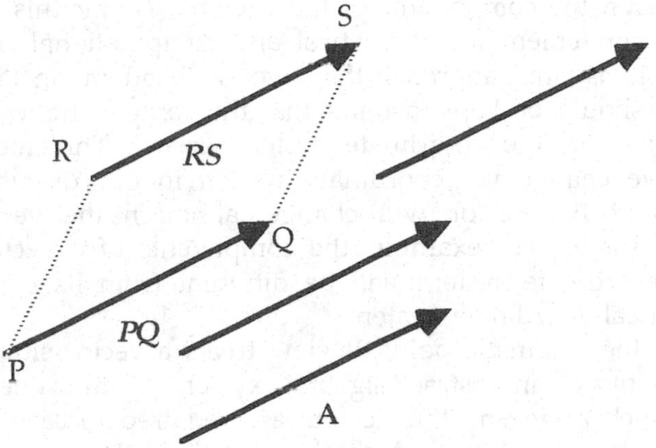

Fig. 1.1 Directed line segments.

line segments, and any two members of the class are said to be equivalent.) The common length of the arrows represents the magnitude of the vector, and the arrowhead indicates the direction of the vector. Further, for the purpose of representation, any one of the arrows in the collection can be used to represent the vector. For example, the collection of arrows shown in Fig. 1.1 defines a vector **A**, and anyone of the arrows can be used to represent the vector **A**. Under this definition, the algebraic operations on vectors are introduced and studied geometrically, making maximum use of our geometrical intuition. One advantage of this approach is that it does not require the introduction of any particular coordinate system and, hence, one can concentrate on the intrinsic vector relations. However, this approach is rather cumbersome and inefficient for computational purposes.

In the analytic approach, a vector in three-dimensional space is defined as an ordered triple of real numbers $[a_1, a_2, a_3]$ relative to a given coordinate system. The real numbers a_1, a_2, a_3 are called the components of the vector. As we shall see later, once a coordinate system is introduced, these components arise naturally from the geometric description of a vector. Likewise, in the two-dimensional space, the xy-plane, a vector is defined as an ordered pair of real numbers $[a_1, a_2]$. Algebraic operations on vectors are then performed through the components of the vectors. By far this approach is the most convenient for theoretical and computational considerations, and it is the approach that we will adopt in this book. However, it should be kept in mind that the components of a vector are dependent on the coordinate system used. This means that whenever we change the coordinate system in our discussion, the components of the vector will change, although the vector itself remains the same. For example, the components of a vector in the cylindrical coordinate system will be different from its components in the spherical coordinate system.

Lastly, the axiomatic point of view treats a vector simply as an undefined entity of an abstract algebraic system called a linear vector space. In such a system, the vectors are required to satisfy certain axioms with respect to two algebraic operations that are undefined concepts. As we shall see later, the sets of axioms for a linear vector space are precisely the properties satisfied by vectors with respect to the vector operations of addition and multiplication by scalars as developed by either the geometric or the analytic approach.

In this chapter, we shall develop the algebra of vectors on the basis of the analytic definition of a vector. We shall use directed line segments or arrows to represent vectors geometrically and to give goemetric interpretations of our results. Accordingly, we assume a coordinate system in our three-dimensional space. As is customary, we assume a right-handed rectangular cartesian coordinate system (x, y, z). The student may recall that such a coordinate system consists of three straight lines that are perpendicular to each other at a common point O, called the origin (Fig.1.2). The lines are designated

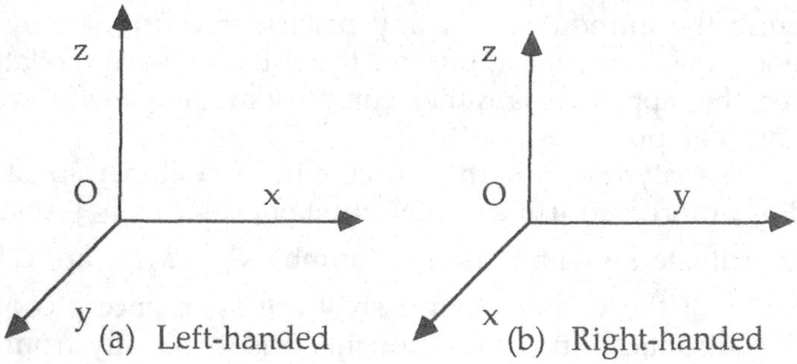

Fig. 1.2 Rectangular cartesian coordinate system.

as the x-, y-, and z-coordinate axes, and a definite direction on each axis is chosen as the positive direction. The coordinate system is then said to be right-handed if the so-called 'right-hand rule' holds; that is, if the index finger and the center finger of the right hand are made to point along the positive x- and y-axes, respectively, the thumb points along the positive z-axis. If we delete the z-axis, we obtain the corresponding cartesian coordinate system in two dimensional space, known as the xy-plane. We now define a three dimensional vector with respect to a right-handed cartesian coordinate system.

Definition 1. A three-dimensional vector **A** is an ordered triple of real numbers a_1, a_2, a_3, written as

$$\mathbf{A} = [a_1, a_2, a_3]$$

a_1 is called the first component, a_2 the second component, and a_3 the third component of the vector.

A vector whose components are all zero is called the zero vector and is denoted by **0**, that is, $\mathbf{0} = [0, 0, 0]$. The negative of a vector **A**, denoted by $-\mathbf{A}$, is defined as the vector whose components are the negative of the components of **A**. That is, if $\mathbf{A} = [a_1, a_2, a_3]$, then $-\mathbf{A} = [-a_1, -a_2, -a_3]$.

Definition 2. The magnitude of a vector $\mathbf{A} = [a_1, a_2, a_3]$, denoted by $|\mathbf{A}|$, is the real number defined by

$$|\mathbf{A}| = \sqrt{a_1^2 + a_2^2 + a_3^2} \tag{1.1}$$

It is clear that $|\mathbf{A}| \geq 0$ and that $|\mathbf{A}| = 0$ if and only if $\mathbf{A} = \mathbf{0}$. If $|\mathbf{A}| = 1$, we call the vector **A** a unit vector. In the sequel, unless stated otherwise, we shall assume our vectors to be non-zero.

1.3 GEOMETRIC REPRESENTATION OF A VECTOR

Let $\mathbf{A} = [a_1, a_2, a_3]$ be a given vector and let P be the point with the coordinates a_1, a_2, a_3. Let us examine the directed line segment or arrow that is drawn from the origin to the point P as shown in Fig. 1.3. From the figure we see that the length of the arrow is equal to $(a_1^2 + a_2^2 + a_3^2)^{1/2}$, which is the magnitude of the vector **A** as defined in (1.1). The direction of the arrow *OP* can be described analytically by the three numbers cos α, cos β, and cos γ, called the direction cosines of the directed line segment. The angles α, β, and γ, known as the direction angles, are the angles made by the directed line segment *OP* with the positive x-, y-, and z-coordinate axes, respectively. Now, from Fig. 1.3, it is clear that

$$\cos\alpha = \frac{a_1}{|\mathbf{A}|}, \quad \cos\beta = \frac{a_2}{|\mathbf{A}|}, \quad \cos\gamma = \frac{a_3}{|\mathbf{A}|} \tag{1.2}$$

Thus the direction cosines of *OP* are proportional to the components of the vector. In this manner, the length and direction of the directed line segment *OP* correspond to the magnitude and direction of the vector **A**. From this discussion it follows that every three dimensional vector is associated with a point in space. Conversely, every point in space determines a three dimensional vector whose components are the coordinates of the point. The same is true of two dimensional vectors and the points in the xy-plane.

Notice that by (1.1) the direction cosines (1.2) satisfy the important relation

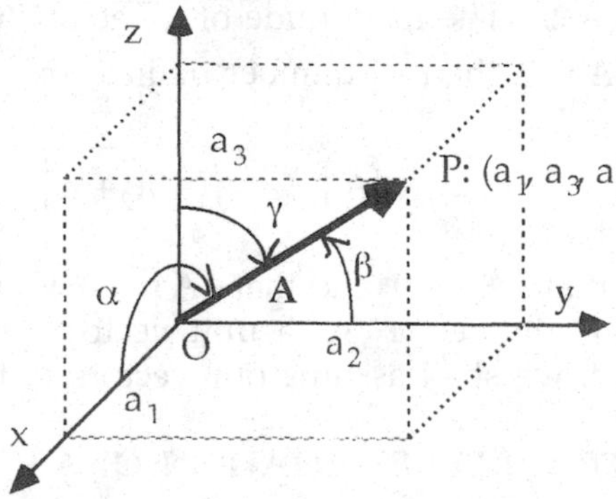

Fig. 1.3 Geometric representation of a vector.

$$\cos^2\alpha + \cos^2\beta + \cos^2\gamma = 1 \tag{1.3}$$

Thus the vector $\mathbf{u} = [a_1/|\mathbf{A}|, a_2/|\mathbf{A}|, a_3/|\mathbf{A}|]$ is a unit vector. This implies that every nonzero vector can be converted into a unit vector. The process of converting a nonzero vector into a unit vector is known as the normalization process. A vector that has been converted into a unit vector is said to have been normalized.

We should point out that the representation of a vector by a directed line segment does not depend on the initial point from which the directed line segment is drawn. In other words, a vector can also be represented by an arrow drawn from an arbitrary point in space so long as the arrow has the same length or magnitude and the same direction as the vector. For example, consider the vector $\mathbf{A} = [a_1, a_2, a_3]$ which is represented geometrically by the arrow ***OP*** in Fig. 1.3. Let Q be an arbitrary point with the coordinates (x_o, y_o, z_o) and let R be the point with the coordinates $(x_o + a_1, y_o + a_2, z_o + a_3)$, see Fig. 1.4. Then the directed line segment drawn from the point Q to the point R also represents geometrically the vector **A**.

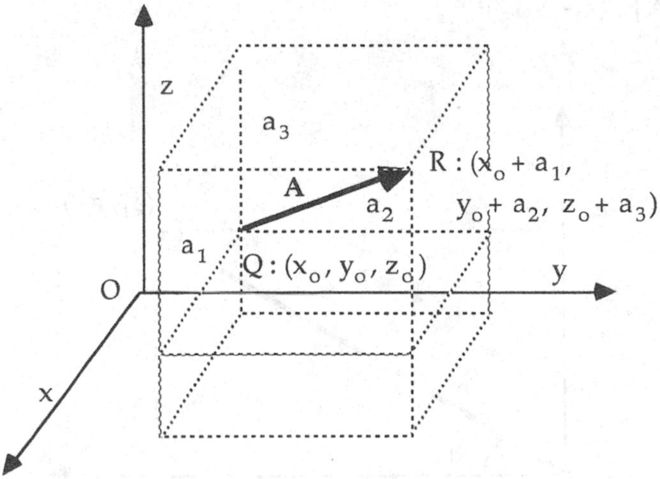

Fig. 1.4 Representation of a vector from an arbitrary point.

This follows immediately from the fact that the directed line segment *QR* (Fig. 1.4) and the directed line segment *OP* (Fig. 1.3) have the same length and the same direction cosines.

From the above discussion, it follows that if a directed line segment *PQ* represents a vector **A**, where *PQ* is drawn from the point P to the point Q whose coordinates are (x_1, y_1, z_1) and (x_2, y_2, z_2), respectively, then the components of the vector **A** are given by

$$a_1 = x_2 - x_1, \quad a_2 = y_2 - y_1, \quad a_3 = z_2 - z_1 \tag{1.4}$$

Similarly, in the xy-plane (a two dimensional space), a vector $\mathbf{A} = [a_1, a_2]$ can be represented geometrically by a directed line segment *OP* drawn from the origin to the point P with the coordinates (a_1, a_2) as shown in Fig. 1.5. The direction of the vector is uniquely defined by the angle $\theta = \arctan(a_2/a_1)$. It is clear that

$$a_1 = |\mathbf{A}| \cos\theta, \quad a_2 = |\mathbf{A}| \sin\theta \tag{1.5}$$

where $|\mathbf{A}| = (a_1^2 + a_2^2)^{1/2}$.

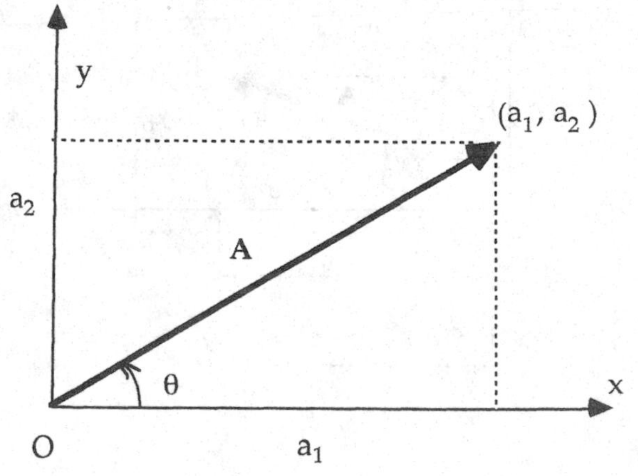

Fig. 1.5 Vector in the plane.

Example 1. Find the components and the magnitude of the vector **A** that is represented by the directed line segment $\overline{PQ}$, where P: (3, -2, 2) and Q: (-1, 1, 4).

Solution: Let a_1, a_2, a_3 denote the components of **A**. By (1.4) we have

$$a_1 = -1 - 3 = -4, \quad a_2 = 1 - (-2) = 3, \quad a_3 = 4 - 2 = 2$$

Hence, by (1.1)

$$|\mathbf{A}| = [(-4)^2 + 3^2 + 2^2]^{1/2} = \sqrt{29}$$

Example 2. Find the direction cosine of a directed line segment which represents the vector **A** = [2, -1, 2].

Solution: The magnitude of the vector is equal to

$$|\mathbf{A}| = [2^2 + (-1)^2 + 2^2]^{1/2} = 3$$

Hence, by (1.2), the direction cosines of a directed line segment representing the vector are

$$\cos\alpha = 2/3, \quad \cos\beta = -1/3, \quad \cos\gamma = 2/3$$

Example 3. If **A** = [-1, 3], what is the length and the direction of an arrow which represents the vector?

Solution: The length of an arrow representing the vector is equal to the magnitude of the vector , which is given by

$$|\mathbf{A}| = [(-1)^2 + 3^2]^{1/2} = \sqrt{10}$$

The direction is determined by the angle θ = arctan (3/-1) = 108.43° measured counterclockwise from the positive x-axis.

Example 4. A vector **A** is represented by the directed line segment $\overline{PQ}$, where P: (2, -1, 3) and Q: (-1, -2, 4). Find the components and the magnitude of the vector, the direction cosines of the line segment, and the unit vector in the same direction as **A**.

Solution: Let **A** = $[a_1, a_2, a_3]$. By (1.4) we have

$$a_1 = -1 - 2 = -3, \quad a_2 = -2 - (-1) = -1, \quad a_3 = 4 - 3 = 1$$

Hence, $|\mathbf{A}| = [(-3)^2 + (-1)^2 + 1^2]^{1/2} = \sqrt{11}$ and, by (1.2),

$$\cos\alpha = -3/\sqrt{11}, \quad \cos\beta = -1/\sqrt{11}, \quad \cos\gamma = 1/\sqrt{11}$$

Thus $\mathbf{u} = [-3/\sqrt{11}, -1/\sqrt{11}, 1/\sqrt{11}]$ is the unit vector in the same direction as **A**.

1.1 EXERCISES

In each of Problems 1 through 6, find the components and the magnitude of the vector that is represented by the directed line segment ***PQ*** whose initial point P and terminal point Q are given.

1. P : (-1, 2), Q : (2, -1).
2. P : (-2, 1), Q : (2, 3).
3. P : (0, 1), Q : (-2, -3).
4. P : (-1, 1, 2), Q : (2, -1, 3).
5. P : (2, 4, 6), Q : (1, -2, 3).
6. P : (1, 0, -2), Q : (3, 1, 0).

In each of Problems 7 through 13, find the coordinates of the initial point P or the the terminal point Q of ***PQ*** which represents the given vector.

7. **A** = [-2, 3], P : (1, -2).
8. **A** = [4, -5], Q : (-1, 1).
9. **A** = [1/2, 2/3], P : (1/2, 0).
10. **A** = [2, 1, 4], P : (1, -1, 1).
11. **A** = [-1, 3, -2], P : (2, -1, -3).
12. **A** = [2, -3, -4], Q : (3, -1, 4).
13. **A** = [3, 1, -2], Q : (-2, 1, 3).
14. If **A** = [-1, -2, 2], find the direction cosines of an arrow representing the vector.
15. Repeat Problem 14 for **A** = [1, - 3, 2].
16. If a plane vector **A** is represented by an arrow whose length is 5 units and whose direction is defined by $\theta = 150^{\circ}$, find the components of the vector.
17. A wind is blowing in the southwesterly direction at 15 mph. Find the x- and the y-component of the wind velocity.

18. Show that the direction cosines defined by equation (1.2) satisfy the relation

$$\cos^2\alpha + \cos^2\beta + \cos^2\gamma = 1$$

19. If **A** = [2, -1, -2], find the unit vector that points in the same direction as **A**; in the opposite direction to **A**. How many vectors are there in either direction?
20. An aircraft takes off from an airport at the air speed of 100 mph at an angle of inclination of 15 deg. How high from the ground and how far from the airport (on the ground) is the aircraft 10 min after takeoff?

1.4 ADDITION AND SCALAR MULTIPLICATION

We now begin the study of the various algebraic operations which can be performed on vectors. First, we introduce the notion of equality between two vectors.

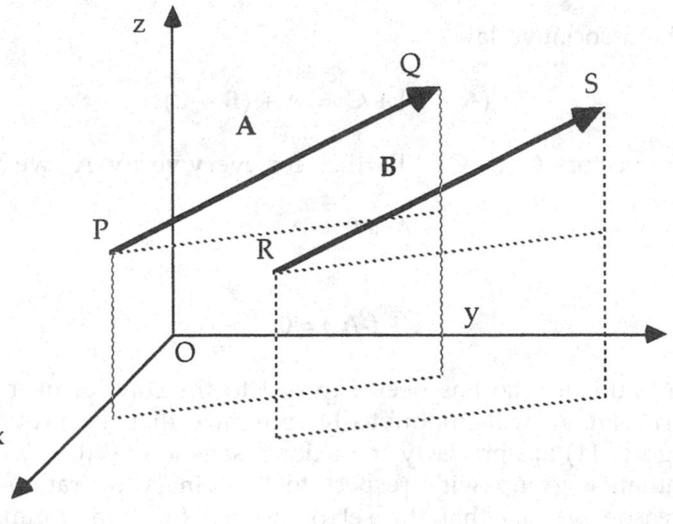

Fig. 1.6 Equality of vectors.

Definition 1. The vectors $\mathbf{A} = [a_1, a_2, a_3]$ and $\mathbf{B} = [b_1, b_2, b_3]$ are said to be equal, written $\mathbf{A} = \mathbf{B}$, if and only if

$$a_1 = b_1, \quad a_2 = b_2, \quad a_3 = b_3 \tag{1.6}$$

Clearly, two vectors that are equal can be represented geometrically either by a single directed line segment or by two directed line segments originating from two distinct points and having the same length and the same direction (see Fig. 1.6).

We now introduce the concept of vector addition.

Definition 2. The sum of two vectors $\mathbf{A} = [a_1, a_2, a_3]$ and $\mathbf{B} = [b_1, b_2, b_3]$, denoted by $\mathbf{A} + \mathbf{B}$, is the vector

$$\mathbf{A} + \mathbf{B} = [a_1 + b_1, a_2 + b_2, a_3 + b_3] \tag{1.7}$$

It follows from the definition (1.7) and the corresponding properties of real numbers that addition of vectors satisfies the commutative law

$$\mathbf{A} + \mathbf{B} = \mathbf{B} + \mathbf{A} \tag{1.8}$$

and the associative law

$$(\mathbf{A} + \mathbf{B}) + \mathbf{C} = \mathbf{A} + (\mathbf{B} + \mathbf{C}) \tag{1.9}$$

for any vectors $\mathbf{A}, \mathbf{B}, \mathbf{C}$. Further, for every vector $\mathbf{A}$, we have

$$\mathbf{A} + \mathbf{0} = \mathbf{A} \tag{1.10}$$

and

$$\mathbf{A} + (-\mathbf{A}) = \mathbf{0} \tag{1.11}$$

A student who has been exposed to the concept of a group in modern algebra will undoubtedly recognize that the properties (1.8) through (1.11) are precisely the axioms satisfied by the elements of a commutative group with respect to the binary operation "+". For this reason we say that the set of vectors forms a commutative group with respect to the operation of vector addition. This constitutes part of the characterization of a linear vector space.

Using the idea of the negative of a vector, we can introduce the operation of subtraction or the difference between two vectors. We define the difference **A** - **B** of two vectors **A** and **B** to be the sum of the vectors **A** and - **B**. That is, if $\mathbf{A} = [a_1, a_2, a_3]$ and $\mathbf{B} = [b_1, b_2, b_3]$, then

$$\mathbf{A} - \mathbf{B} = \mathbf{A} + (-\mathbf{B}) = [a_1 - b_1, a_2 - b_2, a_3 - b_3] \tag{1.12}$$

The addition and subtraction of vectors defined above can be demonstrated geometrically using directed line segments or arrows. For convenience we show this for vectors in the xy-plane. Referring to Fig. 1.7, let the arrows *OP* and *OQ* represent the vectors $\mathbf{A} = [a_1, a_2]$ and $\mathbf{B} = [b_1, b_2]$, respectively, as determined by the points P and Q. Now consider the parallelogram OPRQ with OP and OQ as two of its sides. Since opposite sides of a parallelogram are equal in length and are parallel, it follows that *OP* = *QR* and *OQ* = *PR*.. Hence *QR* also represents the vector **A** and *PR* also represents the vector **B**. Moreover, since $OA = a_1$, $OB = b_1$, we see that $OC = OA + OB = a_1 + b_1$. Likewise, since $OD = b_2$, $OE = a_2$, we have OF = OD

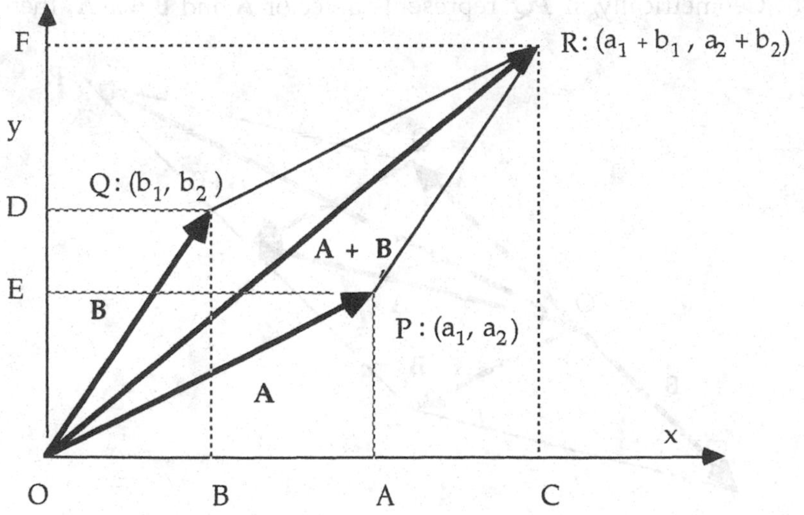

Fig. 1.7 Parallelogram law of addition of vectors.

$+ OE = b_2 + a_2$. Hence, the point R has the coordinates $(a_1 + b_1, a_2 + b_2)$ and, therefore, the arrow *OR* represents the vector sum **A** + **B**. Because *OR* is the diagonal of the parallelogram formed by the vectors **A** and **B**, the rule for the addition of vectors is sometimes called the *parallelogram law of addition.*

The subtraction of two vectors is illustrated geometrically in Fig. 1.8. Notice that in the parallelogram formed by the geometric vectors, one of the diagonals represents the sum of the vectors, while the other represents the difference.

Next we introduce the operation of multiplication of vectors by real numbers or scalars.

Definition 3. Let m be a real number and let $\mathbf{A} = [a_1, a_2, a_3]$ be a vector. The product of m and **A**, denoted by $m\,\mathbf{A}$, is the vector defined by

$$m\,\mathbf{A} = [m\,a_1, m\,a_2, m\,a_3] \tag{1.13}$$

The vector $m\,\mathbf{A}$ is sometimes called a scalar multiple of the vector **A**. This term should not be confused with the scalar product of two vectors, which is an entirely different operation (Sec. 1.6).

Geometrically, if *PQ* represents a vector **A** and $\mathbf{B} = m\,\mathbf{A}$, then

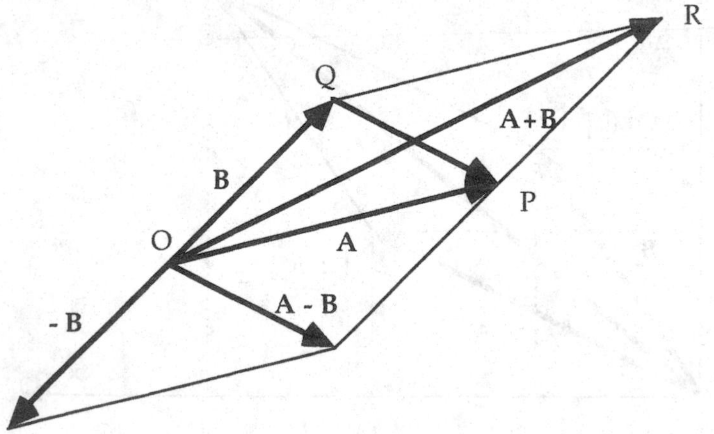

Fig. 1.8 Difference of two vectors.

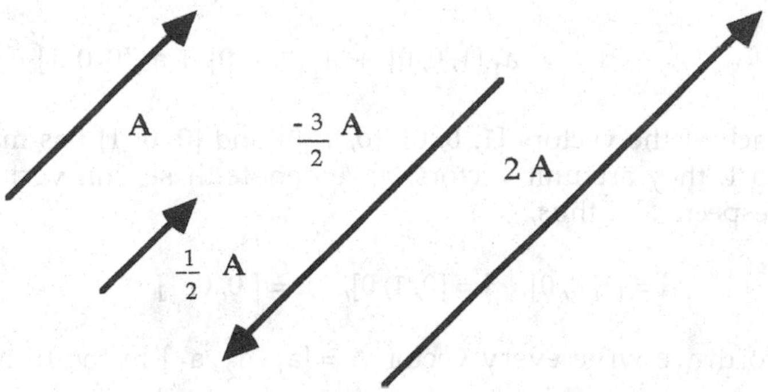

Fig. 1.9 Scalar multiples of a vector.

an arrow representing the vector **B** will have length equal to $|m|$ times the length of PQ; it will point in the same direction as PQ if m is positive and in the opposite direction if m is negative, as shown in Fig. 1.9. From the definition (1.13) and the known properties of real numbers, it is easily verified that for any vectors **A**, **B** and for any scalars m, n, we have

$$m(n\mathbf{A}) = (mn)\mathbf{A} \tag{1.14}$$

$$m(\mathbf{A} + \mathbf{B}) = m\mathbf{A} + m\mathbf{B} \tag{1.15}$$

$$(m+n)\mathbf{A} = m\mathbf{A} + n\mathbf{A} \tag{1.16}$$

$$1\mathbf{A} = \mathbf{A} \tag{1.17}$$

These properties together with those listed in (1.8) through (1.11) constitute the axioms that are satisfied by the elements of a linear vector space with respect to the two binary operations of "addition" and "multiplication by scalars". An element of such a space is then called a vector.

Having introduced the concepts of addition and scalar multiplication of vectors, we now observe that every vector $\mathbf{A} = [a_1, a_2, a_3]$ can be written as

$$[a_1, a_2, a_3] = [a_1, 0, 0] + [0, a_2, 0] + [0, 0, a_3]$$

$$= a_1[1, 0, 0] + a_2[0, 1, 0] + a_3[0, 0, 1]$$

Since each of the vectors [1, 0, 0], [0, 1, 0] and [0, 0, 1] has magnitude equal to 1, they are unit vectors. We denote these unit vectors by **i**, **j**, **k**, respectively; thus,

$$\mathbf{i} = [1, 0, 0], \quad \mathbf{j} = [0, 1, 0], \quad \mathbf{k} = [0, 0, 1] \tag{1.18}$$

Henceforth we write every vector $\mathbf{A} = [a_1, a_2, a_3]$ in the form

$$\mathbf{A} = a_1\mathbf{i} + a_2\mathbf{j} + a_3\mathbf{k} \tag{1.19}$$

and call (1.19) an analytic representation of the vector **A**.

The set of vectors (**i**, **j**, **k**) is called the natural basis for the three dimensional vector space. Geometrically, the unit vectors **i**, **j**, **k** are represented by arrows of unit length along the positive coordinate axes as shown in Fig. 1.10. Assuming that the components of the vector are all positive, we notice that the vectors $a_1\mathbf{i}$, $a_2\mathbf{j}$, $a_3\mathbf{k}$

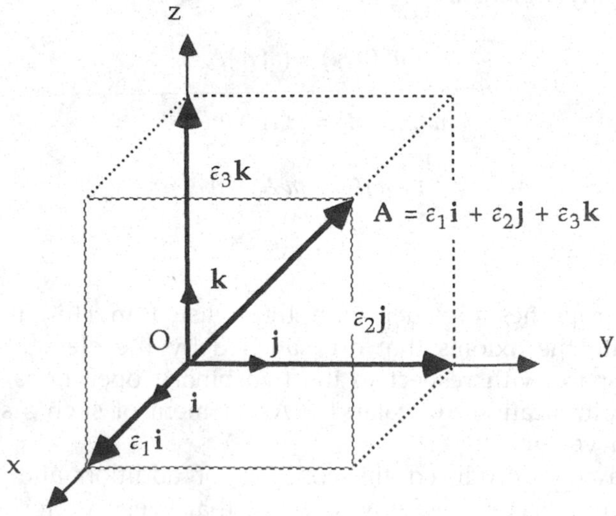

Fig. 1.10 Analytic representation of a vector.

are parallel to the coordinate axes and they form the edges of a rectangular parallelepiped of which **A** is the diagonal.

Position Vector of a Point

Let P be a point with the coordinates (x, y, z). We call the vector $\mathbf{R} = x\,\mathbf{i} + y\,\mathbf{j} + z\,\mathbf{k}$ the position vector of the point. Geometrically, the position vector of a point is the directed line segment drawn from the origin to the point P.

Example 1. Let $\mathbf{A} = \mathbf{i} - 2\mathbf{j} + 2\mathbf{k}$ and $\mathbf{B} = 2\mathbf{i} + 4\mathbf{j} - 5\mathbf{k}$. Find the linear combination $2\mathbf{A} + \mathbf{B}$ and $(1/2)\mathbf{A} - 2\mathbf{B}$.

Solution: We have

$$\begin{aligned} 2\mathbf{A} + \mathbf{B} &= 2(\mathbf{i} - 2\mathbf{j} + 2\mathbf{k}) + 2\mathbf{i} + 4\mathbf{j} - 5\mathbf{k} \\ &= 4\mathbf{i} + 0\mathbf{j} - \mathbf{k} = 4\mathbf{i} - \mathbf{k} \end{aligned}$$

$$\begin{aligned} (1/2)\mathbf{A} - 2\mathbf{B} &= (1/2)(\mathbf{i} - 2\mathbf{j} + 2\mathbf{k}) - 2(2\mathbf{i} + 4\mathbf{j} - 5\mathbf{k}) \\ &= (-7/2)\mathbf{i} - 9\mathbf{j} + 11\mathbf{k} \end{aligned}$$

Example 2. Let $\mathbf{A} = -4\mathbf{i} + 2\mathbf{j}$, $\mathbf{B} = 2\mathbf{i} + \mathbf{j}$, and $\mathbf{C} = 2\mathbf{i} + 3\mathbf{j}$. Find the scalars m and n such that $\mathbf{C} = m\,\mathbf{A} + n\,\mathbf{B}$.

Solution: We have

$$\begin{aligned} m\,\mathbf{A} + n\,\mathbf{B} &= m(-4\mathbf{i} + 2\mathbf{j}) + n(2\mathbf{i} + \mathbf{j}) \\ &= (-4m + 2n)\mathbf{i} + (2m + n)\mathbf{j} = 2\mathbf{i} + 3\mathbf{j} \end{aligned}$$

Equating the corresponding components, we obtain the system of equations

$$-4m + 2n = 2, \quad 2m + n = 3$$

for the scalars m and n. Solving for m and n, we find $m = 1/2$ and $n = 2$. Thus

$$(1/2)\mathbf{A} + 2\mathbf{B} = \mathbf{C}$$

Example 3. Let A, B and C be three distinct points on a straight line and let **A**, **B** and **C** denote their respective position vector relative to a fixed point O (Fig.1.11). If C is the midpoint of the line segment AB, show that $\mathbf{C} = (\mathbf{A} + \mathbf{B})/2$.

Solution: In Fig. 1.11 we note that the arrow **AB** represents the vector $\mathbf{B} - \mathbf{A}$. Since C is the midpoint of the line segment AB, it

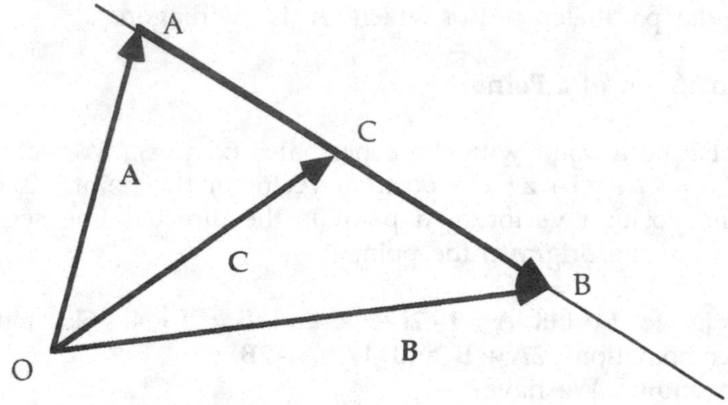

Fig. 1.11 Position vector of the midpoint of a line segment.

follows that the arrow AC represents the vector $(\mathbf{B} - \mathbf{A})/2$. By the law of vector addition, we know that $OC = OA + AC$. Thus we have $\mathbf{C} = \mathbf{A} + (\mathbf{B} - \mathbf{A})/2 = (\mathbf{A} + \mathbf{B})/2$.

Physical Applications

As mentioned previously, physical quantities that have both magnitude and direction are represented and studied by using vectors. In the following examples, we consider some physical problems to illustrate the applications of vectors.

Example 4. Suppose ship A is cruising in the easterly direction at the speed of 15 km/hr while ship B is speeding at 30 km/hr in the northeast direction. Assuming that both ships started from the same place at the same time, find the rate at which the two ships are separating.

Solution: Taking the x-axis in the easterly direction and the y-axis in the northerly direction (Fig. 1.12), we see that the velocity of the ship A is represented by the vector $\mathbf{A} = 15\mathbf{i}$ and the velocity of the ship B by the vector

$$\mathbf{B} = 30(\cos 45^o\mathbf{i} + \sin 45^o\mathbf{j}) = 15\sqrt{2}(\mathbf{i} + \mathbf{j})$$

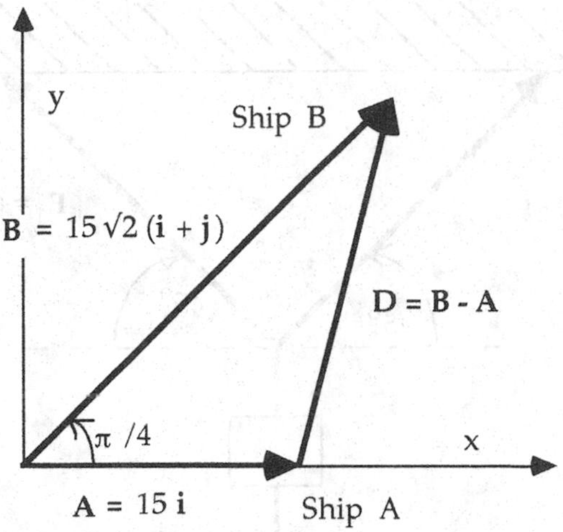

Fig. 1.12 Rate of displacement between two ships.

The displacement of the ship B relative to the ship A is represented by the vector $\mathbf{D} = \mathbf{B} - \mathbf{A}$. Hence the rate at which the ships are separating is equal to the manitude of the displacement vector **D**. Since $\mathbf{D} = \mathbf{B} - \mathbf{A} = 15(\sqrt{2} - 1)\mathbf{i} + 15\sqrt{2}\mathbf{j}$, it follows that

$$|\mathbf{D}| = 15\,[(\sqrt{2} - 1)^2 + 2\,]^{1/2} = 15\,[5 - 2\sqrt{2}]^{1/2} = 22.1 \text{ km/hr}$$

Example 5. A body weighing 100 kg is suspended from two ropes as shown in Fig. 1.13. Find the tension in the ropes.

Solution: Let the tension in the ropes be denoted by $\mathbf{F}_1 = a_1\mathbf{i} + a_2\mathbf{j}$ and $\mathbf{F}_2 = b_1\mathbf{i} + b_2\mathbf{j}$. The weight of the body is represented by the vector $\mathbf{W} = -100\mathbf{j}$. From elementary mechanics, since the system is in equilibrium, the sum of all the forces must be equal to zero. Thus $\mathbf{F}_1 + \mathbf{F}_2 + \mathbf{W} = (a_1 + b_1)\mathbf{i} + (a_2 + b_2 - 100)\mathbf{j} = \mathbf{0}$. This implies

$$a_1 + b_1 = 0, \quad a_2 + b_2 - 100 = 0$$

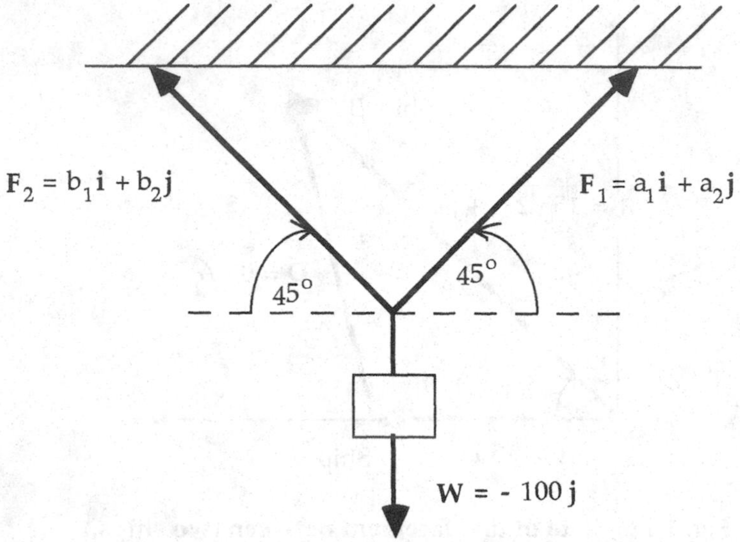

Fig. 1.13 Tension in ropes from which a weight is suspended.

Hence, $a_1 = -b_1$ and $a_2 + b_2 = 100$. By symmetry, we know that $|\mathbf{F}_1| = |\mathbf{F}_2|$, which implies that $a_2 = b_2$. Therefore, $a_2 = b_2 = 50$. Since $a_1 = |\mathbf{F}_1|\cos 45^\circ$ and $a_2 = |\mathbf{F}_1|\sin 45^\circ$, it follows that $|\mathbf{F}_1| = |\mathbf{F}_2| = 50\sqrt{2}$ and $a_1 = -b_1 = 50$. Thus $\mathbf{F}_1 = 50(\mathbf{i} + \mathbf{j})$ and $\mathbf{F}_2 = 50(-\mathbf{i} + \mathbf{j})$. Notice that $\mathbf{F}_1 + \mathbf{F}_2 = 100\mathbf{j}$.

Example 6. An aircraft is cruising at an air speed of $250\sqrt{2}$ mph in the northeast direction. If a wind of 25 mph is blowing from the west (tailwind), what is the effective speed of the aircraft relative to the ground? In what direction is the aircraft headed as a result of the tailwind? Assume that all motions are taking place in a plane.

Solution: If there was no tailwind, the aircraft would remain on its original course flying in the northeast direction. As a result of the tailwind, the aircraft is also moving toward the east at 25 mph. Thus the effective velocity of the aircraft is the sum of its cruising velocity and the tailwind velocity. Let **A** denote the velocity of the aircraft and let **W** denote the velocity of the tailwind. Referring to Fig. 1.14, we note that

$$\mathbf{A} = 250\sqrt{2}(\cos 45^\circ \mathbf{i} + \sin 45^\circ \mathbf{j}) = 250(\mathbf{i} + \mathbf{j}), \quad \mathbf{W} = 25\mathbf{i}$$

Hence the effective velocity of the aircraft is

$$\mathbf{V} = \mathbf{A} + \mathbf{W} = 275\mathbf{i} + 250\mathbf{j}$$

and the effective speed is $|\mathbf{V}| = [(275)^2 + (250)^2]^{1/2} = 25\sqrt{221}$ mph. The direction in which the aircraft is headed is given by the angle

$$\theta = \arctan(250/275) = 42.27 \text{ deg.}$$

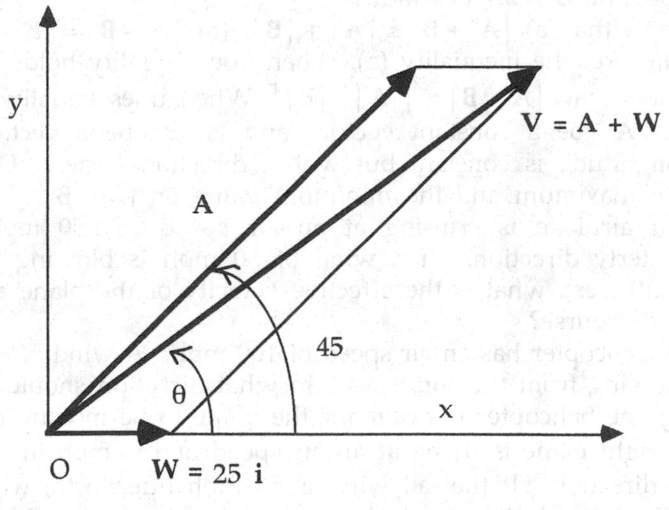

Fig. 1.14 Effective velocity of an aircraft with tailwind.

1.2 EXERCISES

1. Prove the properties (1.8) through (1.11) of vector addition.
2. Prove the properties (1.14) through (1.17) of scalar multiplica-

tion.

3. If $\mathbf{A} = 2\mathbf{i} - \mathbf{j} + 3\mathbf{k}$ and $\mathbf{B} = 3\mathbf{i} + 2\mathbf{j} - 4\mathbf{k}$, find (a) $2\mathbf{A} + \mathbf{B}$, (b) $-3\mathbf{A} + 2\mathbf{B}$, (c) $|\mathbf{A} + \mathbf{B}|$, (d) $|\mathbf{A} - \mathbf{B}|$.
4. If $\mathbf{A} = \mathbf{i} - \mathbf{j}$ and $\mathbf{B} = 2\mathbf{i} - 2\mathbf{j} + 3\mathbf{k}$, find (a) $4\mathbf{A} - 2\mathbf{B}$, (b) $2\mathbf{A} - 3\mathbf{B}$, (c) $|2(\mathbf{A} + \mathbf{B})|$, (d) $|\mathbf{A}| + |\mathbf{B}|$.
5. If $\mathbf{A} = 3\mathbf{i} - \mathbf{j} + 4\mathbf{k}$ and $\mathbf{B} = 2\mathbf{i} + \mathbf{j}$, find the vector $\mathbf{X}$ so that $2\mathbf{A} + 3\mathbf{X} = 5\mathbf{B}$.
6. If $\mathbf{A} = -\mathbf{i} + 2\mathbf{k}$ and $\mathbf{B} = 2\mathbf{i} + \mathbf{j}$, find the vector $\mathbf{X}$ so that $-2\mathbf{X} + 3\mathbf{B} = 2\mathbf{A}$.
7. Let $\mathbf{A} = 3\mathbf{i} - \mathbf{j} + 4\mathbf{k}$ and $\mathbf{B} = 2\mathbf{i} + \mathbf{k}$. If $\mathbf{C} = 4\mathbf{i} - 2\mathbf{j} + 7\mathbf{k}$, find the scalars m and n so that $\mathbf{C} = m\mathbf{A} + n\mathbf{B}$.
8. Let $\mathbf{A} = 4\mathbf{i} - 3\mathbf{j} - 6\mathbf{k}$ and $\mathbf{B} = 2\mathbf{i} + 5\mathbf{j} - 7\mathbf{k}$. If $\mathbf{C} = 8\mathbf{i} - 19\mathbf{j} - 4\mathbf{k}$, find the scalars m and n so that $\mathbf{C} = m\mathbf{A} + n\mathbf{B}$.
9. If $\mathbf{A} = \mathbf{i} - \mathbf{j} + 2\mathbf{k}$, $\mathbf{B} = \mathbf{j} - \mathbf{k}$, $\mathbf{C} = 2\mathbf{i} + \mathbf{k}$, and $\mathbf{D} = 3\mathbf{i} + 2\mathbf{j} - \mathbf{k}$, find the scalars a, b, and c so that $\mathbf{D} = a\mathbf{A} + b\mathbf{B} + c\mathbf{C}$.
10. Repeat Problem 9 if $\mathbf{A} = 2\mathbf{i} - \mathbf{j} + 4\mathbf{k}$, $\mathbf{B} = \mathbf{i} + 2\mathbf{j} - 3\mathbf{k}$, $\mathbf{C} = 2\mathbf{i} - 3\mathbf{j} - 3\mathbf{k}$, and $\mathbf{D} = 3\mathbf{i} - 6\mathbf{j} + 4\mathbf{k}$.
11. Show that (a) $|\mathbf{A} + \mathbf{B}| \leq |\mathbf{A}| + |\mathbf{B}|$, (b) $|\mathbf{A} - \mathbf{B}| = |\mathbf{B} - \mathbf{A}|$. Interpret the inequality (a). When does equality hold?
12. Show that $|\mathbf{A} - \mathbf{B}| \geq ||\mathbf{A}| - |\mathbf{B}||$. When does equality hold?
13. Let $\mathbf{A}$ be a constant vector and let $\mathbf{B}$ be a vector whose magnitude is constant but whose direction varies. Determine the maximum and the minimum value of $|\mathbf{A} - \mathbf{B}|$.
14. An airplane is cruising at an air speed of 250 mph in the easterly direction. If a wind of 50 mph is blowing from the southwest, what is the effective velocity of the plane and what is its course?
15. A helicopter has an air speed of 100 mph. A wind of 30 mph is blowing from the southeast. In what direction should the pilot fly the helicopter in order for the plane to be moving north?
16. A light plane is flying at an air speed of 150 mph in the N 60° E direction. If the tail wind is 50 mph due north, what is the effective velocity of the plane and what is its course?
17. At noon the location and the velocity of ships A and B are sighted from a lighthouse (assumed at the origin) to be $\mathbf{R}_A = 3(\mathbf{i} + \mathbf{j})$ (miles), $\mathbf{V}_A = 4\mathbf{i} + 3\mathbf{j}$ (mph) and $\mathbf{R}_B = 8\mathbf{i} + 3\mathbf{j}$, $\mathbf{V}_B = 2\mathbf{i} + 3\mathbf{j}$, respectively. If the ships continue in their current course, show that the ships will collide. Find the time and the location where the collision occurs.

18. At noon an aircraft takes off from an airport with air speed of 100 mph at an angle of inclination of 15 deg. Ten minutes later, another aircraft takes off from the same airport at the same point with air speed of 110 mph at 20 deg. How far apart are the aircrafts at 12:20 PM?
19. Three forces $\mathbf{F}_1$, $\mathbf{F}_2$ and $\mathbf{F}_3$ act in the direction of the sides AB, BC and CA, respectively, of an equilateral triangle ABC. If $|\mathbf{F}_2| = 2|\mathbf{F}_1|$ and $|\mathbf{F}_3| = 3|\mathbf{F}_1|$, find the resultant of the forces. Show that the resultant force is parallel to the bisector of the angle at the vertex A.
20. A body weighing 100 kg is hanging from two wires that are attached to two pegs 8 m apart on a ceiling as shown in Fig. (i) below. Find the tension in each of the wires.

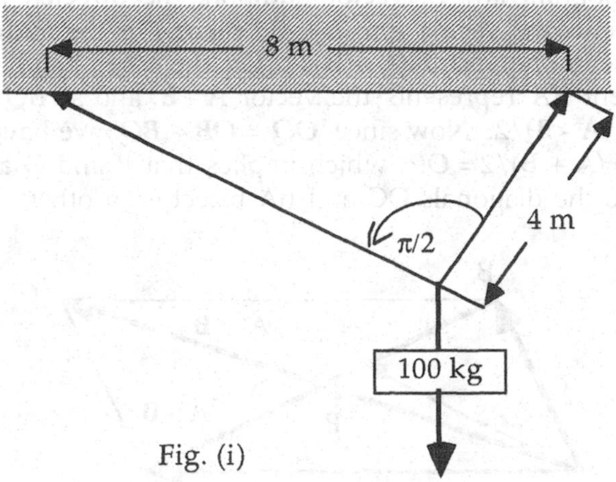

Fig. (i)

21. Suppose **A**, **B** and **C** are three vectors that are represented by the directed line segments *AB*, *BC* and *CA* of a triangle ABC. (a) Show that $\mathbf{A} + \mathbf{B} + \mathbf{C} = \mathbf{0}$. (b) Show that the sum of the vectors represented by *BA*, *CA* and 2(*BC*) has a magnitude equal to 3(BD), where D is a trisecting point on the side AC.
22. A man crossing a river 1500 ft wide rows a boat at the speed of 150 ft/min. Assuming that the man always has the boat pointed perpendicular to the opposite bank and that there is a current of 30 ft/min, what is the velocity of the boat relative to an observer on the ground? How long will it take the man to cross the river and how far downstream on the opposite bank will he land?

1.5 SOME APPLICATIONS IN GEOMETRY

The fact that vectors are mathematical entities independent of any frame of reference makes vector algebra a convenient tool for proving some theorems and establishing certain identities in geometry. In this section we consider some examples to illustrate the technique and procedure.

Example 1. Prove that the diagonals of every parallelogram bisect each other.

Solution: Let OACB be a given parallelogram, and let P and Q be the midpoints of the diagonals OC and BA, respectively (Fig. 1.15). Denote the directed line segments *OA* and *OB* by **A** and **B**. Then by the parallelogram law of vector addition, the directed line segment *OC* represents the vector sum **A** + **B**. Since P is the midpoint of OC, *OP* represents the vector (**A** + **B**)/2. On the other hand, the directed line segment *BA* represents the vector **A** - **B**, and so *BQ* represents the vector (**A** - **B**)/2. Now since *OQ* = *OB* + *BQ*, we have *OQ* = **B** + (**A** - **B**)/2 = (**A** + **B**)/2 = *OP*, which implies that P and Q are the same point. Thus the diagonals OC and BA bisect each other.

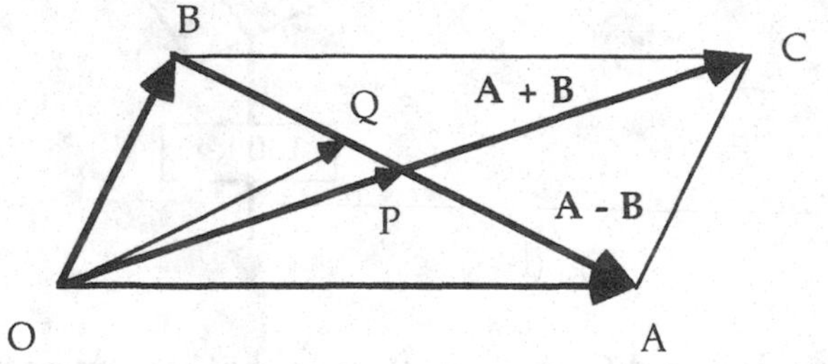

Fig. 1.15 Diagonals of a parallelogram bisect each other.

Example 2. Let **A**, **B**, and **C** denote the position vectors of the points A, B, and C, respectively, relative to a fixed point O, where A, B, and C are collinear (Fig. 1.16). If the point C divides the line segment AB in the ratio $m:n$ (both m and n not equal to zero), show that $\mathbf{C} = (n\mathbf{A} + m\mathbf{B})/(m + n)$.

Solution: Since AC : CB = $m:n$, it follows that *CB* =

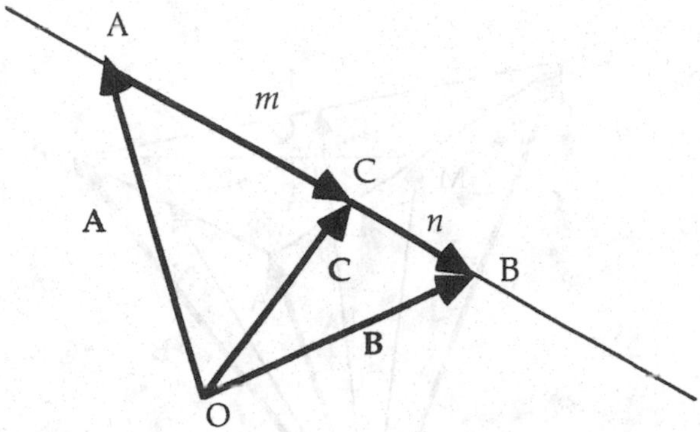

Fig. 1.16 Position vector of a dividing point of a line segment.

$(n/m)AC$. But AC represents the vector **C** - **A** and CB represents the vector **B** - **C**. Hence **B** - **C** = (n/m)(**C** - **A**). Solving for **C**, we obtain the desired result.

When $m = n$, C is the midpoint of AB for which we obtain **C** = (**A**+**B**)/2. This agrees with the result obtained in Example 3, Sec. 1.4.

Example 3. Prove that the medians of a triangle intersect at a point which trisects the medians.

Solution: Let A, B, and C denote the vertices of a triangle, and let **A**, **B**, and **C** denote their respective position vectors relative to a fixed point O (Fig. 1.17). Denote by L, M, and N the respective midpoints of the sides BC, CA, and AB, and let **L**, **M**, and **N** denote their position vectors. Then we have

$$\mathbf{L} = (\mathbf{B} + \mathbf{C})/2, \quad \mathbf{M} = (\mathbf{C} + \mathbf{A})/2, \quad \mathbf{N} = (\mathbf{A} + \mathbf{B})/2$$

Now let P, Q, and R denote the points which divide the respective medians AL, BM, and CN in the ratio 2 : 1 (trisecting points). Then, according to the result of Example 2, we have

$$\begin{aligned} OP &= (\mathbf{A} + 2\mathbf{L})/3 = (\mathbf{A} + \mathbf{B} + \mathbf{C})/3 \\ OQ &= (\mathbf{B} + 2\mathbf{M})/3 = (\mathbf{A} + \mathbf{B} + \mathbf{C})/3 \\ OR &= (\mathbf{C} + 2\mathbf{N})/3 = (\mathbf{A} + \mathbf{B} + \mathbf{C})/3 \end{aligned}$$

This implies that the points P, Q, and R coincide since they have the same position vector.

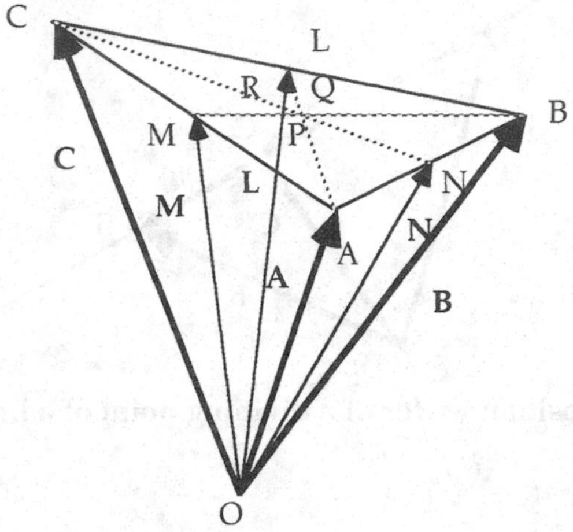

Fig. 1.17 Intersection point of the medians of a triangle.

Example 4. Let A, B, C, and D be four points in space with possition vectors **A, B, C**, and **D**, respectively, relative to a fixed point O (Figure 1.18). If $3\mathbf{A} - 2\mathbf{B} + \mathbf{C} - 2\mathbf{D} = \mathbf{0}$, show that the points A, B, C, and D are coplanar. Determine the point at which the line segments AC and BD intersect and the ratio in which the point divides the line segments.

Solution: If we can show that the line segments AC and BD do intersect, then it will follow that the points A, B, C, and D are coplanar. In Fig. 1.18 we note that the point P is a point on AC if and only if *OP* is a linear combination of **A** and **C**; likewise, Q is a point on BD if and only if *OQ* is a linear combination of **B** and **D**. Now from the given equation, we obtain $3\mathbf{A} + \mathbf{C} = 2(\mathbf{B} + \mathbf{D})$ or $(3\mathbf{A} + \mathbf{C})/4 = (\mathbf{B} + \mathbf{D})/2$. This indicates a point common to the line segment AC and BD. Thus the line segments intersect and, therefore, the points A, B, C, and D are coplanar.

Now $(3\mathbf{A} + \mathbf{C})/4$ denotes a point on AC which divides the line segment in the ratio 1 : 3. On the other hand, $(\mathbf{B} + \mathbf{D})/2$ denotes the midpoint of BD.

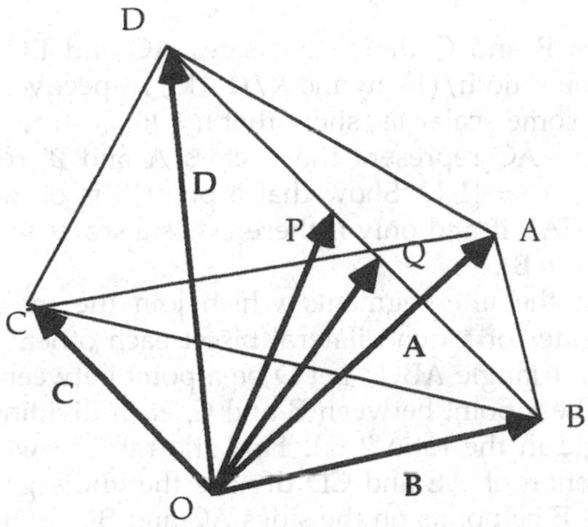

Fig. 1.18 Coplanar points.

1.3 EXERCISES

1. Prove that if the midpoints of the consecutive sides of a quadrilateral are joined by line segments, then the resulting quadrilateral is a parallelogram.
2. Let P, Q, and R be the respective midpoint of the sides AB, BC, and CA of a triangle ABC. Show that the quadrilateral APQR is a parallelogram.
3. Prove that the line segment joining the midpoints of two sides of a triangle is parallel to and is equal in length to half the length of the third side.
4. Show that the line segment joining the midpoints of the two non-parallel sides of a trapezoid is equal in length to half the sum of the length of the parallel sides and is parallel to them.
5. Show that the line which joins one vertex of a parallelogram to the midpoint of an opposite side trisects the diagonal.
6. Let **A** , **B** , **C** , and **D** denote the position vectors of the points A, B, C, and D, respectively, relative to a reference point O. If **B** - **A** = 2(**D** - **C**), show that the line segments AD and BC (extended if necessary) meet in a point which trisects these lines.

7. The points P and Q divide the sides AC and BC of a triangle ABC in the ratio $h/(1 - h)$ and $k/(1 - k)$, respectively. If PQ = m AB, for some scalar m, show that $h = k = 1 - m$.
8. Let AB and AC represent the vectors **A** and **B**, respectively, such that $|\mathbf{A}| = |\mathbf{B}|$. Show that a point P is on the bisector of the angle CAB if and only if there exists a scalar m such that $AP = m(\mathbf{A} + \mathbf{B})$.
9. Show that the line segments which join the midpoints of the opposite sides of a quadrilateral bisect each other.
10. Consider a triangle ABC. Let D be a point between A and B, and let E be a point between B and C, each dividing the sides of the triangle in the ratio 2 : 1. Find the ratio in which the point of intersection of AE and CD divides the line segments.
11. Let D and E be points on the sides AC and BC of a triangle such that the line segments DE is parallel to AB. If the line segments AE and BD meet at a point P, show that the line CP bisects AB.
12. A line from a vertex of a triangle bisects the opposite side. It is bisected by a similar line issuing from another vertex. In what ratio does the latter line intersect the opposite side?
13. Let $\mathbf{A} = OA$, $\mathbf{B} = OB$, $\mathbf{C} = OC$, and $\mathbf{D} = OD$, where A, B, C, and D are four given points, and O is any fixed point. If $\mathbf{A} - \mathbf{B} = \mathbf{C} - \mathbf{D}$, show that the points A, B, C, and D form the vertices of a parallelogram.
14. Let **A**, **B**, **C**, **D**, and **E** denote the vectors represented by the arrows from the center of a regular pentagon to the vertices. Express the sides of the pentagon in terms of the vectors and show that $\mathbf{A} + \mathbf{B} + \mathbf{C} + \mathbf{D} + \mathbf{E} = \mathbf{0}$.

1.6 SCALAR PRODUCT

There are two important operations on vectors which correspond to the operation of multiplication in algebra. One of these operations results in a scalar and the other results in a vector. We consider first the operation that results in a scalar.

Definition 1. Let $\mathbf{A} = a_1\mathbf{i} + a_2\mathbf{j} + a_3\mathbf{k}$ and $\mathbf{B} = b_1\mathbf{i} + b_2\mathbf{j} + b_3\mathbf{k}$ be two vectors. The scalar product of **A** and **B**, denoted by $\mathbf{A}\cdot\mathbf{B}$, is the scalar (real number) given by

$$\mathbf{A}\cdot\mathbf{B} = a_1b_1 + a_2b_2 + a_3b_3 \tag{1.20}$$

The scalar product of two vectors is also known as the dot product (for obvious reason) or the inner product.

From the definition (1.20), it is easy to verify that for any vectors **A**, **B**, **C** and for any scalar m, we have

(a) $\mathbf{A}\cdot\mathbf{B} = \mathbf{B}\cdot\mathbf{A}$ (commutative law)
(b) $\mathbf{A}\cdot(\mathbf{B}+\mathbf{C}) = \mathbf{A}\cdot\mathbf{B} + \mathbf{A}\cdot\mathbf{C}$ (distributive law) (1.21)
(c) $m(\mathbf{A}\cdot\mathbf{B}) = (m\mathbf{A})\cdot\mathbf{B} = \mathbf{A}\cdot(m\mathbf{B})$
(d) $\mathbf{A}\cdot\mathbf{A} = |\mathbf{A}|^2$

These properties are analogous to the properties satisfied by real numbers with respect to ordinary multiplication.

Geometric Interpretation of the Scalar Product of Vectors

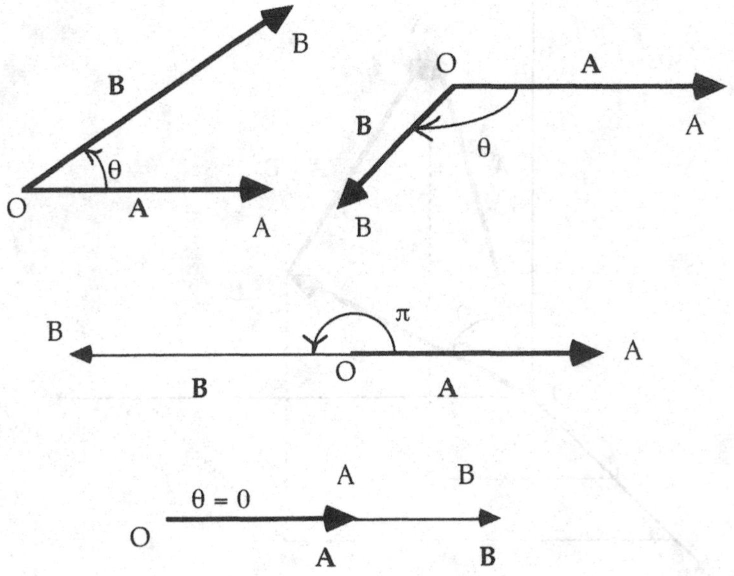

Fig. 1.19 Angle between two vectors.

The scalar product of two vectors has an interesting and important geometric interpretation. Before we present this interpretation, we first introduce the notion of angle between two vectors. Let **A** and **B** be two vectors, and let them be represented by the arrows *OA* and *OB* as shown in Fig. 1.19. By the angle between the vectors **A** and **B**, we mean the angle θ, $0 < \theta < \pi$, formed by the line segmens OA and OB. It follows from this definition that if $\mathbf{B} = m\,\mathbf{A}$, then the angle between **A** and **B** is zero if $m > 0$ and $\theta = \pi$ if $m < 0$. Two vectors are said to be parallel if and only if the angle between them is 0 or π. On the other hand, two vectors are said to be orthogonal or perpendicular if and only if the angle between them is $\pi/2$. These concepts are of course geometrically evident.

Now let **A** and **B** be two vectors represented by the arrows *OA* and *OB* (Fig. 1.20). By the parallelogram law of vector addition, we know that *BA* represents the vector **A** - **B**. Applying the law of cosines to the triangle OAB, we find

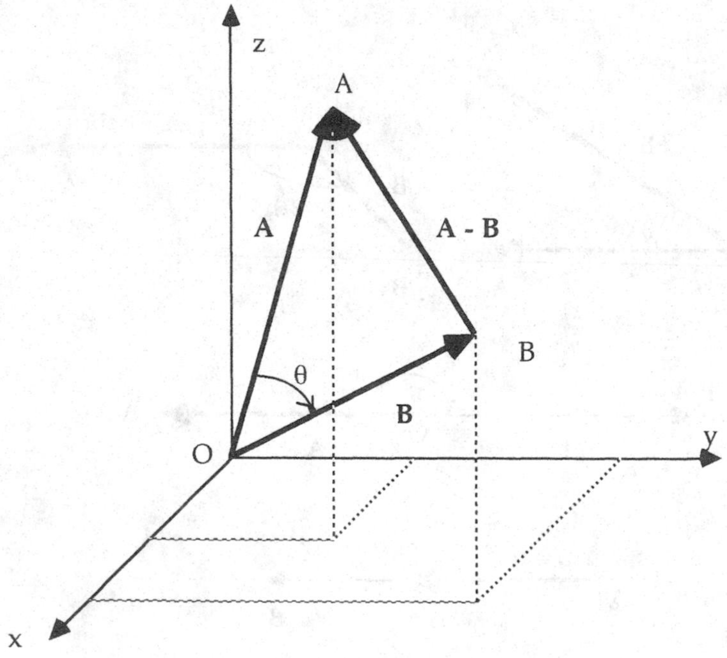

Fig. 1.20 The scalar product of two vectors.

$$|BA|^2 = |OA|^2 + |OB|^2 - 2|OA||OB|\cos\theta$$

where θ is the angle between **A** and **B**. In terms of the magnitude of the vectors **A**, **B**, and **A - B**, this becomes

$$|\mathbf{A} - \mathbf{B}|^2 = |\mathbf{A}|^2 + |\mathbf{B}|^2 - 2|\mathbf{A}||\mathbf{B}|\cos\theta \qquad (1.22)$$

On the other hand, by the properties of scalar product, we have

$$\begin{aligned} |\mathbf{A} - \mathbf{B}|^2 &= (\mathbf{A} - \mathbf{B})\cdot(\mathbf{A} - \mathbf{B}) = \mathbf{A}\cdot\mathbf{A} - \mathbf{A}\cdot\mathbf{B} - \mathbf{B}\cdot\mathbf{A} + \mathbf{B}\cdot\mathbf{B} \\ &= |\mathbf{A}|^2 - 2\mathbf{A}\cdot\mathbf{B} + |\mathbf{B}|^2 \end{aligned} \qquad (1.23)$$

Comparing (1.22) and (1.23), we obtain

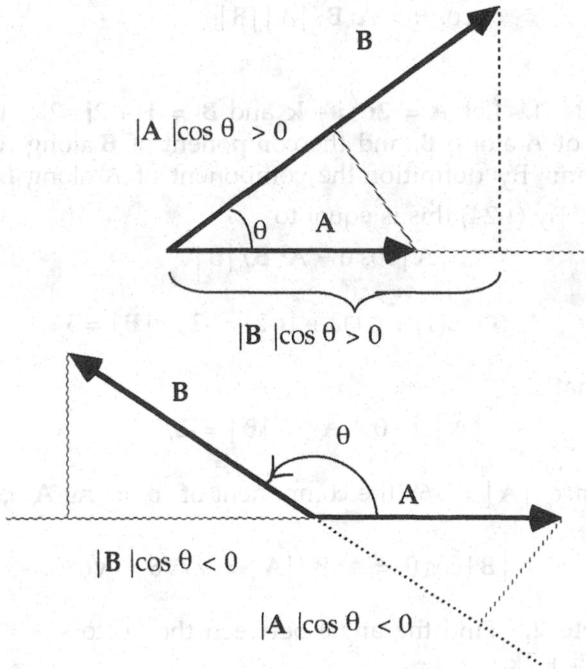

Fig. 1.21 Component of a vector along another vector.

$$\mathbf{A}\cdot\mathbf{B} = |\mathbf{A}||\mathbf{B}|\cos\theta \tag{1.24}$$

Thus the scalar product of two vectors is equal to the product of their magnitudes and the cosine of the angle between then.

If we examine the plane formed by the vectors **A** and **B** in Fig. 1.21, we see that the quantity $|\mathbf{B}|\cos\theta$ is simply the orthogonal projection of the magnitude of **B** onto the vector **A**. We call it the component of **B** along **A**; it is positive, zero, or negative according as $0 < \theta < \pi/2, \theta = 0$, or $\pi/2 < \theta < \pi$. Likewise, the quantity $|\mathbf{A}|\cos\theta$ is called the component of **A** along the vector **B**. Thus the scalar product of **A** and **B** may also be stated as the product of $|\mathbf{A}|$ and the component of **B** along **A**, or the product of $|\mathbf{B}|$ and the component of **A** along **B**.

From (1.24) we obtain a formula for calculating the angle between two vectors, namely,

$$\cos\theta = \mathbf{A}\cdot\mathbf{B}/|\mathbf{A}||\mathbf{B}| \tag{1.25}$$

Example 1. Let $\mathbf{A} = 2\mathbf{i} - \mathbf{j} + \mathbf{k}$ and $\mathbf{B} = \mathbf{i} + 2\mathbf{j} - 2\mathbf{k}$. Find the component of **A** along **B**, and the component of **B** along **A**.

Solution: By definition the component of **A** along **B** is given by $|\mathbf{A}|\cos\theta$. By (1.24) this is equal to

$$|\mathbf{A}|\cos\theta = \mathbf{A}\cdot\mathbf{B}/|\mathbf{B}|$$

Since

$$\mathbf{A}\cdot\mathbf{B} = 2(1) + (-1)2 + 1(-2) = -2, \quad |\mathbf{B}| = 3$$

it follows that

$$|\mathbf{A}|\cos\theta = \mathbf{A}\cdot\mathbf{B}/|\mathbf{B}| = -2/3$$

Similarly, since $|\mathbf{A}| = \sqrt{6}$, the component of **B** along **A** is given by

$$|\mathbf{B}|\cos\theta = \mathbf{A}\cdot\mathbf{B}/|\mathbf{A}| = -2/\sqrt{6} = -\sqrt{6}/3.$$

Example 2. Find the angle between the vectors $\mathbf{A} = 2\mathbf{i} - \mathbf{j} - 2\mathbf{k}$ and $\mathbf{B} = \mathbf{i} + 2\mathbf{j} + 2\mathbf{k}$.

Solution: By (1.20) we have $\mathbf{A}\cdot\mathbf{B} = 2(1) + (-1)2 + (-2)2 = -4$. Since $|\mathbf{A}| = 3$ and $|\mathbf{B}| = 3$, it follows from (1.25) that $\cos\theta = -4/9$.

Thus the angle between the vectors is

$$\theta = \arccos(-4/9) = 116^{\circ}\,23$$

From the formula (1.24) we see immediately that two nonzero vectors **A** and **B** are orthogonal if and only if their scalar product is zero. Indeed, if the vectors are orthogonal, then the angle θ between them is $\pi/2$. Hence $\cos\theta = 0$ and by (1.24), $\mathbf{A}\cdot\mathbf{B} = 0$. Conversly, if $\mathbf{A}\cdot\mathbf{B} = 0$ and **A** and **B** are nonzero vectors, then by (1.24), $\cos\theta = 0$. This implies that $\theta = \pi/2$, which says that the vectors are orthogonal. We state this important result as a theorem.

Theorem 1. Two nonzero vectors **A** and **B** are orthogonal if and only if $\mathbf{A}\cdot\mathbf{B} = 0$.

For the unit base vectors **i**, **j**, and **k**, we have

$$\begin{array}{lll} \mathbf{i}\cdot\mathbf{i} = 1, & \mathbf{i}\cdot\mathbf{j} = 0, & \mathbf{i}\cdot\mathbf{k} = 0 \\ \mathbf{j}\cdot\mathbf{i} = 0, & \mathbf{j}\cdot\mathbf{j} = 1, & \mathbf{j}\cdot\mathbf{k} = 0 \\ \mathbf{k}\cdot\mathbf{i} = 0, & \mathbf{k}\cdot\mathbf{j} = 0, & \mathbf{k}\cdot\mathbf{k} = 1 \end{array}$$

The set of base vectors (**i**, **j**, **k**) is said to form an orthonormal basis.

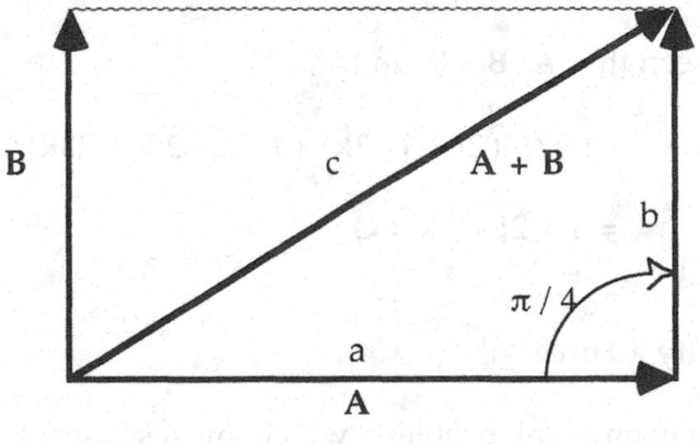

$$c^2 = a^2 + b^2$$

Fig. 1.22 The Pythagorean theorem.

Example 3. Prove that if a and b are the two legs of a right triangle, and c is its hypothenuse, then $a^2 + b^2 = c^2$. (This is the well known Pythagorean theorem in plane geometry.)

Solution: Let the vectors **A** and **B** represent the legs of the right triangle as shown in Fig. 1.22. Then **A** + **B** represents the hypothenuse. Since **A** is perpendicular to **B**, $\mathbf{A}\cdot\mathbf{B} = 0$. Hence

$$c^2 = |\mathbf{A} + \mathbf{B}|^2 = (\mathbf{A} + \mathbf{B})\cdot(\mathbf{A} + \mathbf{B}) = \mathbf{A}\cdot\mathbf{A} + \mathbf{B}\cdot\mathbf{A} + \mathbf{A}\cdot\mathbf{B} + \mathbf{B}\cdot\mathbf{B}$$

$$= |\mathbf{A}|^2 + |\mathbf{B}|^2 = a^2 + b^2$$

Example 3. Let $\mathbf{A} = 2\mathbf{i} + \mathbf{j} - 2\mathbf{k}$ and $\mathbf{C} = \mathbf{i} + 2\mathbf{j} + 3\mathbf{k}$. Express the vector **C** as the sum of a vector that is parallel to **A** and a vector that is orthogonal to **A**.

Solution: A vector that is parallel to **A** is of the form $m\,\mathbf{A}$, for some scalar m. Let **B** be a vector that is orthogonal to **A**. We wish to find m and **B** so that $\mathbf{C} = m\,\mathbf{A} + \mathbf{B}$. Let us take the scalar product of **C** and **A**. Since **B** is orthogonal to **A**, we obtain

$$\mathbf{A}\cdot\mathbf{C} = (m\mathbf{A} + \mathbf{B})\cdot\mathbf{A} = m\mathbf{A}\cdot\mathbf{A} = m\,|\mathbf{A}|^2$$

Since $\mathbf{A}\cdot\mathbf{C} = 2(1) + 1(2) + (-2)3 = -2$ and $|\mathbf{A}| = 3$, we find

$$m = \mathbf{A}\cdot\mathbf{C}\,/\,|\mathbf{A}|^2 = -2/9$$

Hence

$$\mathbf{B} = \mathbf{C} - m\,\mathbf{A} = (13\mathbf{i} + 20\mathbf{j} + 23\mathbf{k})/9.$$

It is readily seen that $\mathbf{A}\cdot\mathbf{B} = 0$ and

$$(-2/9)\mathbf{A} + \mathbf{B} = (-2/9)(2\mathbf{i} + \mathbf{j} - 2\mathbf{k}) + (13\mathbf{i} + 20\mathbf{j} + 23\mathbf{k})/9$$

$$= \mathbf{i} + 2\mathbf{j} + 3\mathbf{k} = \mathbf{C}$$

Work Done By a Force

A simple physical problem which makes use of the scalar product of vectors is the problem of calculating the amount of work done by a force when it displaces an object through a certain distance. Suppose that a constant force **F** acts on an object and

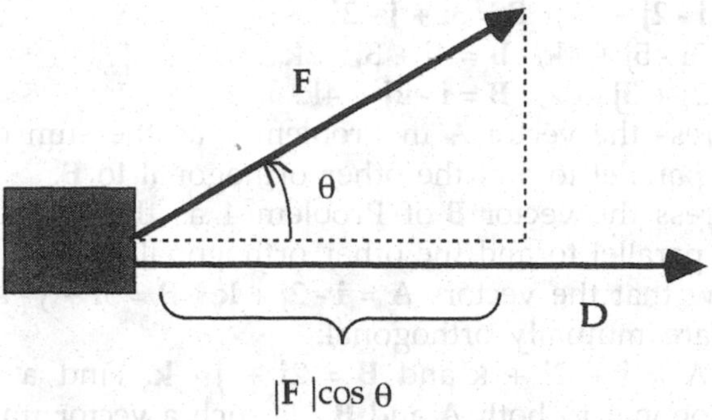

Fig. 1.23 Work done by a force through a displacement.

causes the object to be moved through a distance represented by the displacement vector **D** (Fig. 1.23). In physics, the work done by the force is defined as the product of the displacement and the component of the force along the displacement. Hence if θ is the angle between the force and the displacement, then the component of the force along the displacement is equal to $|\mathbf{F}|\cos\theta$. Thus the work done by the force is given by

$$\text{Work} = |\mathbf{D}|\,|\mathbf{F}|\cos\theta = \mathbf{F}\cdot\mathbf{D}$$

Example 5. If a constant force $\mathbf{F} = 2\mathbf{i} - \mathbf{j} + 3\mathbf{k}$ (in kg), acting on an object, moves the object through a displacement $\mathbf{D} = 3\mathbf{i} + 2\mathbf{j} + 4\mathbf{k}$ (in meters), then the work done by the force is equal to

$$\text{Work} = \mathbf{F}\cdot\mathbf{D} = 2(3) + (-1)2 + 3(4) = 16 \text{ kg-m.}$$

1.4 EXERCISES

In each of Problems 1 through 4, find the orthogonal projection of $|\mathbf{A}|$ along **B**, and determine the cosine of the angle between the vectors.

1. $\mathbf{A} = 2\mathbf{i} - 2\mathbf{j} + \mathbf{k}$, $\mathbf{B} = -\mathbf{i} - 2\mathbf{j} + 2\mathbf{k}$.
2. $\mathbf{A} = \mathbf{i} - 2\mathbf{j} + 3\mathbf{k}$, $\mathbf{B} = 3\mathbf{i} + \mathbf{j} - 2\mathbf{k}$.
3. $\mathbf{A} = 3\mathbf{i} - 5\mathbf{j} + 4\mathbf{k}$, $\mathbf{B} = 4\mathbf{i} + 3\mathbf{j} - 2\mathbf{k}$.
4. $\mathbf{A} = 2\mathbf{i} + 3\mathbf{j} - 6\mathbf{k}$, $\mathbf{B} = \mathbf{i} - 2\mathbf{j} + 4\mathbf{k}$.
5. Express the vector **A** in Problem 1 as the sum of two vectors, one parallel to and the other orthogonal to **B**.
6. Express the vector **B** of Problem 4 as the sum of two vectors, one parallel to and the other orthogonal to **A**.
7. Show that the vectors $\mathbf{A} = \mathbf{i} - 2\mathbf{j} + \mathbf{k}$, $\mathbf{B} = 3\mathbf{i} + \mathbf{j} - \mathbf{k}$, $\mathbf{C} = \mathbf{i} + 4\mathbf{j} + 7\mathbf{k}$ are mutually orthogonal.
8. Let $\mathbf{A} = \mathbf{i} + 2\mathbf{j} + \mathbf{k}$ and $\mathbf{B} = 2\mathbf{i} + \mathbf{j} - \mathbf{k}$. Find a vector that is orthogonal to both **A** and **B**. Is such a vector uniquely determined?
9. Let $\mathbf{A} = \mathbf{i} + 3\mathbf{j} - 4\mathbf{k}$ and $\mathbf{B} = 2\mathbf{i} - 3\mathbf{j} + 5\mathbf{k}$. Find the value of m so that $\mathbf{A} + m\mathbf{B}$ is orthogonal to (a) **A**, (b) **B**.
10. Let $\mathbf{A} = 3\mathbf{i} - \mathbf{j} + 5\mathbf{k}$ and $\mathbf{B} = 2\mathbf{i} - 4\mathbf{j} - 3\mathbf{k}$. Find the value of m so that $m\mathbf{A} + \mathbf{B}$ is orthogonal to (a) **A**, (b) **B**.
11. Let $\mathbf{A} = \mathbf{i} + 3\mathbf{j} + \mathbf{k}$, $\mathbf{B} = 2\mathbf{i} - \mathbf{j} + 2\mathbf{k}$, and $\mathbf{C} = -3\mathbf{i} + 2\mathbf{j} - \mathbf{k}$. Find the values of the scalars m and n so that the vector $\mathbf{C} - m\mathbf{A} - n\mathbf{B}$ is orthogonal to both **A** and **B**.
12. Repeat Problem 11 if $\mathbf{A} = 3\mathbf{i} - 2\mathbf{j} + 5\mathbf{k}$, $\mathbf{B} = 2\mathbf{i} + \mathbf{j} - \mathbf{k}$, and $\mathbf{C} = -\mathbf{i} + 4\mathbf{j} - 6\mathbf{k}$.
13. Let $\mathbf{A} = \mathbf{i} - 2\mathbf{j} + \mathbf{k}$, $\mathbf{B} = \mathbf{i} + \mathbf{j} - 2\mathbf{k}$, and $\mathbf{C} = 2\mathbf{i} - \mathbf{j} + \mathbf{k}$. Find all the unit vectors that are linear combinations of **A** and **B** and that are orthogonal to **C**.
14. A constant force $\mathbf{F} = 3\mathbf{i} + \mathbf{j} - 2\mathbf{k}$ moves an object through a displacement $\mathbf{D} = -2\mathbf{i} + 2\mathbf{j} + \mathbf{k}$. What is the work done by the force?
15. A constant force $\mathbf{F} = 2\mathbf{i} + 5\mathbf{j} - 3\mathbf{k}$ moves an object through consecutive displacements $\mathbf{D}_1 = 3\mathbf{i} - 2\mathbf{j}$ and $\mathbf{D}_2 = \mathbf{i} + 4\mathbf{j} - 2\mathbf{k}$. Find the total work done by the force.
16. Let $\mathbf{A} = \cos\alpha\,\mathbf{i} + \sin\alpha\,\mathbf{j}$ and $\mathbf{B} = \cos\beta\,\mathbf{i} + \sin\beta\,\mathbf{j}$. By taking the dot product of **A** and **B**, derive the trigonometric identity

$$\cos(\alpha - \beta) = \cos\alpha\cos\beta + \sin\alpha\sin\beta$$

17. Show that a set of orthogonal nonzero vectors $(\mathbf{A}_1, \mathbf{A}_2, \ldots, \mathbf{A}_n)$ is linearly independent.

18. Let

$$\mathbf{A} = (\mathbf{i} - \mathbf{j} + \mathbf{k})/\sqrt{3}, \quad \mathbf{B} = (2\mathbf{i} + \mathbf{j} - \mathbf{k})/\sqrt{6}, \quad \mathbf{C} = (\mathbf{j} + \mathbf{k})/\sqrt{2}.$$

Show that the set $(\mathbf{A}, \mathbf{B}, \mathbf{C})$ is an orthonormal basis.

19. Prove the Cauchy-Schwarz inequality $|\mathbf{A} \cdot \mathbf{B}| \le |\mathbf{A}|\,|\mathbf{B}|$.
20. Prove the triangle inequality $|\mathbf{A} + \mathbf{B}| \le |\mathbf{A}| + |\mathbf{B}|$. (Hint: Consider $|\mathbf{A} + \mathbf{B}|^2 = (\mathbf{A}+\mathbf{B}) \cdot (\mathbf{A}+\mathbf{B})$ and use the result of Problem 19.)
21. Deduce $|\mathbf{A} - \mathbf{B}| \ge \big|\,|\mathbf{A}| - |\mathbf{B}|\,\big|$ from the triangle inequality.
22. Show that

$$\text{(a)} \quad |\mathbf{A} + \mathbf{B}|^2 + |\mathbf{A} - \mathbf{B}|^2 = 2|\mathbf{A}|^2 + 2|\mathbf{B}|^2$$
$$\text{(b)} \quad |\mathbf{A} + \mathbf{B}|^2 - |\mathbf{A} - \mathbf{B}|^2 = 4\mathbf{A} \cdot \mathbf{B}$$

23. Prove that the vectors **A** and **B** are orthogonal if and only if $|\mathbf{A} + \mathbf{B}| = |\mathbf{A} - \mathbf{B}|$. What is the geometric interpretation of this?
24. Let **A** and **B** be two vectors and set $m = |\mathbf{A}|$, $n = |\mathbf{B}|$. (a) Show that the vectors $n\mathbf{A} + m\mathbf{B}$ and $n\mathbf{A} - m\mathbf{B}$ are orthogonal. (b) If $\mathbf{C} = (n\mathbf{A} + m\mathbf{B})/(m + n)$, show that the angle between **A** and **C** is the same as the angle between **B** and **C**.
25. Using the scalar product, prove that a triangle inscribed in a semicircle with one of its sides along the diameter is a right triangle.
26. As in Problem 25, prove that the line joining the vertex of an isosceles triangle to the midpoint of its base is perpendicular to the base.
27. Prove **Appolonius** theorem which states that in a triangle ABC,

$$(AB)^2 + (AC)^2 = 2(AD)^2 + (1/2)(BC)^2$$

where D is the midpoint of BC.

1.7 VECTOR PRODUCT

In Sec. 1.6 we defined a product ot two vectors that always yields a scalar or a real number. There is another type of product of two vectors in which the result is another vector. The resulting new vector has the distinct property of being orthogonal to the two given vectors. This product is called the *vector product* or *cross product* and it is defined as follows:

Definition 1. The vector product of two vectors $\mathbf{A} = a_1\mathbf{i} + a_2\mathbf{j} + a_3\mathbf{k}$ and $\mathbf{B} = b_1\mathbf{i} + b_2\mathbf{j} + b_3\mathbf{k}$, denoted by $\mathbf{A} \times \mathbf{B}$ (read **A** cross **B**), is the vector defined by

$$\mathbf{A} \times \mathbf{B} = (a_2 b_3 - a_3 b_2)\mathbf{i} + (a_3 b_1 - a_1 b_3)\mathbf{j} + (a_1 b_2 - a_2 b_1)\mathbf{k} \tag{1.26}$$

Because of its notation, this product is sometimes called the cross product of the vectors **A** and **B**. In contrast to the term inner product for the dot product of two vectors, the product (1.26) is also known as the outer product of **A** and **B**.

Example 1. If $\mathbf{A} = 3\mathbf{i} - 2\mathbf{j} + 4\mathbf{k}$ and $\mathbf{B} = 2\mathbf{i} + 3\mathbf{j} + \mathbf{k}$, then

$$\begin{aligned}\mathbf{A} \times \mathbf{B} &= (3\mathbf{i} - 2\mathbf{j} + 4\mathbf{k}) \times (2\mathbf{i} + 3\mathbf{j} + \mathbf{k})\\ &= (-2 - 4(3))\mathbf{i} + (4(2) - 3)\mathbf{j} + (3(3) - (-2)2)\mathbf{k}\\ &= -14\mathbf{i} + 5\mathbf{j} + 13\mathbf{k}\end{aligned}$$

At first glance, it may appear that the definition (1.26) is too complicated to remember. Actually it is rather easy as all we need to remember is the first component of the vector product, $a_2b_3 - a_3b_2$. We observe that the first term a_2b_3 is the product of the second component a_2 of **A** and the third component b_3 of **B**, while the second term a_3b_2 is just the reverse, being the product of the third component of **A** and the second component of **B**. Thus the first component involves the subscripts 2 and 3, pertaining to the components of **A** and **B**. Knowing the first component, the second component is then obtained by cyclic permutation of 1, 2, 3; that is, the second component involves the subscripts 3, 1, and finally, the third component involves the subscripts 1, 2. Thus, the components of the vector product $\mathbf{A} \times \mathbf{B}$ are:

	1st		2nd		3rd
	$a_2 b_3 - a_3 b_2$	->	$a_3 b_1 - a_1 b_3$	->	$a_1 b_2 - a_2 b_1$
Subscripts:	2 3		3 1		1 2

A more convenient formula for the cross product (1.26) which makes use of the notion of a determinant of a matrix is given in (1.29). For the sake of those who may not be familiar with the concept of a matrix and its determinant, we present briefly the basic ideas below.

Matrices and Determinants

We use capital letters A, B, . . . to denote matrices, and $|A|$ or det A to denote the determinant of a matrix A.

Definition 2. A 2 x 2 (read two by two) matrix A is an array of (real) numbers a_1, a_2, b_1, b_2 written as

$$A = \begin{bmatrix} a_1 & a_2 \\ b_1 & b_2 \end{bmatrix}$$

The numbers a_1, a_2 are called the elements of the first row, and b_1, b_2 the elements of the second row. On the other hand, the numbers a_1, b_1 are called the elements of the first column, and a_2, b_2 the elements of the second column.

The **determinant** of a 2 x 2 matrix A, denoted by $|A|$, is a real number defined by the formula

$$|A| = \begin{vmatrix} a_1 & a_2 \\ b_1 & b_2 \end{vmatrix} = a_1 b_2 - a_2 b_1 \tag{1.27}$$

This is called a determinant of the second order.

For example:

$$\begin{vmatrix} 2 & 1 \\ 3 & 2 \end{vmatrix} = 2(2) - 1(3) = 1$$

$$\begin{vmatrix} -1 & 2 \\ 1 & 1 \end{vmatrix} = -1(1) - 2(1) = -3$$

$$\begin{vmatrix} 1 & 2 \\ 2 & 4 \end{vmatrix} = 1(4) - 2(2) = 0$$

From the definition (1.27), it is easily seen that if we interchange the rows or the columns of the matrix, the determinant of the resulting matrix changes sign. For example, if we interchange the rows, we see that

$$\begin{vmatrix} b_1 & b_2 \\ a_1 & a_2 \end{vmatrix} = a_2b_1 - a_1b_2 = -(a_1b_2 - a_2b_1) = -\begin{vmatrix} a_1 & a_2 \\ b_1 & b_2 \end{vmatrix}$$

If we interchange the columns, we obtain

$$\begin{vmatrix} a_2 & a_1 \\ b_2 & b_1 \end{vmatrix} = a_2b_1 - a_1b_2 = -(a_1b_2 - a_2b_1) = -\begin{vmatrix} a_1 & a_2 \\ b_1 & b_2 \end{vmatrix}$$

Next, we define a 3 x 3 matrix A as an array of nine numbers a_i, b_i, c_i (i = 1, 2, 3), written as

$$A = \begin{bmatrix} a_1 & a_2 & a_3 \\ b_1 & b_2 & b_3 \\ c_1 & c_2 & c_3 \end{bmatrix}$$

The numbers are again arranged in three rows and three columns.

The determinant of a 3 x 3 matrix is defined in terms of determinants of the second order and it is given by the formula

$$|\mathbf{A}| = \begin{vmatrix} a_1 & a_2 & a_3 \\ b_1 & b_2 & b_3 \\ c_1 & c_2 & c_3 \end{vmatrix}$$

$$= a_1\begin{vmatrix} b_2 & b_3 \\ c_2 & c_3 \end{vmatrix} - a_2\begin{vmatrix} b_1 & b_3 \\ c_1 & c_3 \end{vmatrix} + a_3\begin{vmatrix} b_1 & b_2 \\ c_1 & c_2 \end{vmatrix} \qquad (1.28)$$

$$= a_1(b_2c_3 - b_3c_2) - a_2(b_1c_3 - b_3c_1) + a_3(b_1c_2 - b_2c_1)$$

This is called a determinant of the third order .

Without some mnemonic device, formula (1.28) would be difficult to memorize. The rule to remember here is that the right-hand side of (1.28) consists of the algebraic sum of products of an element a_i $(1 \le i \le 3)$ of the first row and the determinant of a 2 x 2 matrix obtained when the elements of the first row and the elements of the i-th column (the column in which a_i lies) are deleted. The signs of the terms alternate beginning with a plus sign in the first term. Thus, the second order determinant in the second term on the right-hand side of (1.28) is the determinant of the resulting matrix when the elements of the first row and the elements of the second column are deleted:

$$\begin{bmatrix} a_1 - - a_2 - - a_3 \\ b_1 \quad b_2 \quad b_3 \\ c_1 \quad c_2 \quad c_3 \end{bmatrix}$$

With the formula (1.28) we say that the determinant is given by expansion by the elements of the first row. The second order determinants appearing in (1.28) together with the algebraic sign are called the *cofactors* of the elements in the first row. That is, the determinants

$$\begin{vmatrix} b_2 & b_3 \\ c_2 & c_3 \end{vmatrix}, \quad -\begin{vmatrix} b_1 & b_3 \\ c_1 & c_3 \end{vmatrix}, \quad \begin{vmatrix} b_1 & b_2 \\ c_1 & c_2 \end{vmatrix}$$

are the cofactor of the elements a_1, a_2, a_3, respectively.

For example:

$$\begin{vmatrix} 2 & 1 & 3 \\ 1 & 0 & 4 \\ 3 & 2 & 0 \end{vmatrix} = 2\begin{vmatrix} 0 & 4 \\ 2 & 0 \end{vmatrix} - 1\begin{vmatrix} 1 & 4 \\ 3 & 0 \end{vmatrix} + 3\begin{vmatrix} 1 & 0 \\ 3 & 2 \end{vmatrix}$$

$$= 2(0 - 8) - (0 - 12) + 3(2 - 0) = 2$$

$$\begin{vmatrix} -3 & 0 & 2 \\ 0 & -2 & 1 \\ 1 & 4 & -5 \end{vmatrix} = -3\begin{vmatrix} -2 & 1 \\ 4 & -5 \end{vmatrix} - 0\begin{vmatrix} 0 & 1 \\ 1 & -5 \end{vmatrix} + 2\begin{vmatrix} 0 & -2 \\ 1 & 4 \end{vmatrix}$$

$$= -3(10 - 4) - 0 + 2(0 + 2) = -14$$

Just like the second order determinant, the determinant of a 3 x 3 matrix also changes sign whenever any two rows or any two columns of the matrix are interchanged. We leave this to the reader to verify. For example, if we interchange the first and the third columns of the last example above, we obtain

$$\begin{vmatrix} 2 & 0 & -3 \\ 1 & -2 & 0 \\ -5 & 4 & 1 \end{vmatrix} = 2\begin{vmatrix} -2 & 0 \\ 4 & 1 \end{vmatrix} - 0\begin{vmatrix} 1 & 0 \\ -5 & 1 \end{vmatrix} + (-3)\begin{vmatrix} 1 & -2 \\ -5 & 4 \end{vmatrix}$$

$$= 2(-2 - 0) + 0 + (-3)(4 - 10) = 14$$

Other important properties of determinants are given in the exercises.

Formula for the Vector Product A x B

Now let us return to the definition of vector product. From the formulas (1.26) and (1.27), we observe that

$$\mathbf{A} \times \mathbf{B} = (a_2 b_3 - a_3 b_2)\mathbf{i} + (a_3 b_1 - a_1 b_3)\mathbf{j} + (a_1 b_2 - a_2 b_1)\mathbf{k}$$

$$= \begin{vmatrix} a_2 & a_3 \\ b_2 & b_3 \end{vmatrix}\mathbf{i} - \begin{vmatrix} a_1 & a_3 \\ b_1 & b_3 \end{vmatrix}\mathbf{j} + \begin{vmatrix} a_1 & a_2 \\ b_1 & b_2 \end{vmatrix}\mathbf{k}$$

Comparing this with the formula (1.28), we deduce the following alternative representation for the vector product

$$\mathbf{A} \times \mathbf{B} = \begin{vmatrix} \mathbf{i} & \mathbf{j} & \mathbf{k} \\ a_1 & a_2 & a_3 \\ b_1 & b_2 & b_3 \end{vmatrix} \tag{1.29}$$

For example, if $\mathbf{A} = 2\mathbf{i} + \mathbf{j} + 3\mathbf{k}$ and $\mathbf{B} = \mathbf{i} - 2\mathbf{j} + \mathbf{k}$, then

$$\mathbf{A} \times \mathbf{B} = \begin{vmatrix} \mathbf{i} & \mathbf{j} & \mathbf{k} \\ 2 & 1 & 3 \\ 1 & -2 & 1 \end{vmatrix} = \begin{vmatrix} 1 & 3 \\ -2 & 1 \end{vmatrix} \mathbf{i} - \begin{vmatrix} 2 & 3 \\ 1 & 1 \end{vmatrix} \mathbf{j} + \begin{vmatrix} 2 & 1 \\ 1 & -2 \end{vmatrix} \mathbf{k}$$

$$= 7\mathbf{i} + \mathbf{j} - 5\mathbf{k}$$

From the definition of vector product, it can be easily shown that for any vectors **A**, **B**, **C** and for any scalar m, the following properties hold:

$$\begin{aligned} &(a)\quad \mathbf{A} \times \mathbf{B} = -(\mathbf{B} \times \mathbf{A}) \\ &(b)\quad \mathbf{A} \times (\mathbf{B} + \mathbf{C}) = (\mathbf{A} \times \mathbf{B}) + (\mathbf{A} \times \mathbf{C}) \\ &(c)\quad m(\mathbf{A} \times \mathbf{B}) = (m\mathbf{A}) \times \mathbf{B} = \mathbf{A} \times (m\mathbf{B}) \\ &(d)\quad \mathbf{A} \times \mathbf{A} = 0 \end{aligned} \tag{1.30}$$

These properties correspond to those listed in equation (1.21) for the scalar product. Unlike the scalar product, however, (1.30a) implies that vector product is not commutative. This property means that the vectors **A** x **B** and **B** x **A** are equal in magnitude but opposite in direction. Thus the order of the factors in a vector product makes a difference. This property should be kept in mind as this may well be the student's first encounter with an operation bearing the name of "product" which is not commutative.

Property (1.30d) says that the cross product of a vector with itself results in the zero vector, even though the vector itself may not be the zero vector. This is quite in contrast to the scalar product of a vector with itself, that is, $\mathbf{A}\cdot\mathbf{A} = |\mathbf{A}|^2$, which is zero if and only if **A** is the zero vector.

Geometric Interpretation of A x B

As in the case of the scalar product, the vector product **A** x **B** also has an important geometric interpretation. First, we show that the vector **A** x **B** is orthogonal to both the vectors **A** and **B**. From the definition of scalar product, we see that

$$\mathbf{A}\cdot(\mathbf{A} \times \mathbf{B}) = a_1(a_2b_3 - a_3b_2) + a_2(a_3b_1 - a_1b_3) + a_3(a_1b_2 - a_2b_1) = 0$$

and similarly, $\mathbf{B}\cdot(\mathbf{A}\times\mathbf{B}) = 0$. Thus $\mathbf{A}\times\mathbf{B}$ is orthogonal to both the vectors **A** and **B**. Further, when the vectors are represented geometrically by arrows, the direction of the vector $\mathbf{A}\times\mathbf{B}$ is related to the direction of **A** and **B** by the right-hand rule (Fig. 1.24). That is, if the index finger of the right hand points along the vector **A** and the center finger points along the vector **B**, then the thumb points in the direction of $\mathbf{A}\times\mathbf{B}$ (assuming that the thumb is perpendicular to the two fingers). If the vectors **A** and **B** are orthogonal, that is, $\mathbf{A}\cdot\mathbf{B} = 0$, then the vectors $\mathbf{A}, \mathbf{B}, \mathbf{A}\times\mathbf{B}$, in that order, form a right-handed triple of vectors that are mutually orthogonal.

For the unit vectors **i**, **j**, **k**, it follows that

$$\begin{array}{lll} \mathbf{i}\times\mathbf{i} = 0, & \mathbf{j}\times\mathbf{j} = 0, & \mathbf{k}\times\mathbf{k} = 0 \\ \mathbf{i}\times\mathbf{j} = \mathbf{k}, & \mathbf{j}\times\mathbf{k} = \mathbf{i}, & \mathbf{k}\times\mathbf{i} = \mathbf{j} \end{array}$$

which of course can also be verified directly from the formula (1.29).

Next, we consider the magnitude of the vector $\mathbf{A}\times\mathbf{B}$. For this purpose, we need the following identity

$$|\mathbf{A}\times\mathbf{B}|^2 = |\mathbf{A}|^2|\mathbf{B}|^2 - (\mathbf{A}\cdot\mathbf{B})^2 \tag{1.31}$$

This identity can be established directly by showing that the two expressions

$$|\mathbf{A}\times\mathbf{B}|^2 = (a_2b_3 - a_3b_2)^2 + (a_3b_1 - a_1b_3)^2 + (a_1b_2 - a_2b_1)^2$$

and

$$|\mathbf{A}|^2|\mathbf{B}|^2 - (\mathbf{A}\cdot\mathbf{B})^2 = (a_1^2 + a_2^2 + a_3^2)(b_1^2 + b_2^2 + b_3^2) - (a_1b_1 + a_2b_2 + a_3b_3)^2$$

are indeed equal. (See also Example 5, Sec. 1.9.) Now let θ denote the angle between **A** and **B**, and let us write $\mathbf{A}\cdot\mathbf{B} = |\mathbf{A}||\mathbf{B}|\cos\theta$. Then the identity (1.31) becomes

$$|\mathbf{A}\times\mathbf{B}|^2 = |\mathbf{A}|^2|\mathbf{B}|^2(1-\cos^2\theta) = |\mathbf{A}|^2|\mathbf{B}|^2\sin^2\theta$$

from which we obtain

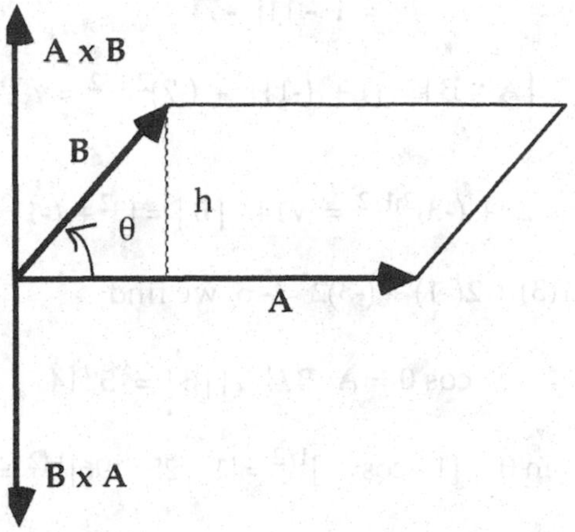

Fig. 1.24 Geometric interpretation of the vector product.

$$|\mathbf{A} \times \mathbf{B}| = |\mathbf{A}||\mathbf{B}| \sin\theta \tag{1.32}$$

This gives the magnitude of the vector **A** x **B**.

Example 2. Find all the unit vectors that are orthogonal to the vectors $\mathbf{A} = \mathbf{i} + 2\mathbf{j} - \mathbf{k}$ and $\mathbf{B} = 2\mathbf{i} - 2\mathbf{j} + \mathbf{k}$.

Solution: We know that the vector

$$\mathbf{A} \times \mathbf{B} = (\mathbf{i} + 2\mathbf{j} - \mathbf{k}) \times (2\mathbf{i} - 2\mathbf{j} + \mathbf{k}) = -3\mathbf{j} - 6\mathbf{k}$$

is orthogonal to the vectors **A** and **B**. By normalizing this, we obtain the unit vectors

$$\mathbf{u} = \pm (\mathbf{j} + 2\mathbf{k})/\sqrt{5}$$

Example 3. Let $\mathbf{A} = \mathbf{i} + 2\mathbf{j} - 3\mathbf{k}$ and $\mathbf{B} = 3\mathbf{i} - \mathbf{j} + 2\mathbf{k}$. Find **A** x **B** and verify the formula (1.32).

Solution: By (1.29) we have

$$\begin{vmatrix} \mathbf{i} & \mathbf{j} & \mathbf{k} \\ 1 & 2 & -3 \\ 3 & -1 & 2 \end{vmatrix} = \begin{vmatrix} 2 & -3 \\ -1 & 2 \end{vmatrix} \mathbf{i} - \begin{vmatrix} 1 & -3 \\ 3 & 2 \end{vmatrix} \mathbf{j} + \begin{vmatrix} 1 & 2 \\ 3 & -1 \end{vmatrix} \mathbf{k}$$

$$= \mathbf{i} - 11\mathbf{j} - 7\mathbf{k}$$

Therefore

$$|\mathbf{A} \times \mathbf{B}| = [1 + (-11)^2 + (-7)^2]^{1/2} = \sqrt{171}$$

Since

$$|\mathbf{A}| = [1^2 + 2^2 + (-3)^2]^{1/2} = \sqrt{14}, \quad |\mathbf{B}| = [3^2 + (-1)^2 + 2^2]^{1/2} = \sqrt{14}$$

and $\mathbf{A} \cdot \mathbf{B} = 1(3) + 2(-1) + (-3)2 = -5$, we find

$$\cos\theta = \mathbf{A} \cdot \mathbf{B} / |\mathbf{A}||\mathbf{B}| = -5/14$$

Hence

$$\sin\theta = [1 - \cos^2\theta]^{1/2} = [1 - 25/196]^{1/2} = \sqrt{171}/14$$

and so

$$|\mathbf{A}||\mathbf{B}| \sin\theta = (\sqrt{14})(\sqrt{14})(\sqrt{171}/14) = \sqrt{171}$$

as we wish to verify.

Referring to Figure 1.24, we see that $|\mathbf{B}|\sin\theta$ represents the altitude of the parallelogram formed by the vectors **A** and **B**, and $|\mathbf{A}|$ represents its base. Since the area of a parallelogram is equal to the product of its base and its altitude, it follows from (1.32) that the magnitude of the vector $\mathbf{A} \times \mathbf{B}$ represents numerically the area of the parallelogram formed by the vectors **A** and **B**. This has an important geometric application as we shall see later.

In Sec. 1.6 we saw that two nonzero vectors **A** and **B** are orthogonal if and only if $\mathbf{A} \cdot \mathbf{B} = 0$. If the vectors are parallel so that the angle between them is either 0 or π, then it follows from (1.32) that $|\mathbf{A} \times \mathbf{B}| = |\mathbf{A}||\mathbf{B}| \sin\theta = 0$. This implies that $\mathbf{A} \times \mathbf{B} = \mathbf{0}$, the zero vector. Conversely, if $\mathbf{A} \times \mathbf{B} = \mathbf{0}$, then $|\mathbf{A} \times \mathbf{B}| = 0$. Since $\mathbf{A} \neq \mathbf{0}$, $\mathbf{B} \neq \mathbf{0}$, we infer from (1.32) that $\sin\theta = 0$. Hence $\theta = 0$ or $\theta = \pi$, which means that the vectors are parallel. We state this result as a theorem.

Theorem 1. Two nonzero vectors **A** and **B** are parallel if and only if $\mathbf{A} \times \mathbf{B} = 0$.

Example 4. Let $\mathbf{A} = 3\mathbf{i} + 2\mathbf{j} - 6\mathbf{k}$ and $\mathbf{B} = \mathbf{i} - 2\mathbf{j} + 2\mathbf{k}$. Find the area of the triangle formed by the vectors **A**, **B**, and **A** - **B**.

Solution: For any vectors **A** and **B**, we know that $|\mathbf{A} \times \mathbf{B}|$

represents the area of the parallelogram formed by the vectors **A** and **B**. Since the area of the triangle formed by the vectors **A**, **B** and **A - B** is one half that of the parallelogram (Fig. 1.25), it follows that

$$\text{area of triangle} = (1/2)\,|\mathbf{A} \times \mathbf{B}|$$

Now

$$\mathbf{A} \times \mathbf{B} = (3\mathbf{i} + 2\mathbf{j} - 6\mathbf{k}) \times (\mathbf{i} - 2\mathbf{j} + 2\mathbf{k}) = -8\mathbf{i} - 12\mathbf{j} - 8\mathbf{k}$$

and

$$|\mathbf{A} \times \mathbf{B}| = [(-8)^2 + (-12)^2 + (-8)^2]^{1/2} = 4\sqrt{17}$$

Hence the area of the triangle is $2\sqrt{17}$ square units.

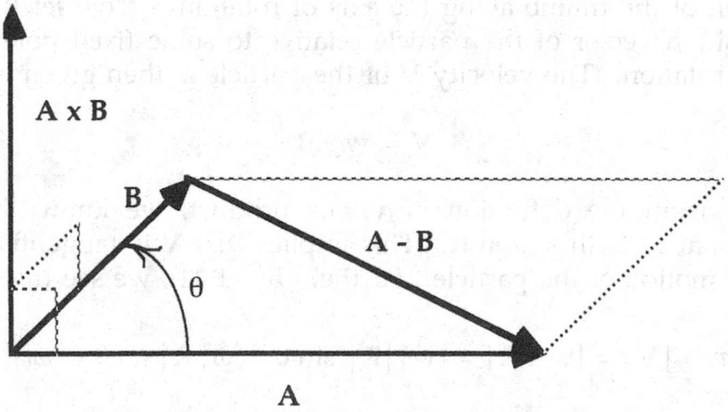

Fig. 1.25 Area of a triangle formed by the vectors A, B, and A - B.

Example 5. Find the area of the triangle with vertices at the points P: (1, 2, 3), Q: (-1, 3, 2), and R: (3, -1, 2).

Solution: Let **A** and **B** denote the vectors represented by the directed line segments *PQ* and *PR*, respectively. Then

$$\mathbf{A} = -2\,\mathbf{i} + \mathbf{j} - \mathbf{k} \quad \text{and} \quad \mathbf{B} = 2\,\mathbf{i} - 3\,\mathbf{j} - \mathbf{k}$$

Taking the cross product of these vectors, we find

$$\mathbf{A} \times \mathbf{B} = \begin{vmatrix} \mathbf{i} & \mathbf{j} & \mathbf{k} \\ -2 & 1 & -1 \\ 2 & -3 & -1 \end{vmatrix} = -4\,\mathbf{i} - 4\,\mathbf{j} + 4\,\mathbf{k}$$

Since $|\mathbf{A} \times \mathbf{B}| = 4\sqrt{3}$, the area of the triangle is $2\sqrt{3}$ square units.

Some Applications of the Vector Product

As an example of the use of vector product, let us consider the motion of a particle P that is revolving at an angular speed ω about a fixed axis (Fig. 1.26). By convention the angular velocity of the particle, which is a vector quantity, is represented by a vector **w** whose magnitude is equal to ω and whose direction is along the axis of rotation as determined by the right-hand rule. (Let the fingers of the right hand point in the direction of rotation. Then **w** points in the direction of the thumb along the axis of rotation). Now let **R** denote the position vector of the particle relative to some fixed point on the axis of rotation. The velocity **V** of the particle is then given by

$$\mathbf{V} = \mathbf{w} \times \mathbf{R} \tag{1.33}$$

In fact, from the definition of vector product, we know that **V** is orthogonal to both **w** and **R**. This implies that **V** is tangential to the path of motion of the particle. Further, by (1.32), we see that

$$v = |\mathbf{V}| = |\mathbf{w} \times \mathbf{R}| = |\mathbf{w}||\mathbf{R}| \sin\theta = \omega|\mathbf{R}|\sin\theta = \omega r$$

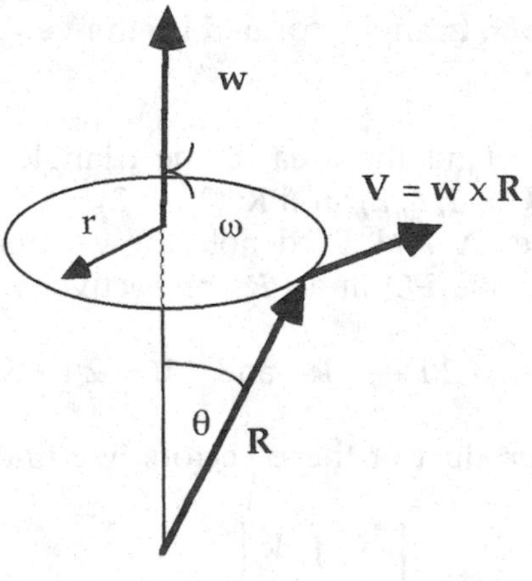

Fig. 1.26 Velocity vector of a particle revolving around an axis.

where r is the radial distance of the particle from the axis. This is the basic formula in elementary physics which relates the tangential speed v of the particle with its angular speed ω.

Example 6. A particle is revolving about the z-axis on a circle of radius 2 cm on the plane z = 5 cm at the rate of 3 revolutions per second. What is the velocity vector and the tangential speed?

Solution: Since the z-axis is represented by the unit vector **k**, the angular velocity is given by $\mathbf{w} = \omega\mathbf{k} = 3\mathbf{k}$. Let **R** denote the position vector of the particle on the plane z = 5. Then $\mathbf{R} = x\,\mathbf{i} + y\,\mathbf{j} + 5\,\mathbf{k}$, where $x^2 + y^2 = 4$. Therefore

$$\begin{aligned} \mathbf{V} = \mathbf{w} \times \mathbf{R} &= 3\mathbf{k} \times (x\,\mathbf{i} + y\,\mathbf{j} + 5\mathbf{k}) = 3x\,(\mathbf{k} \times \mathbf{i}) + 3y\,(\mathbf{k} \times \mathbf{j}) \\ &= 3x\,\mathbf{j} - 3y\,\mathbf{i} = 3(-y\,\mathbf{i} + x\,\mathbf{j}) \end{aligned}$$

and the tangential speed is

$$v = |\mathbf{V}| = 3(x^2 + y^2)^{1/2} = 6 \text{ cm/sec}$$

As another example of a physical problem where vector product is used, consider a force **F** that acts on a particle at a point P. Denote by **R** the position vector of P relative to a fixed point O as shown in Fig. 1.27. The **moment** or **torque** of the force **F** about the point O is defined as the product of the magnitude of the force and the distance from O of the line of action of the force, that is, referring to Fig. 1.27,

$$M_o = |\mathbf{F}|\,|\mathbf{R}|\cos\alpha$$

Since $\alpha = \theta - \pi/2$, where θ is the angle between **F** and **R**, we see that

$$M_o = |\mathbf{F}|\,|\mathbf{R}|\cos(\theta - \pi/2) = |\mathbf{F}|\,|\mathbf{R}|\sin\theta$$

which is the magnitude of the vector product of **F** and **R**. We call the vector

$$\mathbf{M}_o = \mathbf{R} \times \mathbf{F} \tag{1.34}$$

the moment vector or torque vector of a force **F** that acts at a point P, where **R** is the position vector of P relative to a reference point O. Notice that the vector $\mathbf{M}_o$ points in the same direction as the vector

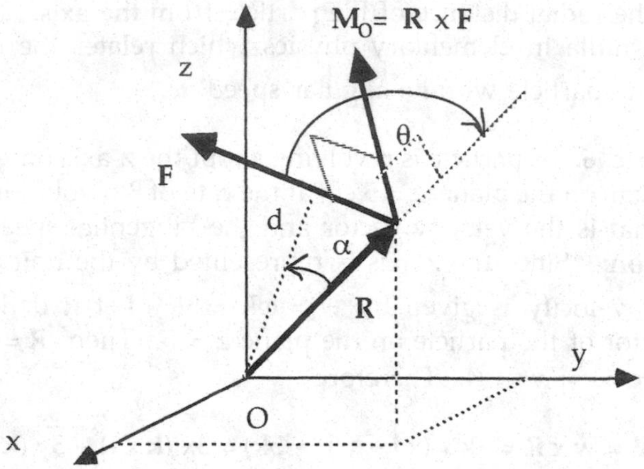

Fig. 1.27 Moment vector $\mathbf{M}_o$.

representing the angular velocity of the point P as caused by the force **F** in accordance with the right-hand rule.

Example 7. A force $\mathbf{F} = -2\mathbf{i} + \mathbf{j} + 3\mathbf{k}$ (in lbs) acts on a body located at the point (-3, 2, 4) (distance in ft). Find the torque vector about the origin and the amount of torque caused by the force.

Solution: By (1.34) the torque vector about the origin is given by

$$\mathbf{M}_o = \mathbf{R} \times \mathbf{F} = \begin{vmatrix} \mathbf{i} & \mathbf{j} & \mathbf{k} \\ -3 & 2 & 4 \\ -2 & 1 & 3 \end{vmatrix} = 2\mathbf{i} + \mathbf{j} + \mathbf{k}$$

Hence the amount of torque is $|\mathbf{M}_o| = \sqrt{6}$ ft-lb.

1.5 EXERCISES

1. Find the determinant of each of the following matrices:

$$\text{(a)} \begin{bmatrix} 2 & 1 \\ 1 & -1 \end{bmatrix}, \quad \text{(b)} \begin{bmatrix} -1 & 3 \\ 2 & -4 \end{bmatrix}, \quad \text{(c)} \begin{bmatrix} 0 & 1 \\ -2 & 5 \end{bmatrix}$$

2. Find the determinant of each of the following matrices:

(a) $\begin{bmatrix} 2 & 1 & 1 \\ -1 & 0 & 1 \\ 1 & -2 & 3 \end{bmatrix}$, (b) $\begin{bmatrix} 1 & -2 & 1 \\ 0 & 2 & 4 \\ 0 & 0 & -1 \end{bmatrix}$

(c) $\begin{bmatrix} a & 0 & 0 \\ 0 & b & 0 \\ 0 & 0 & c \end{bmatrix}$, (d) $\begin{bmatrix} -3 & 0 & 0 \\ 1 & 2 & 4 \\ 3 & 1 & 2 \end{bmatrix}$

3. Show that the determinant of a 3 x 3 matrix changes sign whenever any two rows or two columns are interchanged.
4. Show that if a row or a column of a matrix is multiplied by a constant k, then the determinant of the resulting matrix is k times that of the original matrix.
5. From the results of Problems 3 and 4, deduce that if the elements of any two rows or two columns of a 3 x 3 matrix are proportional, then the determinant of the matrix is zero.
6. Verify each of the following identities:

(a) $$\begin{vmatrix} a_1 & a_2 & a_3 \\ b_1+d_1 & b_2+d_2 & b_3+d_3 \\ c_1 & c_2 & c_3 \end{vmatrix} = \begin{vmatrix} a_1 & a_2 & a_3 \\ b_1 & b_2 & b_3 \\ c_1 & c_2 & c_3 \end{vmatrix} + \begin{vmatrix} a_1 & a_2 & a_3 \\ d_1 & d_2 & d_3 \\ c_1 & c_2 & c_3 \end{vmatrix}$$

(b) $$\begin{vmatrix} a_1 & a_2+r & a_3 \\ b_1 & b_2+s & b_3 \\ c_1 & c_2+t & c_3 \end{vmatrix} = \begin{vmatrix} a_1 & a_2 & a_3 \\ b_1 & b_2 & b_3 \\ c_1 & c_2 & c_3 \end{vmatrix} + \begin{vmatrix} a_1 & r & a_3 \\ b_1 & s & b_3 \\ c_1 & t & c_3 \end{vmatrix}$$

(c) $$\begin{vmatrix} a_1+kb_1 & a_2+kb_2 & a_3+kb_3 \\ b_1 & b_2 & b_3 \\ c_1 & c_2 & c_3 \end{vmatrix} = \begin{vmatrix} a_1 & a_2 & a_3 \\ b_1 & b_2 & b_3 \\ c_1 & c_2 & c_3 \end{vmatrix}$$

7. Prove the properties listed in (1.30).
8. Show in two different ways that the vectors **A** and **B** are

parallel:
(a) $\mathbf{A} = -\mathbf{i} + 2\mathbf{j} - 3\mathbf{k}$, $\mathbf{B} = 2\mathbf{i} - 4\mathbf{j} + 6\mathbf{k}$
(b) $\mathbf{A} = 3\mathbf{i} + 6\mathbf{j} - 9\mathbf{k}$, $\mathbf{B} = \mathbf{i} + 2\mathbf{j} - 3\mathbf{k}$

9. Find a unit vector that is orthogonal to the given two vectors:
(a) $\mathbf{A} = \mathbf{i} - 2\mathbf{j} + 3\mathbf{k}$, $\mathbf{B} = 2\mathbf{i} + \mathbf{j} - \mathbf{k}$
(b) $\mathbf{A} = 3\mathbf{i} - \mathbf{j} + 6\mathbf{k}$, $\mathbf{B} = \mathbf{i} + 4\mathbf{j} + \mathbf{k}$

10. If $\mathbf{A} = 2\mathbf{i} + \mathbf{j} - \mathbf{k}$, $\mathbf{B} = -\mathbf{i} + 3\mathbf{j} + 4\mathbf{k}$, $\mathbf{C} = \mathbf{i} - 3\mathbf{j} + 5\mathbf{k}$, compute the vectors (a) $\mathbf{A} \times \mathbf{B}$, (b) $\mathbf{B} \times \mathbf{C}$, (c) $\mathbf{A} \times \mathbf{B} \cdot \mathbf{C}$, (d) $\mathbf{A} \cdot \mathbf{B} \times \mathbf{C}$, (e) $\mathbf{A} \times (\mathbf{B} \times \mathbf{C})$, (f) $(\mathbf{A} \times \mathbf{B}) \times \mathbf{C}$. What do you notice about $\mathbf{A} \cdot \mathbf{B} \times \mathbf{C}$ and $\mathbf{A} \times \mathbf{B} \cdot \mathbf{C}$; $\mathbf{A} \times (\mathbf{B} \times \mathbf{C})$ and $(\mathbf{A} \times \mathbf{B}) \times \mathbf{C}$?

11. If $\mathbf{A} = 3\mathbf{i} - 6\mathbf{j} + 5\mathbf{k}$, $\mathbf{B} = 2\mathbf{i} - \mathbf{j} + 4\mathbf{k}$, $\mathbf{C} = \mathbf{i} + \mathbf{j} - \mathbf{k}$, compute (a) $\mathbf{A} \times \mathbf{B}$, (b) $\mathbf{B} \times \mathbf{C}$, (c) $(\mathbf{A} + \mathbf{B}) \times (\mathbf{A} - \mathbf{B})$, (d) $(\mathbf{A} \times \mathbf{B}) \times (\mathbf{B} \times \mathbf{C})$, (e) $\mathbf{A} \times (\mathbf{B} \times \mathbf{C})$, and express this vector as a linear combination of $\mathbf{B}$ and $\mathbf{C}$.

12. Use the vector product to compute the area of the triangle with the given vertices:
(a) P: (1, 0, 3), Q: (1,2, -1), R: (-2, 1, 3)
(b) P: (-2, -1, 3), Q: (1, 2, -1), R: (4, 3, -3)
(c) P: (4, -2, 3), Q: (-3,1, 1), R: (1, 1, 1)
(d) P: (-1, -3, 1), Q: (2, 2, -1), R: (-3, 2, -2)

13. A force $\mathbf{F} = 3\mathbf{i} - 2\mathbf{j} + 5\mathbf{k}$ acts on a particle at (4, -2, 2). Find the moment or torque of the force about (a) the origin; (b) the point (1, 2, 1).

14. A force $\mathbf{F} = \mathbf{i} + 4\mathbf{j} - 3\mathbf{k}$ acts on a particle at (-3, 1, 5). Find the torque about (a) the point (1, 0, 2) ; (b) the point (-1, 2, -3).

15. A particle at (2, 5, 0) is acted upon by the forces $\mathbf{F} = -\mathbf{i} + 3\mathbf{j}$ and $\mathbf{G} = -2\mathbf{i} - \mathbf{j}$. Find the total torque about (a) the origin; (b) the point (1, 1, 0).

16. A particle has an angular speed of 2 rad/sec and its axis of rotation passes through the points (1, 0, 2) and (2, 3, -1). Find the velocity of the particle when it is located at the point (3, 4, 5) .

17. A particle has an angular speed of 4 rad/sec with its axis of rotation along the vector $\mathbf{i} + \mathbf{j} + \mathbf{k}$. Find the velocity and the speed of the particle when it is located at the point (-1, 2, 3).

18. If three vertices of a parallelogram are P: (-1, 1, 1), Q: (2, 0, 1), R : (0, 2, 3), find all the possible points D which can be the fourth vertex of the parallelogram.

19. Let $\mathbf{A}$ and $\mathbf{B}$ be two linearly independent vectors.
(a) Show that the angle between $\mathbf{A}$ and $\mathbf{C} = (\mathbf{A} \times \mathbf{B}) - \mathbf{A}$ lies between $\pi/2$ and π.

(b) Show that the angle between **B** and **D** = (**A** x **B**) + **B** lies between 0 and $\pi/2$.

20. Let **A** and **B** be two orthonormal vectors.
(a) Show that the set (**A**, **B**, **A** x **B**) forms an orthonormal basis.
(b) Show that (**A** x **B**) x **A** = **B**, (**A** x **B**) x **B** = - **A**, and give the geometric interpretation.

21. Let **A** = **i** - 2 **j** + **k** and **C** = **i** + **j** + **k**.
(a) Find a vector **B** such that **A** x **B** = **C**. Is the vector **B** uniquely determined?
(b) Find a vector **B** such that **A** x **B** = **C** and **A**· **B** = 1. Is this vector uniquely determined?

22. Prove that |**A** x **B**| = |**A**| |**B**| if and only if **A** and **B** are orthogonal.

23. If **A** + **B** + **C** = 0, show that **A** x **B** = **B** x **C** = **C** x **A**.

24. If **A** + **B** + **C** - **R** = 0 and **A** - **B** + **C** - 3 **R** = 0, show that **B** x **R** = 0, **A** x **R** = **R** x **C**, and **A** x **B** = **B** x **C**.

25. If **A** x **B** = **C** and **A** x **C** = **B**, show that **B** = **0** and **C** = **0**.

26. Show that (a) if **A** x **B** = **0** and **A**·**B** = 0, then **A** = **0** or **B** = **0**;
(b) if **A** $\neq$ **0**, **A** x **B** = **A** x **C**, and **A**· **B** = **B**· **C**, then **B** = **C**.

1.8 LINES AND PLANES IN SPACE

In Sec. 1.5 we saw how vector arithmetic could be used to solve some elementary problems in geometry without reference to any coordinate system. Here we consider applications of the vector operations of scalar product and vector product in the study of lines and planes in space.

Equations of Lines

We begin with the problem of finding an equation of a line. A line is completely determined if we know any two points on the line or if we know a point on the line and the orientation or direction of the line. The direction of a line can be described by means of the direction cosines cos α, cos β and cos γ of the line. Recall that α, β and γ are the angles that the line makes with the positive x-, y- and z-axes, respectively. Equivalently, the direction of a line can also be described by a vector to which the line is parallel, in which case, the components of the vector becomes the direction numbers of the line. So suppose we wish to find an equation of a line that passes through the point P_o: (x_o, y_o, z_o) and is parallel to the vector $\mathbf{A} = a_1\mathbf{i} + a_2\mathbf{j} + a_3\mathbf{k}$. Let $\mathbf{R} = x\,\mathbf{i} + y\,\mathbf{j} + z\,\mathbf{k}$ denote the position vector of an arbitrary point P: (x, y, z) on the line (Fig. 1.28). If $\mathbf{R}_o = x_o\mathbf{i} + y_o\mathbf{j} + z_o\mathbf{k}$ denotes the position vector of the point P_o, then the vector $\mathbf{R} - \mathbf{R}_o$ is parallel to the vector **A**. Hence there exists a scalar t such that $\mathbf{R} - \mathbf{R}_o = t\mathbf{A}$. Thus the position vector **R** of an arbitrary point on the line is given by

$$\mathbf{R} = \mathbf{R}_o + t\mathbf{A}, \qquad (-\infty < t < \infty) \tag{1.35}$$

This is a vector equation of the line. The scalar t is called a parameter. As the parameter ranges from $-\infty$ to ∞, the vector **R** traces the line from one end to the other with $t = 0$ corresponding to the point P_o.

To obtain the equation of the line in cartesian coordinates x, y, z, as we know it in analytic geometry, we equate the corresponding

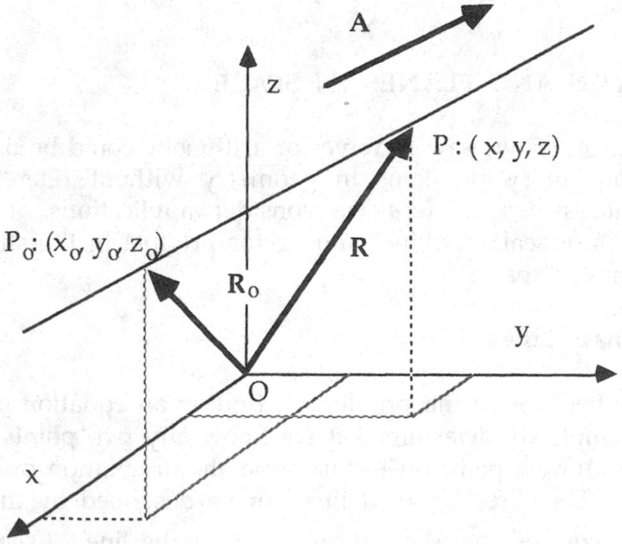

Fig. 1.28 Equation of a line.

components of the vectors in (1.35). We obtain

$$x = x_o + t\,a_1,\quad y = y_o + t\,a_2,\quad z = z_o + t\,a_3 \quad (-\infty < t < \infty) \qquad (1.36)$$

These are called the parametric equations of the line. If we eliminate the parameter t from the equations (1.36), we obtain the *symmetric form* of the equation of the line:

$$\frac{x - x_o}{a_1} = \frac{y - y_o}{a_2} = \frac{z - z_o}{a_3} \qquad (1.37)$$

We point out that equation (1.37) can also be derived from another vector equation of the line, namely, the equation

$$(\mathbf{R} - \mathbf{R}_o) \times \mathbf{A} = \mathbf{0} \qquad (1.38)$$

Notice that the numbers a_1, a_2, a_3 appearing in equations (1.36) and (1.37) are the components of the vector **A**. They are called the direction numbers of the line. Notice also that any vector parallel to **A** may be used in place of **A** in (1.35). This means that any numbers proportional to a_1, a_2, a_3 may also be used as direction numbers of the line.

Example 1. Find a vector equation of the line passing through the points P: (1, -2, 1) and Q: (3, 1, 1). What are the corresponding parametric equations and the equations in symmetric form?

Solution: We note that the line is parallel to the vector

$$\mathbf{A} = (3 - 1)\mathbf{i} + (1 - (-2))\mathbf{j} = 2\mathbf{i} + 3\mathbf{j}$$

determined by the directed line segment PQ. Let $\mathbf{R}_o = \mathbf{i} - 2\mathbf{j} + \mathbf{k}$ and let $\mathbf{R} = x\,\mathbf{i} + y\,\mathbf{j} + z\,\mathbf{k}$ denote the position vector of an arbitrary point (x, y, z) on the line. Then by (1.35) we have

$$\begin{aligned}\mathbf{R} &= \mathbf{R}_o + t\mathbf{A}\\ &= (\mathbf{i} - 2\mathbf{j} + \mathbf{k}) + t\,(2\mathbf{i} + 3\mathbf{j})\\ &= (1 + 2t\,)\mathbf{i} + (-2 + 3t\,)\mathbf{j} + \mathbf{k}\end{aligned}$$

By equating the corresponding components of the vector equation, we obtain the parametric equations

$$x = 1 + 2t\,, \quad y = -2 + 3t\,, \quad z = 1 \qquad (-\infty < t < \infty)$$

By eliminating the parameter t, we obtain the equation in symmetric form

$$(x - 1)/2 = (y + 2)/3\,, \ z = 1$$

The line lies on the plane z = 1.

Example 2. Find the parametric equations of the line that passes through the point (-1, 3, -2) and is perpendicular to the vectors **A** = 3**i** + 4**j** + **k** and **B** = **i** + 2**j**.

Solution: Since the line is perpendicular to the vectors **A** and **B**, it is parallel to the vector product of **A** and **B**. Since

$$\mathbf{A} \times \mathbf{B} = \begin{vmatrix} \mathbf{i} & \mathbf{j} & \mathbf{k} \\ 3 & 4 & 1 \\ 1 & 2 & 0 \end{vmatrix} = -2\mathbf{i} + \mathbf{j} + 2\mathbf{k}$$

it follows that the parametric equations of the line are

$$x = -1 - 2t\,, \quad y = 3 + t\,, \quad z = -2 + 2t \qquad (-\infty < t < \infty)$$

Example 3. Find the point of intersection of the two lines defined by the vector equations

$$\mathbf{R}_1 = \mathbf{i} - \mathbf{j} + 2\mathbf{k} + t\,(\mathbf{i} + \mathbf{j} + \mathbf{k}) \qquad (-\infty < t < \infty)$$
$$\mathbf{R}_2 = 3\mathbf{i} - 3\mathbf{j} + \mathbf{k} + t^*\,(-\mathbf{i} + 3\mathbf{j} + 2\mathbf{k}) \qquad (-\infty < t^* < \infty)$$

and determine the angle between them .

Solution: We need to find a value of the parameter t and a value of the parameter t^* such that $\mathbf{R}_1 = \mathbf{R}_2$. If no such values exist, then the lines do not intersect. Now equating the corresponding components of the equation $\mathbf{R}_1 = \mathbf{R}_2$, we find

$$1 + t = 3 - t^*\,, \quad -1 + t = -3 + 3t^*\,, \quad 2 + t = 1 + 2t^*$$

Solving for t and t^*, we find $t = 1$ and $t^* = 1$. Thus the lines intersect at the point (2, 0, 3).

The angle between the lines is equal to the angle between the two vectors to which the lines are respectively parallel. From the vector equations of the lines, we see that these two vectors are

$$\mathbf{A}_1 = \mathbf{i} + \mathbf{j} + \mathbf{k} \quad \text{and} \quad \mathbf{A}_2 = -\mathbf{i} + 3\mathbf{j} + 2\mathbf{k}$$

Let θ denote the angle between these vectors. Since

$$\mathbf{A}_1 \cdot \mathbf{A}_2 = 4\,,\ |\mathbf{A}_1| = \sqrt{3}\,,\ \text{and}\ |\mathbf{A}_2| = \sqrt{14}$$

we find

$$\cos\theta = \mathbf{A}_1 \cdot \mathbf{A}_2 \,/\, |\mathbf{A}_1|\,|\mathbf{A}_2| = 4/(\sqrt{3}\,\sqrt{14}) = 4/\sqrt{42}$$

Hence the angle between the lines is equal to $\theta = \arccos(4/\sqrt{42}) = 51.9^o$.

Now suppose we wish to find the distance from a point $P : (x_1, y_1, z_1)$ to a line defined by the vector equation (1.35). Using vector method, this problem, which otherwise would involve long and tedious computations, can be easily tackled as follows. Referring to Fig. 1.29, we see that the distance d of the point P from the line is

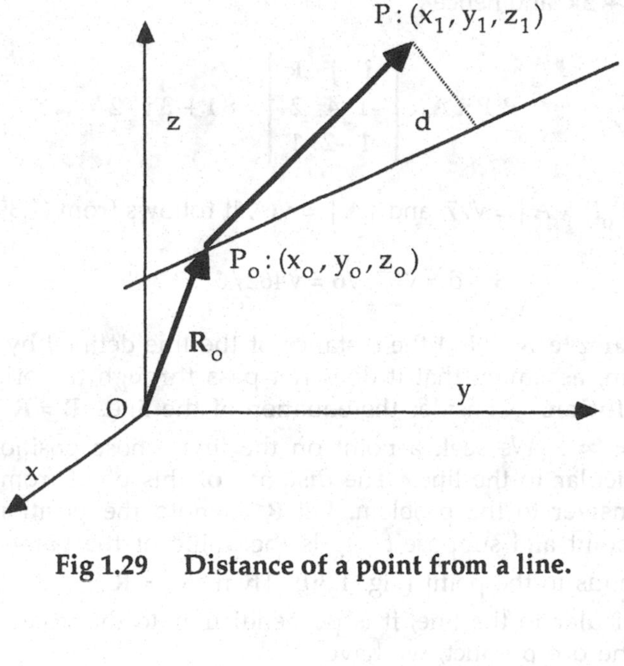

Fig 1.29 Distance of a point from a line.

given by $d = |P_0P| \sin\theta$, where θ is the angle between the given line and the directed line segment P_0P. Notice that P_0 can be chosen as any point on the given line. Since the line is parallel to the vector **A**, we have

$$|P_0P \times \mathbf{A}| = |P_0P|\,|\mathbf{A}| \sin\theta$$

Solving for $|P_0P| \sin\theta$, we obtain the formula

$$d = |P_0P \times \mathbf{A}| / |\mathbf{A}| \tag{1.39}$$

Example 4. Find the distance of the point P: (2, 3, 6) from the line defined by the vector equation

$$\mathbf{R} = (3\mathbf{i} - \mathbf{j} + 4\mathbf{k}) + t(\mathbf{i} - 2\mathbf{j} + \mathbf{k})$$

Solution: We choose the point P_0: (3, -1, 4) on the line corresponding to $t = 0$. Then $P_0P = (2 - 3)\mathbf{i} + (3 - (-1))\mathbf{j} + (6 - 4)\mathbf{k}$ $= -\mathbf{i} + 4\mathbf{j} + 2\mathbf{k}$ and hence

$$\mathbf{P_0P} \times \mathbf{A} = \begin{vmatrix} \mathbf{i} & \mathbf{j} & \mathbf{k} \\ -1 & 4 & 2 \\ 1 & -2 & 1 \end{vmatrix} = 8\mathbf{i} + 3\mathbf{j} - 2\mathbf{k}$$

Since $|P_0P \times \mathbf{A}| = \sqrt{77}$ and $|\mathbf{A}| = \sqrt{6}$, it follows from (1.39)

$$d = \sqrt{77}/\sqrt{6} = \sqrt{462}/6$$

Example 5. Find the distance of the line defined by (1.35) from the origin, assuming that it does not pass through the origin.

Solution: Consider the equation of the line $\mathbf{R} = \mathbf{R}_0 + t\mathbf{A}$, $-\infty < t < \infty$. We seek a point on the line whose position vector is perpendicular to the line. The distance of this point from the origin is the answer to the problem. Let $\mathbf{R}^*$ denote the position vector of such a point and suppose $t = t_0$ is the value of the parameter which corresponds to the point (Fig. 1.30). Then $\mathbf{R}^* = \mathbf{R}_0 + t_0\mathbf{A}$. Since $\mathbf{R}^*$ is perpendicular to the line, it is perpendicular to the vector **A**. Hence, taking the dot product, we have

$$\mathbf{A}\cdot\mathbf{R}^* = \mathbf{A}\cdot\mathbf{R}_o + t_o\mathbf{A}\cdot\mathbf{A} = 0$$

which gives

$$t_o = -\,\mathbf{A}\cdot\mathbf{R}_o/\mathbf{A}\cdot\mathbf{A}$$

Therefore, the position vector of the point we are seeking is

$$\mathbf{R}^* = \mathbf{R}_o - [\mathbf{A}\cdot\mathbf{R}_o/\mathbf{A}\cdot\mathbf{A}]\,\mathbf{A}$$

and so the distance of the line from the origin is given by

$$d = |\,\mathbf{R}^*\,|$$

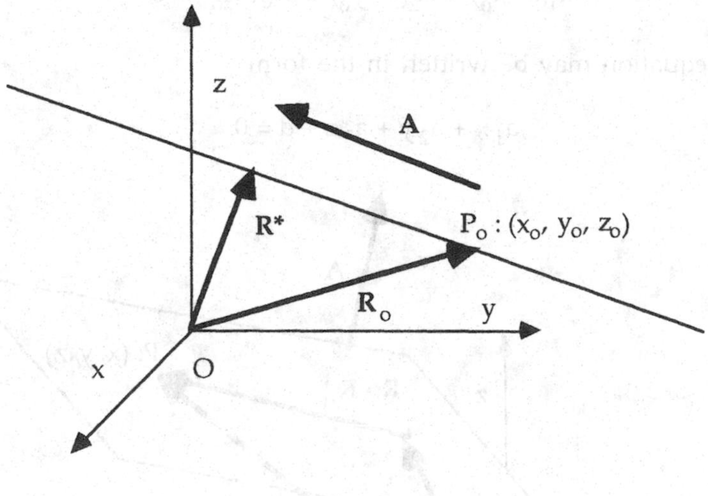

Fig. 1.30 Distance of a line from the origin.

Equations of Planes

We know that a plane is uniquely determined by three non-collinear points or by a point and a line (or a vector) that is perpendicular to the plane. If three non-collinear points, for example, P, Q, and R are given, we can readily find a vector that is perpendicular to the plane determined by the three points. The vector is simply the cross product of any two of the vectors represented by the directed line segments *PQ*, *QR*, and *RP*. Thus it is sufficient to consider only the case where we are given a point and a vector. So let P_o: (x_o, y_o, z_o) be a given point and $\mathbf{A} = a_1\mathbf{i} + a_2\mathbf{j} + a_3\mathbf{k}$ a given vector. We wish to find an equation of the plane that passes through the point P_o and is perpendicular to the vector **A**. Let P: (x, y, z) be an arbitrary point on the plane. Then the vector represented by the directed line segment $P_oP = (x - x_o)\,\mathbf{i} + (y - y_o)\,\mathbf{j} + (z - z_o)\,\mathbf{k}$ lies on the plane, and since **A** is perpendicular to the plane, $|P_oP|$ is orthogonal to the vector **A** (Fig. 1.31). Therefore, $\mathbf{A}\cdot P_oP = 0$, which yields the equation

$$a_1(x - x_o) + a_2(y - y_o) + a_3(z - z_o) = 0 \tag{1.40}$$

This equation may be written in the form

$$a_1 x + a_2 y + a_3 z + d = 0 \tag{1.41}$$

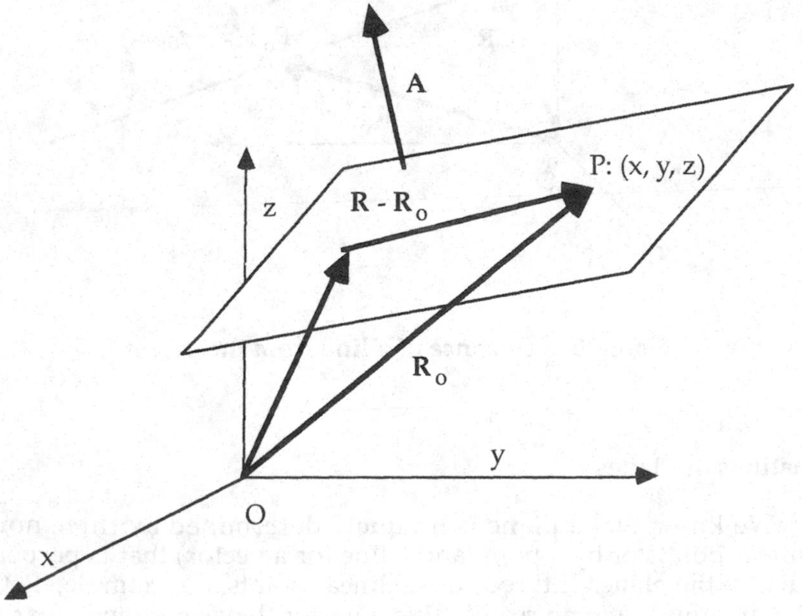

Fig. 1.31 Equation of a plane.

where $d = -(a_1 x_o + a_2 y_o + a_3 z_o)$ is a constant.

We observe that the coefficients in (1.40) and (1.41) are simply the components of the vector **A** that is perpendicular to the plane. The vector **A** is called a normal vector to the plane.

Example 5. Find the equation of the plane that passes through the point (2, -3, 5) and is perpendicular to the line determined by the points (3, 1, 2) and (-2, 3, 4).

Solution: First we find the direction numbers of the line that is determined by the points (3, 1, 2) and (-2, 3, 4). They are given by

$$a_1 = 3 - (-2) = 5, \quad a_2 = 1 - 3 = -2, \quad a_3 = 2 - 4 = -2$$

Hence by (1.40) the equation of the plane is given by

$$5(x - 2) - 2(y + 3) - 2(z - 5) = 0$$

or

$$5x - 2y - 2z - 6 = 0$$

Example 6. Find the equation of the plane that is determined by the points P: (- 2, 1, 3), Q: (1, 2, - 1), and R: (- 3, - 2, 1).

Solution: Let **A** and **B** denote the vectors represented by $\overline{QP}$ and $\overline{QR}$, respectively. Then

$$\mathbf{A} = (-2 - 1)\mathbf{i} + (1 - 2)\mathbf{j} + (3 - (-1))\mathbf{k} = -3\mathbf{i} - \mathbf{j} + 4\mathbf{k}$$

$$\mathbf{B} = (-3 - 1)\mathbf{i} + (-2 - 2)\mathbf{j} + (1 - (-1))\mathbf{k} = -4\mathbf{i} - 4\mathbf{j} + 2\mathbf{k}$$

Since these vectors both lie on the plane in question, it follows that

$$\mathbf{A} \times \mathbf{B} = \begin{vmatrix} \mathbf{i} & \mathbf{j} & \mathbf{k} \\ -3 & -1 & 4 \\ -4 & -4 & 2 \end{vmatrix} = 14\mathbf{i} - 10\mathbf{j} + 8\mathbf{k}$$

is a normal vector to the plane. Hence, using (- 2, 1, 3) as a fixed point, we find

$$14(x + 2) - 10(y - 1) + 8(z - 3) = 0$$

or

$$7x - 5y + 4z + 7 = 0$$

The student should verify that the same equation is obtained when either the point (1, 2, - 1) or the point (- 3, - 2, 1) is used for a fixed point. Alternatively, the equation of the plane may be derived by substituting the coordinates of the points P, Q, and R in (1.41), thus obtaining a system of three equations which can be solved for the unknowns a_1, a_2, a_3.

We now derive a formula analogous to (1.39) for the distance from a given point to a given plane. Let P: (x_1, y_1, z_1) be a given point and let the plane be defined by the equation (1.40):

$$\mathbf{A} \cdot (\mathbf{R} - \mathbf{R}_o) = 0$$

We note that the vector **A** is perpendicular to the plane and the plane passes through the point P_o: (x_o, y_o, z_o) whose position vector is $\mathbf{R}_o$. Referring to Fig. 1.32, we see that the distant from the point P to the plane is given by $d = |P_oP| \cos\theta$, where θ is the angle between the normal vector **A** and the vector represented by P_oP.

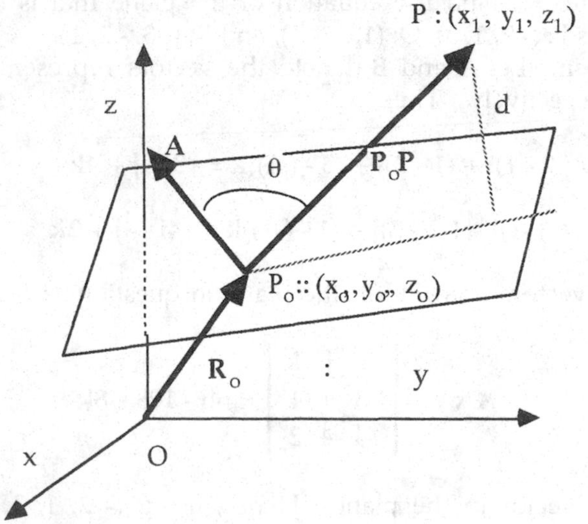

Fig. 1.32 Distance from a point to a plane.

Since $\mathbf{A} \cdot \boldsymbol{P_0P} = |\mathbf{A}|\,|\boldsymbol{P_0P}|\cos\theta$, we obtain the formula

$$d = (\mathbf{A} \cdot \boldsymbol{P_0P})/|\mathbf{A}| \qquad (1.42)$$

Example 7. Find the distance from the point (2, 4, 5) to the plane

$$2x - y + 2z + 6 = 0$$

Solution: First we pick a point P_o on the plane. This is easily done by assigning (arbitrary) values to any two of the variables x, y, z and then solving for the remaining variable. If we set x = -1, y = 2, we find z = -1. Thus the point P_o: (- 1, 2, - 1) lies on the plane, and $\boldsymbol{P_oP}$ represents the vector

$$\mathbf{B} = (2 - (-1))\mathbf{i} + (4 - 2)\mathbf{j} + (5 - (-1))\mathbf{k}$$

$$= 3\mathbf{i} + 2\mathbf{j} + 6\mathbf{k}$$

From the equation of the plane, we deduce the normal vector $\mathbf{A} = 2\mathbf{i} - \mathbf{j} + 2\mathbf{k}$. Hence by (1.42) we find

$$d = \mathbf{A} \cdot \mathbf{B}/|\mathbf{A}| = 16/3$$

1.6 EXERCISES

1. Find a set of parametric equations of the line that passes through point (-1, 3, 2) and is parallel to the vector $\mathbf{A} = 2\mathbf{i} - 3\mathbf{j} + 5\mathbf{k}$.
2. Find the symmetric equations of the line that passes through the points (3, -1, 4) and (-2, 3, 5).
3. Find the parametric equations of the line that passes through the point (1, - 2, 1) and is perpendicular to the plane 2x -3y + 4z = 5.
4. Find the parametric equations of the line that passes through the point (3, 4, 6), perpendicular to the vector $\mathbf{i} + 2\mathbf{j} - \mathbf{k}$ and parallel to the plane 2x - 3y + 5z + 4 = 0.
5. Find an equation of the line of intersection of the planes x - 2y + z - 3 = 0 and 2x + 3y - 5z - 4 = 0.

6. Find the point of intersection of the lines defined by

$$\mathbf{R}_1 = \mathbf{i} + 2\mathbf{j} + \mathbf{k} + (\mathbf{i} - 2\mathbf{j} + \mathbf{k})t$$

and

$$\mathbf{R}_2 = 2\mathbf{i} + \mathbf{j} + 2\mathbf{k} + (\mathbf{i} - 3\mathbf{j} + \mathbf{k})t^* .$$

What is the cosine of the angle between these lines?

7. Find the point of intersection of the lines represented by

$$\frac{x-2}{1} = \frac{y+4}{2} = \frac{z-3}{1} \text{ and } \frac{x-2}{2} = \frac{y-5}{1} = \frac{z-3}{2}$$

and determine the cosine of the angle between them.

8. Find the point of intersection of the line

$$\frac{x-1}{2} = \frac{y-2}{-3} = \frac{z+3}{2}$$

and the plane $3x + 2y - 2z - 5 = 0$.

9. Find an equation of the plane that passes through the point (- 2, 3, 1) and contains the line $\mathbf{R} = \mathbf{i} + 2\mathbf{j} + \mathbf{k} + (3\mathbf{i} - \mathbf{j} + 2\mathbf{k})t$.

10. Find an equation of the plane that passes through the points (1, - 1, 1), (- 2, 3, 4), and (- 3, -2, 1).

11. Find an equation of the plane that contains the two lines given in Problem 6.

12. Find an equation of the plane that contains the two lines given in Problem 7.

13. Find an equation of the plane that passes through the point (- 1, 2, 3) and is perpendicular to the line of intersection of the planes $x + 2y - 3z + 4 = 0$ and $3x - 2y + 5z - 2 = 0$.

14. Find the distance of the point (2, - 2, 3) from the line

$$\frac{x-1}{-1} = \frac{y+2}{2} = \frac{z-1}{-3}$$

15. Find the distance of the line $\mathbf{R} = 2\mathbf{i} - \mathbf{j} + \mathbf{k} + (3\mathbf{i} - \mathbf{j} + 4\mathbf{k})t$ from the origin.

16. Find the distance of the plane $3x - 2y + 5z - 2 = 0$ from the origin.

17. What is the distance of the point (1, 3, 2) from the plane given in Problem 16?
18. Find the distance between the line that joins the points (1, -2, 3) and (2, 4, -1) and the line that joins the points (3, 4, -2) and (2, -1, 5).
19. What is the distance from the point (1, -2, 1) to the plane determined by the three points (2, -2, -1), (3, - 4, -3), and (-1, 2, 0)?
20. Find the distance between the skew lines

$$x = -1 + 4t, \qquad y = 2 - 3t, \qquad z = -3 + 2t$$

and

$$x = 1 + 3t, \qquad y = -3 + t, \qquad z = -2t \qquad (-\infty < t < \infty)$$

21. Find the center and the radius of the sphere that is tangent to the planes x - 2y + z = 4, x - 2y + z = 10 and whose center lies on the planes x - 2y = 2 and y - z = 3. Determine the points at which the sphere is tangent to the given planes?

1.9 SCALAR AND VECTOR TRIPLE PRODUCTS

In this section we study the results of applying the operations of scalar and/or vector product on three vectors. Apart from the or- der in which the vectors appear, the three types of product for three vectors **A** , **B** , **C** are of the form

$$\text{(i)}\ (\mathbf{A}\cdot\mathbf{B})\mathbf{C}, \qquad \text{(ii)}\ \mathbf{A}\cdot(\mathbf{B}\times\mathbf{C}), \qquad \text{(iii)}\ \mathbf{A}\times(\mathbf{B}\times\mathbf{C})$$

The first type is merely the product of the scalar **A** · **B** and the vector **C**, hence it is a scalar multiple of **C**. The second type is the scalar product of **A** and **B** x **C**, which results in a scalar. This product is called the *scalar triple product* of the vectors in the order they appear, and it is denoted by (**ABC**). Thus

$$(\mathbf{ABC}) = \mathbf{A}\cdot\mathbf{B}\times\mathbf{C} \tag{1.43}$$

[Notice that in writing **A** · **B** x **C**, there is no ambiguity as to which operation is to be performed first, since the alternate form (**A** · **B**) x **C** is meaningless. Why?]

If $\mathbf{A} = a_1\,\mathbf{i} + a_2\,\mathbf{j} + a_3\,\mathbf{k}$, $\mathbf{B} = b_1\,\mathbf{i} + b_2\,\mathbf{j} + b_3\,\mathbf{k}$ and $\mathbf{C} = c_1\,\mathbf{i} + c_2\,\mathbf{j} + c_3\,\mathbf{k}$, then in terms of the components of the vectors, we have

$$\mathbf{A}\cdot\mathbf{B}\times\mathbf{C} = \mathbf{A}\cdot\begin{vmatrix} \mathbf{i} & \mathbf{j} & \mathbf{k} \\ b_1 & b_2 & b_3 \\ c_1 & c_2 & c_3 \end{vmatrix}$$

$$= a_1\begin{vmatrix} b_2 & b_3 \\ c_2 & c_3 \end{vmatrix} - a_2\begin{vmatrix} b_1 & b_3 \\ c_1 & c_3 \end{vmatrix} + a_3\begin{vmatrix} b_1 & b_2 \\ c_1 & c_2 \end{vmatrix} \tag{1.44}$$

$$= \begin{vmatrix} a_1 & a_2 & a_3 \\ b_1 & b_2 & b_3 \\ c_1 & c_2 & c_3 \end{vmatrix}$$

Thus the scalar triple product (**ABC**) is simply the determinant of the matrix whose elements in the first, second and third rows are the components of the vectors **A**, **B**, and **C**, respectively.

Example 1. Find the scalar triple product (**ABC**) if **A** = 3**i** + **j** - 2**k**, **B** = 2**i** - **j** - 3**k**, and **C** = 4**i** + 2**j** + **k**.

Solution: By (1.44) we find

$$\mathbf{A}\cdot\mathbf{B}\times\mathbf{C} = \begin{vmatrix} 3 & 1 & -2 \\ 2 & -1 & -3 \\ 4 & 2 & 1 \end{vmatrix} = -15$$

Since the determinant of a matrix changes sign whenever any two of its rows are interchanged, we see that if we interchange in (1.44) the first row and the second row, and then the second row and the third row, the determinant remains the same. Thus, we obtain

$$\mathbf{A}\cdot\mathbf{B}\times\mathbf{C} = \mathbf{B}\cdot\mathbf{C}\times\mathbf{A}$$

If we repeat the process on the the scalar triple product $\mathbf{B}\cdot\mathbf{C}\times\mathbf{A}$, we obtain

$$\mathbf{B}\cdot\mathbf{C}\times\mathbf{A} = \mathbf{C}\cdot\mathbf{A}\times\mathbf{B}$$

Therefore, the scalar triple product of three vectors has the property

$$\mathbf{A} \cdot \mathbf{B} \times \mathbf{C} = \mathbf{B} \cdot \mathbf{C} \times \mathbf{A} = \mathbf{C} \cdot \mathbf{A} \times \mathbf{B} \tag{1.45}$$

Thus in the expression $\mathbf{A} \cdot \mathbf{B} \times \mathbf{C}$, the two operations " · " and " x " are interchangeable, that is, $\mathbf{A} \cdot \mathbf{B} \times \mathbf{C} = \mathbf{A} \times \mathbf{B} \cdot \mathbf{C}$.

The scalar triple product (1.44) provides a convenient way for determining whether or not three given vectors are coplanar; that is, the vectors can be represented by arrows lying on the same plane. The method is contained in the following theorem.

Theorem 1. Three vectors **A, B, C** are coplanar if and only if $(\mathbf{ABC}) = 0$.

Proof: Clearly, the theorem is trivial if one of the vectors is the zero vector. So suppose $(\mathbf{ABC}) = 0$ and that none of the vectors is **0**. By (1.45) it follows that one of the vectors must be orthogonal to the cross product of the other two vectors. Suppose $\mathbf{A} \cdot \mathbf{B} \times \mathbf{C} = 0$. Then **A** is orthogonal to $\mathbf{B} \times \mathbf{C}$. Since $\mathbf{B} \times \mathbf{C}$ itself is orthogonal to **B** and **C**, we deduce that **A** lies on the plane determined by **B** and **C**. This means that the vectors are coplanar.

Conversely, if the vectors are coplanar, then we can express one of the vectors as a linear combination of the other two. Suppose **A** is a linear combination of **B** and **C**. Since $\mathbf{B} \times \mathbf{C}$ is orthogonal to **B** and **C**, it is also orthogonal to **A**, and so $\mathbf{A} \cdot \mathbf{B} \times \mathbf{C} = 0$.

Geometric Interpretaion of Scalar Triple Product

It is interesting to note that the scalar triple product of three nonzero vectors represents in absolute value the volume of the parallelepiped determined by the vectors. To see this, let us consider the parallelepiped formed by the vectors **A, B**, and **C** as shown in Fig. 1.33. From the definition of scalar product, we have

$$\mathbf{A} \cdot \mathbf{B} \times \mathbf{C} = |\mathbf{A}|\,|\mathbf{B} \times \mathbf{C}| \cos\theta$$

where θ is the angle between the vectors **A** and $\mathbf{B} \times \mathbf{C}$. Now from the definition of vector product, we know that $|\mathbf{B} \times \mathbf{C}|$ represents the area of the parallelogram formed by the vectors **B** and **C**. Further, the quantity $h = |\mathbf{A}| \cos\theta$ represents numerically the altitude of the parallelepiped relative to the base formed by the vectors **B** and **C**. Since the volume of a parallelepiped is equal to the product of a base area and its altitude, it follows that the numerical

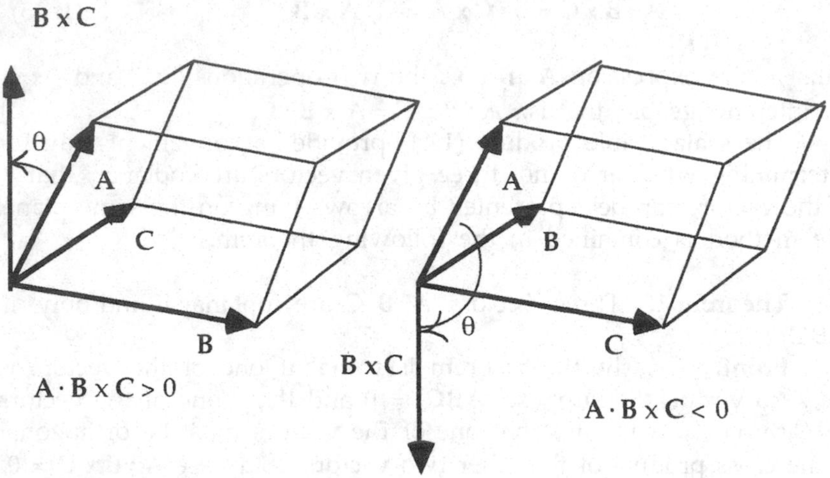

Fig 1.33 Geometric interpretation of the scalar triple product.

value of (1.44) represents the volume of the parallelepiped determined by the vectors.

It is geometrically evident that (**ABC**) is positive or negative according to whether the vectors **A, B, C** form a right-handed or a left-handed triple. When (**ABC**) ≠ 0, the vectors are said to be linearly independent; otherwise, the vectors are said to be linearly dependent. It follows that three nonzero vectors are linearly dependent if and only if they are coplanar. In the case of two vectors, linear dependence implies that the vectors are scalar multiple of each other. Hence two nonzero vectors are linearly independent if and only if the vectors are not parallel.

Example 2. Let P : (1, -2, 3), Q : (2, 1, -2), R : (-2, 1, -1), and S : (2, 2, 3) be four given points. Find the volume of the parallelepiped formed by PQ, PR, and PS.

Solution: Let **A, B,** and **C** denote the vectors represented by *PQ*, *PR*, and *PS*, respectively. Then we have

$$\mathbf{A} = (2 - 1)\mathbf{i} + (1 - (-2))\mathbf{j} + (-2 - 3)\mathbf{k} = \mathbf{i} + 3\mathbf{j} - 5\mathbf{k}$$
$$\mathbf{B} = (-2 - 1)\mathbf{i} + (1 - (-2))\mathbf{j} + (-1 - 3)\mathbf{k} = -3\mathbf{i} + 3\mathbf{j} - 4\mathbf{k}$$
$$\mathbf{C} = (2 - 1)\mathbf{i} + (2 - (-2))\mathbf{j} + (3 - 3)\mathbf{k} = \mathbf{i} + 4\mathbf{j}$$

Thus the volume of the parallelepiped is equal to

$$(\mathbf{ABC}) = \begin{vmatrix} 1 & 3 & -5 \\ -3 & 3 & -4 \\ 1 & 4 & 0 \end{vmatrix} = 79 \text{ cubic units}$$

[Note that if (**ABC**) turns out to be negative, we simply take the absolute value for the volume.]

Example 3. Show that the four points P : (2, 4, 3), Q : (-2, 1, 4), R : (1, 2, 5), and S : (-5, 0, 3) are coplanar.

Solution: The four points are coplanar if and only if the parallelepiped determined by the points degenerate into a planar figure, in which case the volume is then zero. So let **A**, **B**, and **C** denote the vectors represented by ***PQ***, ***PR***, and ***PS***. We have **A** = 4**i** - **j** + **k**, **B** = -**i** - 2**j** + 2**k**, and **C** = -7**i** - 4**j**. Since

$$(\mathbf{ABC}) = \begin{vmatrix} -4 & -3 & 1 \\ -1 & -2 & 2 \\ -7 & -4 & 0 \end{vmatrix} = 0$$

it follows that the points are coplanar.

Finally, let us consider the product **A** x (**B** x **C**). This is called a *vector triple product*. From the definition we know that this vector is orthogonal to the vectors **A** and **B** x **C**. Since **B** x **C** is in turn orthogonal to the vectors **B** and **C**, it follows that **A** x (**B** x **C**), **B**, and **C** are coplanar or linearly dependent (Fig. 1.34). Hence, we can express the vector **A** x (**B** x **C**) in terms of **B** and **C**, that is,

$$\mathbf{A} \times (\mathbf{B} \times \mathbf{C}) = m\,\mathbf{B} + n\,\mathbf{C}$$

for some scalars m and n. We will show that

$$m = \mathbf{C} \cdot \mathbf{A}, \qquad n = -\mathbf{B} \cdot \mathbf{A}$$

so that

$$\mathbf{A} \times (\mathbf{B} \times \mathbf{C}) = (\mathbf{C} \cdot \mathbf{A})\mathbf{B} - (\mathbf{B} \cdot \mathbf{A})\mathbf{C} \tag{1.46}$$

This formula is sometimes referred to as the "CAB minus BAC" formula for the vector triple product **A** x (**B** x **C**).

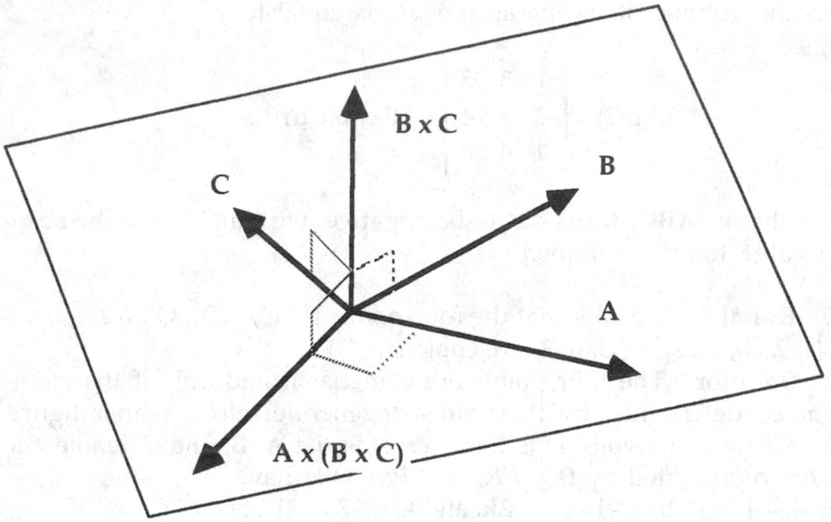

Fig. 1.34 Vector triple product.

We establish the formula (1.46) by showing that the correspon-ding components of the vectors on opposite sides of the equation are equal. So suppose $\mathbf{A} = a_1\mathbf{i} + a_2\mathbf{j} + a_3\mathbf{k}$, $\mathbf{B} = b_1\mathbf{i} + b_2\mathbf{j} + b_3\mathbf{k}$, and $\mathbf{C} = c_1\mathbf{i} + c_2\mathbf{j} + c_3\mathbf{k}$. We note that

$$\mathbf{B} \times \mathbf{C} = \begin{vmatrix} b_2 & b_3 \\ c_2 & c_3 \end{vmatrix} \mathbf{i} + \begin{vmatrix} b_3 & b_1 \\ c_3 & c_1 \end{vmatrix} \mathbf{j} + \begin{vmatrix} b_1 & b_2 \\ c_1 & c_2 \end{vmatrix} \mathbf{k}$$

Thus the first component of the vector $\mathbf{A} \times (\mathbf{B} \times \mathbf{C})$ is given by

$$a_2 \begin{vmatrix} b_1 & b_2 \\ c_1 & c_2 \end{vmatrix} - a_3 \begin{vmatrix} b_3 & b_1 \\ c_3 & c_1 \end{vmatrix}$$

$$= a_2(b_1c_2 - b_2c_1) - a_3(b_3c_1 - b_1c_3)$$

On the other hand, the first component of the vector on the right-hand side of (1.46) is given by

$$\begin{aligned}&(\mathbf{C}\cdot\mathbf{A})b_1 - (\mathbf{B}\cdot\mathbf{A})c_1\\ &= (a_1c_1 + a_2c_2 + a_3c_3)b_1 - (b_1a_1 + b_2a_2 + b_3a_3)c_1\\ &= a_2(b_1c_2 - b_2c_1) - a_3(b_3c_1 - b_1c_3)\end{aligned}$$

which agrees with the first component of the vector $\mathbf{A} \times (\mathbf{B} \times \mathbf{C})$. The corresponding second and third components of the vectors can also be shown to be equal in exactly the same manner, thus establishing the formula (1.46).

Since $\mathbf{A} \times (\mathbf{B} \times \mathbf{C}) = -(\mathbf{B} \times \mathbf{C}) \times \mathbf{A}$, it follows that

$$(\mathbf{A} \times \mathbf{B}) \times \mathbf{C} = (\mathbf{C}\cdot\mathbf{A})\mathbf{B} - (\mathbf{C}\cdot\mathbf{B})\mathbf{A} \tag{1.47}$$

which shows that the vector triple product $(\mathbf{A} \times \mathbf{B}) \times \mathbf{C}$ is also a linear combination of the vectors inside the parentheses.

Example 4. Verify the formula (1.46) for the vectors $\mathbf{A} = \mathbf{i} - \mathbf{j} + 2\mathbf{k}$, $\mathbf{B} = \mathbf{i} - 2\mathbf{k}$ and $\mathbf{C} = -2\mathbf{i} + \mathbf{j} - \mathbf{k}$.

Solution: We have

$$\mathbf{B} \times \mathbf{C} = \begin{vmatrix} \mathbf{i} & \mathbf{j} & \mathbf{k} \\ 1 & 0 & -2 \\ -2 & 1 & -1 \end{vmatrix} = 2\mathbf{i} + 5\mathbf{j} + \mathbf{k}$$

hence

$$\mathbf{A} \times (\mathbf{B} \times \mathbf{C}) = \begin{vmatrix} \mathbf{i} & \mathbf{j} & \mathbf{k} \\ 1 & -1 & 2 \\ 2 & 5 & 1 \end{vmatrix} = -11\mathbf{i} + 3\mathbf{j} + 7\mathbf{k}$$

On the other hand,

$$\begin{aligned}\mathbf{A}\cdot\mathbf{C} &= 1(-2) + (-1)1 + 2(-1) = -5\\ \mathbf{A}\cdot\mathbf{B} &= 1(1) + (-1)0 + 2(-2) = -3\end{aligned}$$

so that

$$-5\mathbf{B} - (-3)\mathbf{C} = -5(\mathbf{i} - 2\mathbf{k}) + 3(-2\mathbf{i} + \mathbf{j} - \mathbf{k}) = -11\mathbf{i} + 3\mathbf{j} + 7\mathbf{k}$$

which equals the vector $\mathbf{A} \times (\mathbf{B} \times \mathbf{C})$, thus verifying the formula (1.46).

Example 5. Prove the identity

$$|\mathbf{A} \times \mathbf{B}|^2 = |\mathbf{A}|^2|\mathbf{B}|^2 - (\mathbf{A} \cdot \mathbf{B})^2$$

for any vectors **A** and **B**.

Solution: We observe that

$$|\mathbf{A} \times \mathbf{B}|^2 = (\mathbf{A} \times \mathbf{B}) \cdot (\mathbf{A} \times \mathbf{B})$$

Set $\mathbf{D} = \mathbf{A} \times \mathbf{B}$ and consider the scalar triple product $\mathbf{D} \cdot (\mathbf{A} \times \mathbf{B})$. By the property of scalar triple product, we see that

$$|\mathbf{A} \times \mathbf{B}|^2 = \mathbf{D} \cdot (\mathbf{A} \times \mathbf{B}) = \mathbf{D} \times \mathbf{A} \cdot \mathbf{B}$$

Since $\mathbf{D} \times \mathbf{A} = (\mathbf{A} \times \mathbf{B}) \times \mathbf{A} = (\mathbf{A} \cdot \mathbf{A})\mathbf{B} - (\mathbf{A} \cdot \mathbf{B})\mathbf{A}$, it follows that

$$\begin{aligned} |\mathbf{A} \times \mathbf{B}|^2 &= [(\mathbf{A} \cdot \mathbf{A})\mathbf{B} - (\mathbf{A} \cdot \mathbf{B})\mathbf{A}] \cdot \mathbf{B} = (\mathbf{A} \cdot \mathbf{A})(\mathbf{B} \cdot \mathbf{B}) - (\mathbf{A} \cdot \mathbf{B})(\mathbf{A} \cdot \mathbf{B}) \\ &= |\mathbf{A}|^2 |\mathbf{B}|^2 - (\mathbf{A} \cdot \mathbf{B})^2 \end{aligned}$$

1.7 EXERCISES

1. Compute the scalar triple product (**ABC**) if
   (a) $\mathbf{A} = 2\mathbf{i} + \mathbf{j}$, $\mathbf{B} = -\mathbf{i} + \mathbf{j} + 2\mathbf{k}$, $\mathbf{C} = \mathbf{i} + \mathbf{j} + 2\mathbf{k}$
   (b) $\mathbf{A} = -\mathbf{i} + \mathbf{j} + 2\mathbf{k}$, $\mathbf{B} = \mathbf{i} + 2\mathbf{j}$, $\mathbf{C} = 2\mathbf{i} - \mathbf{j} + 4\mathbf{k}$
   (c) $\mathbf{A} = 2\mathbf{i} + \mathbf{j} + 3\mathbf{k}$, $\mathbf{B} = -3\mathbf{i} + 2\mathbf{k}$, $\mathbf{C} = 2\mathbf{i} - \mathbf{j} + 4\mathbf{k}$
   (d) $\mathbf{A} = 3\mathbf{i} + \mathbf{j} + 2\mathbf{k}$, $\mathbf{B} = 2\mathbf{i} + 5\mathbf{k}$, $\mathbf{C} = \mathbf{i} + 6\mathbf{j} + 3\mathbf{k}$
2. Find the volume of the parallelepiped determined by the vectors $\mathbf{i} + \mathbf{j}$, $\mathbf{j} + \mathbf{k}$, $\mathbf{i} + \mathbf{k}$.
3. Find the volume of the parallelepiped determined by the line segments PQ, PR, and PS, where P: (1, 2, 0), Q: (3, 5, 0), R: (4, 3, 0), and S: (- 1, - 1, 2).
4. Repeat Problem 3 if the points are P: (2, 0, 2), Q: (4, 3, -1), R: (0, 2, 0) and S: (3, 1, 4).
5. Show that the vectors $\mathbf{A} = 2\mathbf{i} - \mathbf{j} + 4\mathbf{k}$, $\mathbf{B} = \mathbf{i} + 2\mathbf{j} - 3\mathbf{k}$, and $\mathbf{C} = -4\mathbf{i} + 2\mathbf{j} - 8\mathbf{k}$ are linearly dependent and determine which of the vectors is a linear combination of the other two.
6. Show that $\mathbf{A} = \mathbf{i} + \mathbf{j}$, $\mathbf{B} = \mathbf{j} + \mathbf{k}$, $\mathbf{C} = \mathbf{i} + \mathbf{k}$ are linearly independent and express $\mathbf{D} = 4\mathbf{i} - \mathbf{j} - \mathbf{k}$ as a linear combination of these vectors.

7. Repeat Problem 6 if $\mathbf{A} = 2\mathbf{i} + \mathbf{j} + \mathbf{k}$, $\mathbf{B} = \mathbf{i} + 2\mathbf{j} + \mathbf{k}$, $\mathbf{C} = 3\mathbf{i} + \mathbf{j} + 2\mathbf{k}$, and $\mathbf{D} = -\mathbf{i} + 5\mathbf{j}$.
8. Show that the vectors $\mathbf{A} = \mathbf{i} - 2\mathbf{j} + 2\mathbf{k}$ and $\mathbf{B} = 2\mathbf{i} + \mathbf{j} - 3\mathbf{k}$ are linearly independent, and express $\mathbf{D} = 4\mathbf{i} - 3\mathbf{j} + \mathbf{k}$ as a linear combination of these vectors.
9. Repeat Problem 8 if $\mathbf{A} = 2\mathbf{i} + 2\mathbf{j} - 3\mathbf{k}$, $\mathbf{B} = \mathbf{i} - 2\mathbf{j} + \mathbf{k}$, and $\mathbf{D} = 7\mathbf{i} - 2\mathbf{j} - 3\mathbf{k}$.
10. Express $\mathbf{A} = 3\mathbf{i} + 5\mathbf{j}$ as a linear combination of $\mathbf{B} = -\mathbf{i} + \mathbf{j}$ and $\mathbf{C} = 3\mathbf{i} + \mathbf{j}$.
11. Express $\mathbf{A} = -9\mathbf{i} - 2\mathbf{j}$ as a linear combination of $\mathbf{B} = \mathbf{i} + 2\mathbf{j}$ and $\mathbf{C} = 3\mathbf{i} - 2\mathbf{j}$.
12. Show that $\mathbf{A} = 2\mathbf{i} + \mathbf{j} - 3\mathbf{k}$, $\mathbf{B} = \mathbf{i} + 2\mathbf{j} + \mathbf{k}$, and $\mathbf{C} = 2\mathbf{i} + 3\mathbf{j} - 2\mathbf{k}$ are linearly independent.
13. Let $\mathbf{U} = 2\mathbf{A} - \mathbf{B} + \mathbf{C}$, $\mathbf{V} = \mathbf{A} + \mathbf{B} - 2\mathbf{C}$, and $\mathbf{W} = \mathbf{A} - 2\mathbf{B} + 3\mathbf{C}$, where $\mathbf{A} = \mathbf{i} - 2\mathbf{j} + 3\mathbf{k}$, $\mathbf{B} = 2\mathbf{i} + \mathbf{j} + 2\mathbf{k}$, $\mathbf{C} = -\mathbf{i} + \mathbf{j} + \mathbf{k}$. Show that the vectors $\mathbf{U}, \mathbf{V}, \mathbf{W}$ are linearly independent.
14. Let $\mathbf{U} = \mathbf{A} - \mathbf{B}$, $\mathbf{V} = \mathbf{B} + 2\mathbf{C}$, and $\mathbf{W} = 2\mathbf{A} - \mathbf{B} + \mathbf{C}$, where $\mathbf{A} = \mathbf{i} + \mathbf{j} - \mathbf{k}$, $\mathbf{B} = 3\mathbf{i} - \mathbf{j} + 2\mathbf{k}$, $\mathbf{C} = \mathbf{i} - 2\mathbf{j} + 3\mathbf{k}$. Show that $\mathbf{U}$, $\mathbf{V}$, and $\mathbf{W}$ are linearly independent.
15. Let $\mathbf{A} = \mathbf{i} - 2\mathbf{j} - \mathbf{k}$, $\mathbf{B} = 2\mathbf{i} - \mathbf{j} + 3\mathbf{k}$, and $\mathbf{C} = \mathbf{i} + 2\mathbf{j} + 4\mathbf{k}$. Determine in two ways (a) $\mathbf{A} \times (\mathbf{B} \times \mathbf{C})$ and (b) $(\mathbf{A} \times \mathbf{B}) \times \mathbf{C}$.
16. Prove that $(\mathbf{A} \times \mathbf{B}) \times (\mathbf{C} \times \mathbf{D}) = (\mathbf{ABD})\mathbf{C} - (\mathbf{ABC})\mathbf{D} = (\mathbf{ACD})\mathbf{B} - (\mathbf{BCD})\mathbf{A}$.
17. Prove that $(\mathbf{A} \times \mathbf{B}) \cdot (\mathbf{B} \times \mathbf{C}) \times (\mathbf{C} \times \mathbf{A}) = (\mathbf{ABC})^2$.
18. Prove that $\mathbf{A} \times (\mathbf{B} \times \mathbf{C}) + \mathbf{B} \times (\mathbf{C} \times \mathbf{A}) + \mathbf{C} \times (\mathbf{A} \times \mathbf{B}) = \mathbf{0}$.
19. Prove that $\mathbf{D} \times (\mathbf{A} \times \mathbf{B}) \cdot (\mathbf{A} \times \mathbf{C}) = (\mathbf{ABC})(\mathbf{A} \cdot \mathbf{D})$.
20. Show that $\mathbf{A} \times (\mathbf{B} \times \mathbf{C}) = (\mathbf{A} \times \mathbf{B}) \times \mathbf{C}$ if and only if $\mathbf{B} \times (\mathbf{C} \times \mathbf{A}) = \mathbf{0}$.
21. Prove the Lagrange identity

$$(\mathbf{A} \times \mathbf{B}) \cdot (\mathbf{C} \times \mathbf{D}) = (\mathbf{A} \cdot \mathbf{C})(\mathbf{B} \cdot \mathbf{D}) - (\mathbf{A} \cdot \mathbf{D})(\mathbf{B} \cdot \mathbf{C})$$

(Hint: Consider $\mathbf{A} \cdot [\mathbf{B} \times (\mathbf{C} \times \mathbf{D})]$.)

2

DIFFERENTIAL CALCULUS OF VECTOR FUNCTIONS OF ONE VARIABLE

In this chapter we shall extend the vector algebra developed in the previous chapter to vector functions of a real variable. We shall introduce the idea of a derivative of a vector function and consider its geometrical as well as physical significance. Beginning with Sec. 2.8 and through Sec. 2.10, we shall discuss curvilinear motion of a particle in various coordinate systems.

2.1 VECTOR FUNCTIONS OF A REAL VARIABLE

Let us first recall the definition of a real-valued (scalar) function in a domain D. We note that a domain D is an open set of points in which any two points can be joined by line segments lying entirely in the set. Examples are the set of points inside a sphere or a parallelepiped in three dimensional space, the set of points inside a circle or a rectangle in two dimensional space, and the set of points in the interval $a < x < b$ in one dimension. Now we recall that a real-valued function defined in a domain D is a mapping or a rule f that assigns to each point P in D a unique real number f(P) from a set R. We call D the domain of definition of f and R the range of f. In a similar manner, we define a vector-valued function or simply a vector function in a domain D as a rule **F** that assigns to each point P in D a unique vector denoted by **F**(P). In other words, a function defined in a domain D whose range is a set of vectors is called a vector function. In this chapter, we are concerned with vector functions defined in an interval which may contain one or both endpoints or which may be infinite. Such functions are called vector functions of a real variable. The real variable is usually denoted by t, which indicates time in many applications. We denote vector functions by bold-faced letters **F**, **G**, etc., and the value of a function **F** at t by **F**(t).

Following standard notations, we denote the open interval $a < t < b$ by (a, b) and the closed interval $a \le t \le b$ by [a, b]. The semi-infinite interval $-\infty < t \le a$ is denoted by $(-\infty, a]$ and the interval $a \le t < \infty$ by $[a, \infty)$.

Now let **F** be a vector function defined in the interval [a, b]. For each $t \in [a, b]$, **F**(t) is a vector and hence it is uniquely determined by its three components, say, $f_1(t)$, $f_2(t)$ and $f_3(t)$, with respect to a given rectangular cartesian coordinate system. Thus we can associate to each vector function **F** three scalar functions $f_1(t)$, $f_2(t)$ and $f_3(t)$, all defined in the same interval as **F**, whose values at t constitute the components of **F**(t). Therefore, we can represent each vector function **F** as

$$\mathbf{F}(t) = f_1(t)\,\mathbf{i} + f_2(t)\,\mathbf{j} + f_3(t)\,\mathbf{k} \quad (a \le t \le b) \tag{2.1}$$

where the scalar functions $f_1(t)$, $f_2(t)$ and $f_3(t)$ are called the components of **F**(t). This provides an analytic representation of a vector function of one variable. The representation (2.1) is sometimes also written as $\mathbf{F}(t) = [f_1(t), f_2(t), f_3(t)]$.

Example 1. The equation

$$\mathbf{R}(t) = \mathbf{R}_o + t\mathbf{A} \quad (-\infty < t < \infty)$$

of the line that passes through the point $\mathbf{R}_o$ and is parallel to the vector **A** is a vector function of the parameter t. For each value of t, the function defines a vector which is the position vector of a point on the line.

Example 2. The vector function

$$\mathbf{R}(t) = \cos t\,\mathbf{i} + \sin t\,\mathbf{j} \quad (0 \le t < 2\pi)$$

maps each value of t, $0 \le t < 2\pi$, to a vector in the xy-plane. If we represent the vector by an arrow from the origin, then as t ranges over $[0, 2\pi)$, the arrow traces the unit circle (Fig. 2.1). In this case the parameter t denotes the polar angle that the vector makes with the positive x-axis.

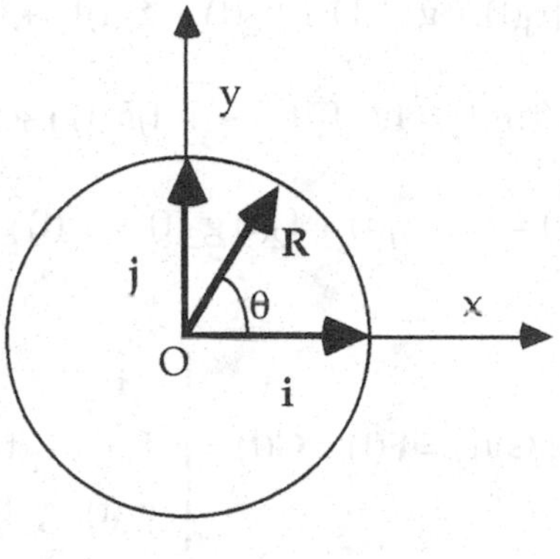

Fig. 2.1 The unit circle.

2.2 ALGEBRA OF VECTOR FUNCTIONS

We now extend the vector algebra developed in Chapter 1 to vector functions. We do this by expressing the vector functions in terms of their components and then performing the operations on the components. First, let **F** be a vector function defined in [a, b], and let f_1, f_2, f_3 denote its components. Then, for each $t \in [a, b]$, the magnitude of **F**(t) is given by

$$|\mathbf{F}(t)| = \sqrt{f_1(t)^2 + f_2(t)^2 + f_3(t)^2} \tag{2.2}$$

This is a scalar function of t in $a \le t \le b$. For some vector functions, however, the magnitude may be a constant. For example, the magnitude of the vector function $\mathbf{F}(t) = a(\cos t\,\mathbf{i} + \sin t\,\mathbf{j})$ is given by $|\mathbf{F}(t)| = a$, which is a constant.

Now let $\mathbf{F}(t) = f_1(t)\mathbf{i} + f_2(t)\mathbf{j} + f_3(t)\mathbf{k}$ and $\mathbf{G}(t) = g_1(t)\mathbf{i} + g_2(t)\mathbf{j} + g_3(t)\mathbf{k}$, $a \le t \le b$, be two vector functions, and h(t) a scalar function all defined in the same interval [a, b]. Then the sum of **F** and **G**, the h scalar multiple of **F**, the scalar and vector products of **F** and **G** are defined as follows:

$$(\mathbf{F} + \mathbf{G})(t) = \mathbf{F}(t) + \mathbf{G}(t)$$

$$= [f_1(t) + g_1(t)]\,\mathbf{i} + [f_2(t) + g_2(t)]\,\mathbf{j} + [f_3(t) + g_3(t)]\,\mathbf{k} \qquad (2.3)$$

$$(h\mathbf{F})(t) = h(t)\mathbf{F}(t) = h(t)f_1(t)\,\mathbf{i} + h(t)f_2(t)\,\mathbf{j} + h(t)f_3(t)\,\mathbf{k} \qquad (2.4)$$

$$(\mathbf{F} \cdot \mathbf{G})(t) = f_1(t)\,g_1(t) + f_2(t)\,g_2(t) + f_3(t)\,g_3(t) \qquad (2.5)$$

and

$$(\mathbf{F} \times \mathbf{G})(t) = \mathbf{F}(t) \times \mathbf{G}(t) = \begin{vmatrix} \mathbf{i} & \mathbf{j} & \mathbf{k} \\ f_1(t) & f_2(t) & f_2(t) \\ g_1(t) & g_2(t) & g_2(t) \end{vmatrix} \qquad (2.6)$$

Notice that the sum **F** + **G**, the scalar multiple h**F**, and the vector product **F** x **G** are all vector valued, whereas the scalar product **F**· **G** is real valued.

The concepts of linear dependence and orthogonality of vectors can likewise be carried over to vector functions. For example, we say that the functions **F**, **G**, **H** defined in the same interval [a, b] are linearly dependent if

$$\mathbf{F}(t) \cdot \mathbf{G}(t) \times \mathbf{H}(t) = 0$$

for all t ε [a, b]. Otherwise, the functions are said to be linearly independent in [a, b]. Similarly, we say that the functions **F** and **G** are orthogonal in [a, b] if $\mathbf{F}(t) \cdot \mathbf{G}(t) = 0$ for all t ε [a, b].

Example 1. If $\mathbf{F}(t) = t\,\mathbf{i} + (1 - t^2)\,\mathbf{j} + \cos t\,\mathbf{k}$ and $\mathbf{G}(t) = \sin t\,\mathbf{i} + \ln t\,\mathbf{j} + t^2\mathbf{k}$ for $t > 0$, then we have

$$\mathbf{F}(t) + \mathbf{G}(t) = (t + \sin t)\mathbf{i} + (1 - t^2 + \ln t)\mathbf{j} + (\cos t + t^2)\mathbf{k}$$

$$\mathbf{F}(t) \cdot \mathbf{G}(t) = t \sin t + (1 - t^2) \ln t + t^2 \cos t$$

and

$$\mathbf{F}(t) \times \mathbf{G}(t) = \begin{vmatrix} \mathbf{i} & \mathbf{j} & \mathbf{k} \\ t & 1-t^2 & \cos t \\ \sin t & \ln t & t^2 \end{vmatrix}$$

$$= [t^2(1 - t^2) - (\cos t) \ln t]\mathbf{i} + (\sin t \cos t - t^3)\mathbf{j} + [t \ln t - (1 - t^2)\sin t]\mathbf{k}$$

The magnitude of **F** and **G** are given by

$$|\mathbf{F}(t)| = \sqrt{1 - t^2 + t^4 + \cos^2 t}$$

and

$$|\mathbf{G}(t)| = \sqrt{\sin^2 t + \ln^2 t + t^4}$$

Example 2. Show that the vector functions $\mathbf{F}(t) = a(\cos t\,\mathbf{i} + \sin t\,\mathbf{j})$ and $\mathbf{G}(t) = b(\sin t\,\mathbf{i} - \cos t\,\mathbf{j})$, $0 \le t < 2\pi$, are orthogonal.

Solution: We show that the scalar product of **F**(t) and **G**(t) is zero. We have

$$\mathbf{F}(t)\cdot \mathbf{G}(t) = ab(\cos t \sin t - \sin t \cos t) = 0$$

for all t in $[0, 2\pi]$. Thus the vector functions are orthogonal.

2.1 EXERCISES

1. Let $\mathbf{F}(t) = t\,\mathbf{i} + (1 + t^2)\,\mathbf{j} - \sin t\,\mathbf{k}$ and $\mathbf{G}(t) = \ln(1 + t)\,\mathbf{i} - e^t\,\mathbf{j} + \mathbf{k}$ for $t \ge 0$. Find (a) $2\mathbf{F} - \mathbf{G}$, (b) $\mathbf{F}\cdot\mathbf{G}$, (c) $\mathbf{F} \times \mathbf{G}$, (d) $|\mathbf{F}|$ and $|\mathbf{G}|$.
2. Let $\mathbf{F}(t) = (3t - 1)\mathbf{i} + (t^2 - 2)\mathbf{j} + t\,\mathbf{k}$ and $\mathbf{G}(t) = 2e^t\,\mathbf{i} - \cos t\,\mathbf{j} + t^2\mathbf{k}$ for $t \ge 0$. Find (a) $3\mathbf{F} - 2\mathbf{G}$, (b) $\mathbf{F}\cdot\mathbf{G}$, (c) $\mathbf{F} \times \mathbf{G}$, (d) $|\mathbf{F} + \mathbf{G}|$.
3. Let $\mathbf{F}(t) = t \cos 2t\,\mathbf{i} + t \sin 2t\,\mathbf{j} + e^t\,\mathbf{k}$ $(t \ge 0)$. Find a scalar function h such that $|h(t)\mathbf{F}(t)| = 1$ for all $t \ge 0$.
4. Repeat Problem 3 if $\mathbf{F}(t) = e^{-t} \cos t\,\mathbf{i} + e^{-t} \sin t\,\mathbf{j} + e^t\,\mathbf{k}$ $(t \ge 0)$.
5. Let $\mathbf{F}(t) = a(\cos t\,\mathbf{i} + \sin t\,\mathbf{j}) + bt\,\mathbf{k}$ and $\mathbf{G}(t) = -\sin t\,\mathbf{i} + \cos t\,\mathbf{j}$ $(0 \le t < 2\pi)$, where a and b are constants. Show that **F** and **G** are orthogonal for all t in $[0, 2\pi)$. Find a unit vector function

that is parallel to $\mathbf{F} \times \mathbf{G}$ for $0 \le t < 2\pi$.

6. Repeat Problem 5 if $\mathbf{F}(t) = (\cos t - \sin t)\,\mathbf{i} + (\sin t + \cos t)\,\mathbf{j} + \mathbf{k}$ and $\mathbf{G}(t) = -(\cos t + \sin t)\,\mathbf{i} + (\cos t - \sin t)\,\mathbf{j}$ $(0 \le t < 2\pi)$.
7. Let $\mathbf{F}(t) = t\,\mathbf{i} + t^2\,\mathbf{j} + t^3\,\mathbf{k}$, $\mathbf{G}(t) = \sin t\,\mathbf{i} + \cos t\,\mathbf{j} + e^t\,\mathbf{k}$, and $\mathbf{H}(t) = (t+1)\,\mathbf{i} - 2t\,\mathbf{j} + e^t\,\mathbf{k}$, $t \ge 0$. Verify that (a) $\mathbf{F}(t)\cdot \mathbf{G}(t) \times \mathbf{H}(t) = \mathbf{F}(t) \times \mathbf{G}(t) \cdot \mathbf{H}(t)$ and (b) $\mathbf{F}(t) \times [\mathbf{G}(t) \times \mathbf{H}(t)] = g(t)\,\mathbf{G}(t) - h(t)\,\mathbf{H}(t)$, where $g(t) = \mathbf{F}(t)\cdot \mathbf{H}(t)$ and $h(t) = \mathbf{F}(t)\cdot \mathbf{G}(t)$ for all t.
8. Repeat Problem 7 if $\mathbf{F}(t) = t^2\,\mathbf{i} - t^3\,\mathbf{j}$, $\mathbf{G}(t) = e^t\,\mathbf{i} + \cosh t\,\mathbf{j}$, and $\mathbf{H}(t) = t\,\mathbf{i} - \ln t\,\mathbf{k}$ $(t > 0)$, and express $[\mathbf{F}(t) \times \mathbf{G}(t)] \times \mathbf{H}(t)$ as a linear combination of $\mathbf{F}(t)$ and $\mathbf{G}(t)$ for each $t > 0$.
9. Let $\mathbf{U}(\theta) = \cos\theta\,\mathbf{i} + \sin\theta\,\mathbf{j}$ and $\mathbf{V}(\phi) = \cos\phi\,\mathbf{i} + \sin\phi\,\mathbf{j}$, where both θ and ϕ range over $[0, 2\pi)$. (a) Show that **U** and **V** are both unit vector functions; (b) by considering the dot product $\mathbf{U}\cdot\mathbf{V}$, obtain the addition formula

$$\cos(\theta - \phi) = \cos\theta\cos\phi + \sin\theta\sin\phi$$

2.3 LIMIT, CONTINUITY, AND DERIVATIVES

The definitions of limit, continuity and derivatives for vector functions bear strong resemblance to the corresponding definitions for scalar functions. As a matter of fact, we may introduce the definitions of limit, continuity and derivative via the components of a vector function since these components are real valued functions. Consequently, many of the theorems on scalar functions concerning limits, continuity and derivatives can be carried over to vector functions in an almost straightforward fashion. We begin with the notion of a limit.

Limit of a function

Definition 1. Let **F** be a vector function defined in $a \le t \le b$, t_0 a point in this interval, and **A** a constant vector. We say that the limit of $\mathbf{F}(t)$ as t approaches t_0 is **A**, written as

$$\lim_{t \to t_0} \mathbf{F}(t) = \mathbf{A} \tag{2.7}$$

if and only if for any given number $\varepsilon > 0$, there is a number $\delta > 0$

such that

$$|\mathbf{F}(t) - \mathbf{A}| < \varepsilon \quad \text{whenever} \quad 0 < |t - t_o| < \delta \tag{2.8}$$

This definition is analogous to the corresponding definition of limit for a scalar function. Note, however, that (2.8) refers to the magnitude of the vector **F**(t) - **A**, whereas in the scalar case it refers to the absolute value of the difference between the scalar function and its limit. Geometrically, the definition implies that relative to a fixed point O the terminal point of the geometric vector **F**(t) lies inside the sphere of radius ε and center at the terminal point of **A** whenever $0 < |t - t_o| < \delta$ (Fig. 2.2).

In terms of components, if $\mathbf{F}(t) = f_1(t)\,\mathbf{i} + f_2(t)\,\mathbf{j} + f_3(t)\,\mathbf{k}$ and $\mathbf{A} = a_1\,\mathbf{i} + a_2\,\mathbf{j} + a_3\,\mathbf{k}$, then (2.7) is equivalent to

$$\lim_{t \to t_o} f_i(t) = a_i \qquad (i = 1, 2, 3) \tag{2.9}$$

Thus we have

$$\lim_{t \to t_o} [f_1(t)\,\mathbf{i} + f_2(t)\,\mathbf{j} + f_3(t)\,\mathbf{k}] = a_1\,\mathbf{i} + a_2\,\mathbf{j} + a_3\,\mathbf{k} \tag{2.10}$$

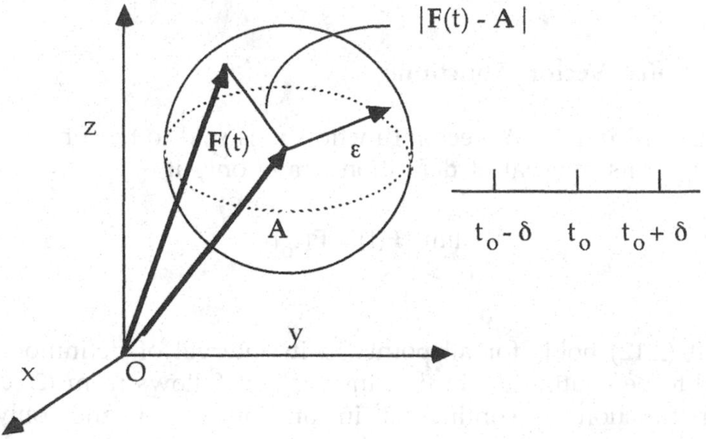

Fig. 2.2 Limit of F(t) as t tends to t_o.

if and only if (2.9) holds. As a result of this, many of the properties of limits for scalar functions also hold for vector functions. We state these properties as a theorem.

Theorem 1. Let **F** and **G** be vector functions and h a scalar function with a common interval of definition. Suppose that

$$\lim_{t \to t_o} \mathbf{F}(t) = \mathbf{A}\,, \qquad \lim_{t \to t_o} \mathbf{G}(t) = \mathbf{B}\,, \qquad \lim_{t \to t_o} h(t) = m$$

where **A** and **B** are constant vectors and m is a scalar. Then we have

$$\begin{aligned}
&\text{(a)} \quad \lim_{t \to t_o} [\mathbf{F}(t) + \mathbf{G}(t)] = \mathbf{A} + \mathbf{B} \\
&\text{(b)} \quad \lim_{t \to t_o} \mathbf{F}(t) \cdot \mathbf{G}(t) = \mathbf{A} \cdot \mathbf{B} \\
&\text{(c)} \quad \lim_{t \to t_o} \mathbf{F}(t) \times \mathbf{G}(t) = \mathbf{A} \times \mathbf{B} \\
&\text{(d)} \quad \lim_{t \to t_o} h(t)\,\mathbf{F}(t) = m\,\mathbf{A}
\end{aligned} \tag{2.11}$$

These properties can, of course, be proved from the definition (2.7) and (2.8) or, more easily, by expressing the vector functions in terms of their components and then using the equivalent definition (2.9).

Continuous Vector Functions

Definition 2. A vector function **F** is said to be continuous at a point t_o in its interval of definition if and only if

$$\lim_{t \to t_o} \mathbf{F}(t) = \mathbf{F}(t_o) \tag{2.12}$$

If (2.12) holds for all points in its interval of definition, then **F** is said to be continuous in that interval. It follows from (2.10) that a vector function is continuous in an interval if and only if its components are continuous in that interval. Moreover, from Theorem 1, we see that if **F, G** and h are continuous in a common interval of definition, then so are the functions $\mathbf{F} + \mathbf{G}$, $\mathbf{F} \cdot \mathbf{G}$, $\mathbf{F} \times \mathbf{G}$ and $h\mathbf{F}$.

Example 1. The following vector functions are continuous in the indicated interval:

(a) $\mathbf{F}(t) = \mathbf{A} \sin t + \mathbf{B} \cos t$, $0 \le t < 2\pi$, $\mathbf{A}$ and $\mathbf{B}$ are constant vectors.

(b) $\mathbf{G}(t) = (1/t)\,\mathbf{i} + (t^2 - 1)\,\mathbf{j} + e^t\,\mathbf{k}$, $t > 0$

(c) $\mathbf{H}(t) = e^{-t} \sin t\,\mathbf{i} + t \ln t\,\mathbf{j} + \cos 2t\,\mathbf{k}$, $t > 0$

Derivative of Vector Functions

Definition 3. Let $\mathbf{F}$ be a vector function defined in [a, b]. The derivative of $\mathbf{F}$ at a point t in [a, b], denoted by $\mathbf{F}'(t)$, is defined as the limit

$$\mathbf{F}'(t) = \lim_{\Delta t \to 0} \frac{\mathbf{F}(t + \Delta t) - \mathbf{F}(t)}{\Delta t} \tag{2.13}$$

provided the limit exists. If $\mathbf{F}'(t)$ exists for all t in (a,b), then we say that $\mathbf{F}$ is differentiable in (a, b).

In terms of components, if $\mathbf{F}(t) = f_1(t)\,\mathbf{i} + f_2(t)\,\mathbf{j} + f_3(t)\,\mathbf{k}$, then (2.13) may be written as

$$\mathbf{F}'(t) = \lim_{\Delta t \to 0} \left[\frac{f_1(t + \Delta t) - f_1(t)}{\Delta t}\,\mathbf{i} + \frac{f_2(t + \Delta t) - f_2(t)}{\Delta t}\,\mathbf{j} + \frac{f_3(t + \Delta t) - f_3(t)}{\Delta t}\,\mathbf{k} \right]$$

Therefore, if the components f_1, f_2, f_3 are all differentiable, then we have

$$\mathbf{F}'(t) = f_1'(t)\,\mathbf{i} + f_2'(t)\,\mathbf{j} + f_3'(t)\,\mathbf{k} \tag{2.14}$$

Thus a vector function $\mathbf{F}$ is differentiable in an interval if and only if its components are all differentiable in that interval. Since a scalar function is continuous whenever it is differentiable, it follows that **F**(t) is continuous in its interval of definition whenever **F**'(t) exists in that interval.

The rules for differentiating vector functions are similar to those for scalar functions. We state these rules as a theorem.

Theorem 2. Let **F**, **G**, and h be differentiable functions in (a, b). Then, in the same interval, we have

$$
\begin{aligned}
&(a) \quad [\mathbf{F}(t) + \mathbf{G}(t)]' = \mathbf{F}'(t) + \mathbf{G}'(t) \\
&(b) \quad [h(t)\mathbf{F}(t)]' = h(t)\mathbf{F}'(t) + h'(t)\mathbf{F}(t) \\
&(c) \quad [\mathbf{F}(t)\cdot \mathbf{G}(t)]' = \mathbf{F}'(t)\cdot \mathbf{G}(t) + \mathbf{F}(t)\cdot \mathbf{G}'(t) \\
&(d) \quad [\mathbf{F}(t) \times \mathbf{G}(t)]' = \mathbf{F}'(t) \times \mathbf{G}(t) + \mathbf{F}(t) \times \mathbf{G}'(t)
\end{aligned}
\qquad (2.15)
$$

These rules are easily established by expressing the vectors in terms of their components and then applying the rules for differentiating scalar functions. Observe that although (2.15d) is similar to the ordinary formula for the derivative of product of scalar functions, the order of the factors must be maintained since cross product of vectors is not commutative.

Formula (2.15c) leads to an important result which is used a great deal in our later discussion. We state this result as a theorem.

Theorem 3. Let **F** be a vector function that is differentiable in (a, b). If **F**(t) has constant magnitude for all t in (a, b) (that is, $|\mathbf{F}(t)|$ = const.), then **F**(t) and **F**'(t) are orthogonal in (a, b).

Proof: To prove the theorem, we need to show that $\mathbf{F}(t)\cdot\mathbf{F}'(t) = 0$ for all t in (a, b). Now, since $|\mathbf{F}(t)|$ is constant in (a, b), we have $|\mathbf{F}(t)|^2 = \mathbf{F}(t)\cdot\mathbf{F}(t) =$ constant, for all t in (a, b). Taking the derivative of both sides of the equation, we find by (2.15c)

$$2\,\mathbf{F}(t)\cdot\mathbf{F}'(t) = 0$$

for all t in (a, b). This implies that **F**(t) and **F**'(t) are orthogonal in (a, b).

As an example, consider the function $\mathbf{F}(t) = \cos t\,\mathbf{i} + \sin t\,\mathbf{j}$, $0 \le t < 2\pi$. Since $|\mathbf{F}(t)| = 1$ for all t in $[0, 2\pi)$, **F**(t) and **F**'(t) are orthogonal on $[0, 2\pi)$. Indeed, we find $\mathbf{F}'(t) = -\sin t\,\mathbf{i} + \cos t\,\mathbf{j}$ so that $\mathbf{F}(t)\cdot\mathbf{F}'(t) = 0$ readily follows.

Example 2. Let $\mathbf{F}(t) = \ln t\,\mathbf{i} + \cos t\,\mathbf{k}$ and $\mathbf{G}(t) = t^2\mathbf{j} + e^t\,\mathbf{k}$ for $t > 0$. Find the derivative of $\mathbf{F}(t)\cdot\mathbf{G}(t)$ in two different ways.

Solution: First, we observe that

$$\mathbf{F}(t)\cdot\mathbf{G}(t) = e^t \cos t$$

Therefore,

$$[\mathbf{F}(t)\cdot\mathbf{G}(t)]' = e^t \cos t - e^t \sin t$$

Alternatively, by (2.15c), we also find

$$\begin{aligned}[\mathbf{F}(t)\cdot\mathbf{G}(t)]' &= [(1/t)\,\mathbf{i} - \sin t\,\mathbf{k}]\cdot[t^2\mathbf{j} + e^t\,\mathbf{k}] \\ &\quad + [\ln t\,\mathbf{i} + \cos t\,\mathbf{k}]\cdot[2t\,\mathbf{j} + e^t\,\mathbf{k}] \\ &= e^t(\cos t - \sin t)\end{aligned}$$

Example 3. Let $\mathbf{F}(t) = t^2\mathbf{i} + e^t\,\mathbf{k}$ and $\mathbf{G}(t) = t\,\mathbf{i} + \sin t\,\mathbf{j} + \mathbf{k}$. Find the derivative of $\mathbf{F}(t) \times \mathbf{G}(t)$ in two different ways.

Solution: First, the cross product of $\mathbf{F}(t)$ and $\mathbf{G}(t)$ is given by

$$\mathbf{F}(t) \times \mathbf{G}(t) = \begin{vmatrix} \mathbf{i} & \mathbf{j} & \mathbf{k} \\ t^2 & 0 & e^t \\ t & \sin t & 1 \end{vmatrix} = -e^t \sin t\,\mathbf{i} + (t\,e^t - t^2)\,\mathbf{j} + t^2 \sin t\,\mathbf{k}$$

Hence

$$[\mathbf{F}(t) \times \mathbf{G}(t)]' = (-\,e^t \sin t - e^t \cos t)\mathbf{i} + (t\,e^t + e^t - 2t)\mathbf{j} + (2t \sin t + t^2 \cos t)\mathbf{k}$$

On the other hand, by (2.15d) we find

$$\mathbf{F}'(t) \times \mathbf{G}(t) = \begin{vmatrix} \mathbf{i} & \mathbf{j} & \mathbf{k} \\ 2t & 0 & e^t \\ t & \sin t & 1 \end{vmatrix} = -\,e^t \sin t\,\mathbf{i} + (t\,e^t - 2t)\,\mathbf{j} + 2t \sin t\,\mathbf{k}$$

and

$$\mathbf{F}(t) \times \mathbf{G}'(t) = \begin{vmatrix} \mathbf{i} & \mathbf{j} & \mathbf{k} \\ t^2 & 0 & e^t \\ 1 & \cos t & 0 \end{vmatrix} = -\,e^t \cos t\,\mathbf{i} + e^t\,\mathbf{j} + t^2 \cos t\,\mathbf{k}$$

The sum of these last two expressions gives the derivative of $\mathbf{F}(t) \times \mathbf{G}(t)$ which agrees with the preceding result.

2.2 EXERCISES

In Problems 1 through 4, let $\mathbf{F}(t) = \mathbf{A}\sin t + \mathbf{B}\cos t$ $(0 \le t < 2\pi)$, where $\mathbf{A} = \mathbf{i} - \mathbf{j}$ and $\mathbf{B} = \mathbf{i} + \mathbf{j}$. Evaluate the following limits.

1. $\lim_{t \to \pi/3} \mathbf{F}(t)$

2. $\lim_{t \to \pi/3} |\mathbf{F}(t)|$

3. $\lim_{t \to \pi/6} (\mathbf{i} + \mathbf{j} + \mathbf{k})\cdot\mathbf{F}(t)$

4. $\lim_{t \to \pi/3} (\tan t)\mathbf{F}(t)$

In Problems 5 through 8, let $\mathbf{F}(t) = t\,\mathbf{i} + t^2\mathbf{j} + t^3\mathbf{k}$ and $\mathbf{G}(t) = 2t\,\mathbf{i} - \mathbf{j} + 3t^3\mathbf{k}$ $(t \ge 0)$. Evaluate the indicated limits.

5. $\lim_{t \to 1} |\mathbf{F}(t) + \mathbf{G}(t)|$

6. $\lim_{t \to 1} \mathbf{F}(t)\cdot\mathbf{G}(t)$

7. $\lim_{t \to 1} \mathbf{F}(t) \times \mathbf{G}(t)$

8. $\lim_{t \to 1} |\mathbf{F}(t) \times \mathbf{G}(t)|$

9. In each case, find the derivative of $\mathbf{F}(t)\cdot\mathbf{G}(t)$ and $\mathbf{F}(t) \times \mathbf{G}(t)$:

   (a) $\mathbf{F}(t) = t\,\mathbf{i} + \mathbf{j} - 2t^2\mathbf{k}$ and $\mathbf{G}(t) = \sin t\,\mathbf{i} + \cos t\,\mathbf{j} + t\,\mathbf{k}$ $(t \ge 0)$
   (b) $\mathbf{F}(t) = \mathbf{i} - t^2\mathbf{j} + t\,\mathbf{k}$ and $\mathbf{G}(t) = t\,\mathbf{i} + (1 + t)\,\mathbf{j} + (t^2 + 1)\,\mathbf{k}$ $(t \ge 0)$.
   (c) $\mathbf{F}(t) = (\mathbf{i} - \mathbf{j} + \mathbf{k})\sin t + (2\mathbf{i} - \mathbf{j} - \mathbf{k})\cos t$ and
   $\mathbf{G}(t) = (2\mathbf{i} + \mathbf{j})e^t - (\mathbf{j} + 2\mathbf{k})\ln t$ $(t > 0)$

10. Let $\mathbf{F}(t) = a(\cos \omega t\,\mathbf{i} + \sin \omega t\,\mathbf{j}) + b\,\mathbf{k}$ $(0 \le t < 2\pi/\omega)$, where a, b and ω are positive constants. Show that $\mathbf{F}$ satisfies Theorem 3.

11. Let $\mathbf{F}(t) = a \sin 2t\, \mathbf{i} + a(1 + \cos 2t)\, \mathbf{j} + 2a \sin t\, \mathbf{k}$ $(0 \le t < \pi)$. Find the derivative of $|\mathbf{F}(t) \cdot \mathbf{F}'(t)|$.
12. Let

$$\mathbf{F}(t) = \frac{2t}{1+t^2}\mathbf{i} + \frac{1-t^2}{1+t^2}\mathbf{j} + \mathbf{k} \qquad (t \ge 0)$$

Show that the angle between $\mathbf{F}(t)$ and $\mathbf{F}'(t)$ is independent of t.
13. Let $\mathbf{F}(t) = \mathbf{A}e^{\omega t} + \mathbf{B}e^{-\omega t}$, where **A** and **B** are constant nonzero vectors and ω is a scalar. Show that $\mathbf{F}(t)$ satisfies the equation $\mathbf{F}''(t) - \omega^2\mathbf{F}(t) = \mathbf{0}$.
14. If $\mathbf{F}(t)$ is a vector function which is twice differentiable, show that $[\mathbf{F}(t) \times \mathbf{F}'(t)]' = \mathbf{F}(t) \times \mathbf{F}''(t)$.
15. Let $\mathbf{F}(t)$ be a differentiable vector function, and set $F(t) = |\mathbf{F}(t)|$. Show that $\mathbf{F}'(t) \cdot \mathbf{F}(t) = F'(t)F(t)$.
16. If $G(t) = \mathbf{F}(t) \cdot \mathbf{F}'(t) \times \mathbf{F}''(t)$, show that $G'(t) = \mathbf{F}(t) \cdot \mathbf{F}'(t) \times \mathbf{F}'''(t)$.
17. Let $\mathbf{F}(t) = \cos \omega t\, \mathbf{i} + \sin \omega t\, \mathbf{j}$ $(0 \le 2\pi/\omega)$. Show that $\mathbf{F}(t)$ satisfies the equation $\mathbf{F}''(t) + \omega^2\mathbf{F}(t) = 0$.
18. Let $\mathbf{F}(t) = \mathbf{A} + \mathbf{B}\, f(t)$, where **A** and **B** are constant vectors and f is a twice differentiable scalar function. Show that $\mathbf{F}'(t) \times \mathbf{F}''(t) = 0$.
19. Let $\mathbf{R}(t) = \mathbf{A} \cos \omega t + \mathbf{B} \sin \omega t$, where **A**, **B**, ω are constant. Show that $\mathbf{R}(t) \times \mathbf{R}'(t) = \omega\, \mathbf{A} \times \mathbf{B}$.
20. Let $\mathbf{r}(t) = \mathbf{R}(t)/|\mathbf{R}(t)|$. Show that $\mathbf{r} \times d\mathbf{r} = (\mathbf{R} \times d\mathbf{R})/|\mathbf{R}|^2$.
21. Show that if $\mathbf{R}(t) \times \mathbf{R}'(t) = \mathbf{0}$, then $\mathbf{R}(t)$ has a constant direction.
22. Show that if $\mathbf{F}'(t) = \mathbf{w} \times \mathbf{F}(t)$ and $\mathbf{G}'(t) = \mathbf{w} \times \mathbf{G}(t)$ for $a \le t \le b$, then $[\mathbf{F}(t) \times \mathbf{G}(t)]' = \mathbf{w} \times [\mathbf{F}(t) \times \mathbf{G}(t)]$, where **w** is a constant vector.

2.4 SPACE CURVES AND TANGENT VECTORS

The study of curves and surfaces in space is greatly facilitated by the use of vector calculus. Beginning in this section, we consider some of the basic geometric properties of space curves and their applications in the study of curvilinear motions.

Equations of Space Curves

We recall that a curve in space may be defined as a set of points (x, y, z) determined by three equations of the form

$$x = x(t), \qquad y = y(t), \qquad z = z(t) \tag{2.16}$$

The functions x(t), y(t), and z(t) are assumed to be continuous functions of the real variable t in some interval [a, b]. As in the case of the equation of a straight line, the equations (2.16) are called parametric equations of the curve, and t is called the parameter. To obtain a vector equation for the curve, we consider the position vector **R**(t) of each point on the curve corresponding to the parameter t. Since the components of **R**(t) are precisely the coordinates of the point, it follows that

$$\mathbf{R}(t) = x(t)\,\mathbf{i} + y(t)\,\mathbf{j} + z(t)\,\mathbf{k}, \quad (a \le t \le b) \tag{2.17}$$

Thus the vector equation of a space curve is a vector function of a

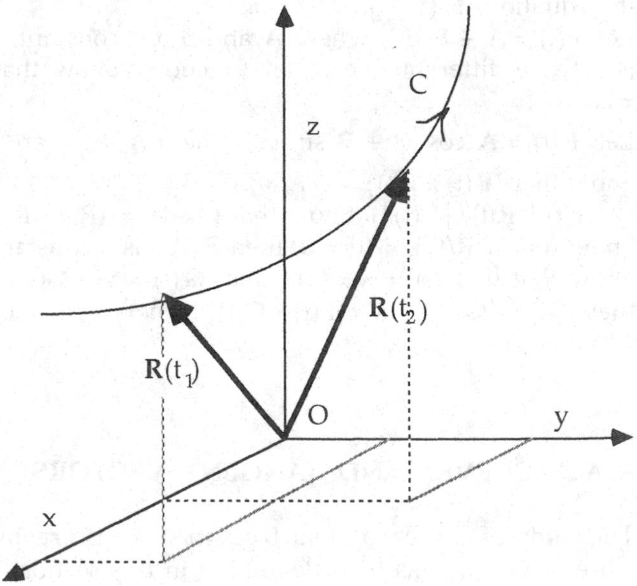

Fig. 2.3 A space curve.

real variable t defined in some interval $a \leq t \leq b$.

Conversely, suppose $\mathbf{R}(t)$ is a vector function that is defined and continuous in [a, b]. If we regard $\mathbf{R}(t)$ as the position vector of a point for each $t \in [a. b]$, then as t ranges over [a, b] the set of points determined by $\mathbf{R}(t)$ describes a curve in space (Fig. 2.3).

Therefore, just as a scalar function of one variable, $y = f(x)$, $(c \leq x \leq d)$ may be represented graphically by a plane curve, so a vector function $\mathbf{R} = \mathbf{R}(t)$ $(a \leq t \leq b)$ of a single variable may be represented graphically by a space curve. Of course, if $z = 0$ identically, then $\mathbf{R}(t) = x(t)\,\mathbf{i} + y(t)\,\mathbf{j}$ represents a curve on the xy-plane. By eliminating the parameter t from the equations $x = x(t)$, $y = y(t)$, it may be possible to obtain an equation of the curve in the form $y = f(x)$.

For example, the vector function

$$\mathbf{R}(t) = (1 + a\cos t)\mathbf{i} + (1 + a\sin t)\mathbf{j}\,, \quad (0 \leq t < 2\pi)$$

represents a circle of radius a and center at the point (1, 1) (Fig. 2.4). If we write the equation in the form

$$\begin{aligned}\mathbf{R}(t) &= (\mathbf{i} + \mathbf{j}) + a(\cos t\,\mathbf{i} + \sin t\,\mathbf{j}) \\ &= \mathbf{R}_0 + \mathbf{R}^*(t)\end{aligned}$$

we see that the position vector of each point on the circle is the vector sum of the position vector $\mathbf{R}_0$ of the center of the circle (relative to the origin) and the position vector $\mathbf{R}^*(t)$ of the point relative

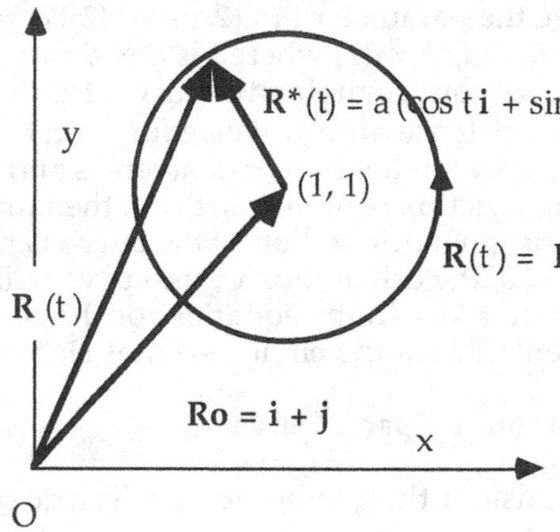

Fig. 2.4 Vector equation of a circle.

to the center of the circle. By eliminating the parameter t from the parametric equations

$$x = 1 + a\cos t, \qquad y = 1 + a\sin t \quad (0 \le t < 2\pi)$$

we obtain

$$(x - 1)^2 + (y - 1)^2 = a^2$$

which is the standard equation of the circle in cartesian coordinates.

Orientation of Space Curves

A curve that is represented in parametric or vector equation can be given one of two possible directions with respect to the parameter in a natural way. There is the direction in which the curve is traced as the parameter t increases from a to b in the interval $a \le t \le b$. It is customary to call this direction the positive direction on the curve; the opposite direction is then called the negative direction. A curve on which a direction has been designated is said to be oriented. Graphically, we denote the orientation of a curve by an arrowhead on the curve. Thus, for example, the circle in Fig. 2.4 has a counterclockwise orientation. This means that the positive direction is counterclockwise, which is the direction in which the circle is traced as t increases from 0 to 2π.

We should point out here that the parametric representation of a space curve is not unique. This means that it is possible, if we so desire, to change the parameter t in (2.16) or (2.17) to another parameter, say s, by setting $t = f(s)$ where f is any differentiable function such that $f' \neq 0$, without changing the curve itself. The condition $f' \neq 0$ ensures the existence of the inverse function f^{-1} of f so that the correspondence between the two parameters s and t is one-to-one. If f is an increasing function of s, that is, $f' > 0$, then under the change of parameters the original orientation of the curve is preserved. On the other hand, if $f' < 0$, the orientation of the curve will be reversed. For example, if we set $t = -s$ in the equation for the circle shown in Fig. 2.4, the orientation of the circle becomes clockwise.

Tangent Vector on a Space Curve

We now consider the geometric significance of the derivative of a vector function with respect to the curve it represents. Let C be a space curve represented by the equation

$$\mathbf{R}(t) = x(t)\,\mathbf{i} + y(t)\,\mathbf{j} + z(t)\,\mathbf{k} \tag{2.18}$$

where the functions x(t), y(t), and z(t) are continuously differentiable in $a < t < b$. From the definition in Sec. 2.3, we recall that

$$\frac{d}{dt}\mathbf{R}(t) = \lim_{\Delta t \to 0} \frac{\Delta\mathbf{R}(t)}{\Delta t} \tag{2.19}$$

where we have set $\Delta\mathbf{R}(t) = \mathbf{R}(t + \Delta t) - \mathbf{R}(t)$. By assumption, this limit exists. Let P and Q denote the points on the curve whose position vectors are given by **R**(t) and **R**(t + Δt), respectively (Fig. 2.5). Then, as $\Delta t \to 0$ the point Q tends to the point P along the curve and the line passing through P and Q approaches a limiting position at P which is the position of the tangent line to the curve at P. Hence, in the limit, the vector represented by the derivative (2.19) (assuming $\mathbf{R}'(t) \neq \mathbf{0}$) is tangent to the curve at P and points in the (positive) direction in which the curve is traced as t increases (see Fig. 2.5). Thus the derivative of a vector function is related to the notion of

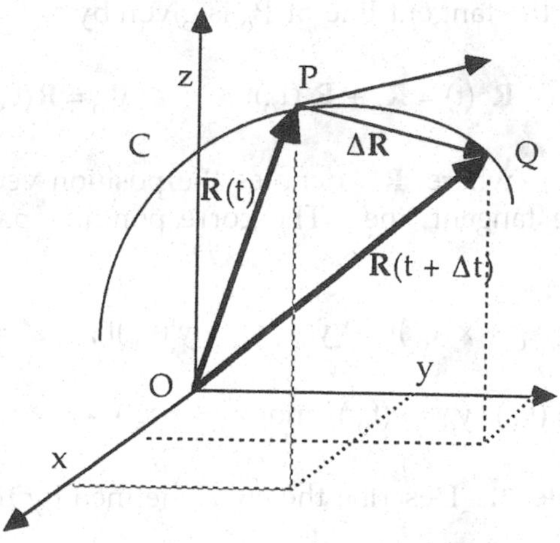

Fig. 2.5 The tangent vector.

tangency as in the case of a scalar function.

We call the vector represented by $d\mathbf{R}(t)/dt$ a tangent vector to the curve at the point P corresponding to a value of the parameter t. The vector

$$\mathbf{t} = \frac{\mathbf{R}'(t)}{|\mathbf{R}'(t)|} \tag{2.20}$$

is called the unit tangent vector. In terms of its components, this unit tangent vector is given by

$$\mathbf{t} = \frac{x'(t)\,\mathbf{i} + y'(t)\,\mathbf{j} + z'(t)\,\mathbf{k}}{\sqrt{x'(t)^2 + y'(t)^2 + z'(t)^2}} \tag{2.21}$$

Knowing the tangent vector at a point of a curve, it is then easy to write down the equation of the tangent line to the curve at that point by the method of Sec. 1.8. In fact, if t_o is the value of the parameter that corresponds to the point P_o on the curve, then a vector equation for the tangent line at P_o is given by

$$\mathbf{R}^*(t) = \mathbf{R}_o + \mathbf{R}'(t_o)t\,, \qquad \mathbf{R}_o = \mathbf{R}(t_o)$$

(see Fig. 2.6), where $\mathbf{R}^*$ denotes the position vector of an arbitrary point on the tangent line. The corresponding parametric equation are

$$x^* = x_o + x'(t_o)t\,, \quad y^* = y_o + y'(t_o)t\,, \quad z^* = z_o + z'(t_o)t$$

where $x_o = x(t_o)$, $y_o = y(t_o)$, and $z_o = z(t_o)$, $-\infty < t < \infty$.

Example 1. Describe the curve defined by the equation

$$\mathbf{R}(t) = e^t \cos t\,\mathbf{i} + e^t \sin t\,\mathbf{j}\,, \quad t \geq 0$$

and find an equation of the tangent line to the curve at the point corresponding to $t = \pi/4$.

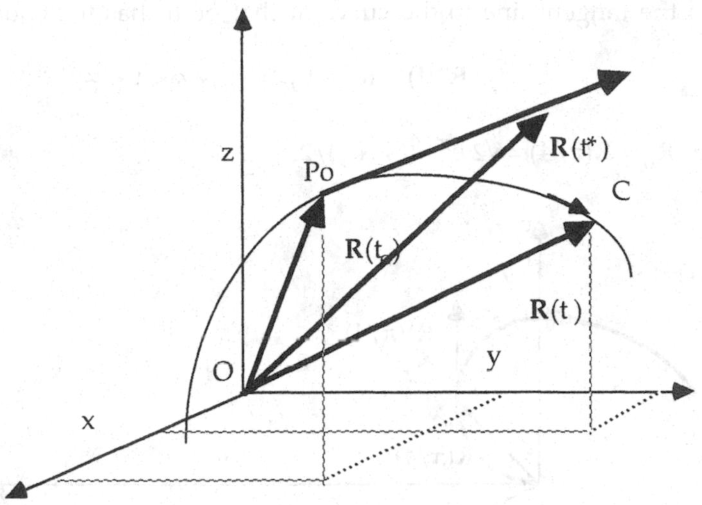

Fig. 2.6 Tangent line to a curve.

Solution: The curve is a plane curve whose parametric equations are given by

$$x = e^t \cos t, \qquad y = e^t \sin t, \quad t \geq 0$$

For each value of t, the distance from the origin of the corresponding point on the curve is equal to

$$r(t) = \sqrt{x^2(t) + y^2(t)} = e^t, \; t \geq 0$$

Thus as t increases the point (x, y) on the curve recedes from the origin. Further, since the sine and cosine functions are periodic, the point (x, y) goes around the origin in the counterclockwise direction. Therefore, as t increases the point (x, y) goes around the origin with increasing radial distance. The curve traced by the points is shown in Fig. 2.7. It is called a spiral.

Calculating $\mathbf{R}'(t)$ at $t = \pi/4$, we find

$$\mathbf{R}'(\pi/4) = e^{\pi/4}\left[(\cos \pi/4 - \sin \pi/4)\,\mathbf{i} + (\sin \pi/4 + \cos \pi/4)\,\mathbf{j}\right] = \sqrt{2}e^{\pi/4}\,\mathbf{j}$$

Hence the tangent vector at $t = \pi/4$ is parallel to the unit vector **j** and so the tangent line to the curve at that point has the equation

$$\mathbf{R}^*(t) = \mathbf{R}_o + t\,\mathbf{j}, \qquad (-\infty < t < \infty)$$

where $\mathbf{R}_o = \mathbf{R}(\pi/4) = \sqrt{2}\, e^{\pi/4}(\mathbf{i} + \mathbf{j})/2$.

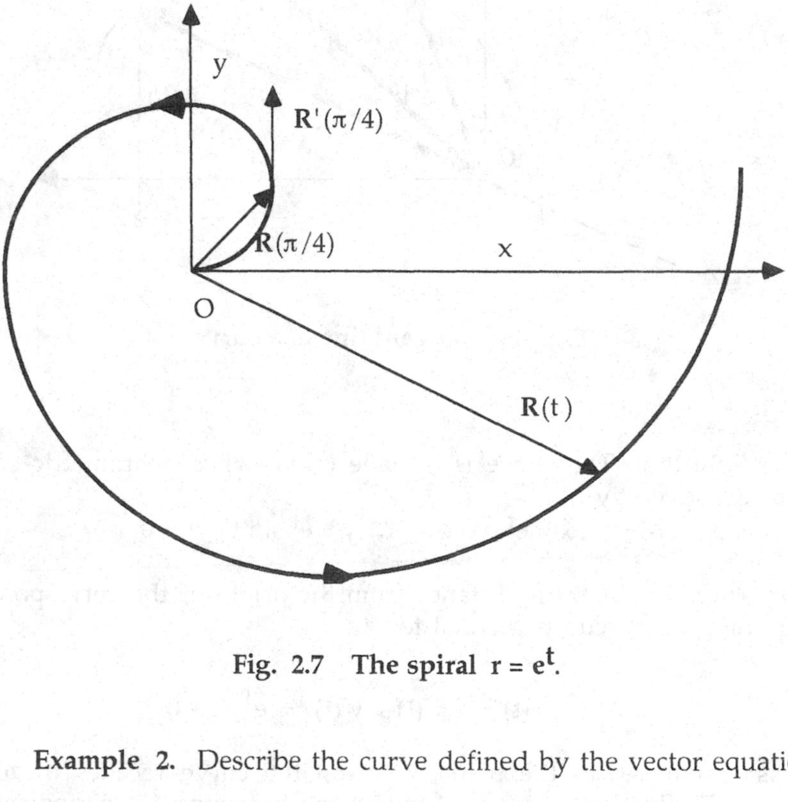

Fig. 2.7 The spiral $r = e^t$.

Example 2. Describe the curve defined by the vector equation

$$\mathbf{R}(t) = a\cos t\,\mathbf{i} + a\sin t\,\mathbf{j} + bt\,\mathbf{k}, \quad (t \geq 0)$$

where a and b are positive constants, and find the unit tangent vector to the curve for any $t > 0$.

Solution: The parametric equations of the curve are given by

$$x = a\cos t, \quad y = a\sin t, \quad z = bt, \quad (t \geq 0)$$

Since

$$x^2 + y^2 = a^2(\cos^2 t + \sin^2 t) = a^2$$

we see that the curve lies on the lateral surface of a circular cylinder of radius a whose axis is the z-axis. Moreover, as t increases the coordinate z increases since $b > 0$. Therefore, as t increases, the curve spirals upward around the cylinder in the counterclockwise direction as viewed from atop the positive z-axis (Fig. 2.8). The curve is called a circular helix.

A tangent vector to the curve for any value of t ($t > 0$) is given by

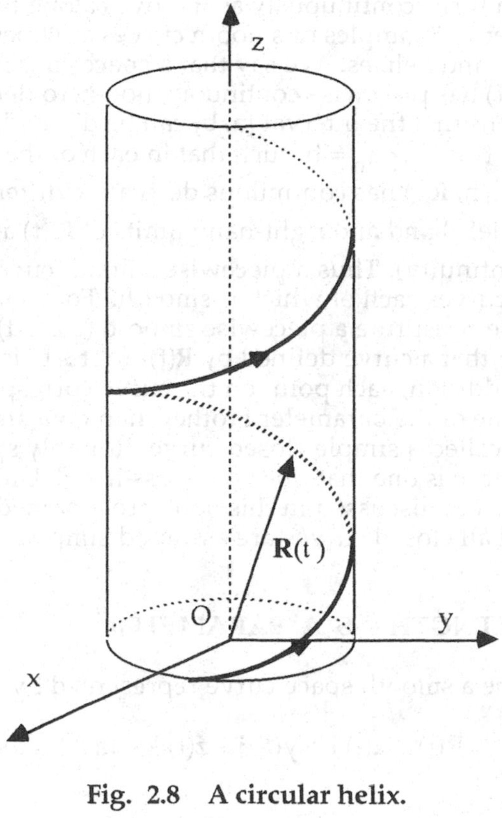

Fig. 2.8 A circular helix.

$$\mathbf{R}'(t) = -a\sin t\,\mathbf{i} + a\cos t\,\mathbf{j} + b\mathbf{k}$$

To obtain the unit tangent vector, we normalize $\mathbf{R}'(t)$ by dividing it by its magnitude, namely, $|\mathbf{R}'(t)| = (a^2 + b^2)^{1/2}$. Hence we have

$$\mathbf{t} = \frac{a(-\sin t\,\mathbf{i} + \cos t\,\mathbf{j}) + b\,\mathbf{k}}{\sqrt{a^2 + b^2}}$$

We close this section with a few remarks concerning the types of curves that we shall be dealing with in this book. We say that a space curve represented by $\mathbf{R}(t)$, $a \le t \le b$, is smooth if $\mathbf{R}(t)$ has continuous derivative in (a, b) such that $\mathbf{R}'(t) \neq \mathbf{0}$. Geometrically, this means that at each point on the curve there is a unique tangent vector which turns continuously as it moves along the curve with the parameter t. Examples of smooth curves are lines, circles, parabolas, spirals and helixes. We say that a space curve is piecewise smooth if $\mathbf{R}(t)$ has piecewise continuous non-zero derivative in (a, b). This means that the interval [a, b] can be divided into subintervals $a = t_o < t_1 < \cdots < t_n = b$ such that in each of the subintervals (t_{i-1}, t_i), $1 \le i \le n$, $\mathbf{R}(t)$ has continuous derivative different from $\mathbf{0}$ and that the left-hand and right-hand limits of $\mathbf{R}'(t)$ at $t = t_i$ exist (jump discontinuity). Thus a piecewise smooth curve is composed of pieces of curves each of which is smooth. For example, the sides of a rectangle constitute a piecewise smooth (closed) curve.

We say that a curve defined by $\mathbf{R}(t)$, $a \le t \le b$, is closed if $\mathbf{R}(a) = \mathbf{R}(b)$. If, in addition, each point on the curve corresponds to one and only one value of the parameter t (other than $t = a$ and $t = b$), then the curve is called a simple closed curve. Roughly speaking, a simple closed curve is one that does not cross itself. Unless otherwise stated, all curves discussed in this book are assumed to be piecewise smooth, and all closed curves are assumed simple.

2.5 ARC LENGTH AS A PARAMETER

Let C be a smooth space curve represented by

$$\mathbf{R}(t) = x(t)\,\mathbf{i} + y(t)\,\mathbf{j} + z(t)\,\mathbf{k} \quad (a \le t \le b) \tag{2.22}$$

Denote by s(t) the arc length of the curve from the point corresponding to t = a to a point corresponding to an arbitrary value of t (Fig. 2.9). Then, as shown in elementary calculus, the arc length is given by the integral

$$s(t) = \int_a^t \sqrt{x'(\tau)^2 + y'(\tau)^2 + z'(\tau)^2}\, d\tau \tag{2.23}$$

Since the integrand in (2.23) is precisely the magnitude of $d\mathbf{R}(t)/dt$, we can write (2.23) as

$$s(t) = \int_a^t \left| \frac{d\mathbf{R}(\tau)}{d\tau} \right| d\tau \tag{2.24}$$

This gives us the relationship between the arc length s of the curve and the parameter t. By the Fundamental Theorem of calculus, it follows that

$$\frac{ds}{dt}(t) = \left| \frac{d\mathbf{R}}{dt}(t) \right| > 0 \tag{2.25}$$

Thus the arc length s is an increasing function of the parameter t. Equation (2.25) implies that

$$\left(\frac{ds}{dt}\right)^2 = \left| \frac{d\mathbf{R}}{dt} \right|^2 = \frac{d\mathbf{R}}{dt} \cdot \frac{d\mathbf{R}}{dt}$$

Hence if we define the vector differential $d\mathbf{R}$ by the equation

$$d\mathbf{R} = dx\,\mathbf{i} + dy\,\mathbf{j} + dz\,\mathbf{k}$$

then we have

$$ds^2 = d\mathbf{R} \cdot d\mathbf{R} = dx^2 + dy^2 + dz^2 \tag{2.26}$$

The differential ds is called the element of arc length.

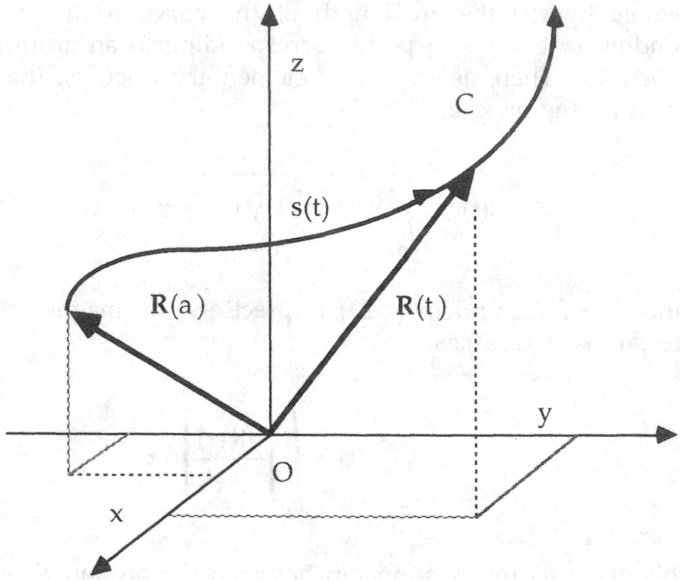

Fig. 2.9 Arc length of a curve.

The study of the various properties of a space curve is greatly facilitated when the curve is represented by parametric equations whose parameter is the arc length s of the curve measured from some initial point on the curve. Indeed, if we take one end of the curve as the initial point (s = 0), then each point P on the curve is uniquely determined by the arc length of the curve measured from the initial point to the point P. With arc length as a parameter, we immediately see from (2.25) that

$$\left|\frac{d\mathbf{R}(s)}{ds}\right| = 1$$

Hence the vector $d\mathbf{R}(s)/ds$ is a unit tangent vector. Therefore, when a curve is represented by an equation $\mathbf{R} = \mathbf{R}(s)$ with its arc length s, $0 \le s \le L$, as a parameter, the unit tangent vector is given by

$$\mathbf{t} = \frac{d\mathbf{R}(s)}{ds} \tag{2.27}$$

For example, the vector equation

$$\mathbf{R}(t) = r(\cos t\,\mathbf{i} + \sin t\,\mathbf{j}), \qquad (0 \le t < 2\pi)$$

represents a circle of radius r with center at the origin. The tangent vector is given by

$$\mathbf{R}'(t) = r(-\sin t\,\mathbf{i} + \cos t\,\mathbf{j})$$

so that $|\mathbf{R}'(t)| = r$. Therefore, by (2.24) the arc length of the circle corresponding to a value of t is given by

$$s = \int_0^t r\,d\tau = rt, \quad 0 \le t < 2\pi$$

Hence, if we replace t by s/r, we obtain the equation

$$\mathbf{R}(s) = r[\cos(s/r)\,\mathbf{i} + \sin(s/r)\,\mathbf{j}]$$

with the arc length s, $0 \le s \le 2\pi r$, as a parameter. The unit tangent vector is given by

$$\mathbf{R}'(s) = -\sin(s/r)\,\mathbf{i} + \cos(s/r)\,\mathbf{j}$$

Example 1. Find the arc length of the curve represented by

$$\mathbf{R}(t) = e^t(\cos t\,\mathbf{i} + \sin t\,\mathbf{j}), \quad (0 \le t \le \pi)$$

What is the parametric equation in terms of its arc length?

Solution: We have

$$\mathbf{R}'(t) = e^t(\cos t - \sin t)\,\mathbf{i} + e^t(\sin t + \cos t)\,\mathbf{j}$$

so that

$$|\mathbf{R}'(t)| = \sqrt{2}\,e^t$$

Hence by (2.24) the arc length of the curve is

$$s = \int_0^\pi \sqrt{2}e^t dt = \sqrt{2}\,(e^\pi - 1)$$

The arc length from the initial point (1, 0) (t = 0) to an arbitrary point (arbitrary value of t) on the curve is given by

$$s = \int_0^t \sqrt{2}e^{\tau}d\tau = \sqrt{2}\,(e^t - 1)$$

Hence if we wish to represent the curve in terms of its arc length as a parameter, we have to express the parameter t in terms of the arc length s. In other words, we need to solve for t in terms of s from the above relation. We readily find

$$t = \ln[s/\sqrt{2} + 1]$$

so that

$$\mathbf{R}(s) = (s/\sqrt{2} + 1)\{\cos[\ln(s/\sqrt{2} + 1)]\,\mathbf{i} + \sin[\ln(s/\sqrt{2} + 1)]\,\mathbf{j}\}$$

Example 2. Represent the helix $\mathbf{R}(t) = a(\cos t\,\mathbf{i} + \sin t\,\mathbf{j}) + bt\,\mathbf{k}$, $0 \le t \le \pi$, in terms of its arc length as a parameter.

Solution: We have

$$\mathbf{R}'(t) = a(-\sin t\,\mathbf{i} + \cos t\,\mathbf{j}) + b\mathbf{k}$$

so that $|\mathbf{R}'(t)| = [a^2 + b^2]^{1/2}$. Hence

$$s = \int_0^t \sqrt{a^2 + b^2}\,d\tau = \sqrt{a^2 + b^2}\,t$$

or

$$t = s/[a^2 + b^2]^{1/2}$$

Replacing t by $s/[a^2 + b^2]^{1/2}$ in the vector equation of the helix, we obtain the desired equation.

In general, it is often very difficult to obtain the arc length s in terms of a parameter t in the sense that the integral (2.24) may not always be expressible by elementary functions. And even when this is possible, it may still be difficult to solve for t in terms of s. The artifice of using arc length as the parameter is, however, very convenient in the theoretical development of geometric properties of curves as we shall see in the next section.

2.3 EXERCISES

1. Obtain the representation of the spiral

$$\mathbf{R}(t) = e^t(\cos t\,\mathbf{i} + \sin t\,\mathbf{j}) \quad (t \geq 0)$$

in terms of the new parameter τ, where $\tau = e^t$, and show that the orientation of the curve is preserved.

2. Obtain another representation of the curve

$$\mathbf{R}(t) = t\,\mathbf{i} + t^2\mathbf{j} + t^3\,\mathbf{k} \qquad (t \geq 0)$$

in terms of the parameter τ, where $\tau = 1 - t^2$. What happens to the orientation of the curve?

3. Show that the curve defined by

$$\mathbf{R}(t) = a(\cos t\,\mathbf{i} + \sin t\,\mathbf{j}) + a(1 + \sin t)\mathbf{k} \qquad (t \geq 0)$$

is a plane curve. (Hint: Show that it is the intersection of a cylinder and a plane.)

4. Show that the curve defined by

$$\mathbf{R}(t) = t(\cos t\,\mathbf{i} + \sin t\,\mathbf{j} + \mathbf{k}) \qquad (t \geq 0)$$

lies on a cone. This curve is called a conical helix.

In each of Problems 5 through 10, describe the curve represented by $\mathbf{R}(t)$, and find the unit tangent vector to the curve at the point corresponding to the given value of t.

5. $\mathbf{R}(t) = t(\cos t\,\mathbf{i} + \sin t\,\mathbf{j})$ $(t \geq 0)$; $t = \pi/6$
6. $\mathbf{R}(t) = a\cos t\,\mathbf{i} + b\sin t\,\mathbf{j}$, a and b are constants $(0 \leq t < 2\pi)$; $t = \pi/4$
7. $\mathbf{R}(t) = (t^2 + 1)\,\mathbf{i} + t\,\mathbf{j}$ $(t \geq 0)$; $t = 2$
8. $\mathbf{R}(t) = \cos t\,\mathbf{i} + \sin t\,\mathbf{j} + (1 + \sin t)\mathbf{k}$ $(0 \leq t < 2\pi)$; $t = \pi/6$
9. $\mathbf{R}(t) = 2(t - \sin t)\mathbf{i} + 2(1 - \cos t)\mathbf{j}$ $(t \geq 0)$; $t = \pi/3$
10. $\mathbf{R}(t) = e^{2t}\,\mathbf{i} + e^{-2t}\,\mathbf{j} + (1 + t^2)\mathbf{k}$ $(t \geq 0)$; $t = 1$
11. Determine the arc length of the curve represented by

$$\mathbf{R}(t) = e^t(\cos t\,\mathbf{i} + \sin t\,\mathbf{j} + \mathbf{k}) \quad (t \geq 0)$$

as a function of t. Express the equation of the curve in terms of the arc length s.

12. Determine the arc length of the curve represented by
$$\mathbf{R}(t) = t(\cos t\,\mathbf{i} + \sin t\,\mathbf{j} + \mathbf{k}) \quad (t \geq 0)$$
as a function of t.

13. Represent the curve defined by
$$\mathbf{R}(t) = (\sin t - t\cos t)\mathbf{i} + (\cos t + t\sin t)\mathbf{j} + (t^2/2)\mathbf{k} \;\; (t \geq 0)$$
in terms of the arc length of the curve.

14. Find the length of the curve defined by
$$\mathbf{R}(t) = \sin t\,\mathbf{i} + t\,\mathbf{j} + (1 - \cos t)\mathbf{k}$$
for $0 \leq t < 2\pi$.

15. Find the length of the circular helix
$$\mathbf{R}(t) = a(\cos t\,\mathbf{i} + \sin t\,\mathbf{j}) + bt\,\mathbf{k}$$
from $t = 0$ to $t = 2\pi$.

16. Find the expression of the arc length ds for the curve given in (a) Problem 3, (b) Problem 4.

17. Let C be a curve in the xy-plane represented in polar coordinates by the equation
$$r = f(\theta), \qquad (x = r\cos\theta, \;\; y = r\sin\theta)$$
where f is a continuously differentiable function. Show that $ds^2 = [f^2(\theta) + f'^2(\theta)]d\theta^2$.

2.6 SIMPLE GEOMETRY OF CURVES

We consider a space curve that is parametrized by $\mathbf{R} = \mathbf{R}(s)$, where s denotes arc length of the curve, $0 \leq s \leq L$. We assume that $\mathbf{R}(s)$ has continuous derivatives up to the second order in the interval (0, L). As we saw in the preceding section, the unit tangent vector to the curve is given by $\mathbf{t} = d\mathbf{R}(s)/ds$. Now let us consider how the unit vector $\mathbf{t}$ changes as it moves along the curve. Since $\mathbf{t}$ has constant magnitude, only its direction can change. The rate of this change is measured by the derivative $d\mathbf{t}/ds$. Since the magnitude of $\mathbf{t}$ is constant, we know from Theorem 3 (Sec. 2.3) that $\mathbf{t}$ and $d\mathbf{t}/ds$ are orthogonal; hence, $\mathbf{t} \cdot d\mathbf{t}/ds = 0$.

Curvature and Principal Normal Vector

If $d\mathbf{t}/ds = \mathbf{0}$, then $\mathbf{t}$ is a constant vector (does not change in magnitude and direction); hence, the curve is a straight line. On the other hand, if $d\mathbf{t}/ds \neq \mathbf{0}$, we can write

$$\frac{d\mathbf{t}}{ds}(s) = \kappa(s)\mathbf{n}(s) \tag{2.28}$$

where **n** is a unit vector in the same direction as $d\mathbf{t}/ds$ and, hence, orthogonal to **t**. It follows from (2.28) that

$$\kappa(s) = \left|\frac{d\mathbf{t}}{ds}(s)\right| \tag{2.29}$$

and so

$$\mathbf{n}(s) = \frac{d\mathbf{t}(s)/ds}{|d\mathbf{t}(s)/ds|} \tag{2.30}$$

The quantity $\kappa(s) \geq 0$ defined by (2.29) is called the *curvature* of the curve at the point corresponding to the parameter s. It is a measure of the rate at which the unit tangent vector **t** changes its direction at a point. A large value of κ at a point implies that the curve is "sharply curved" at that point, whereas a small value of κ means that the curve is almost "straight". Thus for a straight line $\kappa = 0$ as we have observed above.

When $\kappa \neq 0$, its reciprocal $\rho = 1/\kappa$, is called the *radius of curvature.* This is the radius of the circle, called the circle of curvature, that best fits the curve at the point determined by s.

The unit vector **n** defined by (2.30) is called the *principle normal* vector to the curve at the point corresponding to the parameter s. The center of the circle of curvature lies on the line along the principal normal vector.

In order to get an idea of the direction of **n** on the curve, let us examine geometrically the definition of $d\mathbf{t}/ds$, namely,

$$\frac{d\mathbf{t}}{ds}(s) = \lim_{\Delta s \to 0} \frac{\mathbf{t}(s + \Delta s) - \mathbf{t}(s)}{\Delta s}$$

Referring to Fig. 2. 10, we notice that when $\Delta s > 0$ the vector $\Delta\mathbf{t} = \mathbf{t}(s + \Delta s) - \mathbf{t}(s)$ is directed toward the concave side of the curve. Since $\Delta s > 0$, the same is true of the limit of $\Delta\mathbf{t}/\Delta s$ as Δs tends to zero. For $\Delta s < 0$, the vector $\Delta\mathbf{t}$ is directed toward the convex side of the curve; however, since $\Delta s < 0$, the vector $\Delta\mathbf{t}/\Delta s$ points in the opposite direc-

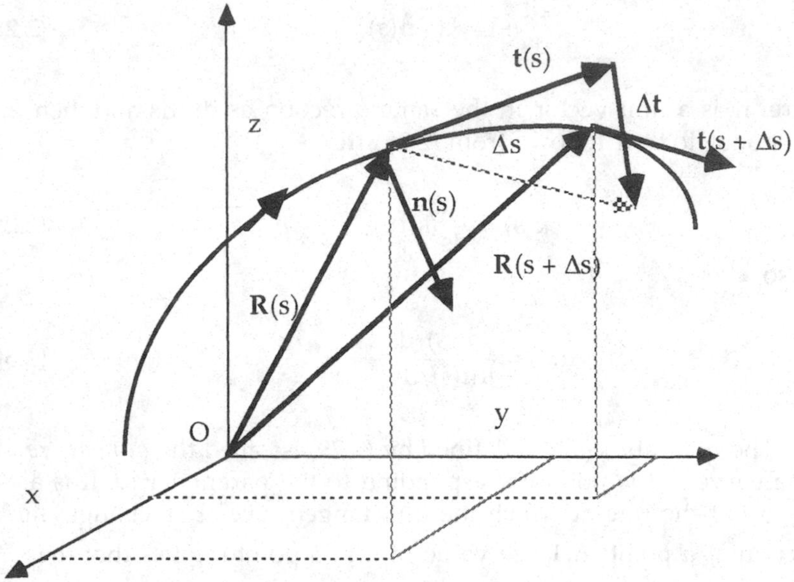

Fig. 2.10 The principal normal vector.

tion to $\Delta\mathbf{t}$, and so is the limit $\Delta\mathbf{t}/\Delta s$ (see Fig. 2.10 and also Example 3). Therefore, at each point on the curve, the principal normal vector $\mathbf{n}$ points toward the concave side of the curve.

Example 1. Consider the circle of radius a described by

$$\mathbf{R}(s) = a(\cos(s/a)\,\mathbf{i} + \sin(s/a)\,\mathbf{j}) \quad (0 \le s \le 2\pi a)$$

We have $\mathbf{t}(s) = \mathbf{R}'(s) = -\sin(s/a)\,\mathbf{i} + \cos(s/a)\,\mathbf{j}$, so that

$$\mathbf{t}'(s) = (1/a)[-\cos(s/a)\,\mathbf{i} - \sin(s/a)\,\mathbf{j}]$$

Since $\mathbf{t}'(s) = \kappa\mathbf{n}$, we deduce that $\kappa = 1/a$ and $\mathbf{n} = -\cos(s/a)\mathbf{i} - \sin(s/a)\mathbf{j}$. Clearly $\mathbf{n} = -\mathbf{R}(s)/a$, which shows that the principal normal vector points toward the center as shown in Fig. 2.11. It follows that the circle of curvature coincides with the circle itself.

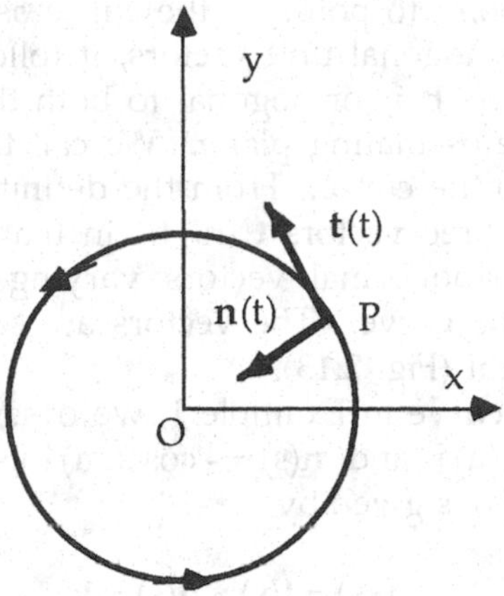

Fig. 2.11 The vectors **t** and **n** on a circle.

Osculating Plane and the Binormal Vector

We observe that at each point on a curve, the two unit vectors **t** and **n** determine a plane containing that point. This plane is known as the *osculating plane*. It is the plane that best fits the curve at that point and thus contains the circle of curvature. For a plane curve (not a straight line), the osculating plane coincides with the plane on which the curve lies. In general, however, the osculating

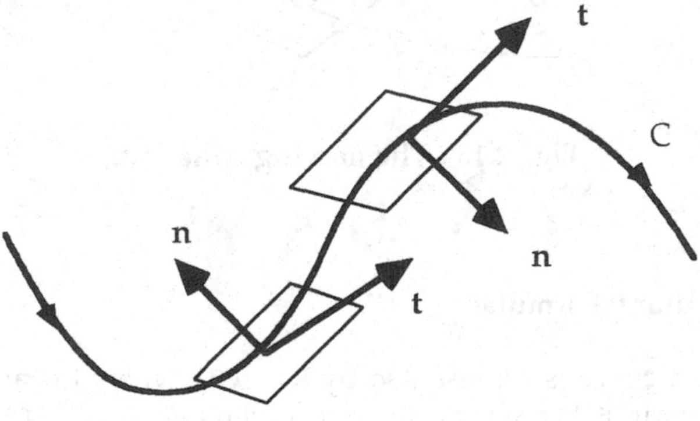

Fig. 2.12 Osculating planes.

plane varies from point to point on the curve as shown in Fig. 2.12. Since **t** and **n** are orthogonal unit vectors, it follows that $|\mathbf{b}| = |\mathbf{t} \times \mathbf{n}| = 1$. Moreover, since **b** is orthogonal to both the vectors **t** and **n**, it is orthogonal to the osculating plane. We call the unit vector **b** the *binormal* vector to the curve. From the definition of cross product, it follows that the three vectors **t**, **n**, **b**, in that order, form a right-handed triple of orthonormal vectors varying in direction from point to point on the curve. The vectors are sometimes referred to as a moving trihedral (Fig. 2.13).

For the plane curve in Example 1, we observe that $\mathbf{t}(s) = -\sin(s/a)\,\mathbf{i} + \cos(s/a)\,\mathbf{j}$ and $\mathbf{n}(s) = -\cos s/a)\,\mathbf{i} - \sin(s/a)\,\mathbf{j}$. Hence the binormal vector $\mathbf{b}(s)$ is given by

$$\mathbf{b}(s) = \mathbf{t}(s) \times \mathbf{n}(s) = \mathbf{k}$$

which is indeed orthogonal to the xy-plane on which the circle lies.

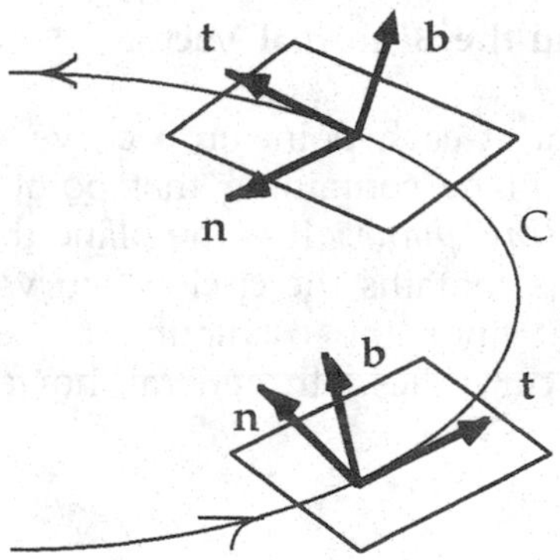

Fig. 2.13 The moving trihedral.

Some Important Formulas

When a curve is represented by $\mathbf{R} = \mathbf{R}(t)$, where t is an arbitrary parameter, its unit tangent vector **t** is given by $\mathbf{t}(t) = \mathbf{R}'(t)/|\mathbf{R}'(t)|$. By the chain rule, the derivative of **t** with respect to arc length s is

$$\frac{d\mathbf{t}}{ds}(t) = \frac{d\mathbf{t}}{dt}\frac{dt}{ds} = \frac{d\mathbf{t}/dt}{ds/dt} = \frac{\mathbf{t}'(t)}{|\mathbf{R}'(t)|} \tag{2.32}$$

Thus the vector $\mathbf{t}'(t)$ has the same direction as $\mathbf{t}'(s)$ and, therefore, the principal normal vector $\mathbf{n}$ may also be given by the formula

$$\mathbf{n}(t) = \frac{\mathbf{t}'(t)}{|\mathbf{t}'(t)|}$$

This shows that the formula (2.30) for the principal normal vector is valid no matter what parameter is used in representing the curve.

From the definition (2.29) of the curvature we deduce from (2.32) that

$$\kappa(t) = \frac{|\mathbf{t}'(t)|}{|\mathbf{R}'(t)|} \tag{2.33}$$

Another formula for the curvature, which may sometimes be more convenient to use, is given by (see Problem 10)

$$\kappa(t) = \frac{|\mathbf{R}'(t) \times \mathbf{R}''(t)|}{|\mathbf{R}'(t)|^3} \tag{2.34}$$

Example 2. Determine the curvature, the principal normal and the binormal vector for the circular helix

$$\mathbf{R}(t) = a(\cos t\,\mathbf{i} + \sin t\,\mathbf{j}) + bt\,\mathbf{k} \qquad (t \geq 0)$$

Solution: Since $\mathbf{R}'(t) = a(-\sin t\,\mathbf{i} + \cos t\,\mathbf{j}) + b\,\mathbf{k}$, we have

$$\mathbf{t}(t) = \frac{\mathbf{R}'(t)}{|\mathbf{R}'(t)|} = \frac{-a\sin t\,\mathbf{i} + a\cos t\,\mathbf{j} + b\mathbf{k}}{\sqrt{a^2 + b^2}}$$

Taking the derivative, we find

$$\mathbf{t}'(t) = \frac{-a\cos t\,\mathbf{i} - a\sin t\,\mathbf{j}}{\sqrt{a^2 + b^2}}$$

and so $|\mathbf{t}'(t)| = a(a^2 + b^2)^{-1/2}$. Hence by (2.33) we obtain

$$\kappa(t) = a/(a^2 + b^2)$$

Thus the helix has a constant curvature as to be expected.

By (2.30) we find

$$\mathbf{n} = -\cos t\,\mathbf{i} - \sin t\,\mathbf{j}$$

which is directed toward the axis (z-axis) of the helix. Taking the cross product of **t** and **n**, we obtain the binormal vector

$$\mathbf{b} = \mathbf{t} \times \mathbf{n} = \frac{1}{\sqrt{a^2 + b^2}} \begin{vmatrix} \mathbf{i} & \mathbf{j} & \mathbf{k} \\ -a \sin t & a \cos t & b \\ -\cos t & -\sin t & 0 \end{vmatrix}$$

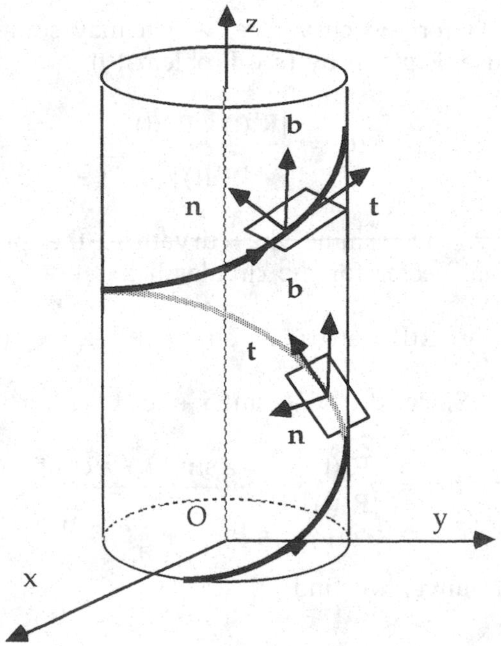

Fig. 2.14 A moving trihedral on a helix.

$$= \frac{b \sin t\, \mathbf{i} - b \cos t\, \mathbf{j} + a\, \mathbf{k}}{\sqrt{a^2 + b^2}}$$

The trihedral $(\mathbf{t}, \mathbf{n}, \mathbf{b})$ on the helix are shown in Fig. 2.14

Curvature for a Plane Curve

Consider a curve in the xy-plane. Since **t** is a unit vector, we may write it in the form

$$\mathbf{t}(t) = \cos \theta(t)\, \mathbf{i} + \sin \theta(t)\, \mathbf{j}$$

where $\theta(t)$ denotes the angle of inclination of the tangent vector at the point corresponding to t (Fig. 2.15). Then, differentiating with respect to t, we obtain

$$\mathbf{t}'(t) = [-\sin \theta(t)\, \mathbf{i} + \cos \theta(t)\, \mathbf{j}]\theta'(t)$$

Hence $|\mathbf{t}'(t)| = |\theta'(t)|$, so that $|\mathbf{t}'(t)|$ is a measure of the absolute

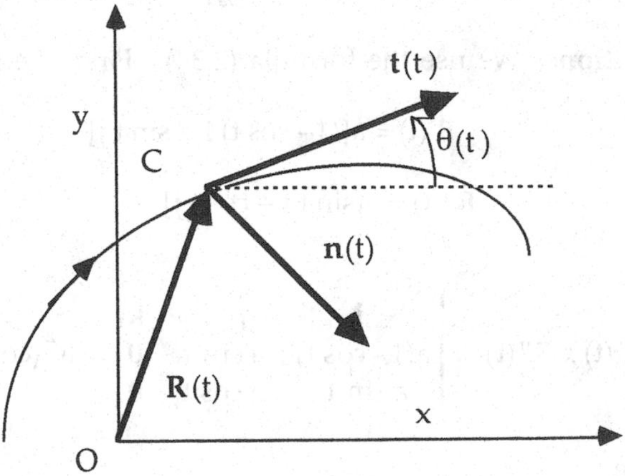

Fig. 2.15 $\kappa(t) = |ds/d\theta|$ **on a plane curve.**

value of the rate of change of the angle of inclination of the tangent vector. If $\theta'(t) > 0$, then $\theta(t)$ is increasing with t and hence the principal normal vector is given by $\mathbf{n}(t) = -\sin\theta(t)\,\mathbf{i} + \cos\theta(t)\,\mathbf{j}$. If $\theta(t) < 0$ so that θ is decreasing with t, then according to (2.30) we have $\mathbf{n}(t) = \sin\theta(t)\mathbf{i} - \cos\theta(t)\mathbf{j}$. In either case, we see that $\mathbf{n}$ is directed toward the concave side of the curve.

Now by (2.33) the curvature κ is given by $\kappa = |\theta'(t)|\,/\,|\mathbf{R}'(t)|$. Since

$$\theta'(t) = \frac{d\theta}{ds}\frac{ds}{dt} = \frac{d\theta}{ds}\,|\mathbf{R}'(t)|$$

it follows that

$$\kappa(t) = \left|\frac{d\theta}{ds}\right|$$

Thus for a plane curve the curvature is a measure of the absolute value of the rate of change of θ, the angle of inclination of the tangent vector, with respect to arc length.

Example 3. Find the curvature of the curve represented by

$$\mathbf{R}(t) = a[(t - \sin t)\mathbf{i} + (1 - \cos t)\mathbf{j}] \qquad 0 \le t \le 2\pi\,,\ a > 0$$

Solution: We use the formula (2.34). First, we have

$$\mathbf{R}'(t) = a[(1 - \cos t)\,\mathbf{i} + \sin t\,\mathbf{j}]$$

$$\mathbf{R}''(t) = a[\sin t\,\mathbf{i} + \cos t\,\mathbf{j}]$$

so that

$$\mathbf{R}'(t) \times \mathbf{R}''(t) = \begin{vmatrix} \mathbf{i} & \mathbf{j} & \mathbf{k} \\ a(1 - \cos t) & a\sin t & 0 \\ a\sin t & a\cos t & 0 \end{vmatrix} = a^2(\cos t - 1)\,\mathbf{k}$$

Therefore

$$|\mathbf{R}'(t) \times \mathbf{R}''(t)| = a^2(\cos t - 1)$$

and so, by (2.34), we find

$$\kappa = \frac{a^2(1 - \cos t)}{|\mathbf{R}'(t)|^3} = \frac{1}{2\sqrt{2(1 - \cos t)}\, a}$$

2.4 EXERCISES

In each of Problems 1 through 5, find the (a) curvature, (b) principal normal vector, and (c) the binormal vector of the space curve represented by the given vector equation.

1. $\mathbf{R}(t) = t^2\mathbf{i} + (1 + t)\,\mathbf{j} + \sqrt{3}\, t\,\mathbf{k}$ $\quad (t \geq 0)$.
2. $\mathbf{R}(t) = e^t \cos t\,\mathbf{i} + e^t \sin t\,\mathbf{j} + e^t\,\mathbf{k}$ $\quad (t \geq 0)$.
3. $\mathbf{R}(t) = t \cos t\,\mathbf{i} + t \sin t\,\mathbf{j} + t\,\mathbf{k}$ $\quad (t \geq 0)$.
4. $\mathbf{R}(t) = \cosh t\,\mathbf{i} + \sinh t\,\mathbf{j} + t\,\mathbf{k}$ $\quad (t \geq 0)$.
5. $\mathbf{R}(t) = (\sin t - t \cos t)\,\mathbf{i} + (\cos t + t \sin t)\,\mathbf{j} + t^2/2\,\mathbf{k}$ $\quad (t \geq 0)$.
6. Find the radius of curvature of the curve described by
   $\mathbf{R}(t) = \cos t\,\mathbf{i} + \sin t\,\mathbf{j} + \cos t\,\mathbf{k}$ $\quad (0 \leq t < 2\pi)$
   and determine the center of the circle of curvature when $t = \pi/4$.
7. Find the center and the radius of the circle of curvature of the curve described by
   $\mathbf{R}(t) = (t + \sin t)\,\mathbf{i} + (1 - \cos t)\,\mathbf{j} + 4 \sin (t/2)\,\mathbf{k}$ when $t = \pi/2$.
8. Find the curvatuve of the curve described by
   $\mathbf{R}(t) = e^t\,\mathbf{i} + e^{-t}\mathbf{j} + \sqrt{2}\, t\,\mathbf{k}$ $\quad (t \geq 0)$.
9. In each case, find an equation of the osculating plane to the given curve at the indicated point:
   (a) the curve given in Problem 2 at $t = \pi/4$;
   (b) the curve given in Problem 3 at $t = \pi$.
10. Let the curve C be represented by $\mathbf{R}(t)$ $(a \leq t \leq b)$ where $\mathbf{R}(t)$ is twice continuously differentiable. Show that the curvature may be given by

$$\kappa(t) = \frac{|\mathbf{R}'(t) \times \mathbf{R}''(t)|}{|\mathbf{R}'(t)|^3}$$

[Hint: Differentiate $\mathbf{R}'(t) = \mathbf{t}\,(ds/dt)$ with respect to s and use (2.28).]

11. Referring to Problem 10, show that also

$$\kappa = \frac{\left[|\mathbf{R}'(t)|^2 |\mathbf{R}''(t)|^2 - \{\mathbf{R}'(t)\cdot\mathbf{R}''(t)\}^2 \right]^{1/2}}{|\mathbf{R}'(t)|^3}$$

[Hint: Differentiate $\mathbf{t} = \mathbf{R}'(t)/|\mathbf{R}'(t)|$ and use (2.33).]

12. If C is a plane curve described by $\mathbf{R}(t) = x(t)\,\mathbf{i} + y(t)\,\mathbf{j}$,
(a) show that the formula in Problem 10 reduces to

$$\kappa = \frac{|x'(t)\,y''(t) - x''(t)\,y'(t)|}{[x'(t)^2 + y'(t)^2]^{3/2}}$$

(b) If C is described by $y = f(x)$, show that

$$\kappa = \frac{|f''(t)|}{[1 + f'(t)^2]^{3/2}}$$

13. Let C be a plane curve described by $\mathbf{R}(s) = x(s)\,\mathbf{i} + y(s)\,\mathbf{j}$, where s denotes arc length. Show that

$$\mathbf{t}(s) = x'(s)\,\mathbf{i} + y'(s)\,\mathbf{j}$$

and

$$\mathbf{n}(s) = -\,y'(s)\,\mathbf{i} + x'(s)\,\mathbf{j}$$

14. Referring to Problem 10, show that **b** and **n** are given by

$$\mathbf{b} = \frac{\mathbf{R}'(t)\times\mathbf{R}''(t)}{|\mathbf{R}'(t)\times\mathbf{R}''(t)|}$$

and

$$\mathbf{n} = \frac{[\mathbf{R}'(t)\times\mathbf{R}''(t)]\times\mathbf{R}'(t)}{\kappa\,|\mathbf{R}'(t)|^4}$$

$$= \frac{[\mathbf{R}'(t)\times\mathbf{R}''(t)]\times\mathbf{R}'(t)}{|\mathbf{R}'(t)\times\mathbf{R}''(t)|\,|\mathbf{R}'(t)|}$$

2.7 TORSION AND FRENET-SERRET FORMULAS

We have seen that the curvature of a curve is related to the rate of change of the unit tangent vector as it moves along the curve. There is another property of a curve, called torsion, that is related to the rate at which the binormal vector **b** changes its direction as it moves along the curve with the other two unit vectors **t** and **n**. This rate of change is measured by the derivative $d\mathbf{b}/ds$. Now since **b** has constant magnitude, we know that $d\mathbf{b}/ds$ is orthogonal to **b**. By differentiating the equation $\mathbf{b}\cdot\mathbf{t} = 0$, we obtain

$$\mathbf{b}\cdot\frac{d\mathbf{t}}{ds} + \frac{d\mathbf{b}}{ds}\cdot\mathbf{t} = \frac{d\mathbf{b}}{ds}\cdot\mathbf{t} = 0$$

since $\mathbf{b}\cdot d\mathbf{t}/ds = \mathbf{b}\cdot\kappa\mathbf{n} = 0$ in view of (2.28). Therefore, the vector $d\mathbf{b}/ds$ is orthogonal to both **b** and **t**.

If $d\mathbf{b}/ds = \mathbf{0}$, then **b** is a constant vector, and hence the curve is a plane curve. On the other hand, if $d\mathbf{b}/ds \neq \mathbf{0}$, then because it is orthogonal to **t** and **b**, it is parallel to the principal normal vector **n**. Hence we can write

$$\frac{d\mathbf{b}}{ds}(s) = -\tau(s)\,\mathbf{n}(s) \tag{2.35}$$

The scalar factor $\tau(s)$ is called the *torsion* of the curve at the point corresponding to the parameter s. The torsion of a curve is a measure of the rate at which the binormal vector or the osculating plane rotates about the tangent vector as it moves along the curve. The negative sign in (2.35) is introduced so that the torsion will come out positive when the right-handed triple **t, n, b** rotates in a right-handed sense about the tangent vector as it moves in the positive direction on the curve.

Just as the curvature of a straight line is zero, we deduce from (2.35) that the torsion of a plane curve is zero.

Equations (2.28) and (2.35) are the results of differentiating the unit vectors **t** and **n** with respect to arc length. To complete the analysis of the orthonormal set of vectors (**t, n, b**), we now consider the derivative of **n**. Since $\mathbf{n} = \mathbf{b}\times\mathbf{t}$, we have

$$\frac{d\mathbf{n}}{ds} = \frac{d\mathbf{b}}{ds}\times\mathbf{t} + \mathbf{b}\times\frac{d\mathbf{t}}{ds}$$

Using (2.28) and (2.35), this becomes

$$\frac{d\mathbf{n}}{ds} = -\tau\,\mathbf{n} \times \mathbf{t} + \mathbf{b} \times \kappa\mathbf{n}$$

$$= -\tau\,\mathbf{b} - \kappa\,\mathbf{t}$$

since $\mathbf{n} \times \mathbf{t} = -\mathbf{b}$ and $\mathbf{b} \times \mathbf{n} = -\mathbf{t}$. Therefore, collecting our results, we have the system of equations

$$\frac{d\mathbf{t}}{ds} = \kappa\,\mathbf{n}$$

$$\frac{d\mathbf{n}}{ds} = -\kappa\,\mathbf{t} + \tau\,\mathbf{b} \qquad (2.36)$$

$$\frac{d\mathbf{b}}{ds} = -\tau\,\mathbf{n}$$

satisfied by the unit vectors $\mathbf{t}, \mathbf{n}, \mathbf{b}$. These equations are known as the Frenet-Serret formulas. They play a central role in the study of differential geometry of curves.

When the curve is represented by $\mathbf{R}(t)$, where t is any parameter, a convenient formula for the torsion is given by

$$\tau(t) = \frac{\mathbf{R}'(t) \times \mathbf{R}''(t) \cdot \mathbf{R}'''(t)}{\kappa^2(t)\,|\mathbf{R}'(t)|^6} \qquad (2.37)$$

If we substitute (2.34) for the curvature κ, (2.37) becomes

$$\tau(t) = \frac{\mathbf{R}'(t) \times \mathbf{R}''(t) \,.\, \mathbf{R}'''(t)}{|\mathbf{R}'(t) \times \mathbf{R}''(t)|^2} \qquad (2.38)$$

To derive the formula (2.37), we note that

$$\mathbf{R}'(t) = \frac{d\mathbf{R}}{ds}\frac{ds}{dt} = \mathbf{t}\,\frac{ds}{dt}$$

$$\mathbf{R}''(t) = \frac{d\mathbf{t}}{ds}\left(\frac{ds}{dt}\right)^2 + \mathbf{t}\frac{d^2s}{dt^2} = \kappa\,\mathbf{n}\left(\frac{ds}{dt}\right)^2 + \mathbf{t}\frac{d^2s}{dt^2}$$

$$\mathbf{R}'''(t) = \mathbf{n}\frac{d}{dt}\left(\kappa\left[\frac{ds}{dt}\right]^2\right) + \kappa\frac{d\mathbf{n}}{ds}\left(\frac{ds}{dt}\right)^3$$

$$+ \frac{d\mathbf{t}}{ds}\frac{ds}{dt}\frac{d^2s}{dt^2} + \mathbf{t}\frac{d^3s}{dt^3}$$

$$= \mathbf{n}\frac{d}{dt}\left(\kappa\left[\frac{ds}{dt}\right]^2\right) + \kappa\,(-\kappa\,\mathbf{t} + \tau\,\mathbf{b})\left(\frac{ds}{dt}\right)^3$$

$$+ \kappa\,\mathbf{n}\frac{ds}{dt}\frac{d^2s}{dt^2} + \mathbf{t}\frac{d^3s}{dt^3}$$

where we have repeatedly used the formulas (2.36). Taking the cross product of $\mathbf{R}'$ and $\mathbf{R}''$, we obtain

$$\mathbf{R}'(t) \times \mathbf{R}''(t) = \kappa\,(\mathbf{t} \times \mathbf{n})\left(\frac{ds}{dt}\right)^3 = \kappa\,\mathbf{b}\left(\frac{ds}{dt}\right)^3$$

Taking the dot product of this expression with $\mathbf{R}'''$, noting that $\mathbf{b}$ is orthogonal to $\mathbf{t}$ and $\mathbf{n}$, we obtain

$$\mathbf{R}'(t)\,\mathbf{R}'(t) \times \mathbf{R}''(t)\cdot\mathbf{R}'''(t) = \kappa^2\,\tau\left(\frac{ds}{dt}\right)^6$$

from which (2.37) follows in view of (2.25).

Example 1. Determine the torsion of the helix described in Example 2, Sec. 2.6.

Solution: From Example 2, Sec. 2.6, we have

$$\mathbf{b}(t) = \frac{b\sin t\,\mathbf{i} - b\cos t\,\mathbf{j} + a\,\mathbf{k}}{\sqrt{a^2 + b^2}}$$

and

$$\frac{ds}{dt} = |\mathbf{R}'(t)| = \sqrt{a^2 + b^2}$$

Hence

$$\frac{d\mathbf{b}}{ds} = \frac{d\mathbf{b}}{dt}\frac{dt}{ds} = \frac{b\cos t\,\mathbf{i} + b\sin t\,\mathbf{j}}{\sqrt{a^2+b^2}} \frac{1}{\sqrt{a^2+b^2}}$$

$$= \frac{-b}{a^2+b^2}\mathbf{n}$$

since $\mathbf{n} = -\cos t\,\mathbf{i} - \sin t\,\mathbf{j}$. Therefore, the torsion of the helix is

$$\tau = \frac{b}{a^2+b^2}$$

which is a constant. This means that the osculating plane rotates clockwise about the tangent vector at a uniform rate as it climbs up the helix. The student should verify that

$$\tau\mathbf{b} - \kappa\mathbf{t} = \frac{\sin t\,\mathbf{i} - \cos t\,\mathbf{j}}{\sqrt{a^2+b^2}} = \frac{d\mathbf{n}}{ds}$$

Example 2. Determine the torsion of the curve represented by

$$\mathbf{R}(t) = a(t - \sin t)\,\mathbf{i} + a(1 - \cos t)\,\mathbf{j} + b\,t\,\mathbf{k} \qquad (t \geq 0)$$

where a and b are positive constants.

Solution: We use the formula (2.38) to determine the torsion. Differentiating **R** three times, we find

$$\mathbf{R}'(t) = a(1 - \cos t)\,\mathbf{i} + a\sin t\,\mathbf{j} + b\,\mathbf{k}$$

$$\mathbf{R}''(t) = a\sin t\,\mathbf{i} + a\cos t\,\mathbf{j}$$

$$\mathbf{R}'''(t) = a\cos t\,\mathbf{i} - a\sin t\,\mathbf{j}$$

Hence

$$\mathbf{R}'(t) \times \mathbf{R}''(t) = \begin{vmatrix} \mathbf{i} & \mathbf{j} & \mathbf{k} \\ a(1-\cos t) & a\sin t & b \\ a\sin t & a\cos t & 0 \end{vmatrix}$$

$$= -ab\cos t\,\mathbf{i} + ab\sin t\,\mathbf{j} + a^2(\cos t - 1)\,\mathbf{k}$$

and so

$$|\mathbf{R}'(t) \times \mathbf{R}''(t)|^2 = a^2b^2 + a^4(\cos t - 1)^2$$

Since

$$\mathbf{R}'(t) \times \mathbf{R}''(t) \cdot \mathbf{R}'''(t) = -a^2b$$

it follows from (2.38) that

$$\tau = \frac{-b}{b^2 + a^2(\cos t - 1)^2}$$

2.5 EXERCISES

In each of Problems 1 through 5, determine the torsion of the given curve.

1. The curve described in Problem 1, Exercise 2.4.
2. The curve described in Problem 2, Exercise 2.4.
3. The curve described in Problem 3, Exercise 2.4.
4. The curve described in Problem 4, Exercise 2.4.
5. The curve described in Problem 5, Exercise 2.4.
6. Find the torsion of the curve described by

$$\mathbf{R}(t) = t\,\mathbf{i} + a\,t^2\mathbf{j} + (2/3)a^2t^3\mathbf{k} \quad (t \geq 0)$$

where a is a positive constant.

7. Show that the torsion of the curve described by

$$\mathbf{R}(t) = t\,\mathbf{i} + (1 + 1/t)\,\mathbf{j} + (1/t - t)\,\mathbf{k} \quad (t \geq 0)$$

is zero.

8. Let C be a curve represented by $\mathbf{R}(t)$, where $\mathbf{R}(t)$ is three times continuously differentiable for $t \geq 0$. Show that the torsion at any point is given by

$$\kappa = \frac{(\mathbf{R}'(t)\,\mathbf{R}''(t)\,\mathbf{R}'''(t))}{\left|\mathbf{R}'(t) \times \mathbf{R}''(t)\right|^2}$$

where the numerator denotes the scalar triple product of $\mathbf{R}'$, $\mathbf{R}''$ and $\mathbf{R}'''$. (Hint: Differentiate $\mathbf{R}''(t)$ and use the Frenet-Serret formula.)

9. Show that if C is represented by $\mathbf{R}(s)$ where s denotes arc length, then

$$\kappa = \left|\mathbf{R}''(s)\right|, \quad \tau = \frac{(\mathbf{R}'(s)\,\mathbf{R}''(s)\,\mathbf{R}'''(s))}{\left|\mathbf{R}'(s)\right|^2}$$

10. Find the curvature and the torsion for the curve C represented by

$$\mathbf{R}(t) = (3t - t^3)\,\mathbf{i} + 3t^2\,\mathbf{j} + (3t + t^3)\,\mathbf{k} \quad (t \geq 0)$$

11. Write the Frenet-Serret formulas for the curve

$$\mathbf{R}(t) = (\sin t - t\cos t)\,\mathbf{i} + (\cos t + t\sin t)\,\mathbf{j} + t^2/2\,\mathbf{k} \quad (t \geq 0)$$

12. Write the Frenet-Serret formulas for the curve

$$\mathbf{R}(t) = \sin t\,\mathbf{i} + t\,\mathbf{j} + (1 - \cos t)\,\mathbf{k} \quad (t \geq 0)$$

2.8 APPLICATIONS TO CURVILINEAR MOTIONS

Let $\mathbf{R}(t)$ denote the position vector of a moving particle at time t relative to a rectangular coordinate system. As t varies over a time interval, the particle traces out a curve in space. This curve is called the path or trajectory of the particle (Fig. 2.16). We shall show that the physical concepts of velocity, speed, and acceleration of the particle can be defined in terms of the derivatives of the position vector $\mathbf{R}(t)$.

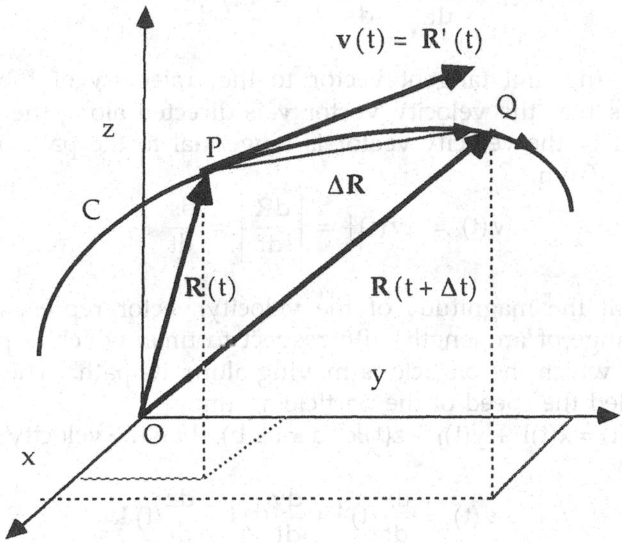

Fig. 2.16 Curvilinear motion of a particle.

Velocity Vector

Suppose that a particle is at a point P at time t and at a point Q at time $t + \Delta t$. Then, as shown in Fig. 2.16, $\Delta \mathbf{R} = \mathbf{R}(t + \Delta t) - \mathbf{R}(t)$ represents the displacement of the particle from P to Q during the time interval Δt. The instantaneous velocity of the particle at time t, which we denote by $\mathbf{v}(t)$, is defined as the limit

$$\mathbf{v}(t) = \lim_{\Delta t \to 0} \frac{\Delta \mathbf{R}}{\Delta t}(t) = \frac{d\mathbf{R}}{dt}(t) \tag{2.39}$$

We call the vector function (2.39) simply the velocity of the particle at time t. Thus the time derivative of the position vector of a moving particle represents the velocity of the particle.

Now let s denote the distance traveled by the particle along its path. By the chain rule we have

$$\mathbf{v}(t) = \frac{d\mathbf{R}}{dt} = \frac{d\mathbf{R}}{ds}\frac{ds}{dt} = \mathbf{t}(t)\frac{ds}{dt} \tag{2.40}$$

where **t** is the unit tangent vector to the trajectory of the particle. This shows that the velocity vector **v** is directed along the tangent vector, that is, the velocity vector is tangential to the path of motion. Moreover, from

$$v(t) = |\mathbf{v}(t)| = \left|\frac{d\mathbf{R}}{dt}\right| = \frac{ds}{dt} \tag{2.41}$$

we see that the magnitude of the velocity vector represents the rate of change of arc length with respect to time, which is precisely the rate at which the particle is moving along its path. The scalar v(t) is called the speed of the particle at time t.

If $\mathbf{R}(t) = x(t)\mathbf{i} + y(t)\mathbf{j} + z(t)\mathbf{k}$ $(a \le t \le b)$, then the velocity (2.40) is given by

$$\mathbf{v}(t) = \frac{dx}{dt}(t)\,\mathbf{i} + \frac{dy}{dt}(t)\,\mathbf{j} + \frac{dz}{dt}(t)\,\mathbf{k} \tag{2.42}$$

so that the speed at time t is

$$v(t) = \sqrt{x'(t)^2 + y'(t)^2 + z'(t)^2} = \frac{ds}{dt} \tag{2.43}$$

Acceleration Vector

The velocity of a moving particle changes when there is change in the speed or the direction of motion. The rate of change of the velocity is called the acceleration, denoted by **a**(t). Thus we have

$$\mathbf{a}(t) = \frac{d\mathbf{v}}{dt}(t) = \frac{d^2\mathbf{R}}{dt^2}(t) \tag{2.44}$$

In terms of its components, the acceleration vector is given by

$$\mathbf{a}(t) = \frac{d^2x}{dt^2}(t)\,\mathbf{i} + \frac{d^2y}{dt^2}(t)\,\mathbf{j} + \frac{d^2z}{dt^2}(t)\,\mathbf{k} \tag{2.45}$$

so that its magnitude is

$$a(t) = |\mathbf{a}(t)| = \sqrt{x''(t)^2 + y''(t)^2 + z''(t)^2} \tag{2.46}$$

Unlike the velocity vector, the accelertation vector is not tangential to the path of motion. In fact, we shall show later that **a**(t) can be expressed as a linear combination of the tangent vector **t** and the principal normal vector **n**.

Example 1. Consider the motion of a particle described by

$$\mathbf{R}(t) = a\,(\cos \omega t\,\mathbf{i} + \sin \omega t\,\mathbf{j}),\ t \geq 0$$

where a and ω are constants. Find the velocity, speed, and acceleration of the particle as a function of time t, $t \geq 0$.

Solution: If we let $\theta = \omega t$ denote the polar angle which the position vector **R** makes with the x-axis, we see that as t increases the particle is moving counterclockwise around the origin on the circle of radius a. The velocity vector is given by

$$\mathbf{v}(t) = \mathbf{R}'(t) = a\omega(-\sin \omega t\,\mathbf{i} + \cos \omega t\,\mathbf{j})$$

so that the speed is

$$v(t) = |\mathbf{v}(t)| = a\omega$$

which is constant. The scalar ω is called the angular speed of the particle, expressed in either revolutions or radians per unit time. It is the rate at which the particle revolves around the origin on the circle.

The corresponding acceleration vector is given by

$$\mathbf{a}(t) = \mathbf{v}'(t) = -a\omega^2(\cos \omega t\,\mathbf{i} + \sin \omega t\,\mathbf{j}) = -\omega^2\mathbf{R}(t)$$

which also has constant magnitude $|\mathbf{a}(t)| = a\omega^2$. This shows that the acceleration vector is opposite in direction to the position vector; hence it is directed toward the origin, as shown in Fig. 2.17. This acceleration is called the centripetal or normal acceleration.

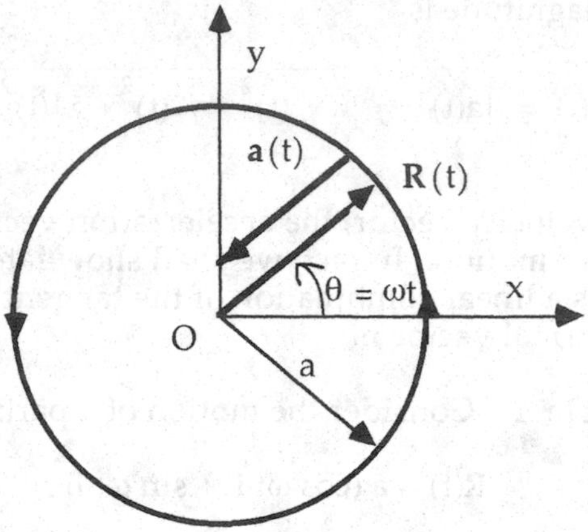

Fig. 2.17 Circular motion.

Tangential and Normal Components of Acceleration

We now show that the acceleration vector can be expressed as a sum of two vectors, one parallel to the unit tangent vector **t** and the other parallel to the principal normal **n**. In view of (2.40) we see that **a** can be written as

$$\mathbf{a} = \frac{d}{dt}\left[\mathbf{t}\frac{ds}{dt}\right] = \frac{d\mathbf{t}}{ds}\left[\frac{ds}{dt}\right]^2 + \mathbf{t}\frac{d^2s}{dt^2}$$

Since $d\mathbf{t}/ds = \kappa\mathbf{n}$ and $v = ds/dt$, this becomes

$$a = \kappa v^2\mathbf{n} + \mathbf{t}\frac{dv}{dt} = \frac{v^2}{\rho}\mathbf{n} + \frac{dv}{dt}\mathbf{t} \tag{2.47}$$

Thus we have resolved the acceleration vector as the sum of two orthogonal vectors, one of magnitude

$$a_t = \frac{dv}{dt} = \frac{d^2s}{dt^2} \tag{2.48}$$

directed along the tangent vector $\mathbf{t}$, and the other of magnitude

$$a_n = \kappa \left[\frac{ds}{dt}\right]^2 = \frac{v^2}{\rho} \tag{2.49}$$

directed along the principal normal vector $\mathbf{n}$, as shown in Fig. 2.18. The scalars (2.48) and (2.49) are called the tangential and the normal components of the acceleration, respectively. It is clears that

$$a^2 = |\mathbf{a}|^2 = a_t^2 + a_n^2 \tag{2.50}$$

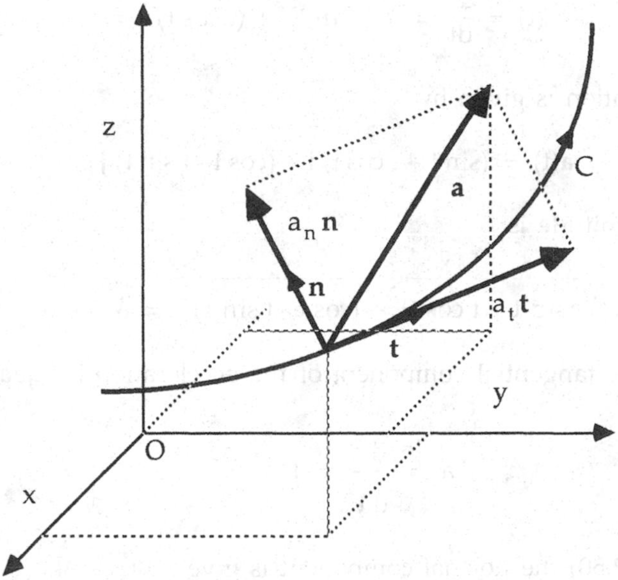

Fig. 2.18 Tangential and normal components of acceleration.

From the formulas (2.48) and (2.49), we see that if a particle is moving around a curve at a constant speed (v = const.), the tangential component of the acceleration is zero so that the only acceleration is the normal acceleration (see Example 1). On the other hand, if the particle is moving along a straight line so that the curvature is zero, then the only acceleration present is the tangential acceleration. These are of course intuitively evident.

Example 2. Consider the motion of a particle described by

$$\mathbf{R}(t) = (\sin t - t\cos t)\,\mathbf{i} + (\cos t + t\sin t)\,\mathbf{j} \qquad (t \geq 0)$$

Find the velocity, the acceleration, and the tangential and normal components of the acceleration.

Solution: The velocity is given by

$$\mathbf{v}(t) = \mathbf{R}'(t) = t\sin t\,\mathbf{i} + t\cos t\,\mathbf{j}$$

so that the speed is

$$v(t) = \frac{ds}{dt} = \sqrt{(t\sin t)^2 + (t\cos t)^2} = t$$

The acceleration is given by

$$\mathbf{a}(t) = (\sin t + t\cos t)\,\mathbf{i} + (\cos t - t\sin t)\,\mathbf{j}$$

and its magnitude is

$$a = \sqrt{(\sin t + t\cos t)^2 + (\cos t - t\sin t)^2} = \sqrt{1 + t^2}$$

By (2.48) the tangential component of the acceleration is equal to

$$a_t = \frac{d^2 s}{dt^2} = 1$$

hence, by (2.50) the normal component is given by

$$a_n = \sqrt{a^2 - a_t^2} = t$$

Example 3. The position vector of a particle at time t is given by $\mathbf{R}(t)$. If $\mathbf{R}(t) \times \mathbf{a}(t) = \mathbf{0}$, where $\mathbf{a}(t)$ is the acceleration, show that the motion takes place on a plane.

Solution: It is sufficient to show that the torsion of the trajectory of the particle is zero, since it implies that the trajectory is a plane curve. Now, by assumption, we have

$$\mathbf{R}(t) \times \mathbf{a}(t) = \mathbf{R}(t) \times \mathbf{R}''(t) = \mathbf{0}$$

Differentiating this, we find

$$\mathbf{R}'(t) \times \mathbf{R}''(t) + \mathbf{R}(t) \times \mathbf{R}'''(t) = \mathbf{0}$$

or

$$\mathbf{R}'(t) \times \mathbf{R}''(t) = - \mathbf{R}(t) \times \mathbf{R}'''(t)$$

Taking the dot product of the last equation with $\mathbf{R}'''(t)$, we obtain

$$\mathbf{R}'(t) \times \mathbf{R}''(t) \cdot \mathbf{R}'''(t) = - \mathbf{R}(t) \times \mathbf{R}'''(t) \cdot \mathbf{R}'''(t) = 0$$

By (2.37) , it follows that $\tau = 0$; hence, the trajectory lies on a plane.

An Energy Theorem in Mechanics

We conclude this section with the derivation of an important theorem in mechanics. Let m denote the mass of a particle in motion whose position vector is given by $\mathbf{R}(t)$. According to Newton's second law of motion, the force $\mathbf{F}$ acting on the particle at time t is given by

$$\begin{aligned} \mathbf{F} &= m\mathbf{a}(t) \\ &= m \frac{d\mathbf{v}}{dt}(t) = m \frac{d^2\mathbf{R}}{dt^2} \end{aligned} \tag{2.51}$$

Thus the force and the acceleration have the same direction. Now since

$$\frac{d}{dt}\left(v^2(t)\right) = \frac{d}{dt}\left(\mathbf{v}(t) \cdot \mathbf{v}(t)\right) = 2\, \mathbf{v}(t) \cdot \frac{d\mathbf{v}(t)}{dt}$$

we have by (2.51)

$$\mathbf{F}(t)\cdot\mathbf{v}(t) = m\frac{d\mathbf{v}}{dt}(t)\cdot\mathbf{v}(t) = \frac{d}{dt}\left(\frac{1}{2}mv^2\right) \tag{2.52}$$

The quantity $(1/2)mv^2(t)$ in (2.52) is known as the kinetic energy of the particle at time t. If T_1 and T_2 denote the kinetic energy of the particle at time t_1 and t_2 ($t_1 < t_2$), respectively, then from (2.52) the change in kinetic energy during the time interval $t_2 - t_1$ is given by

$$T_2 - T_1 = (1/2)[mv^2(t_2) - mv^2(t_1)]$$

$$= \int_{t_1}^{t_2} \mathbf{F}(t)\cdot\mathbf{v}(t)\,dt = \int_{t_1}^{t_2} \frac{d}{dt}\left(\frac{1}{2}mv^2(t)\right)dt \tag{2.53}$$

This is known as the **energy theorem** in mechanics.

For instance, let m denote the mass of the particle in Example 1. Since the acceleration is given by $\mathbf{a} = -\omega^2\mathbf{R}(t)$, the force acting on the particle is equal to $\mathbf{F} = -m\omega^2\mathbf{R}$. This force is directed toward the center of the circular path and it is called the *centripetal* force. The force **-F** which is directed away from the center is called the *centrifugal* force. This is the force that tends to throw a car off its track when it makes a turn or changes its direction. The kinetic energy of the particle is given by

$$T = (1/2)\,m\,(a\omega)^2$$

which is constant. Thus a particle in circular motion at constant angular speed ω has constant energy.

2.6 EXERCISES

In each of Problems 1 through 5, the position vector $\mathbf{R}(t)$ of a moving particle at time t is given. Determine the velocity, speed, and acceleration of the particle at the indicated time.

1. $\mathbf{R}(t) = 3t\,\mathbf{i} + 3t^2\mathbf{j} + 2t^3\mathbf{k}$, $t = 1$.
2. $\mathbf{R}(t) = a(\sinh t - t)\,\mathbf{i} + a(\cosh t - 1)\,\mathbf{j}$, $t = \ln 2$.
3. $\mathbf{R}(t) = t\cos t\,\mathbf{i} + t\sin t\,\mathbf{j} + t^2/2\,\mathbf{k}$, $t = \pi$.
4. $\mathbf{R}(t) = a(1 + \cos t)\,\mathbf{i} + b\sin t\,\mathbf{j} + a\cos t\,\mathbf{k}$, $t = \pi/3$.
5. $\mathbf{R}(t) = \cosh t\cos t\,\mathbf{i} + \cosh t\sin t\,\mathbf{j} + \sinh t\,\mathbf{k}$, $t = \pi$.
6. Consider the motion of a particle on the helix

$$\mathbf{R}(t) = a\cos\omega t\,\mathbf{i} + a\sin\omega t\,\mathbf{j} + b\omega t\,\mathbf{k} \quad (t \geq 0)$$

where a, b, and ω are constants.
(a) Show that the velocity and acceleration both have constant magnitude.
(b) Show that the velocity and acceleration are orthogonal.
(c) Express the acceleration in terms of the principal normal vector to the curve.

7. The motion of a particle in space is described by the equation

$$\mathbf{R}(t) = t\,\mathbf{i} + \sin t\,\mathbf{j} + (1 - \cos t)\,\mathbf{k} \quad (t \geq 0)$$

Find the normal and tangential components of the acceleration. Express the acceleration in terms of **t** and **n**.

8. The motion of a particle in space is described by the equation

$$\mathbf{R}(t) = (t + \sin t)\,\mathbf{i} + (1 - \cos t)\,\mathbf{j} + 4\cos(t/2)\,\mathbf{k} \quad (t \geq 0)$$

Find the acceleration and express it in terms of **t** and **n**.

9. A particle moves in space according to the equation

$$\mathbf{R}(t) = a(1 + \cos t)\,\mathbf{i} + a\sin t\,\mathbf{j} + bt\,\mathbf{k} \quad (t \geq 0)$$

Find the tangential and normal components of the acceleration.

10. The position vector of a moving particle is given by

$$\mathbf{R}(t) = a(1 + \cos t)\,\mathbf{i} + a\sin t\,\mathbf{j} + a(1 + \cos t)\,\mathbf{k} \quad (t \geq 0)$$

(a) Show that the path of motion is an ellipse, and find the equation of the plane on which the ellipse lies.
(b) What are the tangential and normal components of the acceleration?

11. A particle is moving around a circle such that its distance s in feet from a fixed point on the circle is given by $s(t) = t^2 + 1$,

where t is time in seconds. If the magnitude of the acceleration is $(16t^4 + 4)^{1/2}$ ft/sec, find (a) the tangential and normal components of the acceleration, (b) the radius of the circle.

12. The position vector of a moving particle is given by

$$\mathbf{R}(t) = \mathbf{A} \cos \omega t + \mathbf{B} \sin \omega t$$

where $\mathbf{A}$, $\mathbf{B}$, and ω are constants. (a) Find the velocity of the particle. (b) Show that the acceleration is directed toward the origin.

13. Show that if a particle moves in space with constant speed, then the acceleration vector is orthogonal to the velocity vector.

14. Show that if the velocity and acceleration of a moving particle have constant magnitude, then its path has constant curvature.

15. Show that if the force acting on a particle is at all times orthogonal to the direction of motion, then the speed is constant.

16. The position vector of a moving particle is given by $\mathbf{R}(t)$. (a) Show that $\mathbf{R}(t) \times \mathbf{a} = \mathbf{0}$ if and only if $\mathbf{R}(t) \times \mathbf{v} = \mathbf{A}$, where $\mathbf{A}$ is a constant vector and $\mathbf{v}$ and $\mathbf{a}$ denote the velocity and acceleration vectors, respectively. (b) Show that if $\mathbf{R}(t) \times \mathbf{v} = \mathbf{A}$, then the particle lies on a plane.

2.9 CURVILINEAR MOTION IN POLAR COORDINATES

For a particle whose motion takes place on a plane, its trajectory is described by the equation

$$\mathbf{R}(t) = x(t)\,\mathbf{i} + y(t)\,\mathbf{j} \qquad (t \ge 0) \tag{2.54}$$

with respect to a rectangular cartesian coordinate system. Sometimes, however, it is more convenient to study the motion using polar coordinates rather than cartesian coordinates. In this section, we consider this approach and derive the corresponding formulas for the velocity and acceleration.

Since the cartesian coordinates (x, y) are related to the polar coordinates (r, θ) by the equations

$$x = r \cos \theta, \quad y = r \sin \theta$$

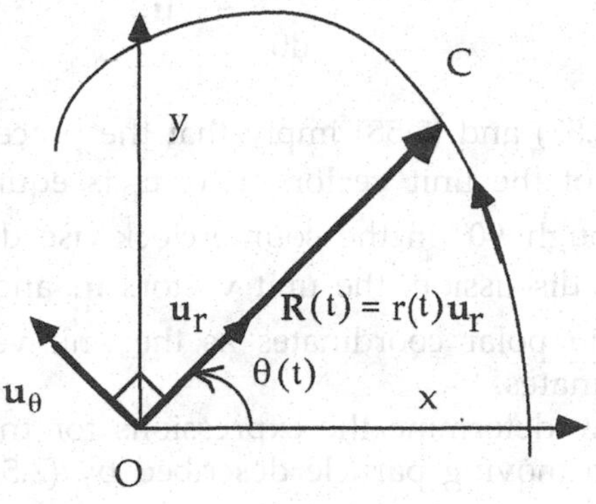

Fig. 2.19 The unit vectors u_r and u_θ in polar coordinates.

we may express (2.54) in polar coordinates as

$$\begin{aligned} \mathbf{R}(t) &= r(t)\,[\cos\theta(t)\,\mathbf{i} + \sin\theta(t)\,\mathbf{j}] \\ &= r(t)\mathbf{u}_r(\theta(t)) \end{aligned} \tag{2.55}$$

where

$$\mathbf{u}_r(\theta(t)) = \cos\theta(t)\,\mathbf{i} + \sin\theta(t)\,\mathbf{j} \tag{2.56}$$

The quantities r(t) and θ(t) denote the radial distance and the polar angle of the particle at time t (Fig. 2.19).

It is clear that $\mathbf{u}_r$ is a unit vector and has the same direction as the position vector **R**. By Theorem 3 (Sec. 2.3) , it follows that

$$\mathbf{u}_\theta = \frac{d\mathbf{u}_r}{d\theta} = -\sin\theta\,\mathbf{i} + \cos\theta\,\mathbf{j} \tag{2.57}$$

is orthogonal to $\mathbf{u}_r$. The vector (2.57) is also a unit vector and it points in the directon of increasing θ (see. Fig. 2.19). Further, by differentiating (2.57), we find

$$\frac{d\mathbf{u}_\theta}{d\theta} = -\mathbf{u}_r \tag{2.58}$$

The formulas (2.57) and (2.58) imply that the process of differentiating either one of the unit vectors $\mathbf{u}_r$ or $\mathbf{u}_\theta$ is equivalent to rotating that vector through 90° in the counterclockwise direction. As we shall see in this discussion, the unit vectors $\mathbf{u}_r$ and $\mathbf{u}_\theta$ play the same important role in polar coordinates as the unit vectors **i** and **j** in cartesian coordinates.

Let us now determine the expressions for the velocity and the acceleration of a moving particle described by (2.55) in polar coordinates. Since θ is a function of time or the parameter t, so is the unit vector $\mathbf{u}_r$. Hence by the chain rule, we have

$$\mathbf{v}(t) = \frac{d\mathbf{R}}{dt} = \frac{dr}{dt}\mathbf{u}_r + r\frac{d\mathbf{u}_r}{d\theta}\frac{d\theta}{dt}$$

which, in view of (2.57), becomes

$$\mathbf{v}(t) = \frac{d\mathbf{R}}{dt} = \frac{dr}{dt}\mathbf{u}_r + r\frac{d\theta}{dt}\mathbf{u}_\theta \tag{2.59}$$

This is the formula for the velocity in polar coordinates. We see from (2.59) that the velocity is the vector sum of two orthogonal vectors, one parallel to $\mathbf{u}_r$ and the other parallel to $\mathbf{u}_\theta$. The scalar factors dr/dt and $rd\theta/dt$ in (2.59) are called the radial and the transverse components of the velocity, respectively (see Fig. 2.20).

Since $\mathbf{u}_r$ and $\mathbf{u}_\theta$ are orthogonal unit vectors, it follows that

$$|\mathbf{v}|^2 = \left(\frac{dr}{dt}\right)^2 + \left(r\frac{d\theta}{dt}\right)^2$$

and so the speed is given by

$$|\mathbf{v}| = \sqrt{(dr/dt)^2 + (r\,d\theta/dt)^2} \tag{2.60}$$

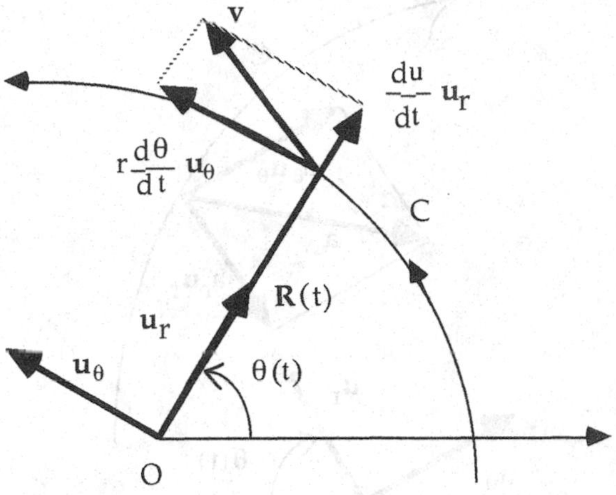

Fig. 2.20 Velocity vector in polar coordinates.

If the particle is moving around a circle with center at the origin, then r = const. and so $dr/dt = 0$. In such a case, the velocity (2.59) reduces simply to $\mathbf{v} = r\,(d\theta/dt)\mathbf{u}_\theta$. The scalar quantity $\omega = d\theta/dt$ is, of course, the angular speed.

To obtain an expression for the acceleration, we differentiate (2.59) and use the relations (2.57) and (2.58). We find

$$\mathbf{a} = \frac{d^2r}{dt^2}\mathbf{u}_r + \frac{dr}{dt}\frac{d\theta}{dt}\frac{d\mathbf{u}_r}{d\theta} + \frac{d}{dt}\left[r\frac{d\theta}{dt}\right]\mathbf{u}_\theta + r\left[\frac{d\theta}{dt}\right]^2\frac{d\mathbf{u}_\theta}{d\theta}$$

$$= \left(\frac{d^2r}{dt^2} - r\left[\frac{d\theta}{dt}\right]^2\right)\mathbf{u}_r + \left(r\frac{d^2\theta}{dt^2} + 2\frac{dr}{dt}\frac{d\theta}{dt}\right)\mathbf{u}_\theta \qquad (2.61)$$

Thus the acceleration vector is also resolved into its radial component $(r'' - r\theta'^2)$ and transverse component $(r\theta'' + 2r'\theta')$, as shown in

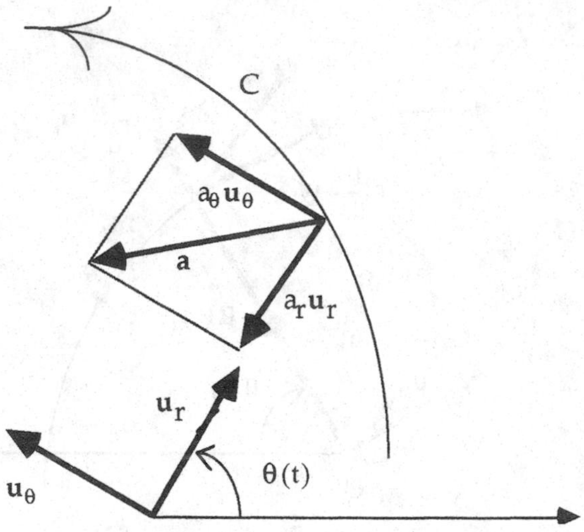

Fig. 2.21 Acceleration vector in polar coordinates.

Fig. 2.21. Unlike the velocity vector, however, the acceleration vector is no longer tangential to the path of motion, as we observed in Sec. 2.8. The term $2r'\theta'\mathbf{u}_\theta$ in (2.61) is sometimes called the Coriolis acceleration.

If a particle is moving around a circle at uniform speed, so that $dr/dt = 0$ and $d\theta/dt$ = const., then $d^2\theta/dt^2 = 0$ and, hence, the acceleration (2.61) reduces to

$$\mathbf{a} = -r\left(\frac{d\theta}{dt}\right)^2 \mathbf{u}_r$$

which is the normal acceleration obtained in Example 1, Sec. 2.8.

Example 1. A particle is moving from the center to the rim along a spoke of a wheel that is rotating at a constant angular speed ω. Its position vector is given by $\mathbf{R}(t) = t\mathbf{u}_r$. Find the velocity and the acceleration of the particle as a function of time.

Solution: By differentiating the vector function **R**, we obtain the velocity

$$\mathbf{v}(t) = \mathbf{u}_r + t\frac{d\mathbf{u}_r}{d\theta}\frac{d\theta}{dt} = \mathbf{u}_r + t\frac{d\theta}{dt}\mathbf{u}_\theta$$

The velocity is composed of the velocity $\mathbf{u}_r$ of the particle relative to the wheel and the additional velocity $t(d\theta/dt)\mathbf{u}_\theta$ due to the rotation of the wheel.

By differentiating $\mathbf{v}$, we obtain the acceleration

$$\mathbf{a} = \frac{d\mathbf{u}_r}{d\theta}\frac{d\theta}{dt} + \frac{d\theta}{dt}\mathbf{u}_\theta + t\left(\frac{d\theta}{dt}\right)^2\frac{d\mathbf{u}_\theta}{d\theta}$$

$$= 2\frac{d\theta}{dt}\mathbf{u}_\theta - t\left(\frac{d\theta}{dt}\right)^2\mathbf{u}_r$$

The Coriolis acceleration is given by $2(d\theta/dt)\mathbf{u}_\theta$ which points in the direction of rotation. The term $-t(d\theta/dt)^2\mathbf{u}_r$ is obviously the centripetal acceleration directed toward the center of the wheel.

Kepler's Law on Planetary Motion

Equation (2.61) leads to an important result when the motion is such that the acceleration is radial, that is, when the transverse component of the acceleration vanishes. This is the case with regard to the motion of the planets around the sun. Here we derive the so-called Kepler's second law on planetary motion. The law states that the position vector from the sun to a planet sweeps out area at a constant rate. In other words, as the planet moves around the sun, its position vector sweeps out equal area during an equal interval of time.

Let us assume that the mass of the sun is M and the mass of a planet is m. Let $\mathbf{R} = r\,\mathbf{u}_r$ denote the position vector from the sun to the planet, and let $\mathbf{F}$ denote the gravitational force with which the planet is attracted to the sun (Fig. 2.22). According to Newton's

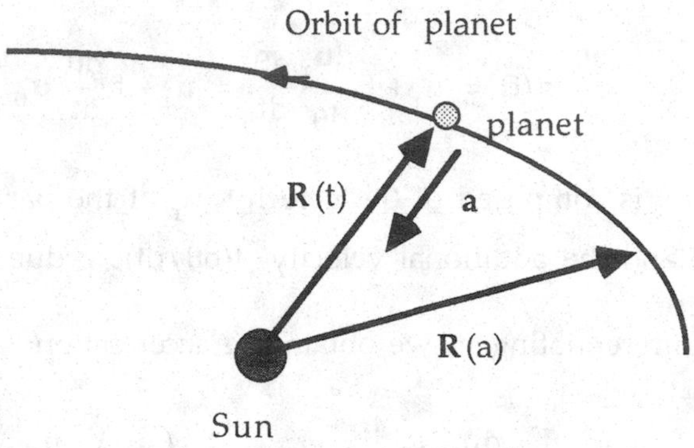

Fig. 2.22 Planetary motion.

universal law of gravitation, we have

$$\mathbf{F} = -G\frac{Mm}{r^2}\mathbf{u}_r$$

where G is the gravitational constant. By Newton's second law of motion this force is equal to the mass of the planet times its acceleration **a**, that is, **F** = m**a**. Hence we deduce

$$\mathbf{a} = -\frac{GM}{r^2}\mathbf{u}_r$$

which is a radial acceleration directed toward the sun (because of the negative sign). Since **R** and **a** are parallel, it follows that **R** x **a** = **0**. From Example 3, Sec. 2.8, we conclude that the orbit of the planet lies on a plane.

Now let A(t) denote the area swept out by the position vector of the planet from a certain time t = a to a subsequent time t (see Fig. 2.22), and let θ_0 and θ denote the corresponding polar angles. From elementary calculus, we know that

$$A(t) = \frac{1}{2}\int_{\theta_o(a)}^{\theta(t)} r^2\, d\phi$$

The rate of change of this area is given by

$$A'(t) = \frac{1}{2}\frac{d}{d\theta}\left(\int_{\theta_o(t)}^{\theta(t)} r^2\, d\phi\right)\frac{d\theta}{dt} = \frac{1}{2} r^2 \frac{d\theta}{dt}$$

Since the acceleration of the planet is radial, the transversal component of its acceleration vanishes. Hence from (2.61) we have

$$r\frac{d^2\theta}{dt^2} + 2\frac{dr}{dt}\frac{d\theta}{dt} = \frac{1}{r}\frac{d}{dt}\left(r^2\frac{d\theta}{dt}\right) = 0$$

which implies that

$$r^2\frac{d\theta}{dt} = \text{const.}$$

Therefore,

$$A'(t) = \frac{1}{2} r^2 \frac{d\theta}{dt} = \text{const.}$$

which says that the area swept out by the position vector of the planet changes at a constant rate. This is Kepler's second law of planetary motion.

2.7 EXERCISES

In each of Problems 1 through 6, the position vector of a moving particle is given. Find the velocity and acceleration of the particle at time t, and point out the Coriolis acceleration, if any.

1. $\mathbf{R}(t) = \cosh \omega t\,(\cos t\,\mathbf{i} + \sin t\,\mathbf{j})$, ω positive constant.
2. $\mathbf{R}(t) = t^2(\cos \omega t\,\mathbf{i} + \sin \omega t\,\mathbf{j})$, $\omega > 0$ constant.
3. $\mathbf{R}(t) = r(t)(\cos 3t\,\mathbf{i} + \sin 3t\,\mathbf{j})$, where $r(t) = a(1 - \cos 3t)$, $a > 0$

constant.

4. $\mathbf{R}(t) = r(t)\,\mathbf{u}_r$, where $r(t) = a(1 + \sin t)$, $\theta = 1 - e^{-t}$.

5. $\mathbf{R}(t) = r(t)(\cos \omega t\, \mathbf{i} + \sin \omega t\, \mathbf{j})$, where $r(t) = a/(1 + \cos t)$, $a > 0$.

6. $\mathbf{R}(t) = r(t)(\cos t\, \mathbf{i} + \sin t\, \mathbf{j})$, where $r(t) = 3a/[2(2 + \cos t)]$, $a > 0$.

7. Let $\mathbf{R}(t) = r(t)\,\mathbf{u}_r$. Express $d^3\mathbf{R}/dt^3$ in terms of the unit vectors $\mathbf{u}_r$ and $\mathbf{u}_\theta$.

8. A particle moves on a plane with constant angular speed ω about the origin. If the acceleration increases at a rate proportional to the vector $\mathbf{u}_r$, (a) find the transverse component of $d^3\mathbf{R}/dt^3$. (b) Show that $d^2r/dt^2 = r\omega^2/3$.

9. A person weighing 160 lb walks at the rate of 2 ft/sec toward the edge of a merry-go-round that is rotating with constant angular speed of $\pi/5$ rad/sec. Find the centrifugal and the Coriolis forces experienced by the man when he is 5 ft from the center.

10. From formula (2.62), deduce that the force acting on the particle of mass m has two components, F_r along $\mathbf{u}_r$ and F_θ along $\mathbf{u}_\theta$. Thus show that the motion of the particle is governed by the quations

$$\text{(i)}\quad F_r = m\frac{d^2r}{dt^2} - m\,r\left[\frac{d\theta}{dt}\right]^2$$

$$\text{(ii)}\quad F_\theta = mr\frac{d^2\theta}{dt^2} + 2m\frac{dr}{dt}\frac{d\theta}{dt}$$

11. (a) From Eq. (ii) of Problem 10, show that

$$r\,F_\theta = \frac{d}{dt}\left[mr^2\frac{d\theta}{dt}\right]$$

The quantity $mr^2(d\theta/dt)$ is called the angular momentum of the particle.

(b) Hence, if the force acting on the particle is always parallel to the radial vector $\mathbf{u}_r$, show that the angular momentum is constant.

12. A particle is rotating counterclockwise around the origin of the xy-plane with angular speed ω.

(a) If $\mathbf{w} = \omega\mathbf{k}$ denotes the angular velocity, show that

$$\mathbf{v} = \mathbf{R}'(t) = \mathbf{w} \times \mathbf{R}(t), \text{ where } \mathbf{R}(t) = r(t)\,\mathbf{u}_r$$

(b) Show that $\mathbf{a} = \mathbf{w}' \times \mathbf{R}(t) - \omega^2\mathbf{R}(t)$, where $\mathbf{w}'$ is the angular aceleration.

(c) Thus deduce that if the angular velocity is constant, the acceleration consists only of the normal acceleration.

2.10 CYLINDRICAL AND SPHERICAL COORDINATES

In studying motions in space it is oftentimes more convenient to use noncartesian coordinate systems. The two most widely used coordinate systems of this kind are the cylindrical and the spherical coordinate systems. In this section we derive the expressions for the velocity and the acceleration in these two coordinate systems.

Cylindrical Coordinates

The equations relating the rectangular coordinates (x, y, z) and the cylindrical coordinates (r, θ, z) of a point are given by

$$x = r\cos\theta, \qquad y = r\sin\theta, \qquad z = z$$

where $r \geq 0$, $0 \leq \theta < 2\pi$, $-\infty < z < \infty$. The inverse transformation is given by

$$r = (x^2 + y^2)^{1/2}, \qquad \theta = \arctan(y/x), \qquad z = z$$

Notice that the coordinates (r, θ) coincides with the polar coordinates of the point (x, y). Referring to Fig. 2.23, we see that the position vector of a point in cylindrical coordinates is given by

$$\mathbf{R}(t) = r(t)\,\mathbf{u}_r + z(t)\,\mathbf{k} \tag{2.62}$$

where $\mathbf{u}_r = \cos\theta\,\mathbf{i} + \sin\theta\,\mathbf{j}$. By taking the first order and the second order derivatives of the vector function (2.62), we obtain the velocity and the acceleration. Since the first term in (2.62) coincides with (2.55), we find from (2.59) and (2.61)

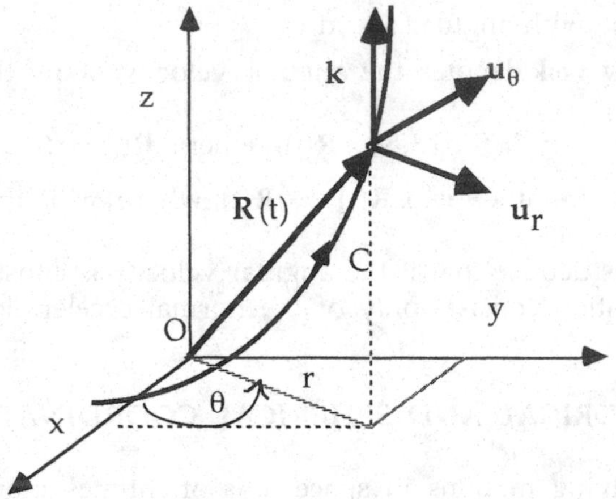

Fig. 2.23 The unit vectors $\mathbf{u}_r$, $\mathbf{u}_\theta$, $\mathbf{k}$ in cylindrical coordinates.

$$\mathbf{v} = \frac{dr}{dt}\mathbf{u}_r + r\frac{d\theta}{dt}\mathbf{u}_\theta + \frac{dz}{dt}\mathbf{k} \tag{2.63}$$

and

$$\mathbf{a} = \left[\frac{d^2r}{dt^2} - r\left(\frac{d\theta}{dt}\right)^2\right]\mathbf{u}_r + \left[r\frac{d^2\theta}{dt^2} + 2\frac{dr}{dt}\frac{d\theta}{dt}\right]\mathbf{u}_\theta + \frac{d^2z}{dt^2}\mathbf{k} \tag{2.64}$$

Example 1. Find the velocity and the acceleration of a moving particle whose position vector is given by

$$\mathbf{R}(t) = a(\cos \omega t\,\mathbf{i} + \sin \omega t\,\mathbf{j}) + b \sin \omega t\,\mathbf{k} \quad (t \geq 0)$$

Solution: We note that

$$\mathbf{u}_r = \cos \omega t\,\mathbf{i} + \sin \omega t\,\mathbf{j} \quad \text{and} \quad \mathbf{u}_\theta = -\sin \omega t\,\mathbf{i} + \cos \omega t\,\mathbf{j}$$

where $\theta = \omega t$. Hence the velocity is given by

$$\mathbf{v}(t) = \mathbf{R}'(t) = a\omega(-\sin \omega t\,\mathbf{i} + \cos \omega t\,\mathbf{j}) + b\omega \cos \omega t\,\mathbf{k}$$
$$= a\omega\,\mathbf{u}_\theta + b\omega \cos \omega t\,\mathbf{k}$$

so that the speed is

$$v(t) = \omega(a^2 + b^2 \cos^2 \omega t)^{1/2}$$

The acceleration is given by

$$\begin{aligned} \mathbf{a}(t) = \mathbf{v}'(t) &= -a\omega^2(\cos \omega t\,\mathbf{i} + \sin \omega t\,\mathbf{j}) - b\omega^2 \sin \omega t\,\mathbf{k} \\ &= -\omega^2 \mathbf{R}(t) \end{aligned}$$

which shows that the acceleration is directed toward the origin.

Spherical Coordinates

The spherical coordinates (ρ, ϕ, θ) of a point P are related to its cartesian coordinates (x, y, z) by the equations

$$x = \rho \sin \phi \cos \theta, \qquad y = \rho \sin \phi \sin \theta, \qquad z = \rho \cos \phi$$

where ρ is the distance of the point from the origin, ϕ is the angle measured from the positive z-axis to the line segment OP, $0 \le \phi \le \pi$, and θ is the polar angle of the point $(x, y, 0)$ on the xy-plane, $0 \le \theta < 2\pi$ (Fig. 2.24). Thus the position vector of a moving particle in spherical coordinates is given by

$$\begin{aligned} \mathbf{R}(t) &= \rho \sin \phi \cos \theta\,\mathbf{i} + \rho \sin \phi \sin \theta\,\mathbf{j} + \rho \cos \phi\,\mathbf{k} \\ &= \rho \mathbf{u}_\rho(\phi, \theta) \end{aligned} \tag{2.65}$$

where

$$\mathbf{u}_\rho(\phi, \theta) = \sin \phi \cos \theta\,\mathbf{i} + \sin \phi \sin \theta\,\mathbf{j} + \cos \phi\,\mathbf{k} \tag{2.66}$$

is the unit vector having the same direction as **R**.

In order to obtain the expressions for the velocity and the acceleration in spherical coordinates, we need to determine two other unit vectors pointing in the direction of increasing ϕ and θ. First, let us differentiate the unit vector $\mathbf{u}_\rho(\phi, \theta)$ with respect to ϕ and set $\mathbf{u}_\phi = \partial \mathbf{u}_\rho / \partial \phi$. We find

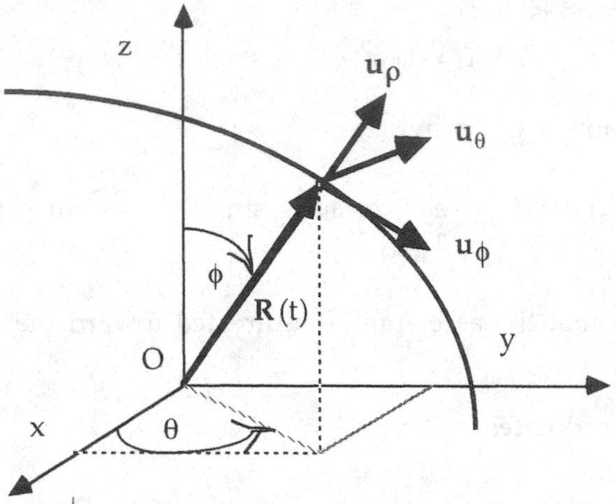

Fig. 2.24 The unit vectors u_ρ, u_ϕ, u_θ in spherical coordinates.

$$\mathbf{u}_\phi = \cos\phi\cos\theta\,\mathbf{i} + \cos\phi\sin\theta\,\mathbf{j} - \sin\phi\,\mathbf{k} \tag{2.67}$$

This is a unit vector orthogonal to $\mathbf{u}_\rho(\phi, \theta)$ and points in the direction of increasing ϕ. Next we differentiate $\mathbf{u}_\rho(\phi, \theta)$ with respect to θ to obtain

$$\partial\mathbf{u}_\rho(\phi, \theta)/\partial\theta = -\sin\phi\sin\theta\,\mathbf{i} + \sin\phi\cos\theta\,\mathbf{j}$$

We know this is orthogonal to $\mathbf{u}_\rho(\phi, \theta)$ and, by direct calculation, it is also orthogonal to $\mathbf{u}_\phi$. However, it is not a unit vector. We therefore normalize it to obtain

$$\mathbf{u}_\theta = -\sin\theta\,\mathbf{i} + \cos\theta\,\mathbf{j}$$

This is the same unit vector we encountered in polar coordinates.

The three mutually orthogonal unit vectors $\mathbf{u}_\rho$, $\mathbf{u}_\phi$, and $\mathbf{u}_\theta$, shown in Fig. 2.24, play similar role in spherical coordinates as the unit vectors $\mathbf{i}$, $\mathbf{j}$, and $\mathbf{k}$ in rectangular cartesian coordinates. They are related to one another by the following formulas, which can be easily verified:

$$\frac{\partial \mathbf{u}_\rho}{\partial \phi} = \mathbf{u}_\phi, \qquad \frac{\partial \mathbf{u}_\phi}{\partial \phi} = -\mathbf{u}_\rho, \qquad \frac{\partial \mathbf{u}_\theta}{\partial \phi} = 0$$

$$\frac{\partial \mathbf{u}_\rho}{\partial \theta} = \sin\phi\, \mathbf{u}_\theta, \quad \frac{\partial \mathbf{u}_\phi}{\partial \theta} = \cos\phi\, \mathbf{u}_\theta \tag{2.68}$$

$$\frac{\partial \mathbf{u}_\theta}{\partial \theta} = -\sin\phi\, \mathbf{u}_\rho - \cos\phi\, \mathbf{u}_\phi$$

Using these formulas, we now determine the expressions for the velocity and the acceleration in spherical coordinates. For the velocity, we differentiate (2.65) to obtain

$$\mathbf{v} = \frac{d\rho}{dt}\mathbf{u}_\rho + \rho\left[\frac{\partial \mathbf{u}_\rho}{\partial \phi}\frac{d\phi}{dt} + \frac{\partial \mathbf{u}_\rho}{\partial \theta}\frac{d\theta}{dt}\right]$$

$$= \frac{d\rho}{dt}\mathbf{u}_\rho + \rho\frac{d\phi}{dt}\mathbf{u}_\phi + \rho\frac{d\theta}{dt}\sin\phi\, \mathbf{u}_\theta \tag{2.69}$$

Differentiating (2.69) we obtain the expression for the acceleration,

$$\mathbf{a} = \frac{d^2\rho}{dt^2}\mathbf{u}_\rho + \frac{d\rho}{dt}\left[\frac{\partial \mathbf{u}_\rho}{\partial \phi}\frac{d\phi}{dt} + \frac{\partial \mathbf{u}_\rho}{\partial \theta}\frac{d\theta}{dt}\right] + \frac{d}{dt}\left[\rho\frac{d\phi}{dt}\right]\mathbf{u}_\phi$$

$$+ \rho\frac{d\phi}{dt}\left[\frac{\partial \mathbf{u}_\phi}{\partial \phi}\frac{d\phi}{dt} + \frac{\partial \mathbf{u}_\phi}{\partial \theta}\frac{d\theta}{dt}\right] + \frac{d}{dt}\left[\rho\frac{d\theta}{dt}\sin\phi\right]\mathbf{u}_\theta$$

$$+ \rho\frac{d\theta}{dt}\sin\phi\left[\frac{\partial \mathbf{u}_\theta}{\partial \phi}\frac{d\phi}{dt} + \frac{\partial \mathbf{u}_\theta}{\partial \theta}\frac{d\theta}{dt}\right]$$

By the formulas in (2.68), this becomes

$$\mathbf{a} = \left[\frac{d^2\rho}{dt^2} - \rho\left(\frac{d\phi}{dt}\right)^2 - \rho\left(\frac{d\theta}{dt}\right)^2 \sin^2\phi\right]\mathbf{u}_\rho$$

$$+\left[2\frac{d\rho}{dt}\frac{d\phi}{dt} + \rho\frac{d^2\phi}{dt^2} - \rho\left(\frac{d\theta}{dt}\right)^2 \sin\phi\cos\phi\right]\mathbf{u}_\phi$$

$$+\left[2\frac{d\rho}{dt}\frac{d\theta}{dt}\sin\phi + \rho\frac{d^2\theta}{dt^2}\sin\phi + 2\rho\frac{d\theta}{dt}\frac{d\phi}{dt}\cos\phi\right]\mathbf{u}_\theta \qquad (2.70)$$

We observe that both the velocity and the acceleration are expressed in terms of their components along the unit vectors $\mathbf{u}_\rho$, $\mathbf{u}_\phi$, and $\mathbf{u}_\theta$.

Example 1. A particle is moving at a constant angular speed γ along a meridian of a sphere of radius a. If the plane of the meridian makes an angle $\theta = \pi/6$ with the x-axis, find the velocity and the acceleration of the particle at any time t.

Solution: Since the particle is revolving at constant angular speed γ, the angle traversed in time t measured from the positive z-axis is equal to $\phi = \gamma t$. Therefore, using spherical coordinates, the position vector of the particle is given by

$$\begin{aligned}\mathbf{R}(t) &= a \sin\gamma t \cos(\pi/6)\,\mathbf{i} + a\sin\gamma t \sin(\pi/6)\,\mathbf{j} + a\cos\gamma t\,\mathbf{k} \\ &= a\sin\gamma t\,\mathbf{u}_r + a\cos\gamma t\,\mathbf{k}\end{aligned}$$

where $\mathbf{u}_r = \cos(\pi/6)\,\mathbf{i} + \sin(\pi/6)\,\mathbf{j}$ (see Fig. 2.25). Hence the velocity is

$$\mathbf{v}(t) = \mathbf{R}'(t) = a\gamma\cos\gamma t\,\mathbf{u}_r - a\gamma\sin\gamma t\,\mathbf{k}$$

and the acceleration is

$$\mathbf{a}(t) = \mathbf{v}'(t) = - a\gamma^2 \sin \gamma t\, \mathbf{u}_r - a\gamma^2 \cos \gamma t\, \mathbf{k} = -\gamma^2 \mathbf{R}(t)$$

It follows that the acceleration is directed toward the center of the sphere.

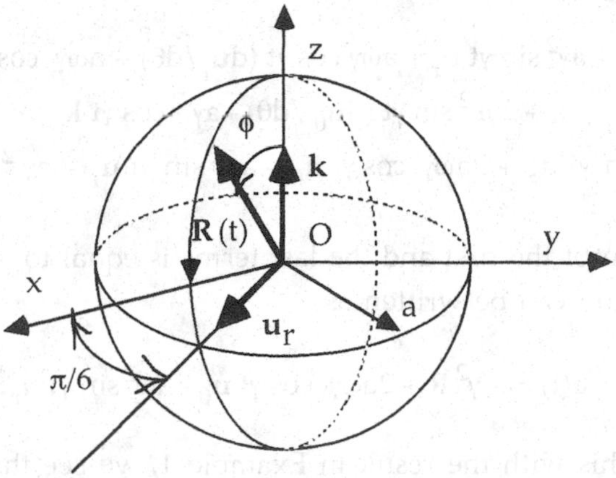

Fig. 2.25 Motion on a sphere.

Example 2. Referring to the problem in Example 1, suppose the sphere is also rotating at a constant angular speed ω about the z-axis, what is the velocity and the acceleration of the particle at any time t?

Solution: Since the plane of the meridian rotates with the sphere, the unit vector $\mathbf{u}_r$ is no longer constant but is of the form

$$\mathbf{u}_r = \cos \omega t\, \mathbf{i} + \sin \omega t\, \mathbf{j}$$

where $\theta = \omega t$ is the angle measured from the positive x-axis. Notice that the x-axis together with the other coordinate axes is fixed in space. By differentiating the position vector

$$\mathbf{R}(t) = a \sin \gamma t\, \mathbf{u}_r + a \cos \gamma t\, \mathbf{k}$$

we obtain the velocity

$$\mathbf{v}(t) = a\,\gamma \cos \gamma t\, \mathbf{u}_r + a\,\omega \sin \gamma t\, (d\mathbf{u}_r/d\theta) - a\,\gamma \sin \gamma t\, \mathbf{k}$$
$$= a\,\gamma \cos \gamma t\, \mathbf{u}_r + a\,\omega \sin \gamma t\, \mathbf{u}_\theta - a\,\gamma \sin \gamma t\, \mathbf{k}$$

The term $a\omega \sin \gamma t\, \mathbf{u}_\theta$ arises from the rotation of the sphere.
Differentiating $\mathbf{v}(t)$, we obtain the acceleration

$$\mathbf{a}(t) = - a\gamma^2 \sin \gamma t\, \mathbf{u}_r + a\omega\gamma \cos \gamma t\, (d\mathbf{u}_r/d\theta) + a\omega\gamma \cos \gamma t\, \mathbf{u}_\theta$$
$$+ \omega^2 \sin \gamma t\, (d\mathbf{u}_\theta / d\theta) - a\gamma^2 \cos \gamma t\, \mathbf{k}$$
$$= - a\gamma^2 \sin \gamma t\, \mathbf{u}_r + 2a\omega\gamma \cos \gamma t\, \mathbf{u}_\theta - a\omega^2 \sin \gamma t\, \mathbf{u}_r - a\gamma^2 \cos \gamma t\, \mathbf{k}$$

Since the sum of the first and the last terms is equal to $-\gamma^2 \mathbf{R}$, the acceleration can be written as

$$\mathbf{a}(t) = -\gamma^2 \mathbf{R} + 2a\omega\gamma \cos \gamma t\, \mathbf{u}_\theta - a\omega^2 \sin \gamma t\, \mathbf{u}_r$$

Comparing this with the result in Example 1, we see that the first term is the radial acceleration directed toward the center of the sphere which is due to the motion of the particle on the meridian. The last term, $a\omega^2 \sin \gamma t\, \mathbf{u}_r$, is the acceleration directed toward the z-axis and is due to the rotation of the sphere. The second term, $2a\omega\gamma \cos \gamma t \mathbf{u}_\theta$, is the Coriolis acceleration, which results from the interaction of the rotation of the sphere and the motion of the particle on the sphere.

2.8 EXERCISES

1. A particle moves in space so that its position at time t has cylindrical coordinates $r = 2$, $\theta = 3t$, $z = 2t$. Find the velocity and acceleration of the particle at time t.
2. A particle moves in space so that its position at time t has cylindrical coordinates $r = t$, $\theta = t$, $z = 2t$. The particle traces out

a curve called a conical helix. Find the velocity and acceleration of the particle at time t.

3. Referring to Problem 2, find a formula for the angle between the velocity vector and the generator of the cone at each point on the curve.

4. A particle moves in space so that its position at time t has cylindrical coordinates $r = \sin t$, $\theta = t$, $z = t$.
(a) Show that the curve traced out by the particle lies on the cylinder $4x^2 + 4(y - 1/2)^2 = 1$.
(b) Find the velocity and acceleration of the particle at time t.
(c) Determine a formula for the angle between the velocity vector and the unit vector **k**.

5. A particle moves in space at constant angular speed ω along the circle of intersection of the plane $z = 1$ with the sphere of radius 2 and center at the origin.
(a) Find the formulas for the velocity and acceleration of the particle at time t. Assume the particle is moving counter-clockwise as viewed atop the positive z-axis.
(b) Suppose the sphere is also rotating at a constant angular speed γ about the z-axis. Find the velocity and acceleration if the sphere rotates in the same direction as the particle; in the opposite direction.

6. Referring to the problem in Example 1, suppose the particle is moving along the meridian at angular speed $\gamma = t$ (so that $\phi = t^2$). Find the transverse and radial components of the acceleration as function of t.

7. Referring to the problem in Example 2, suppose the sphere is rotating at a variable speed such that $\theta = t^2$. Find the acceleration and determine the Coriolis acceleration.

8. A particle moves at constant angular speed α along the circle of intersection of the plane $y = z$ and the sphere of radius a (center at the origin). Show that the position vector of the particle at time t can be expressed as

$$\mathbf{R}(t) = a \cos \alpha t\, \mathbf{i} + a \sin \alpha t\, \mathbf{j}^*$$

where $\mathbf{j}^* = \cos(\pi/4)\,\mathbf{i} + \sin(\pi/4)\,\mathbf{j}$ and αt is the angle made by $\mathbf{R}(t)$ with the x-axis. What are the formulas for the velocity and acceleration?

9. Referring to Problem 8, suppose the sphere is now rotating about the x-axis at constant angular speed γ. Find the formula for the position vector of the particle and determine the velocity and acceleration at time t.

10. Establish the formulas given in (2.70).

11. Show that the scalar coefficient of $\mathbf{u}_\theta$ in (2.70) can be written as

$$\frac{1}{\rho \sin \phi} \frac{d}{dt}\left[\rho^2 \frac{d\theta}{dt} \sin^2 \phi\right]$$

12. A particle slides from (a, 0, 0) down a frictionless circular helix $\mathbf{R}(\theta) = a \cos \theta\, \mathbf{i} + a \sin \theta\, \mathbf{j} + b\theta\, \mathbf{k}$, $\theta \geq 0$, a, b > 0 with constant angular speed $d\theta/dt = \pi$ rad/sec. Find the velocity of the particle and the distance (assumed in meters) it has traveled at t = 2 sec. (Assume the positive z-axis to be pointed downward.)

13. A free falling body starting from rest at (a, 0, 0) slides down a frictionless circular helix $\mathbf{R}(\theta) = a \cos \theta\, \mathbf{i} + a \sin \theta\, \mathbf{j} + b\theta\, \mathbf{k}$, $\theta \geq 0$, a, b > 0 at the speed $v = \sqrt{(2gb\theta)}$, where g is the gravitational acceleration (32 ft/sec^2 or 980 cm/sec^2).

(a) Find θ as a function of time $t \geq 0$. Note that

$$v = |\mathbf{R}'(\theta)| \frac{d\theta}{dt}$$

(b) Find the distance traversed by the body along the helix at t = 2 sec. (Assume the positive z-axis to be pointed downward.)

3

DIFFERENTIAL CALCULUS OF SCALAR AND VECTOR FIELDS

In this chapter, we shall study the differential calculus of scalar and vector fields. We shall generalize the idea of a derivative to that of a directional derivative and consider the various spatial rates of change which lead to the important concepts of the gradient of a scalar field and the divergence and the curl of a vector field. These concepts have significant applications in both geometrical and physical situations.

3.1 SCALAR AND VECTOR FIELDS

Let D be a domain in the three-dimensional space, and let f be a scalar function defined in D. Then at each point P: (x, y, z) in D, f assigns a unique real number f(x, y, z) [or f(P) for brevity], which is the value of f at that point. The domain D, together with the corresponding values of f at each point in D, is called a scalar field. The function f is said to define the scalar field in D. Oftentimes, however, when the domain is clear from the context, the scalar field is simply identified with the scalar function. Thus, for example, if at each point of the atmosphere there is assigned a real number T(P) which represents the temperature at P, then T is a scalar field (also called a temperature field). Other examples of scalar fields are the density of air in the atmosphere, the pressure in a body of fluid, and the gravitational potential in space.

Occasionally, we also consider scalar fields in a plane. Such a field is, of course, defined by a scalar function of two independent variables. A somewhat sophisticated though rather interesting example of a two-dimensional scalar field is that of a "mine field." As we know, a mine field refers to a body of water or a piece of land where mines are placed at randomly chosen points. Such a field can be defined analytically by a function having the value 1 at each point where there is a mine and zero everywhere else. The field is, of course, not continuous.

On the other hand, if at each point P of a domain D there is assigned a unique vector **F**(P), then the domain together with the corresponding vectors at each point constitutes a vector field. The vector function **F** is said to define the vector field in D. In this context, we see that a vector function of a real variable defines a vector field in an interval. As in the case of a scalar field, a vector field is frequently identified with the vector function defining the field. Thus, for example, if at each point P of the atmosphere we assign a vector **v**(P) which represents the wind velocity at P, then **v** defines a vector field known as the velocity field. Other examples of vector fields are the electric field surrounding a power transmission line, the magnetic field around a magnet bar, the velocity field of a fluid in motion, and the gravitational field surrounding the earth.

A vector field may be described graphically by drawing the arrows that represent the vectors at each point in the domain. As an example, the velocity field defined by $\mathbf{F} = -y\mathbf{i}/(x^2+y^2) + x\mathbf{j}/(x^2+y^2)$ on the xy-plane has the graphic representation shown in Fig. 3.1 for points lying on the circles of radii 1 and 2. The field represents a rotational flow about the origin. It approximates, for example, the velocity field of water draining in a bathtab.

Let f define a scalar field in a domain D. The set of points in D at which f assumes the same value will, in general, describe a surface in space. Such a surface is called a level surface of f and has the equation $f(x, y, z) = c$ for some constant c. For example, if $f = x^2 + y^2 + z^2$, then the level surfaces are concentric spheres defined by $x^2 + y^2 + z^2 = c$. In many physical problems, the level surfaces are usually called by special names depending on the nature of the scalar field defined by f. For instance, if f denotes the temperature or pressure in the atmosphere, then the corresponding level surfaces are known as isothermal or isobaric surfaces. If f denotes the gravitational potentential about the earth, then the level surfaces are called equipotential surfaces. These concepts are of great importance in meteorology and physics.

In the two-dimensional case, the set of points at which a scalar field f has the same value describes a plane curve called a level

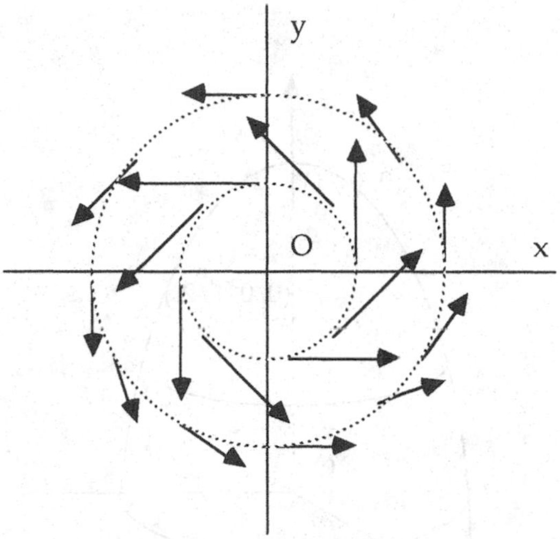

Fig. 3.1 Rotating velocity field.

curve or a contour. It has the representation $f(x,y) = c$. In many applications, it is sometimes necessary to consider level curves on a level surface of a scalar field. For example, suppose $f(x, y, z)$ represents the pressure field of the atmosphere where z denotes the altitude in some suitable scale. The isobaric surfaces $S: f(x,y,z) = c$ consists of points in the atmosphere where the pressure is equal to the constant c. If we set $z = k$ in the equation of an isobaric surface, we then obtain a contour on the surface defined by $f(x, y, k) = c$. This represents the contour of constant pressure at the altitude $z = k$. This contour is known as an isobar in meteorology, and it appears prominently in all weather maps.

Example 1. Discuss the level surfaces for the scalar field defined by $f(x, y, z) = 2x^2 + y^2 + z$. Sketch the level surface for $c = 1$ and the contour at $z = 1/2$.

Solution: The level surfaces are defined by $2x^2 + y^2 + z = c$, which is a family of elliptic paraboloid with vertices on the z-axis. For $c = 1$ the elliptic paraboloid has vertex at $(0, 0, 1)$ as shown in

Fig. 3.2. The contour at $z = 1/2$ is an ellipse whose equation is $4x^2 + 2y^2 = 1$.

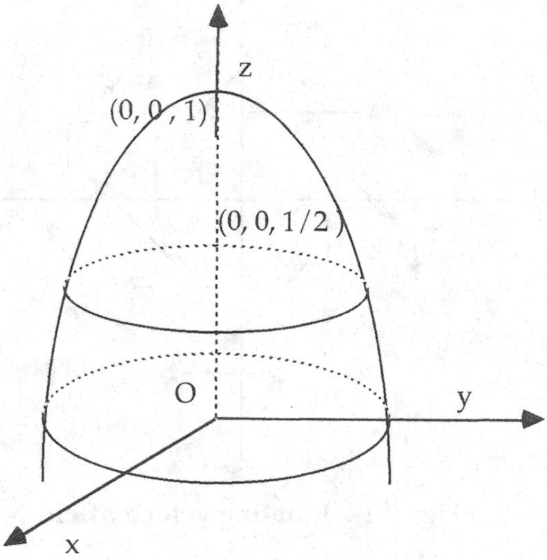

Fig. 3.2 A contour on a level surface.

3.2 ALGEBRA OF VECTOR FIELDS

Like vector functions of a single real variable, a vector field **F** can also be expressed analytically in the form

$$\mathbf{F}(x, y, z) = P(x, y, z)\mathbf{i} + Q(x, y, z)\mathbf{j} + R(x, y, z)\mathbf{k} \qquad (3.1)$$

where P, Q, and R are scalar fields defined in the same domain D. These scalar fields are called the components of the vector field **F**. By performing the operations on the components, the usual operators of vector algebra can be applied to vector fields in exactly the same way it was done in Chapter 2 for vector functions of a single variable. Indeed, suppose $\mathbf{F}(x, y, z) = P(x, y, z)\mathbf{i} + Q(x, y, z)\mathbf{j} + R(x, y, z)\mathbf{k}$ and $\mathbf{G}(x, y, z) = U(x, y, z)\mathbf{i} + Vx, y, z)\mathbf{j} + W(x, y, z)\mathbf{k}$ are two vector fields, and f is a scalar field all defined in the same domain D. Then we have

(i) $(\mathbf{F} + \mathbf{G})(x, y, z) = [P(x, y, z) + U(x, y, z)]\mathbf{i} + [Q(x, y, z) + V(x, y, z)]\mathbf{j} + [R(x, y, z) + W(x, y, z)]\mathbf{k}$

(ii) $(f\,\mathbf{F})(x, y, z) = f(x, y, z)\,[P(x, y, z)\mathbf{i} + Q(x, y, z)\mathbf{j} + R(x, y, z)\mathbf{k}]$

(iii) $(\mathbf{F}\cdot\mathbf{G})(x, y, z) = P(x, y, z)U(x, y, z) + Q(x, y, z)V(x, y, z) + R(x, y, z)W(x, y, z)$ (3.2)

(iv) $$(\mathbf{F} \times \mathbf{G})(x, y, z) = \begin{vmatrix} \mathbf{i} & \mathbf{j} & \mathbf{k} \\ P(x, y, z) & Q(x, y, z) & R(x, y, z) \\ U(x, y, z) & V(x, y, z) & W(x, y, z) \end{vmatrix}$$

The sum $\mathbf{F} + \mathbf{G}$ and the products $f\,\mathbf{F}$ and $\mathbf{F} \times \mathbf{G}$ are again vector fields, whereas the dot product $\mathbf{F}\cdot\mathbf{G}$ is a scalar field.

The notion of orthogonality and linear dependence can also be extended to vector fields in a straightforward manner. Thus, for example, the two vector fields **F** and **G** are orthogonal in D if and only if $\mathbf{F}\cdot\mathbf{G} = 0$ identically in D.

The basic concepts of limit and continuity can also be carried over directly to vector fields. Suppose **F** is a vector field defined in a domain D and **A** is a constant vector. Let $\mathbf{R}_o$ denote the position vector of a point (x_o, y_o, z_o) in D. Then we say that

$$\lim_{\mathbf{R} \to \mathbf{R}_o} \mathbf{F}(x, y, z) = \mathbf{A} \tag{3.3}$$

if for any number $\varepsilon > 0$, there is a number $\delta > 0$ such that

$$|\,\mathbf{F}(x, y, z) - \mathbf{A}\,| < \varepsilon$$

whenever $|\mathbf{R} - \mathbf{R}_o| < \delta$. In this definition, it is not required that **F** be defined at the point (x_o, y_o, z_o) itself.

The inequality $|\mathbf{R} - \mathbf{R}_o| < \delta$ is equivalent to

$$(x - x_o)^2 + (y - y_o)^2 + (z - z_o)^2 < \delta^2$$

which describes the interior of a sphere of radius δ and center at (x_o, y_o, z_o). Thus, geometrically, the definition (3.3) means that for every $\varepsilon > 0$ there is a sphere about (x_o, y_o, z_o) of radius δ such that for each point (x, y, z) inside the sphere the magnitude of the difference between **F**(x, y, z) and **A** is less than ε.

In terms of components, if $\mathbf{F} = F_1\mathbf{i} + F_2\mathbf{j} + F_3\mathbf{k}$ and $\mathbf{A} = a_1\mathbf{i} + a_2\mathbf{j} + a_3\mathbf{k}$, the definition (3.3) implies that

$$\lim_{(x, y, z) \to (x_o, y_o, z_o)} F_i(x, y, z) = a_i \tag{3.4}$$

for i = 1, 2, 3.

The following theorem can be easily established from the definition.

Theorem 1. Let **F** and **G** be vector fields such that

$$\lim_{\mathbf{R} \to \mathbf{R}_o} \mathbf{F}(x, y, z) = \mathbf{A} \quad \text{and} \quad \lim_{\mathbf{R} \to \mathbf{R}_o} \mathbf{G}(x, y, z) = \mathbf{B}$$

Then we have

(a) $\lim_{\mathbf{R} \to \mathbf{R}_o} [\mathbf{F}(x, y, z) + \mathbf{G}(x, y, z)] = \mathbf{A} + \mathbf{B}$

(b) $\lim_{\mathbf{R} \to \mathbf{R}_o} m\,\mathbf{F}(x, y, z) = m\,\mathbf{A}$

(c) $\lim_{\mathbf{R} \to \mathbf{R}_o} \mathbf{F}(x, y, z) \cdot \mathbf{G}(x, y, z) = \mathbf{A} \cdot \mathbf{B}$

(d) $\lim_{\mathbf{R} \to \mathbf{R}_o} \mathbf{F}(x, y, z) \times \mathbf{G}(x, y, z) = \mathbf{A} \times \mathbf{B}$

(e) $\lim_{\mathbf{R} \to \mathbf{R}_o} |\mathbf{F}(x,y,z)| = |\mathbf{A}|$

Thus, for example, if $\mathbf{F} = x^2\mathbf{i} + y^2\mathbf{j} + z^2\mathbf{k}$ and $\mathbf{G} = xy\mathbf{i} + yz\mathbf{j} + xz\mathbf{k}$, then

$$\lim_{(x,y,z) \to (1,2,-1)} \mathbf{F}(x, y, z) \cdot \mathbf{G}(x, y, z)$$

$$= \lim_{(x,y,z) \to (1,2,-1)} (x^3y + y^3z + z^3x) = -7$$

A vector field **F** is said to be continuous at a point (x_o, y_o, z_o) if and only if

$$\lim_{(x, y, z) \to (x_o, y_o, z_o)} \mathbf{F}(x, y, z) = \mathbf{F}(x_o, y_o, z_o) \tag{3.5}$$

The definition requires that the function **F** must be defined at (x_o, y_o, z_o). If **F** is continuous at each point of a domain D, then we say that **F** is continuous in D. From (3.4) it follows that a vector field **F** is continuous in D if and only if each of its components is continuous in D.

From Theorem 1, we see that if **F** and **G** are continuous in a domain D, then so are the sum **F** + **G**, the scalar multiple h**F**, the scalar product $\mathbf{F} \cdot \mathbf{G}$, and the cross product $\mathbf{F} \times \mathbf{G}$.

3.1 EXERCISES

For each of the given scalar fields, sketch the level surface f(x, y, z) = c corresponding to the given value of the constant c.

1. $f(x, y, z) = x + 2y + 2z, \quad c = 1, 2.$
2. $f(x, y, z) = x^2 + y^2 - z, \quad c = -1, 1.$
3. $f(x, y, z) = x^2 + 4y^2 + 9z^2, \quad c = 36.$
4. $f(x, y, z) = y^2 - x^2 - z^2, \quad c = 1.$

For each of the given scalar fields, sketch the contour f(x,y) = c corresesonding to the given value of the constant c.

5. $f(x, y) = xy, \quad c = -2, -1, 1, 2.$
6. $f(x, y) = x^2 - y^2, \quad c = -1, -2, 0, 1, 2.$
7. $f(x, y) = \arctan(y/x), \quad c = -1, 0, 1.$
8. $f(x, y) = e^x \cos y, \quad c = 0, 1.$
9. Sketch the isobars on the isobaric surface $z = (4 - x^2 - y^2)^{1/2}$ at the altitudes $z = 1/2, 1, 3/2$.

9. Sketch the isobars on the isobaric surface $z = (4 - x^2 - y^2)^{1/2}$ at the altitudes $z = 1/2, 1, 3/2$.
10. Sketch the isotherms on the isothermal surface $z = 5 - 2x^2 - 3y^2$ corresponding to $z = 1, 2, 3$.
11. Sketch the vector field defined by each of the following:
(a) $\mathbf{F} = x\mathbf{i} + y\mathbf{j}$; (b) $\mathbf{F} = x^2\mathbf{i} + y\mathbf{j}$
(c) $\mathbf{F} = y\mathbf{i} - x\mathbf{j}$; (d) $\mathbf{F} = x\mathbf{i} + y\mathbf{j} + z\mathbf{k}$
12. Let $\mathbf{F} = \mathbf{A}\sin y + \mathbf{B}\cos x$, where $\mathbf{A} = \mathbf{i} + \mathbf{j}$ and $\mathbf{B} = \mathbf{i} - \mathbf{j}$. Evaluate each of the following:

(a) $\lim_{(x, y) \to (\pi/3, \pi/4)} \mathbf{F}(x, y)$ (b) $\lim_{(x, y) \to (\pi/4, \pi/4)} |\mathbf{F}(x,y)|$

(c) $\lim_{(x, y) \to (\pi/4, \pi/4)} (x + y)\mathbf{F}(x, y)$

13. Let $\mathbf{F} = x\cos z\,\mathbf{i} + x\sin z\,\mathbf{j} + y\,\mathbf{k}$. Evaluate

(a) $\lim_{(x, y, z) \to (1, 2, \pi/2)} \mathbf{F}(x, y, z)$

(b) Show that $\mathbf{F}$ has a constant magnitude on the cylinder $x^2 + y^2 = a^2$, where a is a constant.
(c) Show that $\partial\mathbf{F}/\partial x$, $\partial\mathbf{F}/\partial y$, $\partial\mathbf{F}/\partial z$ are mutually orthogonal.
14. Let $\mathbf{F} = \sin x\,\mathbf{i} + \cos y\,\mathbf{j} + \ln z\,\mathbf{k}$ $(z \neq 0)$, and $\mathbf{G} = e^x \cos y\,\mathbf{i} + e^x \sin y\,\mathbf{j}$. Find (a) $\mathbf{F} + \mathbf{G}$; (b) $\mathbf{F}\cdot\mathbf{G}$; (c) $\mathbf{F} \times \mathbf{G}$.
15. Let $\mathbf{F} = e^x \cos z\,\mathbf{i} + e^x \sin z\,\mathbf{j} + \mathbf{k}$ and $\mathbf{G} = e^x \cos z\,\mathbf{i} + e^x \sin z\,\mathbf{j} - e^{2x}\mathbf{k}$.
(a) Show that $\mathbf{F}$ and $\mathbf{G}$ are orthogonal for all (x, y, z).
(b) Show that the vector field $\mathbf{F} \times \mathbf{G}$ is parallel to the xy-plane.

3.3 DIRECTIONAL DERIVATIVE OF A SCALAR FIELD

Let f be a scalar field defined in a domain D, and suppose that f is differentiable in D. We know that the first partial derivatives $\partial f/\partial x$, $\partial f/\partial y$, and $\partial f/\partial z$ of f represent the rate of change of f along the x, y, and z coordinate axes, respectively. In many applications, it is often necessary to know the rate of change of f in an arbitrary direction. This requires a more general concept of a derivative known as directional derivative.

The definition of a directional derivative is very much like the definition of an ordinary derivative in differential calculus. Let $\mathbf{R}_o$ denote the position vector of a point P_o: (x_o, y_o, z_o) in a domain D and let the unit vector

$$\mathbf{u} = \cos\alpha\,\mathbf{i} + \cos\beta\,\mathbf{j} + \cos\gamma\,\mathbf{k}$$

specify a direction at P_o (Fig. 3.3). We note that points on the line segment issuing from P_o in the direction of $\mathbf{u}$ are given by

$$x = x_o + s\cos\alpha\,,\quad y = y_o + s\cos\beta\,,\quad z = z_o + s\cos\gamma \qquad (3.6)$$

where the parameter s denotes the distance of the point from P_o. We define the derivative of f at P_o in the direction of the vector $\mathbf{u}$ as the limit

$$\frac{df}{ds}(P_o) = \lim_{s \to 0} \frac{f(P) - f(P_o)}{s} \qquad (3.7)$$

$$= \lim_{s \to 0} \frac{f(x_o + s\cos\alpha,\, y_o + s\cos\beta,\, z_o + s\cos\gamma) - f(x_o, y_o, z_o)}{s}$$

If we set

$$g(s) = f(x_o + s\cos\alpha,\ y_o + s\cos\beta,\ z_o + s\cos\gamma) \qquad (3.8)$$

then the definition (3.7) becomes

$$\lim_{s \to 0} \frac{g(s) - g(0)}{s} = g'(0)$$

Now differentiating (3.8) with respect to s and then setting s = 0, we obtain

$$g'(0) = \frac{\partial f}{\partial x}(P_o)\frac{dx}{ds} + \frac{\partial f}{\partial y}(P_o)\frac{dy}{ds} + \frac{\partial f}{\partial z}(P_o)\frac{dz}{ds}$$

From (3.6) we find

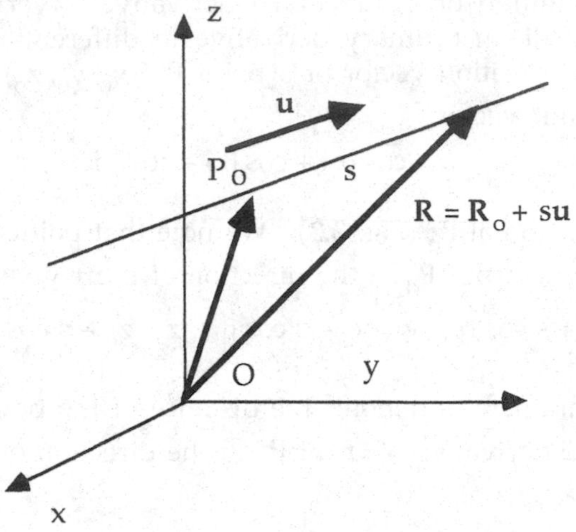

Fig . 3.3 Directional derivative.

$$\frac{dx}{ds} = \cos\alpha\,, \quad \frac{dy}{ds} = \cos\beta\,, \quad \frac{dz}{ds} = \cos\gamma$$

Thus the directional derivative of f at P_o in the direction of the vector **u** is given by the formula

$$\frac{df}{ds}(P_o) = \frac{\partial f}{\partial x}(P_o)\cos\alpha + \frac{\partial f}{\partial y}(P_o)\cos\beta + \frac{\partial f}{\partial z}(P_o)\cos\gamma \qquad (3.9)$$

When $\alpha = 0$, $\beta = \pi/2$, and $\gamma = \pi/2$, so that $\mathbf{u} = \mathbf{i}$, we see that (3.9) reduces to $\partial f(P_o)/\partial x$. Similarly, when $\mathbf{u} = \mathbf{j}$ and $\mathbf{u} = \mathbf{k}$, (3.9) yields $\partial f(P_o)/\partial y$ and $\partial f(P_o)/\partial z$, respectively. Thus the partial derivatives of f are actually directional derivatives in the positive direction of the coordinate axes.

Example 1. Find the directional derivative of $f = 2x^2 + xy + yz^2$ at (1, -1, 2) in the direction of the vector $\mathbf{A} = \mathbf{i} - 2\mathbf{j} + 2\mathbf{k}$.

Solution: We have

$$\frac{\partial f}{\partial x} = 4x + y, \quad \frac{\partial f}{\partial y} = x + z^2, \quad \frac{\partial f}{\partial z} = 2yz$$

so that at (1, -1, 2)

$$\frac{\partial f}{\partial x} = 3, \quad \frac{\partial f}{\partial y} = 5, \quad \frac{\partial f}{\partial z} = -4$$

Since $|\mathbf{A}| = 3$, a unit vector in the direction of **A** is $\mathbf{u} = (1/3)\mathbf{A}$. Hence by (3.9) we have

$$\frac{df}{ds}(1, -1, 2) = 3\left(\frac{1}{3}\right) + 5\left(\frac{-2}{3}\right) - 4\left(\frac{2}{3}\right) = -5$$

Example 2. What is the directional derivative of $f(x,y) = x^2y + 2 \ln y$ at (1, 1) in the direction of the line making 30° with the positive x-axis?

Solution: At (1,1) we have

$$\partial f(1, 1)/\partial x = 2 \quad \text{and} \quad \partial f(1, 1)/\partial y = 3$$

A unit vector in the direction of the given line has the form

$$\mathbf{u} = \cos 30^o\, \mathbf{i} + \sin 30^o\, \mathbf{j}$$

Hence, the directional derivative is equal to

$$\frac{df}{ds}(1, 1) = 2 \cos 30^o + 3 \sin 30^o = \sqrt{3} + \frac{3}{2}$$

We observe that the function defined in (3.8) is actually the value of f at points on the line segment from P_o and extending in the direction of the unit vector **u**. Suppose we now replace the line segment by a smooth curve C passing through P_o. Let the curve be represented by the vector equation

$$\mathbf{R}(s) = x(s)\, \mathbf{i} + y(s)\, \mathbf{j} + z(s)\, \mathbf{k}$$

where s denotes arc length of the curve. Then on C the function f assumes the value

$$g(s) = f\,[x(s), y(s), z(s)]$$

Differentiating this with respect to s, we find by the chain rule

$$g'(s) = \frac{\partial f}{\partial x}(P)\frac{dx}{ds} + \frac{\partial f}{\partial y}(P)\frac{dy}{ds} + \frac{\partial f}{\partial z}(P)\frac{dz}{ds} \tag{3.10}$$

where P is the point corresponding to the parameter s. Since the coefficients dx/ds, dy/ds, and dz/ds in (3.10) are precisely the components of the unit tangent vector $\mathbf{t} = d\mathbf{R}/ds$ to the curve at P, it follows that (3.10) is the directional derivative of f at P on the curve C in the direction of the tangent vector. This derivative is called the tangential derivative of f on the curve. It represents the rate of change of f on the curve with respect to the arc length.

Example 3. Find the tangential derivative of $f = xy$ on the circle $x^2 + y^2 = a^2$ oriented counterclockwise.

Solution: A vector equation that orients the circle in the counterclockwise direction is given by

$$\mathbf{R}(t) = a \cos t\,\mathbf{i} + a \sin t\,\mathbf{j} \qquad (t \geq 0)$$

Hence, the unit tangent vector to the circle is

$$\mathbf{t} = \frac{\mathbf{R}'(t)}{|\mathbf{R}(t)|} = -\sin t\,\mathbf{i} + \cos t\,\mathbf{j}$$

so that $\cos\alpha = -\sin t$ and $\cos\beta = \cos t$. Therefore, by (3.10) the tangential derivative of f on the circle is given by

$$\begin{aligned} df/ds &= y(-\sin t) + x \cos t \\ &= y(-y/a) + x(x/a) \\ &= (x^2 - y^2)/a \end{aligned}$$

Note that if we express f as a function of s on the circle (s = at), we have

$$f = a^2 \sin(s/a) \cos(s/a) = (a^2/2)[\sin 2(s/a)]$$

Differentiating this with respect to s, we find

$$df/ds = a\cos^2(s/a) = a[\cos^2(s/a) - \sin^2(s/a)] = (x^2 - y^2)/a$$

which agrees with the previous result.

3.2 EXERCISES

In each of Problems 1 through 8, find the directional derivative of the given scalar field at the given point in the direction of the given vector.

1. $f(x, y) = x^2 - y^2$, $(2, -1)$; $\mathbf{A} = \mathbf{i} + \mathbf{j}$.
2. $f(x, y) = x^2 - 2xy + 3y^2$, $(-1, 2)$; $\mathbf{A} = -\mathbf{i} + 2\mathbf{j}$.
3. $f(x, y) = e^x \cos y$, $(1, \pi/3)$; $\mathbf{A} = \mathbf{i} - \mathbf{j}$.
4. $f(x, y) = x^3 - 3x^2y - xy + y^2$, $(2, 2)$; $\mathbf{A} = 2\mathbf{i} + \mathbf{j}$.
5. $f(x, y, z) = xy + yz + zx$, $(1, 1, -1)$; $\mathbf{A} = 2\mathbf{i} + \mathbf{j} - 2\mathbf{k}$.
6. $f(x, y, z) = xye^z + z^2$, $(-1, 2, 0)$; $\mathbf{A} = -\mathbf{i} + 2\mathbf{j} + 2\mathbf{k}$.
7. $f(x, y, z) = x\sin z - y\cos z$, $(2, 3, \pi/4)$; $\mathbf{A} = 3\mathbf{i} + 2\mathbf{j} - 2\mathbf{k}$.
8. $f(x, y, z) = \ln(x^2 + y^2 - z)$, $(2, 2, -1)$; $\mathbf{A} = \mathbf{i} + \mathbf{j} - \mathbf{k}$.
9. Show that the directional derivative of a scalar field $f(x, y)$ in the direction of a line making an angle θ with the positive x-axis is given by (see Example 2)

$$\frac{df}{ds}(P) = \frac{\partial f}{\partial x}(P)\cos\theta + \frac{\partial f}{\partial y}(P)\sin\theta$$

   Determine the value of θ for which df/ds is maximum and determine that maximum value. (Note that the point P is fixed.)
10. Apply the result of Problem 9 to find the maximum value of the directional derivative of $f(x, y) = x/(x^2 + y^2)$ at the point (1, 1), and determine the value of θ in which the maximum value occurs.
11. Repeat Problem 10 for $f(x, y) = x^2 - xy + y^2$.
12. The function f has at (1, -1) a directional derivative equal to $\sqrt{2}$ in the direction toward (3, 1), and $\sqrt{10}$ in the direction toward (0, 2). Find the value of $\partial f/\partial x$ and $\partial f/\partial y$ at (1, -1). Determine

the derivative of f at (1, -1) in the direction toward (2, 3).

13. Find the tangential derivative of $f(x, y) = xy^2 - e^x \ln y$ at $(\sqrt{3}, 2)$ on the curve $x = 2 \cos t$, $y = 4 \sin t$ $(t \geq 0)$.

14. Find the tangential derivative of $f(x, y) = x^2 + xy + y^2$ at any point on the curve

$$\mathbf{R}(t) = t\mathbf{i} + (t^2 - t + 2)\mathbf{j} \quad (t \geq 0).$$

15. Let f be a nonconstant scalar field that is twice continuously differentiable on the xy-plane.
(a) Show that the tangential derivative of f is zero on the curve described by $f(x, y) = c$, where c is a constant.
(b) Show that the direction in which the directional derivative of f is maximum is orthogonal to the curve $f(x, y) = c$.

16. Find the tangential derivative of $f(x, y, z) = x^2 - 2xy + z^2$ at $(\sqrt{3}, 1, \pi/3)$ on the helix $x = 2 \sin t$, $y = 2 \cos t$, $z = t$ $(t \geq 0)$.

17. Find the tangential derivative of $f(x, y, z) = e^x y^2 z$ on the curve $\mathbf{R}(t) = (t - 1)\mathbf{i} - t^2\mathbf{j} + 2t\mathbf{k}$ when $t = 1$.

3.4 GRADIENT OF A SCALAR FIELD

Let f be a scalar field defined in a domain D, and suppose that f is differentiable in D. We have seen that the directonal derivative of f at a point P in the direction of a unit vector $\mathbf{u} = \cos\alpha\mathbf{i} + \cos\beta\mathbf{j} + \cos\gamma\mathbf{k}$ is given by

$$\frac{df}{ds}(P) = \frac{\partial f}{\partial x}(P)\cos\alpha + \frac{\partial f}{\partial y}(P)\cos\beta + \frac{\partial f}{\partial z}(P)\cos\gamma \tag{3.11}$$

If we introduce the vector

$$\text{grad } f(P) = \frac{\partial f}{\partial x}(P)\,\mathbf{i} + \frac{\partial f}{\partial y}(P)\,\mathbf{j} + \frac{\partial f}{\partial z}(P)\,\mathbf{k} \tag{3.12}$$

whose components are the partial derivatives of f at P, then the expression (3.11) is simply the dot product of grad f(P) and $\mathbf{u}$, that is,

$$\frac{df}{ds}(P) = \text{grad } f(P)\cdot\mathbf{u} \tag{3.13}$$

The vector field, grad f, is called the gradient of the scalar field f. It is one of the basic concepts in vector analysis that is of great importance in application.

The gradient of a scalar field f is also commonly written as ∇f, where ∇ denotes the vector differential operator

$$\nabla = \mathbf{i}\frac{\partial}{\partial x} + \mathbf{j}\frac{\partial}{\partial y} + \mathbf{k}\frac{\partial}{\partial z} \tag{3.14}$$

In other words, grad f can be interpreted as the result of applying the differential operator ∇ to f. The operator ∇ plays a similar role in vector analysis as the operator $D = d/dx$ in calculus. Indeed, using the operator ∇ together with the familiar rules on differentiation, we can easily establish the following properties:

Theorem 1. Let f and g be scalar fields which are differentiable in a domain D. Then we have

(a) $\nabla(f + g) = \nabla f + \nabla g$

(b) $\nabla(cf) = c\nabla f$, where c is any constant (3.15)

(c) $\nabla(fg) = f\nabla g + g\nabla f$

(d) $\nabla(f/g) = (g\nabla f - f\nabla g)/g^2$, provided $g \neq 0$ in D.

For example, to prove part (c), we proceed as follows:

$$\nabla(fg) = \mathbf{i}\frac{\partial}{\partial x}(fg) + \mathbf{j}\frac{\partial}{\partial y}(fg) + \mathbf{k}\frac{\partial}{\partial z}(fg)$$

$$= \mathbf{i}\left[f\frac{\partial g}{\partial x} + g\frac{\partial f}{\partial x}\right] + \mathbf{j}\left[f\frac{\partial g}{\partial y} + g\frac{\partial f}{\partial y}\right] + \mathbf{k}\left[f\frac{\partial g}{\partial z} + g\frac{\partial f}{\partial z}\right]$$

$$= f\left[\mathbf{i}\frac{\partial g}{\partial x} + \mathbf{j}\frac{\partial g}{\partial y} + \mathbf{k}\frac{\partial g}{\partial z}\right] + g\left[\mathbf{i}\frac{\partial f}{\partial x} + \mathbf{j}\frac{\partial f}{\partial y} + \mathbf{k}\frac{\partial f}{\partial z}\right]$$

$$= f\,\nabla g + g\,\nabla f$$

The operator ∇ has many other applications, and we shall come across them in our discussion later.

Geometric Properties of Grad f

We now turn our attention to some geometric properties of the gradient. Supoose that $\nabla f(P) \neq \mathbf{0}$, and let θ denote the angle between $\nabla f(P)$ and the unit vector $\mathbf{u}$. Then, from the geometric property of scalar product, we have

$$\frac{df}{ds}(P) = \nabla f(P)\cdot \mathbf{u} = |\nabla f(P)|\,|\mathbf{u}|\cos\theta$$
$$= |\nabla f(P)|\cos\theta \tag{3.16}$$

This shows that the directional derivative of a scalar field f at a point P in the direction of a unit vector $\mathbf{u}$ is simply the component of the gradient vector $\nabla f(P)$ at P along $\mathbf{u}$. This component is positive or negative according as $0 \le \theta < \pi/2$ or $\pi/2 < \theta \le \pi$ (see Fig. 3.4).

By examining equation (3.16), we see that the directional derivative is maximum when $\cos\theta = 1$, that is, when $\mathbf{u}$ is in the same direction as $\nabla f(P)$. The maximum value is given by $|\nabla f(P)|$. Since $|\nabla f(P)| > 0$, unless f is identically zero, it follows that f is increasing in the direction of the vector $\nabla f(P)$. In other words, at a point P, the scalar field f experiences its maximum rate of increase in the direction of the gradient vector $\nabla f(P)$.

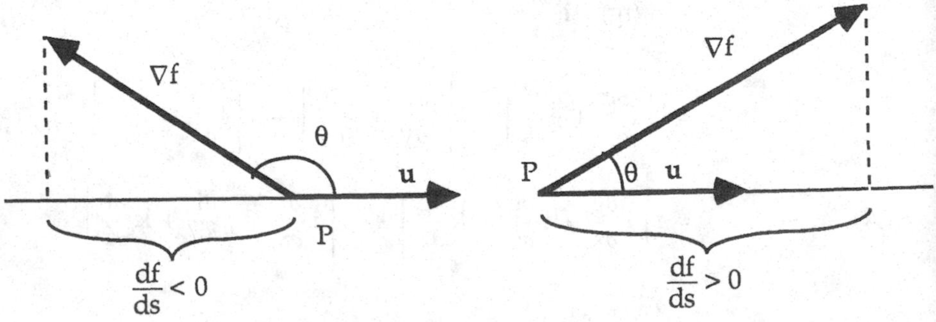

Fig. 3.4 The gradient vector ∇f (P).

Example 1. Let $f(x, y, z) = x^2 y + y^2 z + 1$. Determine the direction in which the directional derivative of f at (2, 1, 3) is maximum, and find that maximum value. What is the minimum value and in what direction does it occur?

Solution: The maximum value of the directional derivative of f at (2, 1, 3) occurs in the direction of the gradient vector $\nabla f(2, 1, 3)$. Since

$$\text{grad } f = 2xy\mathbf{i} + (x^2 + 2yz)\mathbf{j} + y^2\mathbf{k}$$

we find $\nabla f(2, 1, 3) = 4\mathbf{i} + 10\mathbf{j} + \mathbf{k}$. Thus the maximum value of the directional derivative is $|\nabla f(2,1,3)| = \sqrt{117}$.

The minimum value of the directional derivative occurs in the opposite direction to the gradient vector $\nabla f(2, 1, 3)$, that is, in the direction $-4\mathbf{i} - 10\mathbf{j} - \mathbf{k}$ and the minimum value is $-\sqrt{117}$.

Example 2. Suppose the temperature distribution in a metal ball is defined by $T(x,y,z) = a(x^2 + y^2 + z^2)$, where a is a positive constant. What is the direction of maximum cooling?

Solution: The maximum rate of increase of the temperature occurs in the direction of the vector

$$\text{grad } T = 2a(x\mathbf{i} + y\mathbf{j} + z\mathbf{k}) = 2a\mathbf{R}$$

where **R** is the position vector of the point (x, y, z). Hence, maximum cooling occurs in the direction opposite to grad T, that is, in the direction $-\mathbf{R}$, which is toward the origin.

The gradient vector has another geometric significance. Suppose S is a level surface of a scalar field f which is differentiable in a domain D. Let P_0 be a point on this surface, and consider a curve C which lies on S and passes through the point P_0 (Fig. 3.5). Assume that the curve is described by the vector equation $\mathbf{R}(s) = x(s)\mathbf{i} + y(s)\mathbf{j} + z(s)\mathbf{k}$, where s denotes arc length such that $s = s_0$ corresponds to the point P_0. Since C lies on S, we have $f(x(s), y(s), z(s)) = c$, where c is a constant (in fact, $c = f(P_0)$). Hence, taking the derivative of f with respect to s by the chain rule, and setting $s = s_0$, we find

$$\frac{df}{ds}(P_0) = \frac{\partial f}{\partial x}\frac{dx}{ds} + \frac{\partial f}{\partial y}\frac{dy}{ds} + \frac{\partial f}{\partial z}\frac{dz}{ds} = \nabla f(P_0)\cdot\frac{d\mathbf{R}}{ds}(s_0) = 0 \qquad (3.17)$$

This implies that the gradient vector $\nabla f(P_o)$ is orthogonal to the tangent vector $d\mathbf{R}/ds$ at P_o (see Fig. 3. 5). Since (3.17) holds for all curves on S passing through the point P_o, we conclude that $\nabla f(P_o)$ is orthogonal to the tangent vectors of all these curves. These tangent vectors at P_o determine a plane which is called the tangent plane to the surface at P_o. If $\nabla f(P_o) \neq \mathbf{0}$, then $\nabla f(P_o)$ is normal to the tangent plane at P_o. Therefore, given a level surface $f(x, y, z) = c$, we conclude that the gradient of f, $\nabla f(P)$, is orthogonal to the surface at the point P on the surface. The vector

$$\mathbf{n} = \frac{\nabla f(P)}{|\nabla f(P)|} \tag{3.18}$$

is then a unit normal vector to the surface.

If $\mathbf{R}^*$ denotes the position vector of a point on the tangent plane and $\mathbf{R}_o$ is the position vector of the point P_o, then an equation of the tangent plane is given by

$$(\mathbf{R}^* - \mathbf{R}_o)\cdot\nabla f(P_o) = 0$$

In cartesian coordinates, this becomes

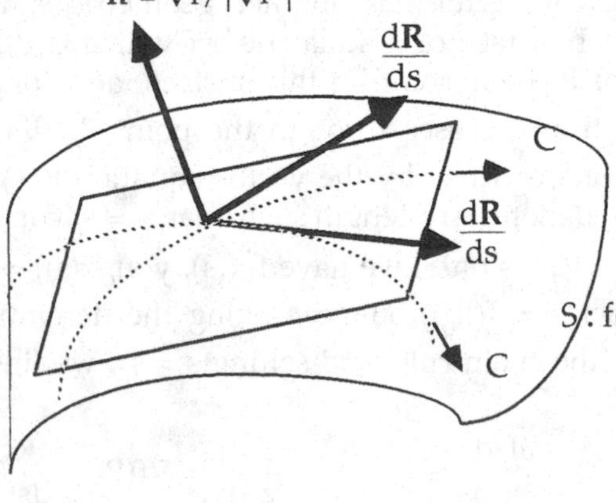

Fig 3.5 Grad f normal to the surface f(x, y, z) = c.

$$\frac{\partial f}{\partial x}(P_o)(x^* - x_o) + \frac{\partial f}{\partial y}(P_o)(y^* - y_o) + \frac{\partial f}{\partial z}(P_o)(z^* - z_o) = 0$$

If a level surface is closed so that it encloses a domain (e. g., a sphere), then the unit normal vector pointing away from the enclosed domain is called the outward unit normal vector on the surface. The directional derivative of a scalar field on a level surface in the direction of the outward unit normal vector is called the *outward normal derivative* of the scalar field on the surface. Thus, if the surface has the representation f(x, y, z) = c so that its outward unit normal vector is $\mathbf{n} = \nabla f/|\nabla f|$, then by definition the outward normal derivative of a scalar field g on the surface is given by

$$\partial g/\partial n = \nabla g \cdot \mathbf{n}$$

Example 3. Find the outward unit normal vector on the sphere $x^2 + y^2 + z^2 = 9$, and determine the equation of the tangent plane at the point (2, 1, 2).

Solution: Let $f(x, y, z) = x^2 + y^2 + z^2$. We find $\nabla f = 2x\mathbf{i} + 2y\mathbf{j} + 2z\mathbf{k}$ so that the outward unit normal vector on the sphere is

$$\mathbf{n} = \frac{x\mathbf{i} + y\mathbf{j} + z\mathbf{k}}{3}$$

This is in the same direction as the position vector **R** of the point (x, y, z). Since $\nabla f(2, 1, 2) = 4\mathbf{i} + 2\mathbf{j} + 4\mathbf{k}$, the equation of the tangent plane is given by

$$4(x - 2) + 2(y - 1) + 4(z - 2) = 0 \quad \text{or} \quad 2x + y + 2z = 9$$

Example 4. Find the outward normal derivative of $g(x, y, z) = xy + yz + zx$ on the surface $x^2 + y^2 + z = 1$.

Solution: Here we take the outward normal vector to be pointing away from the domain bounded by the paraboloid as shown in Fig. 3.6. Letting $f(x, y, z) = x^2 + y^2 + z$, we find

$$\nabla f = 2x\mathbf{i} + 2y\mathbf{j} + \mathbf{k}$$

so that

$$\mathbf{n} = \frac{2x\mathbf{i} + 2y\mathbf{j} + \mathbf{k}}{\sqrt{4(x^2 + y^2) + 1}}$$

Thus, since $\nabla g = (y + z)\mathbf{i} + (z + x)\mathbf{j} + (x + y)\mathbf{k}$, the outward normal derivative of g on the surface is given by

$$\frac{\partial g}{\partial n} = \nabla g \cdot \mathbf{n} = \frac{2x(y + z) + 2y(z + x) + (x + y)}{\sqrt{4(x^2 + y^2) + 1}}$$

In particular, at the point (-1/2, -1/2, 1/2) on the surface, we find

$$\partial g/\partial n = -1/\sqrt{3}$$

The negative sign indicates that g is decreasing along the outward normal vector.

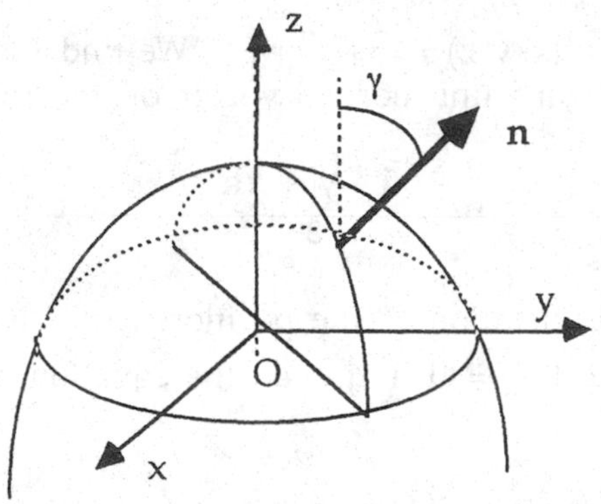

Fig. 3.6 Outward normal vector on a paraboloid.

3.3 EXERCISES

1. Find the gradient vector at the given point for the scalar fields defined by the following functions:

(a) $f(x, y) = x^2 \cos y + y$, $(1, 0)$
(b) $f(x, y) = x \sin y + y \cos x$, $(\pi/4, \pi/4)$
(c) $f(x, y) = \arctan(y/x)$, $(1, -1)$
(d) $f(x, y, z) = x^2 + yz + z^2$, $(1, -1, 1)$
(e) $f(x, y, z) = \ln(x^2 + y^2 - 2z^2)$, $(2, -2, 1)$
(f) $f(x, y, z) = xye^x + y \ln z$, $(-1, -1, 1)$

2. Find the maximum value of the directional derivative of each of the scalar fields at the given point:
(a) $f(x, y) = x^2 y + (x - y)^2$, $(1, 1)$
(b) $f(x, y) = \ln(x^2 + 2y^2)$, $(-1, 1)$
(c) $f(x, y, z) = x^2 + 2y^2 - 3z^2$, $(2, -1, 1)$
(d) $f(x, y, z) = z \arctan(y/x)$, $(-1, 1, 2)$
3. Find the points (x, y) on the unit circle $x^2 + y^2 = 1$ at which the tangential derivative of $f(x, y) = 2x^2 + y^2$ is (a) maximum; (b) minimum. What are these values?
4. Find the outward normal derivative of each of the following scalar fields at the given point on the given surface:
(a) $f(x, y, z) = x^2 + yz + z^2$ at $(-2, 2, 1)$ on the sphere $x^2 + y^2 + z^2 = 9$
(b) $f(x, y, z) = xye^x + y \ln z$ at $(-1, -1, 1)$ on the ellipsoid $2x^2 + 2y^2 + 3z^2 = 7$.
5. Find the directional derivative of $f(x, y, z) = x^2 + y^2 + z^2$ on the surface $x^2 + y^2 = 1 - z$ along the normal vector which makes an acute angle with the positive z-axis.
6. Find the outward normal derivative of $f(x, y, z) = xy + yz + zx$ on the faces of the cube $x = \pm 1, y = \pm 1, z = \pm 1$.
7. Find the cartesian equation of the tangent plane to the surface $x^2 + y^2 = z - 1$ at the point $(-1, 1, 3)$.
8. Find a tangent vector to the curve of intersection of the plane $y - z + 2 = 0$ and the cylinder $x^2 + y^2 = 4$ at the point $(0, 2, 4)$.
9. Find a tangent vector to the curve of intersection of the given surfaces at the indicated point, and determine the angle at which the corresponding tangent planes to the surfaces meet at that point:
(a) $x^2 + y^2 + z^2 = 9$ and $y^2 + z^2 = x + 3$ at $(2, -1, 2)$
(b) $x^2 + z^2 = 4$ and $y^2 + z^2 = 4$ at $(\sqrt{2}, -\sqrt{2}, \sqrt{2})$
10. Find the value of b so that at each point on the curve of intersection of the two spheres

$$(x - 1)^2 + y^2 + z^2 = 1, \qquad x^2 + (y - b)^2 + z^2 = 4$$

the two corresponding tangent planes will be perpendicular.

11. Find the vertex of the cone $x^2 + y^2 = (z - a)^2$ so that at each point of its intersection with the paraboloid $x^2 + y^2 = z$ the two corresponding tangent planes will be perpendicular.
12. Find an equation of the tangent line to the curve of intersection of the surfaces $x^2 y^2 + 2x + z^3 = 16$ and $3x^2 + y^2 - 2z = 9$ at the point (2, 1, 2).
13. Show that the sphere $x^2 + y^2 + z^2 - 8x - 8y - 6z + 24 = 0$ and the ellipsoid $x^2 + 3y^2 + 2z^2 = 9$ are orthogonal at the point (2, 1, 1).
14. Let $u = 1/2 \ln(x^2 + y^2)$ and $v = \arctan(y/x)$. Show that the angle between ∇u and ∇v is a constant.
15. Let $f(x, y, z) = x^2 + y^2 + z^2$, and let $\mathbf{u} = u_1\mathbf{i} + u_2\mathbf{j} + u_3\mathbf{k}$ and $\mathbf{v} = v_1\mathbf{i} + v_2\mathbf{j} + v_3\mathbf{k}$ be two constant unit vectors. If $\phi(u)$ denotes the directional derivative of f in the direction $\mathbf{u}$, show that the directional derivative of ϕ in the direction $\mathbf{v}$ is equal to $2\mathbf{u} \cdot \mathbf{v}$.
16. Let f be a differentiable function of u, where $u = \phi(x, y, z)$, ϕ

being a differentiable scalar field. Show that
17. Let f be a differentiable function of r, where $r = \sqrt{(x^2 + y^2 + z^2)}$.
(a) Show that $\nabla f(r) \times \mathbf{R} = 0$, where $\mathbf{R}$ denotes the position vector of (x, y, z).
(b) Find ∇r^n, where n is a positive integer.
(c) If ∇f is always a scalar multiple of $\mathbf{R}$, show that $f = f(r)$.
18. Let $f(x, y, z) = (\mathbf{R} \times \mathbf{A}) \cdot (\mathbf{R} \times \mathbf{B})$, where $\mathbf{A}$ and $\mathbf{B}$ are constant vectors. Show that
$$\nabla f = \mathbf{B} \times (\mathbf{R} \times \mathbf{A}) + \mathbf{A} \times (\mathbf{R} \times \mathbf{B})$$
19. Prove the rest of Theorem 1.

3.5 DIVERGENCE OF A VECTOR FIELD

There are two basic concepts that arise in connection with the spatial rate of change of a vector field $\mathbf{F}$, namely, the divergence of $\mathbf{F}$, denoted by div $\mathbf{F}$, and the curl of $\mathbf{F}$, denoted by **curl** $\mathbf{F}$. Although these concepts can be defined in terms of their physical interpretations, they can also be defined purely on analytical ground in terms of the del operator

$$\nabla = \frac{\partial}{\partial x}\mathbf{i} + \frac{\partial}{\partial y}\mathbf{j} + \frac{\partial}{\partial z}\mathbf{k}$$

analogous to the definition of the gradient of a scalar field. In this section we shall first introduce the analytic definition of the divergence of a vector field and then consider its physical interpretation. The definition of the curl of a vector field will be considered in the next section.

So let $\mathbf{F}(x, y, z) = P(x, y, z)\mathbf{i} + Q(x, y, z)\mathbf{j} + R(x, y, z)\mathbf{k}$ be a vector field defined in a domain D, where P, Q, and R have continuous first order derivatives. The divergence of **F**, denoted by div **F**, is the scalar field defined by

$$\text{div } \mathbf{F} = \frac{\partial P}{\partial x} + \frac{\partial Q}{\partial y} + \frac{\partial R}{\partial z} \tag{3.19}$$

If we treat the del operator ∇ formally as a vector and take the dot product of ∇ and **F**, we find

$$\nabla \cdot \mathbf{F} = (\frac{\partial}{\partial x}\mathbf{i} + \frac{\partial}{\partial y}\mathbf{j} + \frac{\partial}{\partial z}\mathbf{k}) \cdot (P\,\mathbf{i} + Q\,\mathbf{j} + Q\,\mathbf{k})$$

$$= \frac{\partial P}{\partial x} + \frac{\partial Q}{\partial y} + \frac{\partial R}{\partial z}$$

Thus,

$$\text{div } \mathbf{F} = \nabla \cdot \mathbf{F}$$

It should be noted, however, that since ∇ is actually not a vector, he dot product of ∇ and **F** is not commutative, that is, $\mathbf{F} \cdot \nabla \neq \nabla \cdot \mathbf{F}$. As a matter of fact, the symbol $\mathbf{F} \cdot \nabla$ represents the operator

$$\mathbf{F} \cdot \nabla = P\frac{\partial}{\partial x} + Q\frac{\partial}{\partial y} + R\frac{\partial}{\partial z}$$

which we shall discuss later.

From the definition and by the rules on differentiation, the following properties of the divergence can be easily verified:

$$\begin{aligned} &\text{(a)} \quad \text{div}(\mathbf{F} + \mathbf{G}) = \text{div } \mathbf{F} + \text{div } \mathbf{G} \\ &\text{(b)} \quad \text{div}(f\,\mathbf{F}) = f \text{ div } \mathbf{F} + \text{grad } f \cdot \mathbf{F} \end{aligned} \tag{3.20}$$

where f is any differentiable scalar field. In terms of the del operator ∇, these can be written as

$$\nabla\cdot(\mathbf{F}+\mathbf{G}) = \nabla\cdot\mathbf{F} + \nabla\cdot\mathbf{G}$$
$$\nabla\cdot(f\,\mathbf{F}) = f\nabla\cdot\mathbf{F} + \nabla f\cdot\mathbf{F}$$

If $\mathbf{F} = \nabla f = (\partial f/\partial x)\,\mathbf{i} + (\partial f/\partial y)\,\mathbf{j} + (\partial f/\partial z)\,\mathbf{k}$, for some scalar field f, then

$$\text{div}\,\mathbf{F} = \nabla\cdot\nabla f = \frac{\partial}{\partial x}\left(\frac{\partial f}{\partial x}\right) + \frac{\partial}{\partial y}\left(\frac{\partial f}{\partial y}\right) + \frac{\partial}{\partial z}\left(\frac{\partial f}{\partial z}\right)$$

$$= \frac{\partial^2 f}{\partial x^2} + \frac{\partial^2 f}{\partial y^2} + \frac{\partial^2 f}{\partial z^2}$$

This expression is known as the Laplacian of the function f, denoted by $\nabla^2 f$. Thus, $\nabla^2 f = \text{div}(\text{grad } f)$. The Laplace equation $\nabla^2 f = 0$ is one of the important partial differential equations of mathematical physics.

Example 1. Determine the divergence of (a) $\mathbf{F} = x^2 y\,\mathbf{i} + e^y z\,\mathbf{j} + x \sin z\,\mathbf{k}$ and (b) $\mathbf{F} = \text{grad } f$, where $f = y^2 z + z^2 x$.

Solution: (a) We have

$$\text{div}\,\mathbf{F} = \frac{\partial}{\partial x}(x^2 y) + \frac{\partial}{\partial y}(ze^y) + \frac{\partial}{\partial z}(x \sin z)$$

$$= 2xy + z\,e^y + x\cos z$$

(b) Since

$$\mathbf{F} = \nabla f = \frac{\partial f}{\partial x}\mathbf{i} + \frac{\partial f}{\partial y}\mathbf{j} + \frac{\partial f}{\partial z}\mathbf{k} = z^2\mathbf{i} + 2yz\,\mathbf{j} + (2xz + y^2)\,\mathbf{k}$$

we have

$$\text{div}\,\mathbf{F} = \frac{\partial}{\partial x}(z^2) + \frac{\partial}{\partial y}(2yz) + \frac{\partial}{\partial z}(2xz + y^2) = 2x + 2z$$

Example 2. Let $f(x, y, z) = x^2 y z$ and $\mathbf{F}(x, y, z) = xz\,\mathbf{i} + y^2\mathbf{j} + x^2 z\,\mathbf{k}$. Verify property (b) of (3.20).

Solution: We note that $\nabla f = 2xyz\,\mathbf{i} + x^2 z\,\mathbf{j} + x^2 y\,\mathbf{k}$ and $\text{div}\,\mathbf{F} = z$

$+ 2y + x^2$. Hence

$$f \operatorname{div} \mathbf{F} + \nabla f \cdot \mathbf{F} = (x^2yz)(z + 2y + x^2) + (2xyz)(xz) + (x^2 z)y^2 + x^2y(x^2z)$$

$$= 3x^2yz^2 + 3x^2y^2z + 2x^4yz$$

On the other hand, we find

$$\operatorname{div} (f\, \mathbf{F}) = 3x^2yz^2 + 3x^2y^2z + 2x^4yz$$

thus verifying property (b) of (3.20).

Physical Application of the Divergence

The concepts of the divergence and the curl of vector fields are of fundamental importance in the study of fluid flow. Here we consider the significance of the divergence in a fluid flowing in a domain of our space. Let $\mathbf{V}(x, y, z, t)$ denote the velocity of the fluid at the point (x, y, z) at time t, and let $\rho(x, y, z, t)$ denote the corresponding density (mass per unit volume). Then the vector field $\mathbf{F}(x, y, z, t) = \rho(x, y, z, t)\mathbf{V}(x, y, z, t)$ represents the rate of flow (mass per unit area per unit time) of the fluid. This is called the flux density of the fluid. If the fluid has no sources, that is, points at which fluid is being introduced, or if it has no sinks, that is, points at which fluid is being carried away, then it can be shown (Sec. 4.10) that the divergence of $\mathbf{F}$ satisfies the equation

$$\operatorname{div} \mathbf{F} = \operatorname{div} (\rho\mathbf{V}) = -\partial\rho/\partial t \tag{3.21}$$

This equation is known as the *continuity equation* .

Equation (3.21) says that the divergence of the flux density is related to the rate of change of the density of the fluid. Hence, if $\operatorname{div} \mathbf{F} > 0$ so that $\partial\rho/\partial t < 0$, then the density ρ is diminishing, which means that the fluid is expanding. On the other hand, if $\operatorname{div} \mathbf{F} < 0$ so that $\partial\rho/\partial t > 0$, then the density is increasing, which means that the fluid is being compressed. If the flow is steady, that is, the density does not change with time, then (3.21) becomes

$$\text{div}(\rho\mathbf{V}) = 0 \tag{3.22}$$

If the fluid is incompressible, that is, ρ = const., of which water is a fair example, then (3.21) simplifies to

$$\text{div}\,\mathbf{V} = 0 \tag{3.23}$$

Therefore, for an incompressible fluid flow without source or sink the divergence of its velocity field vanishes. Equation (3.23) is known as the *condition of incompressibility* .

In the case of electrostatic fields, the sources and sinks are electric charges. If a single electric charge of magnitude q (in appropriate units; for example, coulombs) is located at the origin, then the electrostatic field at a point (x, y, z) is given by (assume vacuum medium)

$$\mathbf{E} = \frac{q}{r^3}\mathbf{R}$$

where $\mathbf{R}$ is the position vector of the point (x, y, z) and $r = |\mathbf{R}|$. This can be written as $\mathbf{E} = (q/r^2)\mathbf{u}_r$, where $\mathbf{u}_r = \mathbf{R}/r$ is a unit vector. By direct differentiation, we find

$$\frac{\partial}{\partial x}\left[\frac{x}{r^3}\right] = \left[\frac{1}{r^3} - \frac{3x^2}{r^5}\right], \qquad \frac{\partial}{\partial y}\left[\frac{y}{r^3}\right] = \left[\frac{1}{r^3} - \frac{3y^2}{r^5}\right]$$

$$\frac{\partial}{\partial z}\left[\frac{z}{r^3}\right] = \left[\frac{1}{r^3} - \frac{3z^2}{r^5}\right]$$

so that

$$\text{div}\,\mathbf{E} = \frac{\partial}{\partial x}\left[\frac{x}{r^3}\right] + \frac{\partial}{\partial y}\left[\frac{y}{r^3}\right] + \frac{\partial}{\partial z}\left[\frac{z}{r^3}\right]$$

$$= \frac{3}{r^3} - \frac{3(x^2 + y^2 + z^2)}{r^5} = 0$$

for all (x, y, z) $\neq$ (0, 0, 0). Thus the divergence of an electrostatic field vanishes at each point where no souce or sink is present.

If there are N electric charges of magnitude q_i located at the points (x_i, y_i, z_i), i = 1, . . , N, then the electric field at a point (x, y, z) is given by the sum

$$\mathbf{E} = \sum_{i=1}^{N} \frac{q_i}{|\mathbf{R} - \mathbf{R}_i|^3} (\mathbf{R} - \mathbf{R}_i)$$

where $\mathbf{R}_i$ denotes the position vector of the point (x_i, y_i, z_i), i = 1, . . , N. Again, at each point (x, y, z) $\neq$ (x_i, y_i, z_i), i = 1, . . , N, where there is no electric charge present, it is easily verified that div **E** = 0.

A vector field **F** which satisfies div **F** = 0 is called a *solenoidal* field. A vector field **F** for which there exists a scalar field f such that **F** = grad f is called a conservative field. In such a case, the scalar field f is called a potential function of **F**. Note that if f is a potential function of a conservative field **F**, so is f + C, where C is any constant. This is clear since grad (f + C) = grad f = **F**. Thus, the potential function of a conservative field is unique only up to an additive constant. If a conservative field is also solenoidal, then its potential function satisfies the Laplace equation, that is,

$$\text{div}\,\mathbf{F} = \text{div}(\text{grad}\,f) = \nabla^2 f = 0$$

Later we shall present a method for determining whether or not a given vector field is conservative. In the next example we illustrate a method for finding a potential function for a vector field that is conservative.

Example 3. Determine a potential function for the conservative field

$$\mathbf{F} = (yz + e^x \cos y)\,\mathbf{i} + (xz - e^x \sin y)\,\mathbf{j} + (xy + 2z)\,\mathbf{k}$$

Solution: Let f be a potential function of **F**. Since $\nabla f = \mathbf{F}$, we equate the components of ∇f and **F** to obtain the three equations

$$f_x = yz + e^x \cos y, \qquad f_y = xz - e^x \sin y, \qquad f_z = xy + 2z$$

To detemine f, we first integrate the first equation partially with respect to x (treating all other variables as constants). We find

$$f = xyz + e^x \cos y + h(y, z)$$

where the constant of integration h depends on y and z. Next, we differentiate f with respect to y and equate the result with the second equation above. We find

$$f_y = xz - e^x \sin y + h_y = xz - e^x \sin y$$

This implies $h_y = 0$, so that $h = h(z)$. Finally, differentiating f with respect to z and equating it with the third equation, we obtain

$$xy + h'(z) = xy + 2z$$

Hence $h'(z) = 2z$ or $h(z) = z^2 + C$. Therefore, the potential function is given by

$$f = xyz + e^x \cos y + z^2 + C$$

Example 4. Find a potential function of the electrostatic field

$$\mathbf{E} = \frac{q}{r^3}\mathbf{R}$$

which is known to be a conservative field.

Solution: As in the preceding example, we determine the potential function f from the three equations

$$f_x = \frac{qx}{\left[x^2 + y^2 + z^2\right]^{3/2}}, \quad f_y = \frac{qy}{\left[x^2 + y^2 + z^2\right]^{3/2}}, \quad f_z = \frac{qz}{\left[x^2 + y^2 + z^2\right]^{3/2}}$$

Integrating the first equation with resspect to x , we find

$$f = \frac{q}{\sqrt{x^2 + y^2 + z^2}} + h(y, z)$$

If we differentiate this with respect to y and equate the result with the second equation above, we find that $h_y(y,z) = 0$. Hence h depends only on z, that is, h = h(z). Finally, differentiating f with respect to z and using the third equation, we find h'(z) = 0. Thereore, h is a constant and so the potential function is given by

$$f = \frac{q}{\sqrt{x^2 + y^2 + z^2}} + C = \frac{q}{r} + C$$

On physical ground it is customary to take C = 0 so that f = q/r.

3.4 EXERCISES

Find the divergence of each of the following vector fields.

1. $\mathbf{F}(x, y) = e^x(\cos y\mathbf{i} + \sin y\mathbf{j})$
2. $\mathbf{F}(x, y) = -x/(x^2 + y^2)\mathbf{i} + y/(x^2 + y^2)\mathbf{j}$
3. $\mathbf{F}(x, y, z) = (x^2 + y^2)\mathbf{i} + ze^{xy}\mathbf{j}$
4. $\mathbf{F}(x, y, z) = (1 + y^2)\mathbf{i} + xe^z\mathbf{j}$
5 $\mathbf{F}(x, y, z) = xe^z\mathbf{i} + ye^x\mathbf{j} + ze^y\mathbf{k}$
6. $\mathbf{F}(x, y, z) = x \sin y\mathbf{i} + y \sin (xz)\mathbf{j} + \cos(e^z)\mathbf{k}$
7. $\mathbf{F}(x, y, z) = (x\mathbf{i} + y\mathbf{j} + z\mathbf{k})/(x^2 + y^2 + z^2)^{1/2}$
8. Prove the properties given in (3.20).
9. Let u and v be twice continuously differentiable functions. Show that $\nabla^2(uv) = u\nabla^2 v + 2\nabla u \cdot \nabla v + v\nabla^2 u$.
10. If $\mathbf{F} = r^n \mathbf{R}$, where $\mathbf{R} = x\mathbf{i} + y\mathbf{j} + z\mathbf{k}$ and $r = |\mathbf{R}|$, find div **F** and show that grad (div **F**) = $n(n + 3)r^{n-2}\mathbf{R}$.
11. Find the gradient of the divergence of $\mathbf{F} = xy\mathbf{i} + yz\mathbf{j} + zx\mathbf{k}$.
12. Let $\mathbf{R}_1$ and $\mathbf{R}_2$ denote the position vectors of the fixed points (x_1, y_1, z_1) and (x_2, y_2, z_2), respectively, and let **R** be the position vector of a variable point (x, y, z). If $\mathbf{F} = \mathbf{R} - \mathbf{R}_1$ and $\mathbf{G} = \mathbf{R} - \mathbf{R}_2$, show that $\nabla(\mathbf{F}\cdot\mathbf{G}) = \mathbf{F} + \mathbf{G}$ and $\nabla\cdot(\mathbf{F}\times\mathbf{G}) = 0$.
13. Let f be a differentiable scalar field, and let **F** be the vector field

$$\mathbf{F} = \left(y\frac{\partial f}{\partial z} - z\frac{\partial f}{\partial y}\right)\mathbf{i} + \left(z\frac{\partial f}{\partial x} - x\frac{\partial f}{\partial z}\right)\mathbf{j} + \left(x\frac{\partial f}{\partial y} - y\frac{\partial f}{\partial x}\right)\mathbf{k}$$

Show that $\mathbf{F} = \mathbf{R}\times\nabla f$, $\mathbf{F}\cdot\mathbf{R} = 0$ and $\mathbf{F}\cdot\nabla f = 0$.

For each of the following conservative vector fields, determine a potential function ϕ.

14. $\mathbf{F} = (x\mathbf{i} + y\mathbf{j})/(x^2 + y^2)$
15. $\mathbf{F} = (-y\mathbf{i} + x\mathbf{j})/(x^2 + y^2)$
16. $\mathbf{F} = (2xy + z^2)\mathbf{i} + (2yz + x^2)\mathbf{j} + (2xz + y^2)\mathbf{k}$
17. $\mathbf{F} = e^x \cos z\, \mathbf{i} + 2y\, \mathbf{j} - e^x \sin z\, \mathbf{k}$
18. $\mathbf{F} = (e^y + z^2 \cos x)\mathbf{i} + (xe^y + 2yz \ln z)\mathbf{j} + (y^2 + y^2 \ln z + 2z \sin x)\mathbf{k}$

3.6 CURL OF A VECTOR FIELD

Let $\mathbf{F}(x, y, z) = P(x, y, z)\mathbf{i} + Q(x, y, z)\mathbf{j} + R(x, y, z)\mathbf{k}$ be a vector field defined in a domain D, where P, Q, and R have continuous first order derivatives. The curl of **F**, written as **curl F**, is the vector field defined by the equation

$$\textbf{curl F} = \left(\frac{\partial R}{\partial y} - \frac{\partial Q}{\partial z}\right)\mathbf{i} + \left(\frac{\partial P}{\partial z} - \frac{\partial R}{\partial x}\right)\mathbf{j} + \left(\frac{\partial Q}{\partial x} - \frac{\partial P}{\partial y}\right)\mathbf{k} \tag{3.24}$$

This equation can be easily remembered as the expansion of the determinant

$$\textbf{curl F} = \begin{vmatrix} \mathbf{i} & \mathbf{j} & \mathbf{k} \\ \frac{\partial}{\partial x} & \frac{\partial}{\partial y} & \frac{\partial}{\partial z} \\ P & Q & R \end{vmatrix} \tag{3.25}$$

by the elements of the first row, where each product such as $\partial/\partial y$ and R is to be interpreted as the partial derivative $\partial R/\partial y$. Now since (3.25) resembles the formula for the cross product of two vectors with $\partial/\partial x$, $\partial/\partial y$, $\partial/\partial z$ being the "components" of the "vector" ∇, we can write

$$\textbf{curl F} = \nabla \times \mathbf{F} \tag{3.26}$$

Thus we see that the gradient of a scalar field f, the divergence and the curl of a vector field **F** can be interpreted formally as the result of taking the scalar multiple ∇f, the dot product $\nabla \cdot \mathbf{F}$ and the cross product $\nabla \times \mathbf{F}$, respectively, of the del operator ∇ and the fields f and **F**.

It follows from the definition and the rules on differentiation that the curl satisfies the following properties:

$$\begin{aligned} &\text{(a)}\ \ \mathbf{curl}\,(\mathbf{F}+\mathbf{G}) = \mathbf{curl}\,\mathbf{F} + \mathbf{curl}\,\mathbf{G} \\ &\text{(b)}\ \ \mathbf{curl}\,(f\,\mathbf{F}) = f\,\mathbf{curl}\,\mathbf{F} + \text{grad}\, f \times \mathbf{F} \end{aligned} \tag{3.27}$$

where f is a differentiable scalar field. The student should compare these properties with those listed in (3.20) for the divergence. In terms of the del operator ∇, these properties can be written as

$$\nabla \times (\mathbf{F}+\mathbf{G}) = \nabla \times \mathbf{F} + \nabla \times \mathbf{G}$$

$$\nabla \times (f\,\mathbf{F}) = f\,\nabla \times \mathbf{F} + \nabla f \times \mathbf{F}$$

Example 1. Find the curl of the vector field

$$\mathbf{F} = xy^2\mathbf{i} + x\sin(yz)\mathbf{j} + z^2e^y\mathbf{k}$$

Solution: We have

$$\mathbf{curl}\,\mathbf{F} = \begin{vmatrix} \mathbf{i} & \mathbf{j} & \mathbf{k} \\ \dfrac{\partial}{\partial x} & \dfrac{\partial}{\partial y} & \dfrac{\partial}{\partial z} \\ xy^2 & x\sin(yz) & z^2e^y \end{vmatrix}$$

$$= [z^2e^y - xy\cos(yz)]\mathbf{i} + (0)\mathbf{j} + [\sin(yz) - 2xy]\mathbf{k}$$

$$= [z^2e^y - xy\cos(yz)]\mathbf{i} + [\sin(yz) - 2xy]\mathbf{k}$$

Example 2. Show that the curl of $\mathbf{F} = (-y\mathbf{i} + x\mathbf{j})/(x^2+y^2)$ vanishes for all $(x, y) \neq (0, 0)$.

Solution: Since **F** is independent of the variable z and its third component is zero (R= 0), we readily find

$$\mathbf{curl}\,\mathbf{F} = \left[\frac{\partial}{\partial x}\left(\frac{x}{x^2+y^2}\right) - \frac{\partial}{\partial y}\left(\frac{-y}{x^2+y^2}\right)\right]\mathbf{k}$$

$$= \left[\frac{(x^2+y^2)-2x^2}{(x^2+y^2)^2} + \frac{(x^2+y^2)-2y^2}{(x^2+y^2)^2} \right] \mathbf{k}$$

$$= 0$$

Alternatively, we can perform the above calculation by using (3.27b) with $\mathbf{G} = -y\,\mathbf{i} + x\,\mathbf{j}$ and $f = 1/(x^2+y^2)$. In this case we see that $\mathbf{curl}\ \mathbf{G} = 2\mathbf{k}$, grad $f = -(2x\,\mathbf{i} + 2y\,\mathbf{j})/(x^2+y^2)^2$ and

$$\text{grad } f \times \mathbf{G} = [-2(x^2+y^2)/(x^2+y^2)^2]\mathbf{k} = [-2/(x^2+y^2)]\mathbf{k}$$

Thus

$$\begin{aligned} \mathbf{curl}\,(f\,\mathbf{G}) &= f\,\mathbf{curl}\,\mathbf{G} + \text{grad } f \times \mathbf{G} \\ &= [2/(x^2+y^2)]\mathbf{k} - [2/(x^2+y^2)]\mathbf{k} = 0 \end{aligned}$$

Example 3. Let $\mathbf{A}$ be a constant vector and $\mathbf{R}$ the position vector of the point (x, y, z). Show that $\mathbf{curl}\,(\mathbf{A} \times \mathbf{R}) = 2\mathbf{A}$.

Solution: Let a_1, a_2, a_3 denote the components of the vector $\mathbf{A}$. Since

$$\mathbf{A} \times \mathbf{R} = (a_2 z - a_3 y)\mathbf{i} + (a_3 x - a_1 z)\mathbf{j} + (a_1 y - a_2 x)\mathbf{k}$$

we find

$$\mathbf{curl}\,(\mathbf{A} \times \mathbf{R}) = \begin{vmatrix} \mathbf{i} & \mathbf{j} & \mathbf{k} \\ \dfrac{\partial}{\partial x} & \dfrac{\partial}{\partial y} & \dfrac{\partial}{\partial z} \\ a_2 z - a_3 y & a_3 x - a_1 z & a_1 y - a_2 x \end{vmatrix}$$

$$= 2a_1\mathbf{i} + 2a_2\mathbf{j} + 2a_3\mathbf{k} = 2\mathbf{A}$$

Physical Application of the Curl

Roughly speaking the curl can be interpreted in fluid motion as a measure of the tendency of the velocity field to cause a rotation at a point. For example, let us consider the motion of a fluid that is in pure rotation about a fixed axis with an angular speed ω. As we saw in Sec.1.7, the rotation can be described by a vector **w** of magnitude ω pointing in the direction of the axis of rotation determined by the right-hand rule. Assuming that the rotation is about the z-axis, we then have **w** = ω**k**. Hence the velocity field **V** is given by

$$\mathbf{V} = \omega\mathbf{k} \times \mathbf{R} = \omega(-y\,\mathbf{i} + x\,\mathbf{j})$$

Now let us take the curl of the velocity field **V**. We find

$$\text{curl}\,\mathbf{V} = \begin{vmatrix} \mathbf{i} & \mathbf{j} & \mathbf{k} \\ \frac{\partial}{\partial x} & \frac{\partial}{\partial y} & \frac{\partial}{\partial z} \\ -\omega y & \omega x & 0 \end{vmatrix}$$

$$= 2\omega\mathbf{k} = 2\mathbf{w}$$

Thus the curl of the velocity field of a fluid in pure rotation about the z-axiz is twice the angular velocity **w** . Moreover, we have div **V** = 0, so that the velocity field is solenoidal.

The graphical view of the velocity field **V** as shown in Fig. 3.7 suggests that we may describe the velocity field geometrically as "curling".

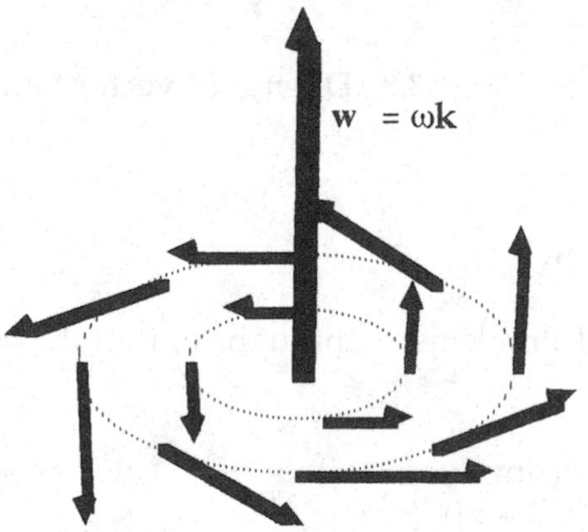

Fig. 3.7 Graphic view of the velocity field $\mathbf{V} = \omega(-y\,\mathbf{i} + x\,\mathbf{j})$.

In contrast suppose the velocity field of a fluid flow is described by $\mathbf{V} = (x\mathbf{i} + y\mathbf{j} + z\mathbf{k})/r = \mathbf{R}/r$, where $r = \sqrt{(x^2 + y^2 + z^2)}$. Then, since $\mathbf{curl\ R} = \mathbf{0}$, grad $(1/r) = -\mathbf{R}/r^3$ and $\mathbf{R} \times \mathbf{R} = \mathbf{0}$, we find

$$\mathbf{curl\ V} = (1/r)\,\mathbf{curl\ R} + \text{grad}\,(1/r) \times \mathbf{R} = \mathbf{0}$$

Thus the vector field does not "curl". However, if we take its divergence, we find

$$\text{div}\,\mathbf{V} = (1/r)\,\text{div}\,\mathbf{R} + \text{grad}\,(1/r)\cdot\mathbf{R} = 3/r - 1/r = 2/r,\ r > 0$$

The graphical view of the velocity field **V** is shown in Fig. 3.8, which suggests that we may describe the field geometrically as "diverging".

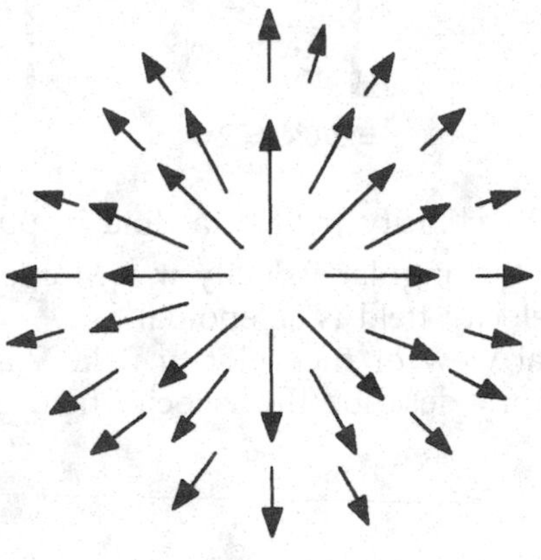

Fig. 3.8 Divergent vector field.

3.5 EXERCISES

In each of Problems 1 through 6, find the curl of the given vector field.

1. $\mathbf{F}(x, y) = e^x(\sin y\mathbf{i} + \cos y\mathbf{j})$
2. $\mathbf{F}(x, y) = (-y\mathbf{i} + x\mathbf{j})/(x^2 + y^2)^{1/2}$
3. $\mathbf{F}(x, y, z) = x^2yz^3(\mathbf{i} + \mathbf{j} + \mathbf{k})$
4. $\mathbf{F}(x, y, z) = (yz^2 - 2x)\mathbf{i} + (xz^2 + 2y)\mathbf{j} + (2xy + x^2)\mathbf{k}$

5 $F(x, y, z) = xyz^2i + 2xy^3j - x^2yzk$
6. $F(x, y, z) = (4xy - z^3)i + 2x^2j - 3xz^2k$
7. Prove the formulas in (3.27).

In each of Problems 8 through 11, let $\mathbf{F}(x, y, z) = y\mathbf{i} + z\mathbf{j} + x\mathbf{k}$ and $\mathbf{G}(x, y, z) = x^2z\mathbf{i} + y^2x\mathbf{j} + z^2y\mathbf{k}$.

8. Calculate div(**F** x **G**).
9. Calculate **curl** (**F** x **G**).
10. Calculate **F** x **curl G**.
11. Calculate **curl F** x **curl G**.
12. Let f be a differentiable function of r, where $r = (x^2 + y^2 + z^2)^{1/2}$. Show that **curl** (f(r)**R**) = **0**, where **R** is the position vector of (x, y, z).
13. Suppose that u and v are differentiable scalar fields satisfying a functional relation f(u,v) = 0. Show that $\nabla u \times \nabla v = 0$, assuming that f is differentiable.
14. Let $\mathbf{F}(x, y, z) = x^2yz\,\mathbf{i} - x^3y^3\mathbf{j} + xyz^3\mathbf{k}$. Calculate div (**curl F**).
15. Let $\mathbf{F}(x, y, z) = 2xyz\,\mathbf{i} + x^2z\,\mathbf{j} + x^2y\,\mathbf{k}$. Show that **curl F** = **0**, and thus find all scalar fields f so that grad f = **F**.
16. Show that **curl** [(**A** · **R**)**R**] = **A** x **R** for any constant vector **A**.
17. Show that **curl** [(**A** · **R**)**B**] = **A** x **B** for any constant vectors **A** and **B**.

3.7 OTHER PROPERTIES OF THE DIVERGENCE AND THE CURL

Let f be a scalar field with continuous second-order derivatives. We have already seen that

$$\text{div}(\text{grad } f) = \nabla^2 f \tag{3.28}$$

so that if f is a harmonic function, then $\nabla^2 f = 0$. On the other, if we take the curl of the gradient of f, we find

$$\mathbf{curl}\,(\mathrm{grad}\, f) = \nabla \times \nabla f = \begin{vmatrix} \mathbf{i} & \mathbf{j} & \mathbf{k} \\ \frac{\partial}{\partial x} & \frac{\partial}{\partial y} & \frac{\partial}{\partial z} \\ \frac{\partial f}{\partial x} & \frac{\partial f}{\partial y} & \frac{\partial f}{\partial z} \end{vmatrix}$$

$$= \left[\frac{\partial^2 f}{\partial y\,\partial z} - \frac{\partial^2 f}{\partial z\,\partial y}\right]\mathbf{i} + \left[\frac{\partial^2 f}{\partial z\,\partial x} - \frac{\partial^2 f}{\partial x\,\partial z}\right]\mathbf{j} + \left[\frac{\partial^2 f}{\partial x\,\partial y} - \frac{\partial^2 f}{\partial y\,\partial x}\right]\mathbf{k}$$

Since the second-order derivatives of f are continuous, the mixed derivatives are independent of the order of differentiation, that is, $f_{zy} = f_{yz}$, etc. Therefore, we have

$$\mathbf{curl}\,(\mathrm{grad}\, f) = \nabla \times \nabla f = \mathbf{0} \tag{3.29}$$

for any twice continuously differentiable scalar field f. Thus the gradient of a scalar field is irrotational.

Equation (3.28) implies that if a vector **F** is conservative, so that **F** = ∇f, for some scalar field f, then **curl F** = **0**. Hence a necessary condition for a vector field to be conservative is that its curl must vanish. For a special type of domain, it turns out that the condition **curl F** = **0** is also a sufficient condition for **F** to be a conservative field, that is, there exists a scalar field f such that grad f = **F**. A more detailed discussion of this topic is given in the next chapter.

A counterpart of equation (3.28) for the divergence is that if **F** is twice continuously differentiable, then

$$\mathrm{div}\,(\mathbf{curl\,F}) = \nabla \cdot (\nabla \times \mathbf{F})$$

$$= \frac{\partial}{\partial x}\left(\frac{\partial R}{\partial y} - \frac{\partial Q}{\partial z}\right) + \frac{\partial}{\partial y}\left(\frac{\partial P}{\partial z} - \frac{\partial R}{\partial x}\right) + \frac{\partial}{\partial z}\left(\frac{\partial Q}{\partial x} - \frac{\partial P}{\partial y}\right)$$

$$= 0 \tag{3.30}$$

Thus the curl of a vector field is solenoidal. Under certain condition on the domain where **F** is defined (for example, a spherical domain), it turns out that a solenoidal field is the curl of some vector field (see Problem 1). That is, suppose div $\mathbf{F} = 0$, say in a spherical domain D, then there exists a vector field $\mathbf{V}$ in D such that $\mathbf{F} =$ **curl V**. Moreover, it can be shown that $\mathbf{V}$ is determined up to a term ∇f, where f is an arbitrary differentiable function (see Problem 3). Therefore, for certain types of domains a solenoidal field is the curl of some vector field just as a conservative field is the gradient of some scalar field.

In (3.28) and (3.29) we calculated the divergence and the curl of the vector field grad f. In (3.30) we have the divergence of the vector field **curl F**. To complete the analysis, we now compute the curl of $\nabla \times \mathbf{F}$, that is,

$$\textbf{curl}\,(\textbf{curl}\,\mathbf{F}) = \nabla \times (\nabla \times \mathbf{F})$$

Although this can be evaluated strictly from the definition of a curl, we introduce here a method which makes use of the algebraic properties of the vector operations of dot product, cross product and vector triple product. The method consists in treating the del operator ∇ as though it were a vector and applying known vector identities. In doing this, however, we have to make sure that the operator ∇ applies only to the factor it is supposed to operate on and that it produces the appropriate kind of field (a scalar or a vector field).

We illustrate the procedure by considering the expression $\nabla \times \nabla \times \mathbf{F})$ bearing in mind that this results in a vector field. By applying the vector identity

$$\begin{aligned}\mathbf{A} \times (\mathbf{B} \times \mathbf{C}) &= (\mathbf{A} \cdot \mathbf{C})\mathbf{B} - (\mathbf{A} \cdot \mathbf{B})\mathbf{C} \\ &= \mathbf{B}\,(\mathbf{A} \cdot \mathbf{C}) - (\mathbf{A} \cdot \mathbf{B})\mathbf{C}\end{aligned}$$

identifying $\mathbf{A}$ and $\mathbf{B}$ as ∇, we obtain

$$\nabla \times \nabla \times \mathbf{F}) = \nabla(\nabla \cdot \mathbf{F}) - \nabla \cdot \nabla)\mathbf{F} = \text{grad}\,(\text{div}\,\mathbf{F}) - \nabla^2\mathbf{F} \qquad (3.31)$$

Here $\nabla^2\mathbf{F}$ denotes the Laplacian of $\mathbf{F}$ which is given by

$$\nabla^2\mathbf{F} = \nabla^2(P\mathbf{i} + Q\mathbf{j} + R\mathbf{k}) = (\nabla^2 P)\mathbf{i} + (\nabla^2 Q)\mathbf{j} + (\nabla^2 R)\mathbf{k}$$

Notice that $\nabla(\nabla \cdot \mathbf{F}) \neq (\nabla \cdot \mathbf{F})\nabla$ since ∇ is not a vector. The student should verify this result from the definition of the curl of a vector field (see Problem 4).

We now list a few other important properties of the gradient, divergence, and curl, expressing them in terms of the differential operator ∇:

$$\begin{aligned}
&\text{(a)} \quad \nabla(\mathbf{F}\cdot\mathbf{G}) = (\mathbf{F}\cdot\nabla)\mathbf{G} + (\mathbf{G}\cdot\nabla)\mathbf{F} + \mathbf{F}\times(\nabla\times\mathbf{G}) + \mathbf{G}\times(\nabla\times\mathbf{F}) \\
&\text{(b)} \quad \nabla\cdot(\mathbf{F}\times\mathbf{G}) = \mathbf{G}\cdot\nabla\times\mathbf{F} - \mathbf{F}\cdot\nabla\times\mathbf{G} \qquad (3.32) \\
&\text{(c)} \quad \nabla\times(\mathbf{F}\times\mathbf{G}) = (\mathbf{G}\cdot\nabla)\mathbf{F} - (\mathbf{F}\cdot\nabla)\mathbf{G} + (\nabla\cdot\mathbf{G})\mathbf{F} - (\nabla\cdot\mathbf{F})\mathbf{G}
\end{aligned}$$

where **F** and **G** are differentiable vector fields.

These identities can be proved by the same method we used in establishing the identity (3.31). For example, to prove (3.32b) we write

$$\nabla\cdot(\mathbf{F}\times\mathbf{G}) = \nabla_F\cdot(\mathbf{F}\times\mathbf{G}) + \nabla_G\cdot(\mathbf{F}\times\mathbf{G}) \qquad (3.33)$$

where the symbol ∇_F means that the operator ∇ applies only to the factor **F**, and **G** is to be treated as a constant. A similar interpretation holds for the symbol ∇_G. This is in accordance with the rule for differentiating product of functions. Now, to calculate the terms on the right-hand side of (3.33), we make judicious use of the vector identity

$$\mathbf{A}\cdot\mathbf{B}\times\mathbf{C} = \mathbf{B}\cdot\mathbf{C}\times\mathbf{A} = \mathbf{C}\cdot\mathbf{A}\times\mathbf{B} \qquad (3.34)$$

Since ∇_F is to operate only on the factor **F**, we adopt the last member of (3.34) to obtain

$$\nabla_F\cdot(\mathbf{F}\times\mathbf{G}) = \mathbf{G}\cdot\nabla\times\mathbf{F}$$

Similarly, we have

$$\nabla_G\cdot(\mathbf{F}\times\mathbf{G}) = -\nabla_G\cdot(\mathbf{G}\times\mathbf{F}) = -\mathbf{F}\cdot\nabla\times\mathbf{G}$$

Substituting these on the right-hand side of (3.33), we obtain the identity (3.32b). The student should try to prove the rest of (3.32) by following the same procedure.

3.6 EXERCISES

1. Let $\mathbf{V} = x\mathbf{i} + 2y\mathbf{j} - 3z\mathbf{k}$. Verify that div $\mathbf{V} = 0$ for all (x, y, z) and hence find a vector field **F** so that **curl F** = **V**. [Hint: Assume $\mathbf{F}(x, y, z) = P(x, y, z)\mathbf{i} + Q(x, y, z)\mathbf{j}$ and then solve the

system $Q_z = -x$, $P_z = 2y$, $Q_x - P_y = -3z$.]

2. Repeat Problem 1 when $\mathbf{V} = yz\mathbf{i} + zx\mathbf{j} + xy\mathbf{k}$.
3. Let $\mathbf{V}$ be a given vector field for which div $\mathbf{V} = 0$ in a spherical domain D. If $\mathbf{F}$ is a vector field such that $\mathbf{curl\ F} = \mathbf{V}$, show that $\mathbf{F}$ is determined up to a term ∇f, where f is a differentiable scalar field; that is, if $\mathbf{curl\ F_1} = \mathbf{curl\ F_2} = \mathbf{V}$, then $\mathbf{F_1} = \mathbf{F_2} + \nabla f$.
4. Verify the formula (3.31) from the definition of the curl.
5. Using the vector identity for $\mathbf{A} \times (\mathbf{B} \times \mathbf{C})$, show that

   (a) $\nabla_F(\mathbf{F} \cdot \mathbf{G}) = \mathbf{G} \times \nabla \times \mathbf{F}) + (\mathbf{G} \cdot \nabla)\mathbf{F}$

   (b) $\nabla_F \times (\mathbf{F} \times \mathbf{G}) = (\mathbf{G} \cdot \nabla)\mathbf{F} - (\nabla \cdot \mathbf{F})\mathbf{G}$
6. Prove part (a) of (3.32). [Hint : Use part (a) of Problem 5.]
7. Prove part (c) of (3.32). [Hint : Use part (b) of Problem 5.]
8. Show that $\nabla \cdot (\nabla f \times \nabla g) = \mathbf{0}$.
9. Show that $(\mathbf{F} \cdot \nabla)\mathbf{F} = (1/2)\nabla(\mathbf{F} \cdot \mathbf{F}) - \mathbf{F} \times (\nabla \times \mathbf{F})$
10. Show that div(f grad g - g grad f) $= f\nabla^2 g - g\nabla^2 f$
11. If $f(x, y, z) = (\mathbf{R} \times \mathbf{A}) \cdot (\mathbf{R} \times \mathbf{B})$, where $\mathbf{R}$ is the position vector of (x, y, z) and $\mathbf{A}, \mathbf{B}$ are constant vectors, show that

$$\text{grad } f = \mathbf{B} \times (\mathbf{R} \times \mathbf{A}) + \mathbf{A} \times (\mathbf{R} \times \mathbf{B})$$

12. Compute $\nabla(\mathbf{F} \cdot \mathbf{G} \times \mathbf{H})$.
13. Compute $\nabla \cdot [\mathbf{F} \times (\mathbf{G} \times \mathbf{H})]$.
14. Compute $\nabla \times [\mathbf{F} \times (\mathbf{G} \times \mathbf{H})]$.

3.8 CURVILINEAR COORDINATE SYSTEMS

So far we have studied the various field properties - gradient, divergence, and curl - and their analytic representations in a rectangular cartesian coordinate system. In many applications of vector analysis, however, it often becomes necessary to use coordinate systems other than the cartesian. In such situations, the representations of the gradient of a scalar field, the divergence and the curl of a vector field take on completely different forms. In this section we shall discuss the idea of a curvilinear coordinate system in preparation for the task of deriving the corresponding expressions of the gradient, the divergence, and the curl in such a coordinate system.

Let u_1, u_2, u_3 denote new coordinates, and suppose that they are related to the cartesian coordinates x, y, z by the equations

$$u_1 = \phi_1(x, y, z), \quad u_2 = \phi_2(x, y, z), \quad u_3 = \phi_3(x, y, z) \tag{3.35}$$

or briefly, $u_i = \phi_i(x, y, z)$, $i = 1, 2, 3$. We assume that the functions $\phi_i(x, y, z)$, $1 \leq i \leq 3$, have continuous first-order derivatives in some domain D of the xyz-space, and that at some point P_o: (x_o, y_o, z_o) in D the condition

$$\frac{\partial(u_1, u_2, u_3)}{\partial(x, y, z)} = \begin{vmatrix} \frac{\partial \phi_1}{\partial x} & \frac{\partial \phi_1}{\partial y} & \frac{\partial \phi_1}{\partial z} \\ \frac{\partial \phi_2}{\partial x} & \frac{\partial \phi_2}{\partial y} & \frac{\partial \phi_2}{\partial z} \\ \frac{\partial \phi_3}{\partial x} & \frac{\partial \phi_3}{\partial y} & \frac{\partial \phi_3}{\partial z} \end{vmatrix} \neq 0 \tag{3.36}$$

holds. The determinant on the right-hand side of (3.36) is called the Jacobian of the functions ϕ_1, ϕ_2, ϕ_3.

The condition (3.36) ensures that in some neighborhood of the point P_o it is possible to determine x, y, z in terms of the coordinates u_1, u_2, u_3; that is, there exist functions f, g, h such that

$$x = f(u_1, u_2, u_3), \qquad y = g(u_1, u_2, u_3),$$
$$z = h(u_1, u_2, u_3) \tag{3.37}$$

where f, g, h are defined in a neighborhood of the point $(u^o{}_1, u^o{}_2, u^o{}_3)$ determined from (3.35) by the point (x_o, y_o, z_o). Moreover, in that neighborhood f, g and h also have continuous first derivatives for which

$$\frac{\partial(x, y, z)}{\partial(u_1, u_2, u_3)} \neq 0 \tag{3.38}$$

at $(u^o{}_1, u^o{}_2, u^o{}_3)$. The functions f, g, and h define the inverse transformation of (3.35). It is important to note that the Jacobians (3.36) and (3.38) satisfy the relation (see Problem 4)

$$\frac{\partial(u_1, u_2, u_3)}{\partial(x, y, z)} \cdot \frac{\partial(x, y, z)}{\partial(u_1, u_2, u_3)} = 1 \tag{3.39}$$

Now let P be any point in D with the coordinates (x, y, z), and let the numbers u_1, u_2, u_3 be determined by (3.35). We call the ordered triple of numbers (u_1, u_2, u_3) the curvilinear coordinates of the point P. The equations given in (3.35) are called the coordinate transformation, and they are said to define a curvilinear coordinate system in D. It follows from (3.39) that the Jacobian of a coordinate transformation is the reciprocal of the Jacobian of its inverse.

Example 1. Consider the transformation from the rectangular cartesian coordinates (x, y) on a plane to the polar coordinates (r, θ) defined by

$$r = \sqrt{x^2 + y^2}$$

$$\theta = \arccos \frac{x}{\sqrt{x^2 + y^2}} = \arcsin \frac{y}{\sqrt{x^2 + y^2}}$$

[Thus θ is the unique angle $0 \le \theta < 2\pi$ such that $\cos\theta = x/\sqrt{(x^2 + y^2)}$ and $\sin\theta = y/\sqrt{(x^2 + y^2)}$]. The transformation is defined for all (x, y) except at (0, 0) where θ is not defined. The Jacobian of this transformation is given by

$$\frac{\partial(r, \theta)}{\partial(x, y)} = \begin{vmatrix} \dfrac{x}{\sqrt{x^2+y^2}} & \dfrac{y}{\sqrt{x^2+y^2}} \\ \dfrac{-y}{x^2+y^2} & \dfrac{x}{x^2+y^2} \end{vmatrix} = \frac{1}{r}$$

Hence, except at the origin, the transformation has an inverse, and it is given by

$$x = r\cos\theta \ , \quad y = r\sin\theta$$

This inverse transformation is defined, however, for all (r, θ).

By direct calculation, we find

$$\frac{\partial(x, y)}{\partial(r, \theta)} = \begin{vmatrix} \cos\theta & -r\sin\theta \\ \sin\theta & r\cos\theta \end{vmatrix} = r$$

It is clear that

$$\frac{\partial(r, \theta)}{\partial(x, y)} \cdot \frac{\partial(x, y)}{\partial(r, \theta)} = 1$$

in accord with (3.39).

Coordinate Surfaces and Coordinate Curves

Let us now examine the nature of a curvilinear coordinate system from a geometric point of view. Suppose P_o is a point in space with curvilinear coordinates $(u^o{}_1, u^o{}_2, u^o{}_3)$ as defined by the coordinate transformation (3.35). Then the equations

$$\phi_1(x, y, z) = u^o{}_1, \quad \phi_2(x, y, z) = u^o{}_2, \quad \phi_3(x, y, z) = u^o{}_3 \tag{3.40}$$

define three surfaces in space each of which passes through the point P_o (Fig. 3.9). These surfaces correspond to the three coordinate planes $x = x_o$, $y = y_o$, $z = z_o$ in the cartesian coordinate system, where (x_o, y_o, z_o) are the cartesian coordinates of P_o. Recall that by the transformation (3.35), we have $u^o{}_i = \phi_i(x_o, y_o, z_o)$, $i = 1, 2, 3$. Accordingly, we call the three surfaces described by (3.40) the coordinate surfaces intersecting at the point P_o. The curves of intersection of these coordinate surfaces correspond to the coordinate axes in cartesian coordinates. Thus, for example, the surfaces $\phi_2(x, y, z) = u^o{}_2$ and $\phi_3(x, y, z) = u^o{}_3$ intersect in the curve on which only the coordinate u_1 varies. Hence we call this curve the u_1-coordinate curve or simply the u_1-curve. The u_2- and u_3-curves are defined similarly and they are shown in Fig. 3.9.

Using the inverse transformation (3.37), the position vector of a point in curvilinear coordinates now has the representation

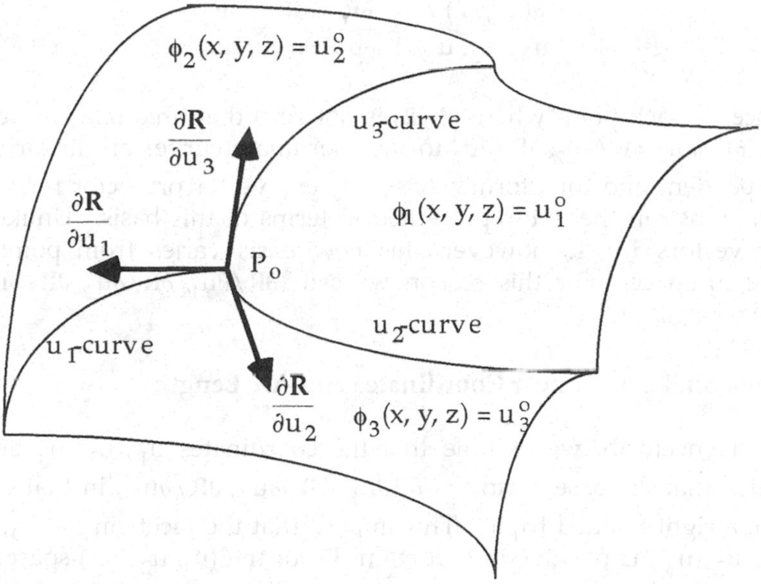

Fig. 3.9 Curvilinear coordinate surfaces and curves.

$$\mathbf{R}(u_1, u_2, u_3) = f(u_1, u_2, u_3)\mathbf{i} + g(u_1, u_2, u_3)\mathbf{j} + h(u_1, u_2, u_3)\mathbf{k} \quad (3.41)$$

If we set $u_2 = c_2$ and $u_3 = c_3$, where c_2, c_3 are constants, then the resulting vector function $\mathbf{R}(u_1, c_2, c_3)$ represents the u_1-curve in which the two coordinate surfaces $\phi_2(x, y, z) = c_2$, $\phi_3(x, y, z) = c_3$ intersect. On this curve u_1 is the parameter. It follows that the derivative $\partial\mathbf{R}/\partial u_1$ represents the tangent vector to this curve. Likewise, we have $\partial\mathbf{R}/\partial u_2$ and $\partial\mathbf{R}/\partial u_3$ representing the tangent vectors to the u_2- and u_3-curves, respectively (see Fig. 3.9).

Since the determinant of a matrix is the same as the determinant of its transpose, it follows from the definition of the Jacobian (3.36) and that of the scalar triple product that

$$\frac{\partial(x, y, z)}{\partial(u_1, u_2, u_3)} = \frac{\partial \mathbf{R}}{\partial u_1} \cdot \frac{\partial \mathbf{R}}{\partial u_2} \times \frac{\partial \mathbf{R}}{\partial u_3} \tag{3.42}$$

Hence, at each point where (3.42) is not zero the three tangent vectors $\partial \mathbf{R}/\partial u_1$, $\partial \mathbf{R}/\partial u_2$, $\partial \mathbf{R}/\partial u_3$ to the coordinate curves are linearly independent and thus form a basis. Every vector or vector field at each point can then be represented in terms of this basis. Unlike the unit vectors **i**, **j**, **k**, however, this new basis varies from point to point in space. For this reason we call $(\partial \mathbf{R}/\partial u_1, \partial \mathbf{R}/\partial u_2, \partial \mathbf{R}/\partial u_3)$ a local basis.

Orthogonal Curvilinear Coordinates and Arc Length

Henceforth, we assume that the coordinates u_1, u_2, u_3 are so labeled that the base vectors $\partial \mathbf{R}/\partial u_1$, $\partial \mathbf{R}/\partial u_2$, $\partial \mathbf{R}/\partial u_3$, in that order, form a right-handed triple. This implies that the Jacobian $\partial(x, y, z)/\partial(u_1, u_2, u_3)$ is positive in a domain D^* of the (u_1, u_2, u_3)-space.

A curvilinear coordinate system defined by (3.35) or (3.37) is said to be *orthogonal* if at each point the base vectors $\partial \mathbf{R}/\partial u_1$, $\partial \mathbf{R}/\partial u_2$, $\partial \mathbf{R}/\partial u_3$ are mutually orthogonal. This means that although the vectors may vary in magnitude and direction from point to point, they remain orthogonal to each other at each point. When this is the case, the formulas for the gradient, the divergence, and the curl appear very much like their representations in rectangular cartesian coordinate system. We shall derive these representations in the next section. As we shall see the polar coordinates in two dimensions and the cylindrical and spherical coordinates in three dimensions are all orthogonal curvilinear coordinate systems.

We conclude this section by deriving the formula for arc length in a curvilinear coordinate system. Let $\mathbf{R}(u_1, u_2, u_3)$ denote the position vector of a point in curvilinear coordinates. Then by definition and by the chain rule, we have

$$\begin{aligned} ds^2 &= d\mathbf{R} \cdot d\mathbf{R} \\ &= \left[\sum_{i=1}^{3} \frac{\partial \mathbf{R}}{\partial u_i} du_i\right] \cdot \left[\sum_{j=1}^{3} \frac{\partial \mathbf{R}}{\partial u_j} du_j\right] \end{aligned} \tag{3.43}$$

$$= \sum_{i,j=1}^{3} g_{ij}\, du_i\, du_j$$

where

$$g_{ij} = \frac{\partial \mathbf{R}}{\partial u_i} \cdot \frac{\partial \mathbf{R}}{\partial u_j} \qquad (i, j = 1, 2, 3) \qquad (3.44)$$

The quantities g_{ij} may be written in the matrix form

$$(g_{ij}) = \begin{bmatrix} g_{11} & g_{12} & g_{13} \\ g_{21} & g_{22} & g_{23} \\ g_{31} & g_{32} & g_{33} \end{bmatrix} \qquad (3.45)$$

This is known as the metric of the space with respect to the curvilinear coordinates u_1, u_2, u_3. If the coordinate system is orthogonal, then (3.44) reduces to

$$g_{ij} = \begin{cases} h_i^2, & i = j \\ 0, & i \neq j \end{cases}$$

where $h_i = |\partial \mathbf{R}/\partial u_i|$, i = 1, 2, 3. In this case formula (3.43) simplifies to the form

$$ds^2 = (h_1 du_1)^2 + (h_2 du_2)^2 + (h_3 du_3)^2 \qquad (3.46)$$

This reduces to the familiar form $ds^2 = dx^2 + dy^2 + dz^2$ in rectangular cartesian coordinates.

Example 2. The transformation relating the cylindrical coordinates r, θ, z to the rectangular cartesian coordinates x, y, z is defined by the equations

$$r = \sqrt{x^2 + y^2}$$

$$\theta = \cos^{-1} \frac{x}{\left(x^2 + y^2\right)^{1/2}} = \sin^{-1} \frac{y}{\left(x^2 + y^2\right)^{1/2}}$$

$$z = z$$

The transformation is defined for all (x, y, z) except for points on the z-axis where θ is not defined. By the same calculation as in Example 1, we find

$$\frac{\partial(r, \theta, z)}{\partial(x, y, z)} = \begin{vmatrix} \frac{\partial r}{\partial x} & \frac{\partial r}{\partial y} & \frac{\partial r}{\partial z} \\ \frac{\partial \theta}{\partial x} & \frac{\partial \theta}{\partial y} & \frac{\partial \theta}{\partial z} \\ 0 & 0 & 1 \end{vmatrix} = \frac{1}{r}$$

The inverse transformation is given by

$$x = r\cos\theta, \quad y = r\sin\theta, \quad z = z$$

which is valid for all $r \geq 0$, θ, z.

The coordinate surfaces intersecting at a point (x_o, y_o, z_o) consist of a cylinder of radius $r_o = \sqrt{(x_o^2 + y_o^2)}$ with axis along the z-axis, a plane containing the z-axis and making an angle $\theta_o = \cos^{-1}(x_o/r_o) = \sin^{-1}(y_o/r_o)$ with the xz coordinate plane, and the plane $z = z_o$. The coordinate curves comprise of a circle and two straight lines as shown in Fig. 3.10.

The position vector in cylindrical coordinates assumes the form

$$\mathbf{R}(r, \theta, z) = r\cos\theta\,\mathbf{i} + r\sin\theta\,\mathbf{j} + z\,\mathbf{k}$$

so that the tangent vector to the coordinate curves are

$$\frac{\partial \mathbf{R}}{\partial r} = \cos\theta\,\mathbf{i} + \sin\theta\,\mathbf{j}, \quad \frac{\partial \mathbf{R}}{\partial \theta} = -r\sin\theta\,\mathbf{i} + r\cos\theta\,\mathbf{j}, \quad \frac{\partial \mathbf{R}}{\partial z} = \mathbf{k}$$

these vectors are mutually orthogonal for all $r \geq 0$ and θ. Therefore, the cylindrical coordinate system is an orthogonal curvilinear coordinate system.

Letting $u_1 = r$, $u_2 = \theta$, and $u_3 = z$, it follows that $h_1 = 1$, $h_2 = r$, $h_3 = 1$ so that from (3.46) we have

$$ds^2 = dr^2 + (r\,d\theta)^2 + dz^2$$

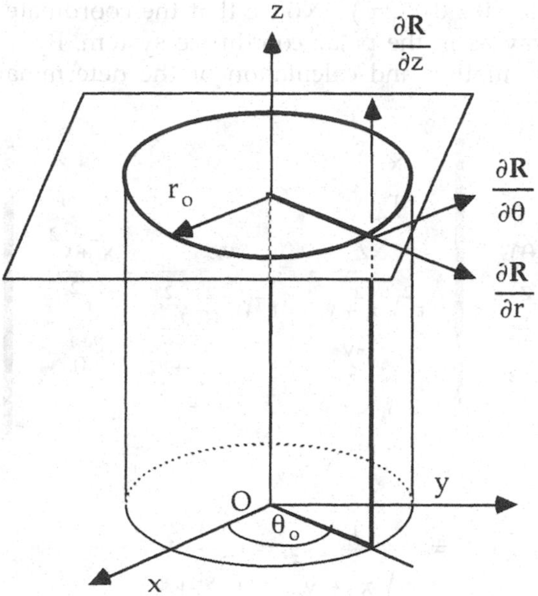

Fig . 3. 10 Cylindrical coordinate system.

This is the expression for *arc length* in cylindrical coordinate system.

Obviously, when we drop the third variable z from the above discussion, we obtain the polar coordinate system. It follows that the polar coordinate system is a two dimensional curvilinear coordinate system.

Example 3. The spherical coordinates $u_1 = r, u_2 = \phi, u_3 = \theta$ are defined by the equations

$$r = \sqrt{x^2 + y^2 + z^2}\ , \quad \phi = \arccos \frac{z}{r}$$

$$\theta = \arccos \frac{x}{\left(x^2 + y^2\right)^{1/2}} = \arcsin \frac{y}{\left(x^2 + y^2\right)^{1/2}}$$

($r \geq 0,\ 0 \leq \phi < \pi,\ 0 \leq \theta < 2\pi$). Notice that the coordinate θ is defined in the same way as in the polar coordinate system. By a straightforward differentiation and calculation of the determinant, we find

$$\frac{\partial(r, \phi, \theta)}{\partial(x, y, z)} = \begin{vmatrix} \dfrac{x}{r} & \dfrac{y}{r} & \dfrac{z}{r} \\ \dfrac{xz}{r^2\sqrt{x^2+y^2}} & \dfrac{yz}{r^2\sqrt{x^2+y^2}} & -\dfrac{x^2+y^2}{r^2} \\ \dfrac{-y}{\sqrt{x^2+y^2}} & \dfrac{x}{\sqrt{x^2+y^2}} & 0 \end{vmatrix}$$

$$= \frac{1}{r\sqrt{x^2+y^2}} = \frac{1}{r^2 \sin\phi}$$

The student is asked to verify this result in Prob. 5. Hence, except for points on the z-axis, the transformation has an inverse given by

$$x = r \sin\phi \cos\theta, \quad y = r \sin\phi \sin\theta, \quad z = r\cos\phi$$

valid for $r \geq 0,\ 0 \leq \phi \leq \pi,\ 0 \leq \theta < 2\pi$. From (3.39) or by direct calculation, we find

$$\frac{\partial(x, y, z)}{\partial(r, \phi, \theta)} = r^2 \sin\phi$$

The coordinate surfaces intersecting at a point (x_o, y_o, z_o) consist of a sphere of radius $r_o = \sqrt{(x_o^2 + y_o^2 + z_o^2)}$ with center at the origin, a "cone" with vertex at the origin, axis along the z-axis, and generating angle $\phi_o = \arccos(z_o/r_o)$, and the plane containing the z-axis making an angle

$$\theta_o = \arccos[x_o/(x_o^2 + y_o^2)^{1/2}] = \arcsin[y_o/(x_o^2 + y_o^2)^{1/2}]$$

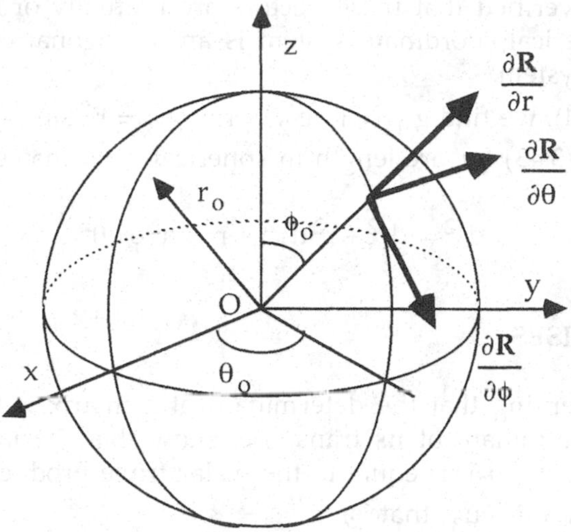

Fig. 3.11 Coordinates surfaces in spherical coordinates.

with the xz-coordinate plane (see Fig. 3.11). In this connection we must broaden the interpretation of the word "cone" to include the xy-plane for which $\phi = \pi/2$ and the cones with generating angles ϕ greater that $\pi/2$.

The position vector in spherical coordinates is given by

$$\mathbf{R}\,(r, \phi, \theta) = r \sin\phi \cos\theta\, \mathbf{i} + r \sin\phi \sin\theta\, \mathbf{j} + r \cos\phi\, \mathbf{k}$$

Thus the tangent vectors to the coordinate curves are

$$\frac{\partial \mathbf{R}}{\partial r} = \sin\phi \cos\theta\, \mathbf{i} + \sin\phi \sin\theta\, \mathbf{j} + \cos\phi\, \mathbf{k}$$

$$\frac{\partial \mathbf{R}}{\partial \phi} = r \cos\phi \cos\theta\, \mathbf{i} + r \cos\phi \sin\theta\, \mathbf{j} - r \sin\phi\, \mathbf{k}$$

$$\frac{\partial \mathbf{R}}{\partial \theta} = - r \sin\phi \sin\theta\, \mathbf{i} + r \sin\phi \cos\theta\, \mathbf{j}$$

It is readily verified that these vectors are mutually orthogonal so that the spherical coordinate system is an orthogonal curvilinear coordinate system.

By (3.44), we find $g_{11} = 1$, $g_{22} = r^2$, $g_{33} = r^2 \sin^2 \phi$ and hence the formula (3.43) for arc length in spherical coordinates is given by

$$ds^2 = dr^2 + r^2 d\phi^2 + r^2 \sin^2 \phi \, d\theta^2$$

3.7 EXERCISES

1. By observing that the determinant of a matrix is the same as the determinant of its transpose, show that the Jacobian $\partial(x, y, z)/\partial(u_1, u_2, u_3)$ is equal to the scalar triple product of $\partial\mathbf{R}/\partial u_1$, $\partial\mathbf{R}/\partial u_2$, $\partial\mathbf{R}/\partial u_3$, that is,

$$\frac{\partial(x, y, z)}{\partial(u_1, u_2, u_3)} = \frac{\partial\mathbf{R}}{\partial u_1} \cdot \frac{\partial\mathbf{R}}{\partial u_2} \times \frac{\partial\mathbf{R}}{\partial u_3}$$

2. Show that

$$\frac{\partial(x, y, z)}{\partial(u_1, u_2, u_3)} = \nabla u_1 \cdot \nabla u_2 \times \nabla u_3$$

3. Recall that the product of two $n \times n$ matrices $A = (a_{ij})$ and $B = (b_{ij})$ is defined as the matrix $AB = (c_{ij})$, where

$$c_{ij} = \sum_{k=1}^{n} a_{ik} b_{kj} \qquad (i, j = 1, .., n)$$

Thus show that

$$\begin{bmatrix} \frac{\partial u_1}{\partial x} & \frac{\partial u_1}{\partial y} & \frac{\partial u_1}{\partial z} \\ \frac{\partial u_2}{\partial x} & \frac{\partial u_2}{\partial y} & \frac{\partial u_2}{\partial z} \\ \frac{\partial u_3}{\partial x} & \frac{\partial u_3}{\partial y} & \frac{\partial u_3}{\partial z} \end{bmatrix} \begin{bmatrix} \frac{\partial x}{\partial u_1} & \frac{\partial x}{\partial u_2} & \frac{\partial x}{\partial u_3} \\ \frac{\partial y}{\partial u_1} & \frac{\partial y}{\partial u_2} & \frac{\partial y}{\partial u_3} \\ \frac{\partial z}{\partial u_1} & \frac{\partial z}{\partial u_2} & \frac{\partial z}{\partial u_3} \end{bmatrix} = \begin{bmatrix} 1 & 0 & 0 \\ 0 & 1 & 0 \\ 0 & 0 & 1 \end{bmatrix}$$

[Hint: $\delta_{ij} = \partial u_i/\partial u_j = (\partial u_i/\partial x)(\partial x/\partial u_j) + (\partial u_i/\partial y)(\partial y/\partial u_j) + (\partial u_i/\partial z)(\partial z/\partial u_j)$]

4. Deduce the relation (3.39) from the result of Problem 3.
5. Verify the calculation of the Jacobian $\partial(r, \phi,\theta)/\partial(x, y, z)$ in Example 3.
6. The transformation relating the cartesian coordinates x, y, z to the parabolic cylindrical coordinates u, v, z is given by the equations

$$x= (u^2 - v^2)/2, \quad y = uv, \quad z = z \quad (-\infty < u < \infty, v \geq 0).$$

(a) Determine the curves on the xy-plane (z = 0) which correspond to u = const.; and v = const. In particular, what are the curves which correspond to u = 0; v = 0; u = v; u = - v?
(b) Find the tangent vectors $\partial\mathbf{R}/\partial u$, $\partial\mathbf{R}/\partial v$, $\partial\mathbf{R}/\partial z$, and show that the new coordinate system is orthogonal.
(c) What is the expression of the arc length in the new coordinate system?

7. The transformation relating the cartesian coordinates x, y, z to the elliptic cylindrical coordinates u, v, z is given by

$$x = a \cosh u \cos v, \quad y = a \sinh u \sin v, \quad z = z$$

($u \geq 0$, $0 \leq v < 2\pi$, $a > 0$ const.)
(a) Show that in the xy-plane a curve u = const. represents an ellipse, while a curve v = const. represents half of one branch of an hyperbola.
(b) Sketch the curves on the xy-plane conrresponding to the values u = 0; v = 0 and $v = \pi$; $v = \pi/2$ and $v = 3\pi/2$.
(c) Verify that the new coordinate system is orthogonal.
(d) Show that the arc length in the new coordinates is given by

$$ds^2 = a^2 (\cosh^2 u - \cos^2 v)(du^2 + dv^2) + dz^2$$

8. The transformation relating the cartesian coordinates x, y, z to the parabolic coordinates u, v, θ is given by

$$x = uv \cos\theta, \quad y = uv \sin\theta, \quad z = (u^2 - v^2)/2$$

($u \geq 0$, $v \geq 0$). Verify that the parabolic coordinate system is orthogonal and find the expression for the arc length in that coordinate system.

9. Let u, v, w denote the new coordinates defined by the equations

$$x = uv, \quad y = (u^2 + v^2)/2, \quad z = w$$

Find the Jacobian of u, v, w and show that the new coordinate system is not orthogonal.

10. Consider the new coordinates u, v, w defined by

$$u = x - y, \quad v = y + z, \quad w = x - z$$

(a) Find the inverse transformation.
(b) Show that the coordinate curves are straight lines.
(c) Show that the coordinate system (u, v, w) is not orthogonal. We call this an oblique cartesian coordinate system.
(d) Show that the u, v, w - coordinate axes are left handed.
(e) Find the expression of the arc length in the coordinates u, v, w.

3.9 GRADIENT, DIVERGENCE AND CURL IN ORTHOGONAL CURVILINEAR COORDINATE SYSTEMS

In this section we derive the expressions of the various vector concepts in an orthogonal curvilinear coordinate system. Let u_1, u_2, u_3 denote such a coordinate system defined by the transformation

$$u_i = \phi_i(x, y, z) \qquad (i = 1, 2, 3) \tag{3.47}$$

and let

$$x = f(u_1, u_2, u_3), \ y = g(u_1, u_2, u_3), \ z = h(u_1, u_2, u_3) \tag{3.48}$$

denote the inverse transformation. As in the preceding section, we assume the same differentiability conditions on both the functions (3.47) and (3.48). By relabeling the coordinates, if necessary, we may assume that the Jacobian $\partial(x, y, z)/\partial(u_1, u_2, u_3)$ is positive so that the base vectors $\partial\mathbf{R}/\partial u_1, \partial\mathbf{R}/\partial u_2, \partial\mathbf{R}/\partial u_3$, in that order, form a right-handed triple. To obtain an orthonormal basis, we normalize the base vectors by multiplying each vector by the reciprocal of its magnitude. We denote these unit base vectors by

$$\mathbf{u}_i = \frac{1}{h_i}\frac{\partial \mathbf{R}}{\partial u_i} \qquad (i = 1, 2, 3) \tag{3.49}$$

where $h_i = |\partial \mathbf{R}/\partial u_i|$. It follows that $\mathbf{u}_i = \mathbf{u}_j \times \mathbf{u}_k$, where i, j, k are cyclic permutations of 1, 2, 3; that is,

$$\mathbf{u}_1 = \mathbf{u}_2 \times \mathbf{u}_3, \quad \mathbf{u}_2 = \mathbf{u}_3 \times \mathbf{u}_1, \quad \mathbf{u}_3 = \mathbf{u}_1 \times \mathbf{u}_2 \tag{3.50}$$

Gradient of a scalar field

Now suppose f is a differentiable scalar field given in a rectangular cartesian coordinate system x, y, z. Under the transformation of coordinates (3.48), f becomes a function of u_1, u_2, u_3. The total differential of f is given by

$$df = \frac{\partial f}{\partial u_1} du_1 + \frac{\partial f}{\partial u_2} du_2 + \frac{\partial f}{\partial u_3} du_3 \tag{3.51}$$

Let

$$\nabla f = f_1 \mathbf{u}_1 + f_2 \mathbf{u}_2 + f_3 \mathbf{u}_3 \tag{3.52}$$

be the expression of grad f in the new coordinate system. We shall determine the components f_1, f_2, f_3. Since $df = \nabla f \cdot d\mathbf{R}$ and by (3.49),

$$d\mathbf{R} = \frac{\partial \mathbf{R}}{\partial u_1} du_1 + \frac{\partial \mathbf{R}}{\partial u_2} du_2 + \frac{\partial \mathbf{R}}{\partial u_3} du_3$$

$$= \mathbf{u}_1 h_1 du_1 + \mathbf{u}_2 h_2 du_2 + \mathbf{u}_3 h_3 du_3$$

we find

$$df = \nabla f \cdot d\mathbf{R} = f_1 h_1 du_1 + f_2 h_2 du_2 + f_3 h_3 du_3$$

Comparing this with (3.51), we deduce

$$f_i = \frac{1}{h_i}\frac{\partial f}{\partial u_i} \qquad (i = 1, 2, 3)$$

Substituting these in (3.52), we finally have

$$\nabla f = \frac{1}{h_1}\frac{\partial f}{\partial u_1}\mathbf{u}_1 + \frac{1}{h_2}\frac{\partial f}{\partial u_2}\mathbf{u}_2 + \frac{1}{h_3}\frac{\partial f}{\partial u_3}\mathbf{u}_3 \tag{3.53}$$

This is the representation of the gradient of a scalar field in an orthogonal curvilinear coordinate system (u_1, u_2, u_3).

From (3.53) we deduce the expression

$$\nabla = \mathbf{u}_1\frac{1}{h_1}\frac{\partial}{\partial u_1} + \mathbf{u}_2\frac{1}{h_2}\frac{\partial}{\partial u_2} + \mathbf{u}_3\frac{1}{h_3}\frac{\partial}{\partial u_3} \tag{3.54}$$

for the del operator ∇. Thus we see that the gradient of f may again be interpreted as the result of applying the operator (3.54) to the function f.

Example 1. Find the expression of ∇ and determine the gradient of $f = xyz$ in cylindrical coordinates.

Solution: In cylindrical coordinates the function becomes $f = r^2 z \sin\theta\cos\theta$. Let $u_1 = r$, $u_2 = \theta$, and $u_3 = z$. Then, by (3.54), we have

$$\nabla = \mathbf{u}_1\frac{\partial}{\partial r} + \mathbf{u}_2\frac{1}{r}\frac{\partial}{\partial\theta} + \mathbf{u}_3\frac{\partial}{\partial z}$$

where (see Example 2, Sec. 3.8)

$$\mathbf{u}_1 = \cos\theta\,\mathbf{i} + \sin\theta\,\mathbf{j}, \quad \mathbf{u}_2 = -\sin\theta\,\mathbf{i} + \cos\theta\,\mathbf{j}, \quad \mathbf{u}_3 = \mathbf{k}$$

Applying ∇ to the given scalar field f, we find

$$\nabla f = 2rz\sin\theta\cos\theta\,\mathbf{u}_1 + rz(\cos^2\theta - \sin^2\theta)\mathbf{u}_2 + r^2\sin\theta\cos\theta\,\mathbf{u}_3$$

It is instructive to note that if we substitute the expressions of $\mathbf{u}_1, \mathbf{u}_2, \mathbf{u}_3$ in the above formula for ∇f and collect the terms involving $\mathbf{i}, \mathbf{j}, \mathbf{k}$, we find

$$\begin{aligned}\nabla f &= rz \sin\theta\, \mathbf{i} + rz\cos\theta\, \mathbf{j} + r^2 \sin\theta \cos\theta\, \mathbf{k} \\ &= yz\mathbf{i} + xz\mathbf{j} + xy\mathbf{k}\end{aligned}$$

which is precisely the gradient of f in rectangular coordinates. This shows the invariant property of ∇f under any coordinate transformation.

Divergence of a Vector Field

Next we derive the expression of the divergence of a vector field. First, we observe that by applying the del operator (3.53) to the functions defined in (3.47) and noting that $\partial u_i / \partial u_j = \delta_{ij}$, where δ_{ij} is the Kronecker delta, we obtain

$$\nabla u_i = \nabla f_i = \frac{\mathbf{u}_i}{h_i} \qquad (i = 1, 2, 3)$$

Since the curl of the gradient of a scalar field is zero, it follows that

$$\nabla \times \nabla u_i = \nabla \times \left[\frac{\mathbf{u}_i}{h_i}\right] = \mathbf{0} \qquad (i = 1, 2, 3) \qquad (3.54)$$

Moreover, by (3.32b), (3.50) and (3.53), we have

$$\begin{aligned}\nabla \cdot \left[\frac{\mathbf{u}_i}{h_j h_k}\right] &= \nabla \cdot \left[\frac{\mathbf{u}_j}{h_j} \times \frac{\mathbf{u}_k}{h_k}\right] \\ &= \frac{\mathbf{u}_k}{h_k} \cdot \nabla \times \left[\frac{\mathbf{u}_j}{h_j}\right] - \frac{\mathbf{u}_j}{h_j} \cdot \nabla \times \left[\frac{\mathbf{u}_k}{h_k}\right] = 0 \qquad (3.55)\end{aligned}$$

where i, j, k assume the cyclic permutations 123, 231, 312.

Now let

$$\mathbf{F} = F_1 \mathbf{u}_1 + F_2 \mathbf{u}_2 + F_3 \mathbf{u}_3 \tag{3.56}$$

be a vector in the orthogonal curvilinear coordinate system u_i with basis $\mathbf{u}_i$, $i = 1, 2, 3$. By definition, we have

$$\operatorname{div} \mathbf{F} = \nabla \cdot \mathbf{F} = \nabla \cdot (F_1 \mathbf{u}_1) + \nabla \cdot (F_2 \mathbf{u}_2) + \nabla \cdot (F_3 \mathbf{u}_3) \tag{3.57}$$

To calculate each of the terms on the right-hand side of (3.57), we write

$$F_i \mathbf{u}_i = (F_i h_j h_k) \left[\frac{\mathbf{u}_i}{h_j h_k} \right]$$

for ijk = 123, 231, 312. By the property of divergence, we find

$$\nabla \cdot (F_i \mathbf{u}_i) = \nabla (F_i h_j h_k) \cdot \left[\frac{\mathbf{u}_i}{h_j h_k} \right] + F_i h_j h_k \nabla \cdot \left[\frac{\mathbf{u}_i}{h_j h_k} \right]$$

$$= \nabla (F_i h_j h_k) \cdot \left[\frac{\mathbf{u}_i}{h_j h_k} \right] \tag{3.58}$$

But from (3.52) we have

$$\nabla (F_i h_j h_k) = \frac{\mathbf{u}_1}{h_1} \frac{\partial}{\partial u_1} (F_i h_j h_k) + \frac{\mathbf{u}_2}{h_2} \frac{\partial}{\partial u_2} (F_i h_j h_k) + \frac{\mathbf{u}_3}{h_3} \frac{\partial}{\partial u_3} (F_i h_j h_k)$$

Since the vectors $\mathbf{u}_1, \mathbf{u}_2, \mathbf{u}_3$ are orthonormal, (3.58) yields

$$\nabla \cdot (F_i \mathbf{u}_i) = \frac{1}{h_1 h_2 h_3} \frac{\partial}{\partial u_i} (F_i h_j h_k)$$

for ijk = 123, 231, 312. Substituting this in (3.57) for i = 1, 2, 3, we finally obtain the formula

$$\operatorname{div} \mathbf{F} = \frac{1}{h_1 h_2 h_3}\left[\frac{\partial}{\partial u_1}(F_1 h_2 h_3) + \frac{\partial}{\partial u_2}(F_2 h_3 h_1) + \frac{\partial}{\partial u_3}(F_3 h_1 h_2)\right] \tag{3.59}$$

This expresses the divergence of a vector field in an orthogonal curvilinear coordinate system.

In particular, if $\mathbf{F} = \nabla f$, where ∇f is given by (3.52), then we have

$$\nabla^2 f = \operatorname{div} \operatorname{grad} f \tag{3.60}$$

$$= \frac{1}{h_1 h_2 h_3}\left[\frac{\partial}{\partial u_1}\left(\frac{h_2 h_3}{h_1}\frac{\partial f}{\partial u_1}\right) + \frac{\partial}{\partial u_2}\left(\frac{h_3 h_1}{h_2}\frac{\partial f}{\partial u_2}\right) + \frac{\partial}{\partial u_3}\left(\frac{h_1 h_2}{h_3}\frac{\partial f}{\partial u_3}\right)\right]$$

This is the expression of the Laplacian of a scalar field in an orthogonal curvilinear coordinate system.

Example 2. Find the divergence in cylindrical coordinates of the vector field $\mathbf{F} = r^3\,\mathbf{u}_1 + r^2 \sin\theta\,\mathbf{u}_2 + z^2\,\mathbf{u}_3$.

Solution: In cylindrical coordinates we have $h_1 = 1$, $h_2 = r$, and $h_3 = 1$. Hence by (3.59) we have

$$\operatorname{div} \mathbf{F} = \frac{1}{r}\left[\frac{\partial}{\partial r}(r\,r^3) + \frac{\partial}{\partial \theta}(r^2 \sin\theta) + \frac{\partial}{\partial z}(r\,z^2)\right]$$

$$= \frac{1}{r}(4r^3 + r^2\cos\theta + 2\,r\,z) = 4r^2 + r\cos\theta + 2\,z$$

If we change back to cartesian coordinates, this result becomes

$$\operatorname{div} \mathbf{F} = 4(x^2 + y^2) + x + 2z$$

which is precisely the divergence of $\mathbf{F}$ when $\mathbf{F}$ is expressed in cartesian form.

Curl of a Vector Field

Finally, we derive the expression for the curl of a vector field **F** which we assume in the form (3.56). By defintion, we have

$$\mathbf{curl\ F} = \nabla \times \mathbf{F} = \nabla \times (F_1 \mathbf{u}_1) + \nabla \times (F_2 \mathbf{u}_2) + \nabla \times (F_3 \mathbf{u}_3) \tag{3.61}$$

To calculate each of the terms on the right-hand side of (3.61), we write

$$F_i \mathbf{u}_i = (F_i h_i) \frac{\mathbf{u}_i}{h_i} \qquad (i = 1, 2, 3)$$

Then by the property of the curl and by (3.54), we find

$$\nabla \times (F_i \mathbf{u}_i) = \nabla(F_i h_i) \times \frac{\mathbf{u}_i}{h_i} + (F_i h_i) \nabla \times \frac{\mathbf{u}_i}{h_i}$$

$$= \nabla(F_i h_i) \times \frac{\mathbf{u}_i}{h_i} \tag{3.62}$$

For i = 1, 2, 3 , we know that

$$\nabla(F_i h_i) = \frac{\mathbf{u}_1}{h_1} \frac{\partial}{\partial u_1}(F_i h_i) + \frac{\mathbf{u}_2}{h_2} \frac{\partial}{\partial u_2}(F_i h_i) + \frac{\mathbf{u}_3}{h_3} \frac{\partial}{\partial u_3}(F_i h_i)$$

and from (3.50) we have $\mathbf{u}_1 = \mathbf{u}_2 \times \mathbf{u}_3$, $\mathbf{u}_2 = \mathbf{u}_3 \times \mathbf{u}_1$, $\mathbf{u}_3 = \mathbf{u}_1 \times \mathbf{u}_2$. Hence, by taking the cross product in (3.62) for i = 1, 2, 3, we find

$$\nabla \times (F_1 \mathbf{u}_1) = \frac{\mathbf{u}_2 \times \mathbf{u}_1}{h_2 h_1} \frac{\partial}{\partial u_2}(F_1 h_1) + \frac{\mathbf{u}_3 \times \mathbf{u}_1}{h_3 h_1} \frac{\partial}{\partial u_3}(F_1 h_1)$$

$$= -\frac{\mathbf{u}_3}{h_2 h_1} \frac{\partial}{\partial u_2}(F_1 h_1) + \frac{\mathbf{u}_2}{h_3 h_1} \frac{\partial}{\partial u_3}(F_1 h_1)$$

$$\nabla \times (F_2 \mathbf{u}_2) = \frac{\mathbf{u}_1 \times \mathbf{u}_2}{h_1 h_2} \frac{\partial}{\partial u_1}(F_2 h_2) + \frac{\mathbf{u}_3 \times \mathbf{u}_2}{h_3 h_2} \frac{\partial}{\partial u_3}(F_2 h_2)$$

$$= \frac{\mathbf{u}_3}{h_1 h_2}\frac{\partial}{\partial u_1}(F_2 h_2) - \frac{\mathbf{u}_1}{h_3 h_2}\frac{\partial}{\partial u_3}(F_2 h_2)$$

and

$$\nabla \times (F_3 \mathbf{u}_3) = \frac{\mathbf{u}_1 \times \mathbf{u}_3}{h_1 h_3}\frac{\partial}{\partial u_1}(F_3 h_3) + \frac{\mathbf{u}_2 \times \mathbf{u}_3}{h_2 h_3}\frac{\partial}{\partial u_2}(F_3 h_3)$$

$$= -\frac{\mathbf{u}_2}{h_1 h_3}\frac{\partial}{\partial u_1}(F_3 h_3) + \frac{\mathbf{u}_1}{h_2 h_3}\frac{\partial}{\partial u_2}(F_3 h_3)$$

Substituting these results on the right-hand side of (3.61), we finally obtain

$$\begin{aligned} \textbf{curl F} = {} & \frac{\mathbf{u}_1}{h_2 h_3}\left[\frac{\partial}{\partial u_2}(F_3 h_3) - \frac{\partial}{\partial u_3}(F_2 h_2)\right] \\ & + \frac{\mathbf{u}_2}{h_3 h_1}\left[\frac{\partial}{\partial u_3}(F_1 h_1) - \frac{\partial}{\partial u_1}(F_3 h_3)\right] \\ & + \frac{\mathbf{u}_3}{h_1 h_2}\left[\frac{\partial}{\partial u_1}(F_2 h_2) - \frac{\partial}{\partial u_2}(F_1 h_1)\right] \end{aligned} \tag{3.63}$$

This can be written in the more convenient form

$$\textbf{curl F} = \frac{1}{h_1 h_2 h_3}\begin{vmatrix} h_1 \mathbf{u}_1 & h_2 \mathbf{u}_2 & h_3 \mathbf{u}_3 \\ \frac{\partial}{\partial u_1} & \frac{\partial}{\partial u_2} & \frac{\partial}{\partial u_3} \\ F_1 h_1 & F_2 h_2 & F_3 h_3 \end{vmatrix} \tag{3.64}$$

which resembles the corresponding form in rectangular cartesian coordinate system.

3.8 EXERCISES

1. Find the expressions of $\nabla^2 f$ and $\nabla \times \mathbf{F}$ in cylindrical coordinate system.
2. Find the expressions of $\nabla^2 f$, $\nabla \cdot \mathbf{F}$, and $\nabla \times \mathbf{F}$ in spherical coordinates.
3. Find ∇f and $\nabla^2 f$ in cylindrical and spherical coordinates when $f(x, y, z) = xy + yz + zx$; (b) $f(x, y, z) = x^2 + y^2 + z^2$.
4. Consider the unit base vectors $\mathbf{u}_1 = \cos\theta\,\mathbf{i} + \sin\theta\,\mathbf{j}$,

   $\mathbf{u}_2 = -\sin\theta\,\mathbf{i} + \cos\theta\,\mathbf{j}$, $\mathbf{u}_3 = \mathbf{k}$ in cylindrical coordinates.

   Treating the equations as algebraic system, solve for the unit vectors $\mathbf{i}, \mathbf{j}, \mathbf{k}$ in terms of the unit vectors $\mathbf{u}_1, \mathbf{u}_2, \mathbf{u}_3$. Thus express the vector field $\mathbf{F} = z\mathbf{i} - 2x\mathbf{j} + 2y\mathbf{k}$ in cylindrical coordinates .
5. Find the divergence and the curl in cylindrical coordinates of the vector field given in Problem 4.
6. Repeat Problem 5 when $\mathbf{F} = y\mathbf{i} + z^2\mathbf{j} + (x^2 + y^2)\mathbf{k}$.
7. (a) From Example 3, Sec. 3.9, determine the unit base vectors

   $\mathbf{u}_r = \mathbf{u}_1$, $\mathbf{u}_\phi = \mathbf{u}_2$, $\mathbf{u}_\theta = \mathbf{u}_3$ in the spherical coordinate system.

   (b) Determine $\mathbf{i}, \mathbf{j}, \mathbf{k}$ in terms of $\mathbf{u}_r, \mathbf{u}_\phi, \mathbf{u}_\theta$.
8. Express the vector field $\mathbf{F} = z\mathbf{i} + 2y\mathbf{j} + 3z\mathbf{k}$ in spherical coordinate system, and find the divergence and curl in that coordinate system.
9. Repeat Problem 8 when $\mathbf{F} = z\mathbf{i} + 2y\mathbf{j} + 3z\mathbf{k}$.
10. Find the expressions of ∇f, $\nabla \cdot \mathbf{F}$, and $\nabla \times \mathbf{F}$ in the parabolic cylindrical coordinate system given in Problem 6, Exercise 3.7.
11. Find the expressions of ∇f, $\nabla \cdot \mathbf{F}$, and $\nabla \times \mathbf{F}$ in the parabolic cordinate system given in Problem 8, Exercise 3.7.
12. Using the formulas (3.53), (3.59), and (3.63), show that (a) div **curl** $\mathbf{F} = 0$ and (b) **curl** grad $f = 0$ in any orthogonal curvilinear coordinate system.

4

INTEGRAL CALCULUS OF SCALAR AND VECTOR FIELDS

In this chapter we shall study line and surface integrals of scalar and vector fields. These integrals are the natural generalization of ordinary single and multiple integrals. In the ordinary single integral

$$\int_a^b f(x)\,dx$$

the function f is defined in the interval [a, b] of the x-axis, while in the double integral

$$\iint_D f(x, y)\,dx\,dy$$

f is defined in the domian D of the xy-plane. In contrast, in a line integral (the term curvilinear integral would seem more appropriate), the function is defined on a space curve and the integration is performed with respect to arc length of the curve. In a surface integral the function is defined on a surface and the integration is performed with respect to the area of the surface. In actual computation of these integrals, however, they are generally converted to an ordinary single or multiple integrals where the integration is performed in the usual manner.

For this reason, the student is presumed to be already familiar with the usual definition of ordinary single and multiple integrals as well as with the techniques for evaluating such integrals.

4.1 LINE INTEGRALS OF SCALAR FIELDS

Let f be a scalar field defined in a domain D of the xyz-space and let $\mathbf{R}(t) = x(t)\mathbf{i} + y(t)\mathbf{j} + z(t)\mathbf{k}$, $a \le t \le b$, represent a space curve C that lies in D (Fig. 4.1). We assume that C is a smooth curve and that f is continuous on C. From Sec. 2.5 we know that the arc length of C is given by

$$s(t) = \int_a^t |\mathbf{R}'(\tau)|\, d\tau \qquad (a \le t \le b)$$

so that $ds = |\mathbf{R}'(t)|\, dt$. Thus we define the line integral of f on C with respect to arc length as follows:

Definition 1: The line integral of f on C with respect to arc length, denoted by $\int_C f\, ds$, is the integral

$$\int_C f(x, y, z)ds = \int_a^b f\,[x(t), y(t), z(t)]\,|\mathbf{R}'(t)|dt \tag{4.1}$$

where

$$|\mathbf{R}'(t)| = \sqrt{x'(t)^2 + y'(t)^2 + z'(t)^2}$$

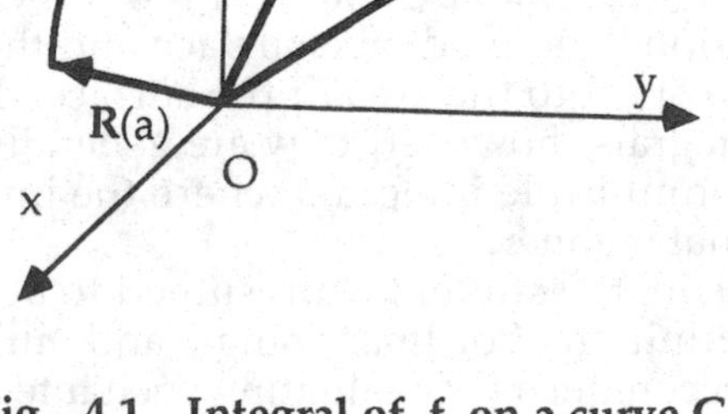

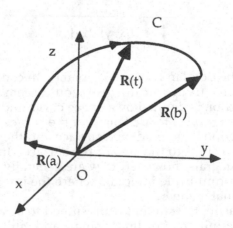

Fig. 4.1 Integral of f on a curve C.

Notice that when f = 1, the integral (4.1) gives the length of the curve C. Also, the value of the integral (4.1) depends only on the function and the geometry of the curve C. It does not depend on the parametrization of the curve C. This means that if $\mathbf{R} = \mathbf{R}(t)$, $a \le t \le b$, and $\mathbf{R}^* = \mathbf{R}^*(\tau)$, $\alpha \le \tau \le \beta$, are two different representations of the curve C such that they orient the curve in the same direction, that is, $\mathbf{R}(a) = \mathbf{R}^*(\alpha)$ and $\mathbf{R}(b) = \mathbf{R}^*(\beta)$, then

$$\int_C f\,ds = \int_a^b f[x(t), y(t), z(t)]\,|\mathbf{R}'(t)|\,dt$$

$$= \int_\alpha^\beta f[x(\tau), y(\tau), z(\tau)]\,|\mathbf{R}^{*\prime}(\tau)|\,d\tau$$

Example 1. Evaluate the line integral $\int_C f ds$ of $f(x,y) = x + 2y$ on the straight line $y = 2x$ from the origin to the point (1, 2).

Solution: We represent the straight line by the equation $\mathbf{R}(t) = t\mathbf{i} + 2t\mathbf{j}$, $0 \le t \le 1$. Then $|\mathbf{R}'(t)| = \sqrt{5}$ so that

$$\int_C (x + 2y)\,ds = \int_0^1 (t + 4t)\sqrt{5}\,dt = \frac{5\sqrt{5}}{2}$$

Example 2. Evaluate the line integral $\int_C xyz\,ds$, where C is the helix $\mathbf{R}(t) = \cos t\,\mathbf{i} + \sin t\,\mathbf{j} + t\mathbf{k}$, $0 \le t \le 2\pi$.

Solution: First we note that $\mathbf{R}'(t) = -\sin t\,\mathbf{i} + \cos t\,\mathbf{j} + \mathbf{k}$ so that $|\mathbf{R}'(t)| = \sqrt{2}$. Thus

$$\int_C xyz\,ds = \sqrt{2}\int_0^{2\pi} t\cos t\sin t\,dt$$

$$= \sqrt{2}\left|\frac{t}{2}\sin^2 t\right|_0^{2\pi} - \frac{\sqrt{2}}{2}\int_0^{2\pi} \sin^2 t\,dt$$

$$= -\frac{\sqrt{2}}{2}\int_0^{2\pi} \frac{1 - \cos 2t}{2}\,dt = -\frac{\sqrt{2}}{2}\pi$$

Applications of Line Integrals

Line integrals occur in many problems in mathematical physics. For example, if f represents a force that moves a body along the curve C and that f remains tangential to the curve at each point, then the line integral $\int_C f ds$ represents the work done by the force in moving the body along the curve. We shall consider this point in more detail in the next section. In mechanics the calculation of the mass of a body, the center of gravity, and the moment of inertias of geometric figures lying on curves, all make use of line integrals. For example, if a piece of wire is in the shape of a space curve C and its density (mass per unit length) is given by $\rho(x, y, z)$, then the total mass of the wire is given by the line integral

$$M = \int_C \rho(x, y, z)\, ds$$

The center of mass of the wire refers to the point whose coordinates (x_C, y_C, z_C) are given by the equations

$$x_C = \frac{1}{M}\int_C x\rho(x, y, z)\, ds \ , \ y_C = \frac{1}{M}\int_C y\rho(x, y, z)\, ds$$

$$z_C = \frac{1}{M}\int_C z\rho(x, y, z)\, ds$$

all involving line integrals. The moment of inertia of the wire with respect to a line L is given by the line integral

$$I_L = \int_C \delta^2(x, y, z)\, ds$$

where δ denotes the perpendicular distance of each point (x, y, z) on the wire from the line L. Thus, for example, if L is the z-axis, then $\delta^2 = x^2 + y^2$ so that

$$I_z = \int_C (x^2 + y^2)\rho(x, y, z)\, ds$$

On the other hand, if f(x, y, z) denotes the temperature of the wire at the point (x, y, z), then the integral

$$\frac{1}{L}\int_C f(x, y, z)\, ds$$

where L denotes the length of the wire, represents the average temperature of the wire. We consider a few examples below.

Example 3. A piece of wire is in the shape of a parabola $y = x^2$, $-2 \le x \le 2$, and its density is given by $\rho(x,y) = |x|$. Find the mass of the wire.

Solution: Using x as a parameter, we represent the parabola by the vector equation $\mathbf{R}(x) = x\mathbf{i} + x^2\mathbf{j}$, $-2 \le x \le 2$. Then

$$ds = |\mathbf{R}'(x)|\, dx = \sqrt{(1 + 4x^2)}dx$$

and so the mass is given by

$$M = \int_{-2}^{2} |x| \sqrt{1 + 4x^2}\, dx$$

$$= 2\int_{0}^{2} x\sqrt{1 + 4x^2}\, dx$$

$$= \left|\frac{1}{6}(1 + 4x^2)^{3/2}\right|_0^2 = \frac{17\sqrt{17} - 1}{6}$$

Example 4. A piece of wire is in the shape of a semicircle of radius a, $y = \sqrt{(a^2 - x^2)}$, $-a \le x \le a$, and its density is given by $\rho(x, y) = y$. Find the center of mass of the wire, and its moment of inertia with respect to the x-axis; the y axis.

Solution: Let us represent the semicircle by the equation

$$\mathbf{R}(\theta) = a(\cos\theta\, \mathbf{i} + \sin\theta\, \mathbf{j}), \qquad 0 \le \theta \le \pi$$

Then, $ds = a\, d\theta$ and so the mass of the wire is given by

$$M = \int_0^{\pi} (a \sin \theta)\, a\, d\theta = a^2(-\cos \theta)\,|_0^{\pi} = 2a^2$$

Let (x_c, y_c) denote the center of mass of the wire. By symmetry we deduce that $x_c = 0$, (this can of course be verified by actual calculation). The ordinate y_c satisfies the equation

$$My_c = \int_0^{\pi} (a \sin \theta)(a \sin\theta) a\, d\theta = a^3 \int_0^{\pi} \sin^2 \theta\, d\theta$$

$$= \frac{a^3}{2} \int_0^{\pi} (1 - \cos 2\theta)\, d\theta = \frac{\pi a^3}{2}$$

Hence, $y_c = (\pi a^3/M) = \pi a/4$ and so the center of mass is at the point $(0, \pi a/4)$. The moment of inertia of the wire with respect to the x-axis is given by the integral

$$I_x = \int_C y^2 \rho\, ds = \int_0^{\pi} (a \sin \theta)^2 (a \sin \theta)\, a\, d\theta$$

$$= a^4 \int_0^{\pi} \sin^3 \theta\, d\theta = a^4 \int_0^{\pi} \sin \theta\, (1 - \cos^2 \theta)\, d\theta$$

$$= a^4 \left[-\cos \theta + \frac{\cos^3 \theta}{3} \right]_0^{\pi} = \frac{4}{3} a^4$$

The moment of inertia with respect to the y-axis is given by

$$I_y = \int_C x^2 \rho\, ds = \int_0^{\pi} (a \cos \theta)^2 (a \sin \theta)\, a\, d\theta$$

$$= a^4 \left[-\frac{\cos^3 \theta}{3} \right]_0^{\pi} = \frac{2}{3} a^4$$

Line Integrals with Respect to x, y, z

In a similar manner we define the line integral of f on a space curve C: $\mathbf{R}(t) = x(t)\mathbf{i} + y(t)\mathbf{j} + z(t)\mathbf{k}$, $a \le t \le b$, with respect to the variables x, y, and z as the ordinary integrals

$$\int_C f(x, y, z)\, dx = \int_a^b f[x(t), y(t), z(t)]\, x'(t)\, dt$$

$$\int_C f(x, y, z)\, dy = \int_a^b f[x(t), y(t), z(t)]\, y'(t)\, dt \qquad (4.2)$$

$$\int_C f(x, y, z)\, dz = \int_a^b f[x(t), y(t), z(t)]\, z'(t)\, dt$$

Notice that when $f(x, y, z) = 1$, these integrals give $x(b) - x(a)$, $y(b) - y(a)$, and $z(b) - z(a)$, which are the projections of the space curve on the x-, y-, and z-coordinate axes, respectively. As with integrals with respect to arc length, the integrals (4.2) do not depend on the parametrization of the curve C.

Example 5. Evaluate the line integrals of $f(x, y, z) = xyz$ with respect to x, y, and z along the curve $\mathbf{R}(t) = \cos t\,\mathbf{i} + \sin t\,\mathbf{j} + t\mathbf{k}$, $0 \le t \le \pi$.

Solution: We have

$$\int_C xyz\, dx = \int_0^\pi t \cos t \sin t\,(-\sin t)\, dt$$

$$= \frac{1}{3}\left[-t \sin^3 t - \cos t + \frac{1}{3}\cos^3 t\right]_0^\pi = \frac{4}{9}$$

$$\int_C xyz\, dy = \int_0^\pi t \cos t \sin t\,(\cos t)\, dt$$

$$= \frac{1}{3}\left[-t \cos^3 t + \sin t - \frac{1}{3}\sin^3 t\right]_0^\pi = \frac{\pi}{3}$$

and

$$\int_C xyz\,dz = \int_0^{\pi} t\cos t\sin t\,dt = \frac{1}{2}\left[t\sin^2 t + \frac{1}{2}t - \frac{\sin 2t}{4}\right]_0^{\pi} = \frac{\pi}{4}$$

Example 6. Calculate the line integrals $\int_C (x^2 + y)dx$ and $\int_C (x^2 + y)dy$ from (0, 0) to (4, 2) along the parabola $y = \sqrt{x}$ $(0 \le x \le 4)$.

Solution: Using x as a parameter, we find

$$\int_C (x^2 + y)\,dx = \int_0^4 (x^2 + \sqrt{x}\,)dx = \frac{80}{3}$$

and

$$\int_C (x^2 + y)\,dy = \int_0^4 (x^2 + \sqrt{x}\,)\frac{dx}{2\sqrt{x}}$$

$$= \frac{1}{2}\int_0^4 (x^{3/2} + 1)\,dx = \frac{42}{5}$$

If we use y as a parameter, we find

$$\int_C (x^2 + y)dx = \int_0^2 (y^4 + y)2ydy = \frac{80}{3}$$

and

$$\int_C (x^2 + y)dy = \int_0^2 (y^4 + y)dy = \frac{42}{5}$$

This verifies the fact that the line integral along the parabola does not dependent on the manner we parametrize the curve.

Geometric Interpretation of Line Integrals in the Plane

The student may recall from elementary calculus that when $f(x) \ge 0$ the integral $\int_a^b f(x)dx$ represents the area bounded by the curve defined by $y = f(x)$ and the lines $x = a$, $x = b$. It is interesting to note that an analogous geometric interpretation also holds for line integrals along curves in the xy-plane. Indeed, suppose $f(x, y)$ is a non-negative function defined on a curve C in the xy-plane. Then we may regard $z = f(x, y)$ as the height of a vertical cylindrical surface at each point (x, y) on the curve C (see Fig. 4.2); that is, $z = f(x, y)$ defines a "fence" whose base is the curve C. Then, the line integral $\int_C f(x, y)ds$ represents the area of the "fence" along C whose height

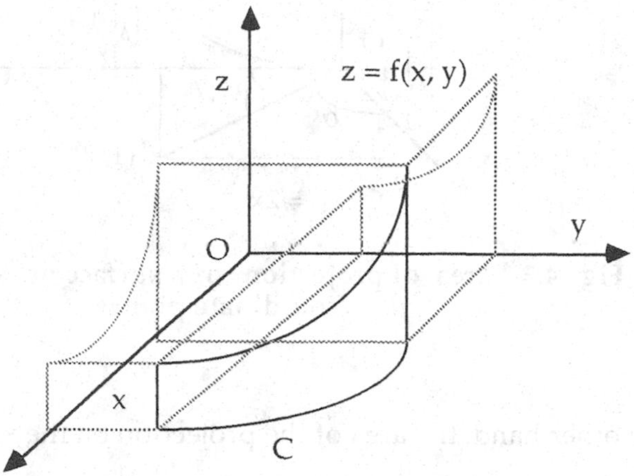

Fig. 4.2 Geometric interpretation of a line integral.

at each point (x, y) is given by $z = f(x, y)$. In this context, the integrals $\int_C f(x, y)dx$ and $\int_C f(x, y)dy$ represent the area of the projection of the "fence" on the xz- and yz-coordinate planes, respectively, provided there is no overlapping in the projection. As a simple illustration, we consider the following example.

Example 7. Find the area of the fence whose base is the line $y = 2x$, $0 \leq x \leq 1$ (ft), and whose height is defined by $z = xy$ (ft), see Fig. 4.3. Verify the area of the projection of the fence on the coordinate planes by using the "area cosine principle".

Solution: Using x as a parameter, we have $ds = \sqrt{(1 + y'^2)}\, dx = \sqrt{5}\, dx$, so that the area is

$$A \text{ (area)} = \int_C xy\, ds = \int_0^1 2\sqrt{5}\, x^2 dx = \frac{2\sqrt{5}}{3} \text{ ft}^2$$

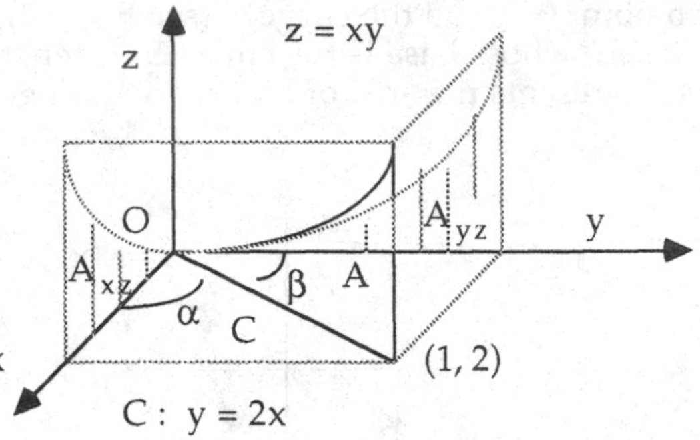

Fig. 4.3 Area of projections of a surface on xz- and yz-coordinate planes.

On the other hand, the area of the projection on the xz-plane is

$$A_{xz} = \int_0^1 2x^2 dx = \frac{2}{3}\ \text{ft}^2$$

and the area of the projection on the yz-plane is

$$A_{yz} = \int_0^2 \frac{1}{2} y^2 dy = \frac{4}{3}\ \text{ft}^2$$

Now by the area cosine principle, the area of the projection on the xz-coordinate plane is given by the product of the area and the cosine of the angle α which the line makes with the x-axis. Likewise, the area of the projection on the yz-coordinate plane is the product of the area and the cosine of the angle β which the line makes with the y-axis. From Fig. 4.3 we see that $\cos\alpha = 1/\sqrt{5}$ and $\cos\beta = 2/\sqrt{5}$. Thus we have

$$A_{xz} = \frac{2\sqrt{5}}{3} \cdot \frac{1}{\sqrt{5}} = \frac{2}{3}$$

$$A_{yz} = \frac{2\sqrt{5}}{3} \cdot \frac{2}{\sqrt{5}} = \frac{4}{3}$$

which agree with the results given by the integrals.

Example 8. Find the surface area of a wall that is erected on the curve $x^2 + 4y^2 = 1, x \geq 0, y \geq 0$, whose height at each point (x, y) is given by $f(x, y) = xy$. Assume the measurement to be in meters.

Solution: Let us parametrize the curve by setting $x = 2\cos\theta$, $y = \sin\theta$ $(0 \leq \theta \leq \pi/2)$. Then we have

$$\mathbf{R}(\theta) = 2\cos\theta\,\mathbf{i} + \sin\theta\,\mathbf{j}, \quad \mathbf{R}'(\theta) = -2\sin\theta\,\mathbf{i} + \cos\theta\,\mathbf{j}$$

so that

$$|\mathbf{R}'(\theta)| = \sqrt{4\sin^2\theta + \cos^2\theta} = \sqrt{3\sin^2\theta + 1}$$

Hence the surface area of the wall is equal to

$$\int_C xy\,ds = \int_0^{\pi/2} 2\cos\theta\sin\theta\sqrt{3\sin^2\theta + 1}\,d\theta$$

$$= \frac{2}{9}(3\sin^2\theta + 1)^{3/2}\Big|_0^{\pi/2} = \frac{14}{9}\ m^2$$

The area of the projection of the wall on the xz-plane is given by $\int_C xy\,dx$. To ensure a positive area, we need to integrate in the direction of increasing values of x. Thus we find

$$\int_C xy\,dx = \int_{\pi/2}^{0} 2\cos\theta\sin\theta\,d(2\cos\theta)$$

$$= \int_{\pi/2}^{0} -4\cos\theta\sin^2\theta\,d\theta = -\frac{4}{3}\sin^3\theta\,\Big|_{\pi/2}^{0} = \frac{4}{3}$$

The area of the projection on the yz-plane is given by

$$\int_C xy\, dy = \int_0^{\pi/2} 2 \cos\theta \sin\theta\, d(\sin\theta)$$

$$= \int_0^{\pi/2} 2 \cos^2\theta \sin\theta\, d\theta = -\frac{2}{3} \cos^3\theta \Big|_0^{\pi/2} = \frac{2}{3}$$

4.1 EXERCISES

In each of Problems 1 through 5, calculate the line integral along the given curve.

1. $\int_C (x^2 - y^2)\, ds$, C: $\mathbf{R}(t) = a \cos t\, \mathbf{i} + a \sin t\, \mathbf{j}$ $(0 \le t < 2\pi)$ traced counterclockwise.
2. $\int_C (x - y)\, ds$, C: quadrilateral with vertices at (1, 0), (0, 1), (-1, 0), (0, -1) traced counterclockwise.
3. $\int_C (z + 1)\, ds$, C: $\mathbf{R}(t) = t \cos t\, \mathbf{i} + t \sin t\, \mathbf{j} + t\, \mathbf{k}$ $(0 \le t \le \pi)$ from (0, 0, 0) to $(-\pi, 0, \pi)$.
4. $\int_C (xy - z)\, ds$, C: $\mathbf{R}(t) = a \cos t\, \mathbf{i} + a \sin t\, \mathbf{j} + bt\, \mathbf{k}$ $(0 \le t \le 3\pi/4)$ from (a, 0, 0) to $(-\sqrt{2}a/2, \sqrt{2}a/2, 3\pi b/4)$.
5. $\int_C xz\, ds$, C: $\mathbf{R}(t) = 2 (\cos t\, \mathbf{i} + \sin t\, \mathbf{j} + \sin t\, \mathbf{k})$ $(0 \le t \le \pi/2)$ from (2, 0, 0) to (0, 2, 2)
6. A piece of wire is in the shape of an equilateral triangle with vertices at (1, 0, 0), (0, 1, 0), (0, 0, 1). If the density is given by $f(x, y, z) = x + y + z$, find the mass and the center of mass of the wire.
7. The density of a ring of wire with radius a is defined by $\rho = a/\pi + |x|$. Find (a) the mass of the wire, (b) the center of mass, (c) the moment of inertia with respect to the x-axis, (d) the moment of inertia with respect to the y-axis.
8. Find the average temperature of a piece of wire that is in the shape of a curve defined by $x = t$, $y = \sqrt{5}\, t^2/2$, $z = 2t$ $(0 \le t \le 1)$ if the temperature at each point (x, y, z) is given by $f(x, y, z) = yz$.

9. Repeat Problem 8 for a piece of wire in the shape of a curve defined by $x = t\cos t$, $y = t\sin t$, $z = 2t$ $(0 \le t \le \pi)$ whose temperature is given by $f(x, y, z) = z$.

In each of Problems 10 through 15, calculate the line integral along the given curve.

10. $\int_C (x - y^2)\,dx$ along the parabola $y = x - x^2/4$ from (0, 0) to (4, 0).

11. $\int_C (x - y^2)\,dy$ along the curve of Problem 10.

12. $\int_C (x^2 - y^2)\,dx$ around a circle of radius a traced counterclockwise.

13. $\int_C (x^2 + y^2)\,dy$ around the ellipse $\dfrac{x^2}{a^2} + \dfrac{y^2}{b^2} = 1$ traced counterclockwise.

14. $\int_C (xy + z^2)\,dz$ along the helix $x = 2\cos t$, $y = 2\sin t$, $z = 2t$ from (2, 0, 0) to $(-\sqrt{2}, \sqrt{2}, 3\pi/2)$.

15. $\int_C (xz + yz + xy)\,dy$ around the intersection of the cylinder $x^2 + y^2 = 1$ and the plane $y + z = 1$ in the clockwise direction as viewed from the origin.

16. Let C_1 and C_2 be two simple smooth curves represented by the equations

$$C_1:\ y = f(x) \quad (a \le x \le b)$$

$$C_2:\ y = g(x) \quad (a \le x \le b)$$

where $f(a) = g(a)$, $f(b) = g(b)$, and $f(x) < g(x)$ for $a \le x \le b$. Let C denote the closed curve composed of C_1 and C_2 as shown in Fig. 4.4. Show that the integral $\int_C y\,dx$ gives the area of the domain bounded by C.

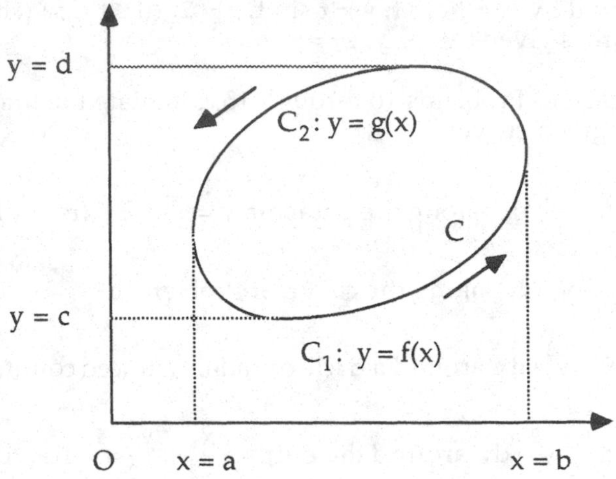

Fig. 4.4 Area of a domain bounded by a closed curve.

4.2 LINE INTEGRALS OF VECTOR FIELDS

We now extend the definition of a line integral to vector fields To motivate the definition, we consider the notion of work in physics. Suppose **F** represents a force field in space and C represents the smooth path of a particle moving under the influence of **F**. We want to calculate the amount of work done by the force in moving the particle along C. If **F** is a constant force and C is a line segment represented by the vector **D**, then, by definition, the work done by **F** in moving the particle through the displacement **D** is given by

$$\text{Work} = \mathbf{F} \cdot \mathbf{D}$$

More generally, if C is a space curve defined by $\mathbf{R}(t) = x(t)\mathbf{i} + y(t)\mathbf{j} + z(t)\mathbf{k}$ $(a \leq t \leq b)$, we calculate the work done by **F** by decomposing C into a finite number of paths C_i and determining the work done by **F** on each path, and then taking the sum. So let us divide the interval [a, b] into *n* subintervals with endpoints

$$a = t_o < t_1 < \ldots . < t_{n-1} < t_n = b$$

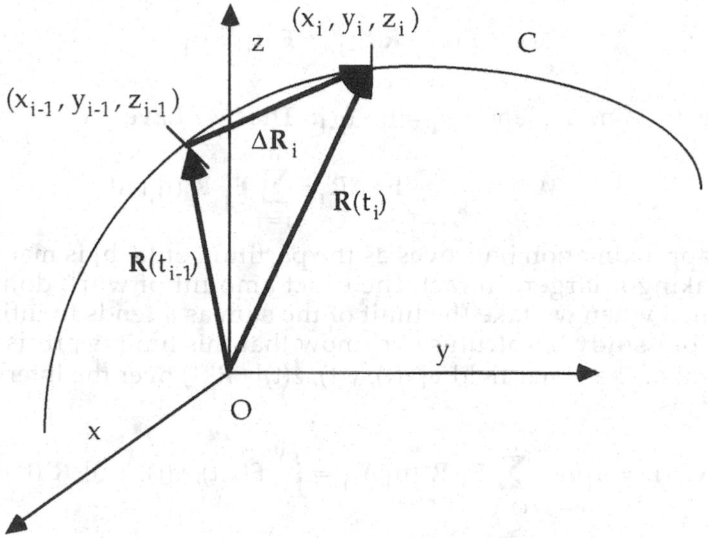

Fig. 4.4 Work done by a force field F along a curve C.

Let $\Delta\mathbf{R}_i = \mathbf{R}(t_i) - \mathbf{R}(t_{i-1})$ and denote the length of the arc between $\mathbf{R}(t_{i-1})$ and $\mathbf{R}(t_i)$ by Δs_i (Fig. 4.5). When n is large, the arc length Δs is small and may be approximated by $|\Delta\mathbf{R}_i$. Further, the force $\mathbf{F}$ is approximately constant on C_i, say it is equal to $\mathbf{F}_i = \mathbf{F}\,[x(\xi_i), y(\xi_i), z(\xi_i)]$, for some ξ_i, $t_{i-1} \le \xi_i \le t_i$ $(1 \le i \le n)$. Then the work done by $\mathbf{F}_i$ in moving the particle through Δs_i is approximately equal to

$$\Delta W_i = \mathbf{F}_i \cdot \Delta\mathbf{R}_i$$

Hence the work done by F in moving the particle along the curve C is approximately equal to the sum

$$\text{Work} = \sum_{i=1}^{n} \Delta W_i = \sum_{i=1}^{n} \mathbf{F}_i \cdot \Delta\mathbf{R}_i$$

Now by the mean value theorem of differential calculus, we can write

$$\Delta\mathbf{R}_i = \mathbf{R}(t_i) - \mathbf{R}(t_{i-1}) = \mathbf{R}'(\eta_i)\,\Delta t_i$$

where $t_{i-1} \le \eta_i \le t_i$ and $\Delta t_i = t_i - t_{i-1}$. Thus we have

$$\text{Work} = \sum_{i=1}^{n} \mathbf{F}_i \cdot \Delta\mathbf{R}_i = \sum_{i=1}^{n} \mathbf{F}_i \cdot \mathbf{R}'(\eta_i)\,\Delta t_i$$

This approximation improves as the partition of [a, b] is made finer by making n larger. In fact, the exact amount of work done is obtained when we take the limit of the sum as n tends to infinity. From our study of calculus, we know that this limit is precisely the integral of the scalar field $\mathbf{F}[x(t), y(t), z(t)] \cdot \mathbf{R}'(t)$ over the interval [a, b], that is,

$$\text{Work} = \lim_{n \to \infty} \sum_{i=1}^{n} \mathbf{F}_i \cdot \mathbf{R}'(\eta_i)\Delta t_i = \int_a^b \mathbf{F}[x(t), y(t), z(t)] \cdot \mathbf{R}'(t)\,dt$$

We call this integral the line integral of $\mathbf{F}$ on the curve $\mathbf{R}(t) = x(t)\mathbf{i} + y(t)\mathbf{j} + z(t)\mathbf{k}$ $(a \le t \le b)$. Thus, we define the line integral of a vector field on a curve as follows:

Definition 1. Let $\mathbf{F}(x, y, z) = P(x, y, z)\mathbf{i} + Q(x, y, z)\mathbf{j} + R(x, y, z)\mathbf{k}$ be a vector field defined and continuous in a domain D, and let C be a smooth curve in D represented by $\mathbf{R}(t) = x(t)\mathbf{i} + y(t)\mathbf{j} + z(t)\mathbf{k}$, $a \le t \le b$. The line integral of $\mathbf{F}$ on C, denoted by $\int_C \mathbf{F}\cdot d\mathbf{R}$, is the integral

$$\int_C \mathbf{F}\cdot d\mathbf{R} = \int_a^b \mathbf{F}[x(t), y(t), z(t)] \cdot \frac{d\mathbf{R}(t)}{dt}\,dt \tag{4.3}$$

Notice that the integral (4.3) is taken along the positive direction on the curve C. If the curve is traversed in the opposite direction, the integral changes sign. Further, as in the case of a line integral of a scalar field, the integral (4.3) does not depend on the parametrization of the curve C. That is,

$$\int_C \mathbf{F}\cdot d\mathbf{R} = \int_a^b \mathbf{F}[x(t), y(t), z(t)] \cdot \mathbf{R}'(t)\,dt$$

$$= \int_{\alpha}^{\beta} \mathbf{F}[x(\tau), y(\tau), z(\tau)] \cdot \mathbf{R}^{*\prime}(\tau)d\tau = \int_C \mathbf{F} \cdot d\mathbf{R}^*$$

where $\mathbf{R} = \mathbf{R}(t)$, $a \le t \le b$, and $\mathbf{R}^* = \mathbf{R}^*(t)$, $\alpha \le t \le \beta$, are two different representations of the curve maintaining the same orientation.

If $\mathbf{F}(x, y, z) = P(x, y, z)\mathbf{i} + Q(x, y, z)\mathbf{j} + R(x, y, z)\mathbf{k}$, the integral (4.3) can also be written as

$$\int_C \mathbf{F} \cdot d\mathbf{R} = \int_C P\,dx + Q\,dy + R\,dz$$

$$= \int_a^b \{ P[x(t), y(t), z(t)]\,x'(t) + Q[x(t), y(t), z(t)]\,y'(t) \qquad (4.4)$$

$$+ R[x(t), y(t), z(t)]\,z'(t)\}dt$$

Notice that the functions P, Q, and R are integrated along the curve Cwith respect to the variables x, y, and z, respectively, in accordance with the definition given in the preceding section.

Example 1. Calculate the line integral of $\mathbf{F}(x, y) = x^2 y\mathbf{i} + (x^2 + y)\mathbf{j}$ from the origin to the point (2, 4) along the parabola $y = x^2$.

Solution: We represent the parabola by $\mathbf{R}(t) = t\mathbf{i} + t^2\mathbf{j}$, $0 \le t \le 2$. Then on the curve we have

$$\mathbf{F}(t) \cdot \mathbf{R}'(t) = (t^4\mathbf{i} + 2t^2\mathbf{j}) \cdot (\mathbf{i} + 2t\mathbf{j}) = t^4 + 4t^3$$

so that

$$\int_C \mathbf{F} \cdot d\mathbf{R} = \int_0^2 (t^4 + 4t^3)\,dt = \frac{112}{5}$$

Alternatively, if we write

$$\int_C \mathbf{F} \cdot d\mathbf{R} = \int_C x^2 y\,dx + (x^2 + y)\,dy$$

then, using x as the parameter, we find

$$\int_C x^2y\,dx + (x^2 + y)\,dy = \int_0^2 x^4\,dx + (x^2 + x^2)\,2x\,dx$$

$$= \int_0^2 (x^4 + 4x^3)\,dx = \frac{112}{5}$$

Thus if **F** represents a force field that moves a particle from the origin to the point (2, 4) along the parabola, the value of the integral (in appropriate units, e. g., ft-lb) represents the amount of work done by the force on the particle.

Example 2. Calculate the line integral of $\mathbf{F}(x, y, z) = y\mathbf{i} - x\mathbf{j} + z\mathbf{k}$ along the helix C: $\mathbf{R}(t) = \cos t\,\mathbf{i} + \sin t\,\mathbf{j} + t\mathbf{k}$, $0 \le t \le \pi$.

Solution: We note that $\mathbf{R}'(t) = -\sin t\,\mathbf{i} + \cos t\,\mathbf{j} + \mathbf{k}$ so that on the helix we have

$$\mathbf{F}[\cos t, \sin t, t] \cdot \mathbf{R}'(t) = -\sin^2 t - \cos^2 t + t = -1 + t$$

Hence,

$$\int_C \mathbf{F} \cdot d\mathbf{R} = \int_0^\pi (-1 + t)\,dt = -\pi + \frac{\pi^2}{2}$$

Other Notations

We recall that the unit tangent vector on a curve defined by $\mathbf{R}(t)$, $a \le t \le b$, is given by

$$\mathbf{t} = \frac{\mathbf{R}'(t)}{|\mathbf{R}'(t)|}$$

Thus the line intgeral (4.3) can also be written as

$$\int_C \mathbf{F} \cdot d\mathbf{R} = \int_a^b \mathbf{F} \cdot \frac{\mathbf{R}'(t)}{|\mathbf{R}'(t)|}\,|\mathbf{R}'(t)|\,dt = \int_a^b \mathbf{F} \cdot \mathbf{t}\,|\mathbf{R}'(t)|\,dt$$

$$= \int_C \mathbf{F} \cdot \mathbf{t}\, ds$$

This says that the integral $\int_C \mathbf{F} \cdot d\mathbf{R}$ is simply the integral of the tangential component of the vector field **F** along the curve C. If **F** represents a force field, then indeed only the tangential component of the force along the curve contributes to the work done by **F**.

When C is a closed curve, it is customary to write the integral in the form

$$\oint_C \mathbf{F} \cdot d\mathbf{R}$$

The integration around the closed curve C is taken in the positive direction which is, by convention, counterclockwise.

Example 3. Evaluate the line integral of $\mathbf{F}(x, y) = x^2\,\mathbf{i} + \mathbf{j}$ around the circle $x^2 + y^2 = 4$.

Solution: We represent the circle in polar coordinates

$$x = 2\cos\theta, \quad y = 2\sin\theta, \quad 0 \le \theta < 2\pi$$

Then we have

$$\oint_C \mathbf{F} \cdot d\mathbf{R} = \int_0^{2\pi} [(4\cos^2\theta)(-2\sin\theta)\, d\theta + 2\sin\theta\,(2\cos\theta)\, d\theta]$$

$$= \int_0^{2\pi} (-8\cos^2\theta\sin\theta + 4\sin\theta\cos\theta)\, d\theta = 0$$

Example 4. Evaluate the integral of $\mathbf{F}(x, y, z) = (x + y^2)\mathbf{i} + (x + z)\mathbf{j} + xy\mathbf{k}$ from the origin to the point (1, -1, 1) along (a) the straight line joining the points; (b) the curve $\mathbf{R}(t) = t\,\mathbf{i} - t^2\,\mathbf{j} + t^3\,\mathbf{k}$ $(0 \le t \le 1)$.

Solution: (a) Let us represent the line segment that joins the origin and the point (1, -1, 1) by the parametric equations

$$L: \quad x = t, \quad y = -t, \quad z = t, \; 0 \le t \le 1$$

Then

$$\int_L \mathbf{F}\cdot d\mathbf{R} = \int_L (x + y^2)\,dx + (x + z)\,dy + xy\,dz$$

$$= \int_0^1 [t + (-t)^2 + (t + t)(-1) + t(-t)]\,dt = \int_0^1 -t\,dt = -\frac{1}{2}$$

(b) Along the curve C: $\mathbf{R}(t) = t\,\mathbf{i} - t^2\,\mathbf{j} + t^3\,\mathbf{k}$, $0 \le t \le 1$, we have

$$\int_C \mathbf{F}\cdot d\mathbf{R} = \int_C (x + y^2)\,dx + (x + z)\,dy + xy\,dz$$

$$= \int_0^1 [(t + t^4) + (t + t^3)(-2\,t) + t(-t^2)(3t^2)]\,dt$$

$$= \int_0^1 (t - 2t^2 - t^4 - 3t^5)\,dt = -\frac{13}{15}$$

This example shows that the value of the line integral from (0, 0, 0) to (1, -1, 1) depends on the curve that joins the points. This is true in general. Later we shall study vector fields whose line integrals from one point to another are independent of the curve joining the points.

Example 5. A particle of mass m is moving along a curve $\mathbf{R} = \mathbf{R}(t)$ under the action of a force $\mathbf{F} = \mathbf{F}(t)$ $(t > 0)$. If $\mathbf{v}(t)$ denotes the velocity of the particle at time t, show that

$$\int_{t_1}^{t_2} \mathbf{F}\cdot\mathbf{v}\,dt = \frac{1}{2}mv^2(t_2) - \frac{1}{2}mv^2(t_1)$$

where $v(t) = |\mathbf{v}(t)|$.

Solution: According to Newton's second law of motion, we have

$$\mathbf{F} = m\mathbf{a} = m\frac{d\mathbf{v}}{dt}$$

Then

$$\mathbf{F}\cdot\mathbf{v} = m\frac{d\mathbf{v}}{dt}\cdot\mathbf{v} = \frac{1}{2}m\frac{d}{dt}(\mathbf{v}\cdot\mathbf{v}) = \frac{1}{2}m\frac{d}{dt}(v^2)$$

hence

$$\int_{t_1}^{t_2} \mathbf{F}\cdot\mathbf{v}\,dt = \int_{t_1}^{t_2} \frac{1}{2}m\frac{d}{dt}(v^2)\,dt = \left|\frac{1}{2}m(v^2)\right|_{t_1}^{t_2}$$

which yields the desired result.

The quantity $(1/2)mv^2(t)$ is known as the kinetic energy of the particle at time t. Thus the formula says that the work done by **F** during the time interval $[t_1, t_2]$ is equal to the change in kinetic energy of the particle. This is a basic law in mechanics.

4.3 PROPERTIES OF LINE INTEGRALS

We have defined line integrals of scalar and vector fields in terms of ordinary integrals of a function of a single variable. As a result of this line integrals satisfy many algebraic properties which are similar to those satisfied by ordinary integrals. For example, suppose f and g are scalar fields and **F** and **G** are vector fields, all defined and continuous in a domain containing the smooth curve C: $\mathbf{R} = \mathbf{R}(t)$ $(a \le t \le b)$. Then for any constants c_1 and c_2, we have

$$\int_C (c_1 f + c_2 g)\,ds = c_1\int_C f\,ds + c_2\int_C g\,ds \tag{4.5}$$

and

$$\int_C (c_1\mathbf{F} + c_2\mathbf{G})\cdot d\mathbf{R} = c_1\int_C \mathbf{F}\cdot d\mathbf{R} + c_2\int_C \mathbf{G}\cdot d\mathbf{R} \tag{4.6}$$

This property is known as the linearity property of line integrals.

Next, suppose the curve C is comprised of two connected curves $C_1 : \mathbf{R} = \mathbf{R}_1(t)$ $(a \le t \le b)$ and $C_2 : \mathbf{R} = \mathbf{R}_2(t)$ $(c \le t \le d)$. Then

$$\int_C f\,ds = \int_{C_1} f\,ds + \int_{C_2} f\,ds \tag{4.7}$$

and

$$\int_C \mathbf{F}\cdot d\mathbf{R} = \int_{C_1} \mathbf{F}\cdot d\mathbf{R}_1 + \int_{C_2} \mathbf{F}\cdot d\mathbf{R}_2 \tag{4.8}$$

This is known as the additive property of line integrals. It corresponds to the familiar formula

$$\int_a^b g(x)\,dx = \int_a^c g(x)\,dx + \int_c^b g(x)\,dx \quad , a \le c \le b$$

of ordinary integrals.

More generally, if C is composed of *n* smooth curves $C_1, C_2, \ldots, C_n$, as shown in Fig. 4.6 for n = 5, then

$$\int_C f\,ds = \int_{C_1} f\,ds + \ldots + \int_{C_n} f\,ds \tag{4.9}$$

and

$$\int_C \mathbf{F}\cdot d\mathbf{R} = \int_{C_1} \mathbf{F}\cdot d\mathbf{R} + \ldots + \int_{C_n} \mathbf{F}\cdot d\mathbf{R} \tag{4.10}$$

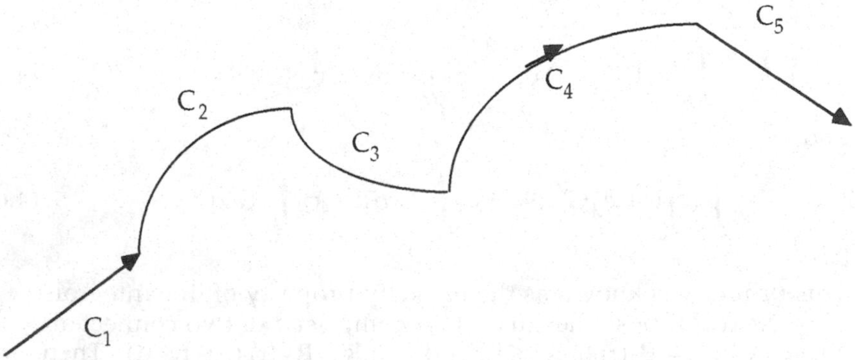

Fig. 4.6 Line integral along a piecewise smooth curve.

We observe that each of the integrals on the right-hand side of (4.9) and (4.10) may be evaluated using different parametrization so long as each parametrization of a curve preserves the original orientation of the curve.

Example 1. Evaluate the integral $\int_C xdx - zdy + 2ydz$, where C consists of the line segments from the origin to the point (1, 1, 0) and from (1, 1, 0) to the point (1, 1, 2) as shown in Fig. 4.7.

Solution: Referring to Fig. 4.7, we note that on C_1, $x = y$, $z = 0$, so that $dx = dy$, $dz = 0$, and on C_2, $x = 1$, $y = 1$ so that $dx = 0$, $dy = 0$. Hence we have

$$\int_C x\,dx - z\,dy + 2y\,dz = \int_{C_1} x\,dx + \int_{C_2} 2y\,dz$$

$$= \int_0^1 x\,dx + \int_0^2 2\,dz = \frac{9}{2}$$

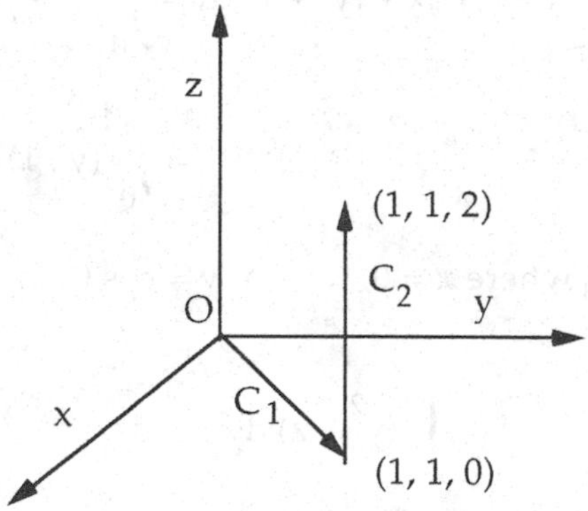

Fig. 4.7 Integrals along line segements.

Example 2. Evaluate the line integral

$$I = \int_C (x^2 + y)\,dx + (y^2 + z)\,dy + (z^2 + x)\,dz$$

around the closed curve C consisting of the line segments C_1: $x + z = 1$ $(0 \le x \le 1, y = 0)$ and C_2: $x + y = 1$ $(0 \le y \le 1, z = 0)$, and the quarter circle C_3: $y^2 + z^2 = 1$ $(y \ge 0, z \ge 0)$ in the direction indicated in Fig. 4.8.

Solution: We calculate the line integral on each of the line segments and on the quarter circle. On C_1 we have $y = 0$, $z = 1 - x$ so that $dy = 0$ and $dz = dx$. Hence, using x as the parameter, we find

$$I_1 = \int_{C_1} (x^2 + y)\,dx + (z^2 + x)\,dz = \int_0^1 x^2\,dx + [(1 - x)^2 + x](-\,dx)$$

$$= \int_0^1 (x - 1)\,dx = -\frac{1}{2}$$

On C_2, using y as the parameter, we have $x = 1 - y$, $z = 0$, so that $dx = -\,dy$, $dz = 0$. Hence

$$I_2 = \int_{C_2} (x^2 + y)\,dx + (y^2 + z)\,dy = \int_0^1 [(1 - y)^2 + y](-\,dy) + y^2\,dy$$

$$= \int_0^1 (y - 1)\,dy = -\frac{1}{2}$$

Finally, on C_3 where $x = 0$, we set $y = \cos\theta$, $z = \sin\theta$ $(0 \le \theta \le \pi/2)$. Then

$$I_3 = \int_{C_3} (y^2 + z)\,dy + z^2\,dz$$

$$= \int_0^{\pi/2} [(\cos^2\theta + \sin\theta)(-\sin\theta) + \sin^2\theta\cos\theta]\,d\theta$$

$$= \int_0^{\pi/2} (-\cos^2\theta\sin\theta - \sin^2\theta + \sin^2\theta\cos\theta)\,d\theta$$

$$= \frac{\cos^3\theta}{3} - \left(\frac{\theta}{2} - \frac{\sin 2\theta}{4}\right) + \frac{\sin^3\theta}{3}\Big|_0^{\pi/2} = -\left(\frac{\pi}{4}\right)$$

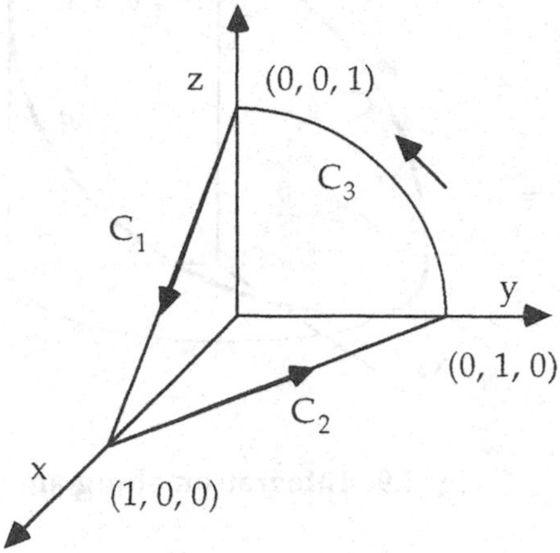

Fig. 4.8 Integration around a closed curve.

Thus the line integral around the closed curve is equal to the sum

$$I_1 + I_2 + I_3 = -1 - \pi/4$$

Example 3. Evaluate the line integral $\int_C ydx - xdy + zdz$ around the curve of intersection of the cylinder $x^2 + y^2 = a^2$ and the plane $z - y = a$ taken in the counterclockwise direction as shown in Fig. 4.9.

Solution: First, we need to find a parametrization of the curve of intersection of the cylinder and the plane. We note that every point on the cylinder $x^2 + y^2 = a^2$ may be represented in cylindrical coordinates by the equations

$$x = a\cos\theta, \quad y = a\sin\theta, \quad z = z \quad (0 \le \theta < 2\pi)$$

Since the curve of intersection consists of points lying on the cylinder whose third coordinate z is given by $z = a + y$, we have

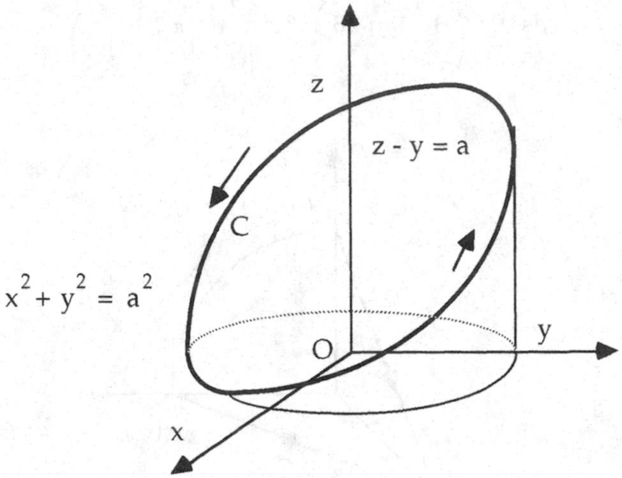

Fig. 4.9 Integration along an ellipse.

$$x = a\cos\theta\,,\quad y = a\sin\theta\,,\quad z = a + a\sin\theta \quad (0 \le \theta < 2\pi)$$

a parametric representation of the curve of intersection. Observe that the curve is an ellipse. Thus we have

$$\oint_C y\,dx - x\,dy + z\,dz =$$

$$\int_0^{2\pi} [a\sin\theta\,(-\,a\sin\theta) - a\cos\theta(a\cos\theta) + a(1+\sin\theta)(a\cos\theta)]d\theta$$

$$= \int_0^{2\pi} [-a^2 + a^2(1+\sin\theta)\cos\theta]\,d\theta$$

$$= \left|-a^2\theta + a^2\frac{(1+\sin\theta)^2}{2}\right|_0^{2\pi} = -2\pi a^2$$

4.2 EXERCISES

In each of Problems 1 through 8, compute the line integral of the vector field **F** along the curve described.

1. $\mathbf{F}(x, y) = (x^2 - y)\mathbf{i} + (y^2 - x)\mathbf{j}$ from the origin to the point (1, 2) along the parabola $y = \sqrt{2}\, x$.
2. $\mathbf{F}(x, y) = (x^2 + y^2)\mathbf{i} - xy\mathbf{j}$ from the point (1, 1) to (-2, 4) along the parabola $y = x^2$.
3. $\mathbf{F}(x, y) = 2xy\mathbf{i} + (x^2 - y^2)\mathbf{j}$ from (2, 0) to (0, 2) along the first quadrant arc of $x^2 + y^2 = 4$.
4. $\mathbf{F}(x, y) = (y^2 - 2xy)\mathbf{i} + (x^2 + 2xy)\mathbf{j}$ from (-1, 0) to (1, 0) along the upper semi-circle of radius 1.
5. $\mathbf{F}(x, y, z) = y(x - z)\mathbf{i} + z(y - x)\mathbf{j} + x(z - y)\mathbf{k}$ from (1, 1, 1) to (2, -2, 3) along the line segment joining the points.
6. $\mathbf{F}(x, y, z) = (x - z)\mathbf{i} + (y - x)\mathbf{j} + (z - y)\mathbf{k}$ from the origin to (1, 1, 1) along the curve $\mathbf{R}(t) = t\mathbf{i} + t^2\mathbf{j} + t^3\mathbf{k}$ $(0 \le t \le 1)$.
7. $\mathbf{F}(x, y, z) = (y^2 + z)\mathbf{i} + (z^2 + x)\mathbf{j} + (x^2 + y)\mathbf{k}$ from (1, 0, 0) to (0, 1, 0) to (0, 0, 1) along the line segments joining the points.
8. $\mathbf{F}(x, y, z) = (y - 1)\mathbf{i} + (x + 1)\mathbf{j} + 2z\mathbf{k}$ from (0, a, 0) to (a, 0, π) along the helix $\mathbf{R}(t) = a(\sin t)\mathbf{i} + a(\cos t)\mathbf{j} + 2t\mathbf{k}$ $(0 \le t \le \pi/2)$.
9. Evaluate the line integral

$$\int_C \frac{-y\,dx + x\,dy}{x^2 + y^2}$$

in each of the two cases:
(a) when C is the sides of the square $|x| = 1$, $|y| = 1$ oriented counterclockwise.
(b) when C is the unit circle $x^2 + y^2 = 1$ oriented counterclockwise.

10. Evaluate the integral

$$\int_C \frac{(x + y)\,dx - (x - y)\,dy}{x^2 + y^2}$$

around the circle $x^2 + y^2 = a^2$ in the counterclockwise direction.

11. Evaluate the integral $\int_C y\,dx - x\,dy + z\,dz$ around the intersection of the cylinder $x^2 + y^2 = 1$ and the plane $y + z = 1$ in the clockwise direction as viewed from the origin.

12. Evaluate the integral $\int_C y dx - y(x-1)dy + y^2 z dz$ around the intersection of the hemisphere $x^2 + y^2 + z^2 = 4$ $(z \geq 0)$ and the cylinder $(x-1)^2 + y^2 = 1$ in the clockwise direction as viewed from the origin.
13. Calculate the work done by the force field $\mathbf{F}(x, y) = (x^2 - y^2)\mathbf{i} + 2xy\mathbf{j}$ in moving a particle once around the (a) sides of the square $|x| = 1$, $|y| = 1$, (b) unit circle $x^2 + y^2 = 1$, both in the counterclockwise direction.
14. A force field is given by $\mathbf{F}(x, y, z) = (y - z)\mathbf{i} + (z - x)\mathbf{j} - (x - y)\mathbf{k}$. Find the work done by this force field in moving a particle once around the intersection of the hemisphere $x^2 + y^2 + z^2 = 4$ $(z \geq 0)$ and the cylinder $x^2 + (y-1)^2 = 1$ traced in the clockwise direction as viewed from the origin.
15. A particle moves along a path C: $\mathbf{R}(t) = \sin t\,\mathbf{i} + \cos t\,\mathbf{j} + t\,\mathbf{k}$ $(t \geq 0)$ under the influence of a force $\mathbf{F}(x, y, z) = (y - 1)\mathbf{i} + z\mathbf{j} + x\mathbf{k}$. Suppose the particle leaves its path and moves along the tangent at $t = \pi/2$, calculate the work done by the force on the particle in the time interval $[\pi/2, \pi]$.
16. Evaluate $\int_C f\, ds$, where $f(x, y, z) = x^2 - y + z$ and C is the broken line segments joining the origin to the point (1, 1, 0) and (1, 1, 0) to the point (1, 1, 2).
17. Evaluate $\int_C (xy - z) ds$ from (1, 0, 0) to (0, 0, 1) along the line segment $x + y = 1$ joining (1, 0, 0) to (0, 1, 0) and then along the quarter circle $y^2 + z^2 = 1$ $(y > 0, z > 0)$ joining (0, 1, 0) to (0, 0, 1).
18. Let $\mathbf{R}(t)$ $(a \leq t \leq b)$ describe a smooth curve C, and let $t = \phi(\tau)$ $(\alpha \leq \tau \leq \beta)$, where ϕ is a differentiable function such that $\phi' > 0$. If $\phi(\alpha) = a$ and $\phi(\beta) = b$, show that $\mathbf{R}^*(\tau) = \mathbf{R}(\phi(\tau))$ describes the same curve and preserves the orientation of the curve. HINT: Show that $\mathbf{R}'(t)$ points in the same direction as $\mathbf{R}^{*\prime}(\tau)$. Thus establish the validity of the integral

$$\int_C \mathbf{F} \cdot d\mathbf{R} = \int_C \mathbf{F} \cdot d\mathbf{R}^*$$

19. Let $\mathbf{R}(t)$ $(a \leq t \leq b)$ describes a smooth curve C, and suppose that $\mathbf{F}$ is parallel to the tangent vector at each point on C. Show that

$$\int_C \mathbf{F} \cdot d\mathbf{R} = \int_C |\mathbf{F}|\, ds$$

20. Show that

$$\left| \int_C \mathbf{F} \cdot d\mathbf{R} \right| \le ML$$

where $M = \max |\mathbf{F}|$ on C and L is the length of the curve.

4.4 LINE INTEGRALS INDEPENDENT OF PATH

Let **F**(x, y, z) be a vector field that is defined and continuous in a domain D, and let A and B be any two points in D. The line integral of **F** from the point A to the point B will generally depend on the curve C that joins the points. If the value of the integral from A to B remains the same however we choose the curve that connects the points, then we say that the integral of **F** is independent of path in D. The integral depends only on the points A and B. In such a case, we write the integral as

$$\int_A^B \mathbf{F} \cdot d\mathbf{R}$$

An immediate result arising from an integral being independent of path is contained in the following theorem.

Theorem 1. Let **F** be continuous in a domain D. The integral of **F** is independent of path in D if and only if

$$\oint_C \mathbf{F} \cdot d\mathbf{R} = 0$$

for any piecewise smooth simple closed curve C in D.

Proof: We note that a simple closed curve is a closed curve that does not cross itself. So suppose that the integral of **F** around any simple closed curve in D is zero. Let A and B be any two points in D, and let C_1 and C_2 be any two curves that join A to B (Fig. 4.10). Then we can form a closed curve C consisting of C_1 from A to B and $\sim C_2$ from B to A, opposite in direction to C_2. Since the line integral around C is zero, we then have

$$\oint_C \mathbf{F} \cdot d\mathbf{R} = \int_{C_1} \mathbf{F} \cdot d\mathbf{R} + \int_{\sim C_2} \mathbf{F} \cdot d\mathbf{R}$$

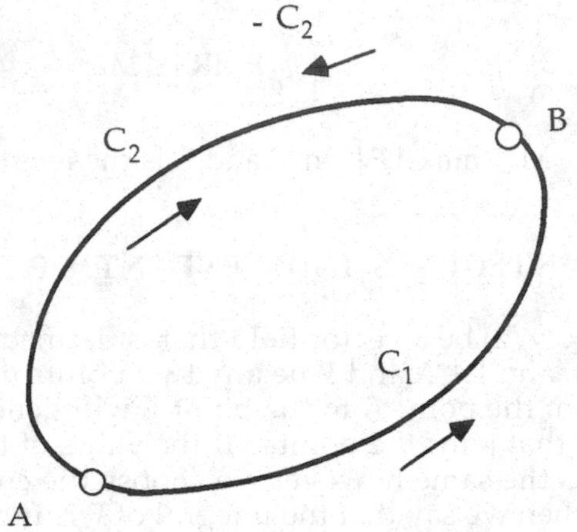

Fig. 4.10 Integral around a closed curve.

$$= \int_{C_1} \mathbf{F} \cdot d\mathbf{R} - \int_{C_2} \mathbf{F} \cdot d\mathbf{R} = 0$$

This implies

$$\int_{C_1} \mathbf{F} \cdot d\mathbf{R} = \int_{C_2} \mathbf{F} \cdot d\mathbf{R}$$

Thus the line integral of **F** from A to B along C_1 is the same as that along C_2. Therefore, $\int \mathbf{F} \cdot d\mathbf{R}$ is independent of path in D.

On the other hand, suppose $\int \mathbf{F} \cdot d\mathbf{R}$ is independent of path in D. If C is a simple closed curve in D, we can choose two points, say A and B, on C and split C into two parts: C_1 from A to B and C_2 from B to A. Then we have

$$\oint_C \mathbf{F} \cdot d\mathbf{R} = \int_{C_1} \mathbf{F} \cdot d\mathbf{R} + \int_{C_2} \mathbf{F} \cdot d\mathbf{R}$$

$$= \int_A^B \mathbf{F} \cdot d\mathbf{R} + \int_B^A \mathbf{F} \cdot d\mathbf{R}$$

$$= \int_A^B \mathbf{F}\cdot d\mathbf{R} - \int_A^B \mathbf{F}\cdot d\mathbf{R} = 0$$

Caution: Notice that the theorem requires that the integral $\int_C \mathbf{F}\cdot d\mathbf{R}$ must be zero on every closed curve in D in order for the integral to be independent of path in D. In other words, the line integral may be zero for a particular closed curve or even for infinitely many closed curves, and still the integral may not be independent of path (see Example 5). For this reason, Theorem 1 is only useful for establishing that an integral is not independent of path by showing that $\int_C \mathbf{F}\cdot d\mathbf{R} \neq 0$ for a certain closed curve C.

Necessary Condition for Independence of Path - Conservative Field

We now discuss a necessary and a sufficient condition for the line integral of a vector field to be independent of path. First, we show that if the line integral of **F** is independent of path in a domain D, then **F** must be a conservative field; that is, there exists in D a scalar field ϕ, called a potential function, such that $\mathbf{F} = \nabla\phi$. Thus **F** being a conservative field is a necessary condition for $\int \mathbf{F}\cdot d\mathbf{R}$ to be independent of path in D. To prove this, we define a function ϕ by the line integral of **F** from any fixed point (x_o, y_o, z_o) to an arbitrary point (x, y, z) in D, that is,

$$\phi(x, y, z) = \int_{(x_o,y_o,z_o)}^{(x,y,z)} \mathbf{F}\cdot d\mathbf{R} = \int_{(x_o,y_o,z_o)}^{(x,y,z)} P\,d\xi + Q\,d\eta + R\,d\zeta \qquad (4.11)$$

We shall show that $\nabla\phi = \mathbf{F}$, so that the function (4.11) is a potential function for **F**.

Indeed, since the integral is independent of path, we may choose a path from (x_o, y_o, z_o) to (x, y, z) to consist of a smooth curve from (x_o, y_o, z_o) to (x_1, y, z) and the line segment from (x_1, y, z) to (x, y, z) as indicated in Fig. 4.11. By the additive property of line integrals, we have

$$\phi(x, y, z) = \int_{(x_o,y_o,z_o)}^{(x_1,y,z)} \mathbf{F}\cdot d\mathbf{R} + \int_{(x_1,y,z)}^{(x,y,z)} \mathbf{F}\cdot d\mathbf{R}$$

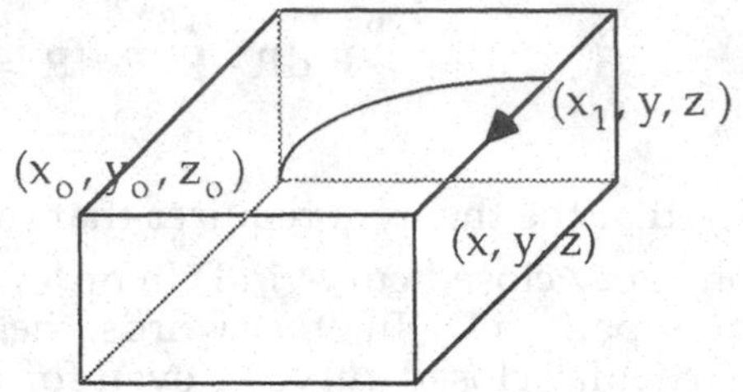

Fig. 4.11 An integral independent of path.

$$= \phi(x_1, y, z) + \int_{(x_1,y,z)}^{(x,y,z)} \mathbf{F}\cdot d\mathbf{R}$$

$$= \phi(x_1, y, z) + \int_{x_1}^{x} P(\xi, y, z)\, d\xi$$

Notice that along the line segment joining (x_1, y, z) and (x, y, z), the coordinates y and z remain fixed so that $dy = 0$, and $dz = 0$. Now, differentiating $\phi(x, y, z)$ with respect to x, we obtain

$$\frac{\partial \phi}{\partial x}(x, y, z) = P(x, y, z)$$

Similarly, if we integrate along a smooth curve from (x_o, y_o, z_o) to (x, y_1, z) and then along a line segment from (x, y_1, z) to (x, y, z), we obtain

$$\phi(x, y, z) = \phi(x, y_1, z) + \int_{(x,y_1,z)}^{(x,y,z)} \mathbf{F}\cdot d\mathbf{R}$$

$$= \phi(x, y_1, z) + \int_{y_1}^{y} Q(x, \eta, z)\, d\eta$$

Differentiating this with respect to y, we find

$$\frac{\partial \phi}{\partial y}(x, y, z) = Q(x, y, z)$$

The final result

$$\frac{\partial \phi}{\partial z}(x, y, z) = R(x, y, z)$$

is established in the same way. Thus we have shown that $\nabla\phi = \mathbf{F}$, so that **F** is a conservative field.

Since $\nabla \times \nabla\phi = \mathbf{curl}\ (\text{grad}\ \phi) = 0$, an equivalent *necessary condition* for $\int \mathbf{F}\cdot d\mathbf{R}$ to be independent of path is $\mathbf{curl\ F} = \mathbf{0}$. Therefore, if $\mathbf{curl\ F} \neq \mathbf{0}$, we can immediately conclude that the line integral of **F** is path dependent.

Sufficient Condition - Conservative Field and Connectivity

The converse of the preceding result is not true unless we make some restriction about the domain D in which $\mathbf{curl\ F} = \mathbf{0}$. In other words, we may have $\mathbf{curl\ F} = \mathbf{0}$ in D and yet the line integral of **F** is not independent of path in D (see Examples 1, 2 below). In order for the converse to hold, we require that D be *simply connected*. By this we mean that D has the property that whenever a simple closed curve lies in D, all its interior also lies in D. Thus, for example, a domain bounded by a circle or a polygon is simply connected, whereas a domain bounded by two concentric circles is not. However, the domain bounded by two concentric spheres is simply connected, whereas the interior of a torus (think of a donut) is not (see Fig. 4. 12).

So suppose $\mathbf{F} = \nabla\phi$, or equivalently, $\mathbf{curl\ F} = 0$ in a domain D, where D is simply connected. Let A and B be any two points in D and let C: $\mathbf{R}(t) = x(t)\,\mathbf{i} + y(t)\,\mathbf{j} + z(t)$, $a \le t \le b$, be a piecewise smooth curve joining A and B. We will show that the line integral $\int \mathbf{F}\cdot d\mathbf{R}$ from A to B does not depend on C. Now on the curve C we have

$$\mathbf{F}\cdot\frac{d\mathbf{R}}{dt} = \nabla\phi\cdot\frac{d\mathbf{R}}{dt} = \frac{\partial\phi}{\partial x}\frac{dx}{dt} + \frac{\partial\phi}{\partial y}\frac{dy}{dt} + \frac{\partial\phi}{\partial y}\frac{dy}{dt} = \frac{d\phi}{dt}$$

Hence

$$\int_A^B \mathbf{F}\cdot d\mathbf{R} = \int_a^b \mathbf{F}\cdot\frac{d\mathbf{R}}{dt}\,dt = \int_a^b \frac{d\phi}{dt}dt$$

$$= \phi[x(b), y(b), z(b)] - \phi[x(a), y(a), z(a)]$$

$$= \phi(B) - \phi(A)$$

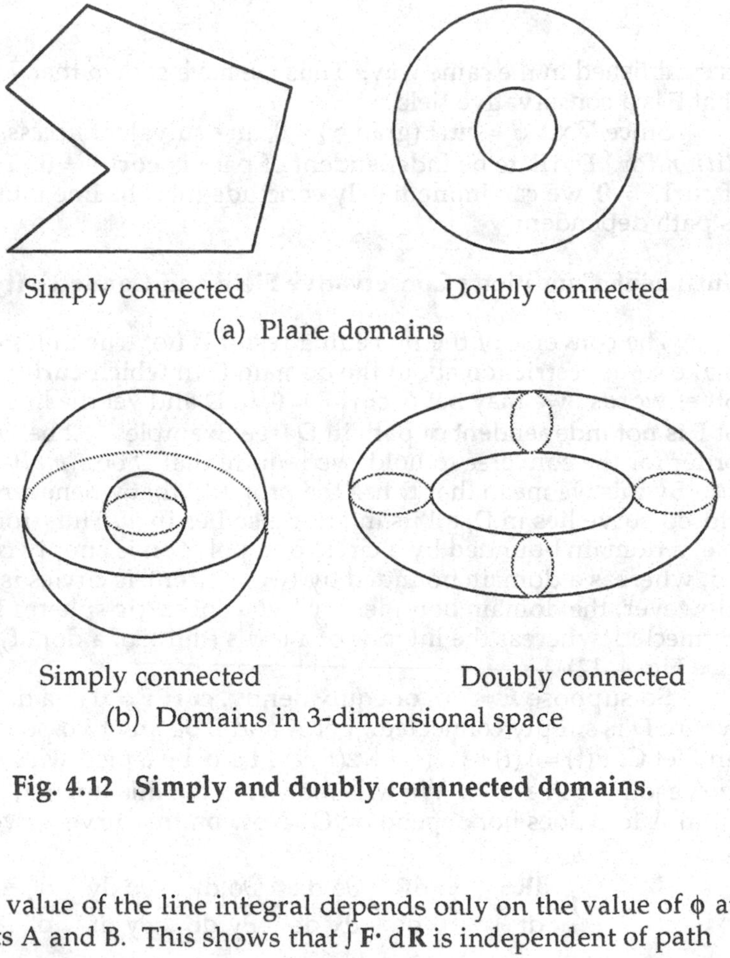

Fig. 4.12 Simply and doubly connnected domains.

Thus the value of the line integral depends only on the value of ϕ at the points A and B. This shows that $\int \mathbf{F} \cdot d\mathbf{R}$ is independent of path in D.

We summarize the results obtained above in the following theorem.

Theorem 2. Let $\mathbf{F} = P\mathbf{i} + Q\mathbf{j} + R\mathbf{k}$ be a continuous vector field in a domain D, and let A, B be any two points in D. If the line integral

$$\int_A^B \mathbf{F}\cdot d\mathbf{R} = \int_A^B P\,dx + Q\,dy + R\,dz \tag{4.12}$$

is independent of the path joining A and B, then **curl F = 0** or, equivalently, **F** is a conservative field, $\mathbf{F} = \nabla\phi$. Conversely, if **curl F = 0** in D, and D is simply connected, then the line integral of **F** in D is independent of path, and

$$\int_A^B \mathbf{F}\cdot d\mathbf{R} = \int_A^B \nabla\phi\cdot d\mathbf{R} = \phi(B) - \phi(A) \tag{4.13}$$

for any two points A and B in D, where ϕ is a function such that $\nabla\phi = \mathbf{F}$.

Example 1. Consider the vector field

$$\mathbf{F}(x, y) = \frac{-y}{x^2+y^2}\mathbf{i} + \frac{x}{x^2+y^2}\mathbf{j}, \qquad (x, y) \neq (0, 0)$$

It is easily verified that $\mathbf{F}(x, y) = \nabla\phi$, where $\phi = \arctan(y/x)$, so that **F** is a conservative field in the domain $D = \{(x, y) \mid (x, y) \neq (0, 0)\}$. However, since the origin has to be excluded, the domain has a "hole" and, therefore, it is not simply connected. Indeed let C be any closed curve C that encloses the origin. Then we see that

$$\oint_C \mathbf{F}\cdot d\mathbf{R} = \oint_C \nabla\phi\cdot d\mathbf{R} = \int_0^{2\pi} d\phi = 2\pi$$

Hence, by Theorem 1, the integral $\int \mathbf{F}\cdot d\mathbf{R}$ is not independent of path in D. However, in any simply connected domain that does not contain the origin, $\int_C \mathbf{F}\cdot d\mathbf{R} = 0$ for every closed curve C in D, which shows that the integral is independent of path. For example, if we introduce a "cut" in the xy-plane by deleting the entire non-positive x-axis, then the resulting domain D* becomes simply connected (no closed curve can now enclose the origin), and so $\int \mathbf{F}\cdot d\mathbf{R} = \int d\phi$ will vanish on every closed curve in D*.

Example 2. Consider next the vector field

$$\mathbf{F}(x, y) = \frac{x}{x^2 + y^2}\mathbf{i} + \frac{y}{x^2 + y^2}\mathbf{j} \qquad (x, y) \neq (0, 0)$$

Again it is easily verified that $\mathbf{F}(x, y) = \nabla\phi$, where $\phi = \ln r$, $r^2 = x^2 + y^2$. Thus **F** is a conservative field in the doubly connected domain D = { (x, y) | (x, y) ≠ (0, 0) }. For any two points A and B in D, we find

$$\int_A^B \mathbf{F}\cdot d\mathbf{R} = \int_A^B \nabla\phi\cdot d\mathbf{R} = \int_A^B d[\ln r]$$

$$= \ln r_B - \ln r_A$$

where r_A are r_B denote the respective distance of the points A and B from the origin. Thus the value of the line integral depends only on the points A and B. Hence the line integral is independent of path although the domain {(x, y) | (x, y) ≠ (0, 0)} is not simply connected.

This example shows that in Theorem 2 the condition that D be simply connected is merely a sufficient condition and not a necessary one.

Example 3. Let $\mathbf{F} = \nabla\phi$, where $\phi = x^2 y + yz + z^2$. Calculate the line integral of **F** from (1, -1, 1) to (2, 1, 3) along any path.

Solution: Since **F** is a conservative field, its line integral depends only on the endpoints (1, -1, 1) and (2, 1, 3). Hence, we have

$$\int_{(1,-1,1)}^{(2,1,3)} \mathbf{F}\,.\,d\mathbf{R} = \phi(2, 1, 3) - \phi(1, -1, 1) = \Big| x^2 y + yz + z^2 \Big|_{(1,-1,1)}^{(2,1,3)} = 17$$

Example 4. Given that $\mathbf{F}(x, y, z) = (\sin y + z)\,\mathbf{i} + (x\cos y + e^z)\,\mathbf{j} + (x + ye^z)\,\mathbf{k}$ is a conservative vector field. Calculate the integral

$$\int_{(0,0,0)}^{(1,\pi,2)} \mathbf{F}\cdot d\mathbf{R}$$

Solution: Since the vector field is conservative, the integral is independent of path. So we may choose any path from the origin to the point (1, π, 2). We choose the path consisting of the line segments joining (0, 0, 0) to (1, 0, 0), (1, 0, 0) to (1, π, 0), and finally (1, π ,

0) to $(1, \pi, 2)$. On each of these line segments the integral reduces to just one term. Thus we obtain

$$\int_{(0,0,0)}^{(1,\pi,2)} \mathbf{F} \cdot d\mathbf{R} = \int_0^{\pi} (\cos y + 1)dy + \int_0^2 (1 + \pi e^z)dz$$

Example 5. Show that the line integral of $\mathbf{F} = xy\,\mathbf{i} + y\,\mathbf{j}$ is zero for any circle with center at the origin, but that $\mathbf{F}$ is not a conservative vector field.

Solution: Consider any circle of radius r with center at the origin and represent it by the parametric equations

$$C: \; x = r\cos\theta, \; y = r\sin\theta, \quad 0 \le \theta < 2\pi$$

Then we find

$$\oint_C \mathbf{F}\cdot d\mathbf{R} = \oint_C xy\,dx + y\,dy$$

$$= \int_0^{2\pi} (-r^3\cos\theta\sin^2\theta + r^2\sin\theta\cos\theta)\,d\theta = 0$$

On the other hand, if we take C to be the sides of the square with vertices at (0, 0), (1, 0), (1, 1), and (0, 1), taken counterclockwise, we find

$$\oint_C \mathbf{F}\cdot d\mathbf{R} = \int_0^1 y\,dy + \int_1^0 x\,dx + \int_1^0 y\,dy = -\frac{1}{2}$$

Therefore, by Theorem 1, the integral is not independent of path and hence $\mathbf{F}$ is not a conservative field.

Example 6. Show that the integral of $\mathbf{F} = (2xy + z^2)\,\mathbf{i} + x^2\,\mathbf{j} + 2xz\,\mathbf{k}$ around the intersection of the plane $x = y$ and the sphere $x^2 + y^2 + z^2 = a^2$ (great circle) is zero.

Solution: To obtain a parametric representation of the intersection of the plane $x = y$ and the sphere, we note that the plane is defined in spherical coordinates by $\theta = \pi/4$, and the sphere is defined by $r = a$. Hence their intersection has the parametric representation

$$C: \; x = \frac{\sqrt{2}}{2} a\sin\phi, \; y = \frac{\sqrt{2}}{2} a\sin\phi, \; z = a\cos\phi, \quad 0 \le \phi \le 2\pi$$

Therefore,

$$\oint_C (2xy + z^2)\, dx + x^2 dy + 2xz\, dz$$

$$= \int_0^{2\pi} \left[2\left(\frac{\sqrt{2}}{2} a \sin\phi\right)^2 + (a\cos\phi)^2 + \left(\frac{\sqrt{2}}{2} a \sin\phi\right)^2 \right] \frac{\sqrt{2}}{2} a \cos\phi \, d\phi$$

$$+ \int_0^{2\pi} 2\left(\frac{\sqrt{2}}{2} a \sin\phi\right) a\cos\phi\,(- a \sin\phi)\, d\phi$$

$$= \int_0^{2\pi} \left(\frac{\sqrt{2}}{2} a^3 \cos\phi + \frac{\sqrt{2}}{4} a^3 \sin^2\phi \cos\phi - \sqrt{2} a^3 \sin^2\phi \cos\phi \right) d\phi$$

$$= 0$$

Principle of Conservation of Energy

Suppose we have a continuous conservative force field **F** having ϕ as its potential function. Then, according to Theorem 2, the work done by **F** in moving a particle from a point A to a point B is independent of the path joining A and B, and it is given by

$$W = \int_A^B \mathbf{F} \cdot d\mathbf{R} = \phi(B) - \phi(A) \tag{4.14}$$

But according to Example 5 of Sec. 4.2, this work is also equal to the change in the kinetic energy of the particle at the points A and B. Therefore, if we denote by K(P) the kinetic energy of the particle at the point P, then

$$K(B) - K(A) = \phi(B) - \phi(A)$$

or

$$K(B) - \phi(B) = K(A) - \phi(A) \tag{4.15}$$

The scalar quantity $-\phi(P)$ is called the potential energy of the particle at the point P. Thus equation (4.17) says that in a conservative force field the sum of the kinetic and potential energies of a particle moving in that field is constant. This is the statement of the well known principle of conservation of energy in mechanics. For this reason a force field in which this principle holds is called a conservative field. In this context, Theorem 2 implies that in a conservative field when a particle is moved around a closed curve back to its starting point, the net work done is zero.

A very important example of a conservative field is, of course, the gravitational force field considered in Example 4 of Sec. 3.6.

4.3 EXERCISES

1. Let **F** be a continuous vector field with continuous first derivative in a domain D. If the line integral of **F** is independent of path, show that **curl F** = **0** at each point in D.
2. Using the result of Problem 1, verify that for each of the following vector fields the line integral is not independent of path.
   (a) $\mathbf{F}(x, y) = (x^2 + y)\mathbf{i} + (y^2 - x)\mathbf{j}$
   (b) $\mathbf{F}(x, y) = (y + x \cos y)\mathbf{i} + (x + y \sin x)\mathbf{j}$
   (c) $\mathbf{F}(x, y, z) = (x^2 + \sin y + z)\mathbf{i} + (y \cos y + 1)\mathbf{j} + (z^2 - xy)\mathbf{k}$
   (d) $\mathbf{F}(x, y, z) = (2xy - z^2)\mathbf{i} + (y^2 + x \cos z)\mathbf{j} + x^2 z\mathbf{k}$

In each of Problems 3 through 9, the given vector field is a conservative vector field. Verify that **curl F** = **0** and determine a potential function for each of them.

3. $\mathbf{F}(x, y) = [x/(x^2 + y^2)]\mathbf{i} + [y/(x^2 + y^2)]\mathbf{j}$
4. $\mathbf{F}(x, y) = (y^2 \cos x + 2xe^y)\mathbf{i} + (2y \sin x + x^2e^y)\mathbf{j}$
5. $\mathbf{F}(x, y, z) = (y z^2 e^{xy} - y \sin x)\mathbf{i} + (\cos x + x z^2 e^{xy})\mathbf{j} + (2ze^{xy} + 1)\mathbf{k}$
6. $\mathbf{F}(x, y, z) = (ye^z - z)\mathbf{i} + (xe^z + 2y)\mathbf{j} + (xye^z - x)\mathbf{k}$
7. $\mathbf{F}(x, y, z) = (2xz - ye^{-x})\mathbf{i} + (\cos(z - 1) + e^{-x})\mathbf{j} + [x^2 - y \sin(z - 1)]\mathbf{k}$
8. $\mathbf{F}(x, y, z) = (x\mathbf{i} + y\mathbf{j} + z\mathbf{k})/(x^2 + y^2 + z^2)$, $(x, y, z) \neq (0, 0, 0)$
9. $\mathbf{F}(x, y, z) = (3e^z + 2xy)\mathbf{i} + (x^2 + z \sin y)\mathbf{j} + (3xe^z - \cos y)\mathbf{k}$
10. Show that the line integral $\int_C \mathbf{F} \cdot d\mathbf{R}$ of the vector field $\mathbf{F}(x, y) = (x^2 + xy)\mathbf{i} + (x^2 + y^2)\mathbf{j}$ is zero when C is (a) any circle with center at the origin; (b) the sides of the square $|x| + |y| = 1$; (c) the sides of the triangle with vertices at (-a, 0), (0, b), (a, 0). Can you generalize this result for arbitrary C? Is **F** a conservative field? Does this contradict Theorem 1 or Theorem 2?
11. Show that for the vector field $\mathbf{F}(x, y) = (x + y^2)\mathbf{i} + (xy - 1)\mathbf{j}$, $\int_C \mathbf{F} \cdot d\mathbf{R} = 0$ when (a) C is any circle with center on the x-axis,

(b) C is the sides of the square with vertices at (0, 0), (1, -1), (2, 0), (1, 1). Is **F** is a conservative field?

12. Show that $\mathbf{F} = (2xy + z^3)\mathbf{i} + (x^2 + 2y)\mathbf{j} + 3xz^2\mathbf{k}$ is a conservative field and calculate its line integral from (0, 1, 1) to (1, 2, 3) along any path.
13. Show that $\mathbf{F} = (2xy + e^z\cos x)\mathbf{i} + (x^2 + z)\mathbf{j} + (e^z\sin x + y)\mathbf{k}$ is a conservative field and calculate its line integral from the origin to the point $(\pi/2, 4, 1)$ along any path.
14. Evaluate the integral $\int_C \mathbf{F}\cdot d\mathbf{R}$, where $\mathbf{F} = y\mathbf{i} + z\mathbf{j} + x\mathbf{k}$ and C is the helix $\mathbf{R}(t) = \cos t\,\mathbf{i} + \sin t\,\mathbf{j} + t\mathbf{k}$, $0 \le t \le \pi$.
15. Evaluate the integral

$$\int_{(1,\,0,\,0)}^{(2,\,1,\,3)} \frac{x\,dx + y\,dy}{x^2 + y^2} + z\,dz$$

along the line segment joining the points.

4.5 GREEN'S THEOREM IN THE PLANE

For a vector field defined in a domain of the xy-plane, there is a fundamental theorem which converts its line integral around a simple closed curve into a double integral over the domain enclosed by the curve. The theorem is known as Green's theorem, in honor of the English mathematician G. Green of the 19th century. The theorem is one of the great theorems of mathematical analysis. Its generalizations to three dimensions provide the foundation for important theorems in applied mathematics. In this section we state the theorem in its general form and prove it for special domains having certain properties.

Theorem 1. Let D be a domain in the xy-plane bounded by a piecewise smooth simple closed curve C. If P, Q and their first partial derivatives are continuous in D and on C, then

$$\oint_C P(x, y)dx + Q(x, y)dy = \int\int_D \left(\frac{\partial Q}{\partial x} - \frac{\partial P}{\partial y} \right) dx\,dy \qquad (4.16)$$

where the line integral around C is taken in the positive (counterclockwise) direction.

Proof: We observe that it is sufficient to establish the following two identities:

$$\oint_C P(x, y)dx = -\int\int_D \frac{\partial P}{\partial y} dx\, dy \tag{4.17}$$

and

$$\oint_C Q(x, y)dy = \int\int_D \frac{\partial Q}{\partial x} dx\, dy \tag{4.18}$$

since their sum leads to the formula (4.16). We establish these identities in the special case where the boundary C of D can be split into two parts with two different representations. We assume that C can be represented by:

$$C_1:\ y = f(x),\quad C_2:\ y = g(x),\quad f(x) < g(x),\quad a \le x \le b$$

and by

$$C^*_1:\ x = h(y),\ C^*_2:\ x = k(y),\ h(y) < k(y),\ c \le y \le d$$

as shown in Fig. 4.13.

To prove (4.17) we consider the decomposition of C as shown in Fig. 4.13(a). Evaluating the double integral over D, we find

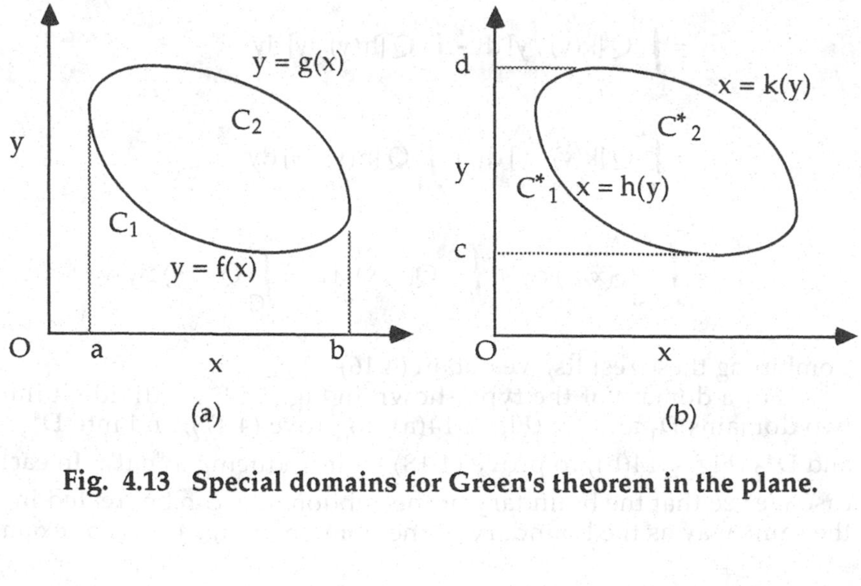

Fig. 4.13 Special domains for Green's theorem in the plane.

$$-\int\int_D \frac{\partial P}{\partial y} dx\, dy = -\int_a^b \int_{f(x)}^{g(x)} \frac{\partial P}{\partial y} dy\, dx$$

$$= -\int_a^b \{P[x, g(x)] - P[x, f(x)]\}\, dx$$

$$= \int_a^b P[x, f(x)]\, dx + \int_b^a P[x, g(x)]\, dx$$

$$= \int_{C_1} P(x, y)dx + \int_{C_2} P(x, y)dx$$

$$= \oint_C P(x, y)dx$$

Similarly, decomposing C into the two curves shown in Fig. 4.13(b), and performing the double integral (4.18) over D, we find

$$\int\int_D \frac{\partial Q}{\partial x} dx\, dy = \int_c^d \int_{h(y)}^{k(y)} \frac{\partial Q}{\partial x} dx\, dy$$

$$= \int_c^d Q\,[k(y)\,,\, y]\, dy - \int_c^d Q\,[h(y)\,,\, y]\, dy$$

$$= \int_c^d Q\,[k(y)\,,\, y]\, dy + \int_d^c Q\,[h(y)\,,\, y]\, dy$$

$$= \int_{C^*_2} Q(x, y)\, dy + \int_{C^*_1} Q(x, y)\, dy = \oint_C Q(x, y)\, dy$$

Combining these results, we obtain (4.16).

For a domain of the type shown in Fig. 4.14, we divide it into two domains D_1 and D_2 (Fig. 4.14(a)) to prove (4.17), and into D^*_1 and D^*_2 (Fig. 4.14(b)) to prove (4.18) by introducing a "cut". In each case we see that the boundary of the subdomains can be treated in the same way as the boundary of the domain in Fig. 4.13. For example, to establish (4.17) we perform the double integral of $\partial P/\partial y$ over D_1 and D_2 and add the results. When the integrals along the boundaries are combined, the line integral along the "cut" C_1, C_2 cancels out since it is raced in opposite direction. The integrals along the other parts of the boundary add up to the integral along the boundary C of the domain. On the other hand, the sum of the double integrals over D_1 and D_2 is exactly the double integral over the entire domain. Thus (4.17) holds for the domain considered. To establish (4.18), we divide the domain as indicated in Fig. 4.14(b). The boundary of each of the subdomains can be treated as the domain of Fig. 4.13(b). Integrating $\partial Q/\partial x$ over D^*_1 and D^*_2 and adding the results, we obtain (4.18). Thus we have established Green's theorem for the domain of the type shown in Fig. 4.14.

This technique can be used in extending Green's theorem to more general type of domains that are frequently encountered in applications.

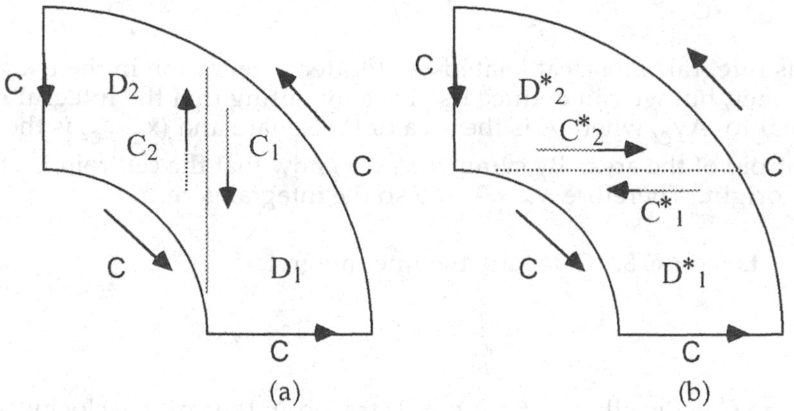

Fig. 4.14 Other types of domain for Green's theorem.

Example 1. By Green's theorem, evaluate the integral

$$\oint_C (x + y)dx + 2xdy$$

where C is the unit circle $x^2 + y^2 = 1$ traced in the counterclockwise direction.

Solution: Here $P(x,y) = x + y$ and $Q(x,y) = 2x$, so that $\partial P/\partial y = 1$ and $\partial Q/\partial x = 2$. Hence, we have

$$\oint_C (x + y)dx + 2xdy = \int\int_D (2 - 1)dx\,dy = \pi$$

Example 2. Calculate the line integral

$$\oint_C (x + y^2)dx - xy\,dy$$

where C is the sides of the squre with vertices at (1, 0), (0, 1), (-1, 0), (0, -1), traced in the counterclockwise direction.

Solution: By Green's theorem, we have

$$\oint_C (x + y^2)dx - xy\,dy = \int\int_D (-y - 2y)dx\,dy = -3\int\int_D y\,dx\,dy$$

This integral can be calculated by iterated integration in the usual manner, but we can deduce its value by noting that the integral is equal to Ay_c, where A is the area of the square and (x_c, y_c) is the centroid of the area. By symmetry, we know that the centroid is at the origin. Therefore $y_c = 0$ and so the integral is zero.

Example 3. Calculate the line integral

$$\oint_C (x + 2y)dx + (3x - y)dy$$

where C is the ellipse $x^2 + 4y^2 = 4$, traced in the counterclockwise direction.

Solution: By Green's theorem, we have

$$\oint_C (x + 2y)dx + (3x - y)dy = \int\int_D (3 - 2)dx\,dy = \int\int_D dx\,dy$$

$$= \text{area of the ellipse}$$
$$= \pi ab$$

where a and b are the major and minor axes of the ellipse. From the equation of the ellipse, we deduce a = 2, b = 1. Thus

$$\oint_C (x + 2y)dx + (3x - y)dy = 2\pi$$

Example 4. Show that

$$\oint_C \frac{-y\,dx + x\,dy}{x^2 + y^2} = 2\pi$$

where C is any piecewise smooth simple closed curve that encloses the origin (see Fig. 4.15). (Cf. Example 1, Sec. 4.4).

Solution: Since the domain enclosed by C contains the origin at which the integrand is not defined, we exclude the origin by deleting the interior of a circle C' of radius ε, 0 < ε < 1, about the origin. The doubly connected domain D* is now bounded outside by C, traced in the counterclockwise (positive) direction, and inside by C', traced in the clockwise direction. Now in the domain D*

$$\frac{\partial}{\partial y}\left(\frac{-y}{x^2 + y^2}\right) = \frac{\partial}{\partial x}\left(\frac{x}{x^2 + y^2}\right)$$

and so, by Green's theorem , we have

$$\oint_{C+C} \frac{-y\,dx + x\,dy}{x^2 + y^2} = 0$$

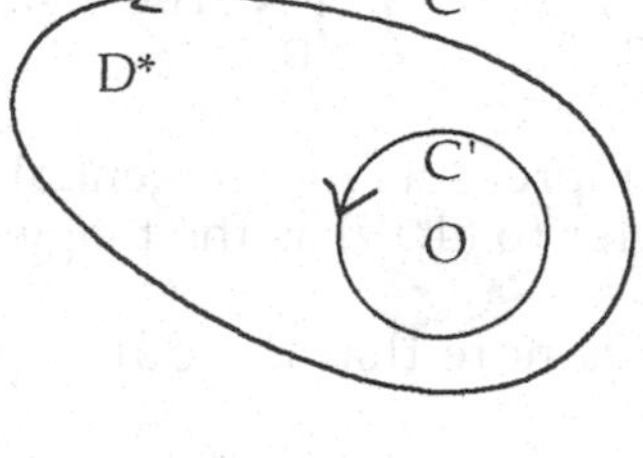

Fig. 4.15 Green's theorem in a doubly connected domain.

Hence we have

$$\oint_C \frac{-y\,dx + x\,dy}{x^2 + y^2} = -\oint_{C'} \frac{-y\,dx + x\,dy}{x^2 + y^2}$$

Note that the circle C' is traced in the clockwise direction. Setting $x = \varepsilon \cos\theta$, $y = \varepsilon \sin\theta$. we find

$$-\oint_{C'} \frac{-y\,dx + x\,dy}{x^2 + y^2} = \int_0^{2\pi} \frac{\varepsilon^2(\sin^2\theta + \cos^2\theta)d\theta}{\varepsilon^2} = 2\pi$$

Tangential and Normal Forms of Green's Theorem

We can write Green's theorem in different forms by making use of the unit tangent vector and the unit outward normal vector on the curve. For this purpose, let $\mathbf{F}(x,y) = P(x, y)\mathbf{i} + Q(x, y)\mathbf{j}$, and let the curve C be represented by $\mathbf{R}(s) = x(s)\mathbf{i} + y(s)\mathbf{j}$, where s denotes arc length of C, $s \geq 0$. Since the unit tangent vector on the curve is given by

$$\mathbf{t} = \frac{dx}{ds}\mathbf{i} + \frac{dy}{ds}\mathbf{j}$$

and since

$$\mathbf{curl\ F} = \left(\frac{\partial Q}{\partial x} - \frac{\partial P}{\partial y}\right)\mathbf{k}$$

equation (4.16) can be written in vector form as

$$\oint_C \mathbf{F}\cdot\mathbf{t}\,ds = \int\int_D \mathbf{curl\ F}\cdot\mathbf{k}\,dx\,dy \tag{4.19}$$

We note that $\mathbf{F}\cdot\mathbf{t}$ represents the tangential component of F on C. For this reason we refer to (4.19) as the tangential form of Green's theorem.

Similarly, we note that the outward normal vector **n** on the curve is given by

$$\mathbf{n} = \frac{dy}{ds}\mathbf{i} - \frac{dx}{ds}\mathbf{j}$$

so that

$$\mathbf{F}\cdot\mathbf{n} = P(x, y)\frac{dy}{ds} - Q(x, y)\frac{dx}{ds}$$

Hence, applying Green's theorem, we obtain

$$\oint_C \mathbf{F}\cdot\mathbf{n}\, ds = \int_C -Q(x, y)dx + P(x, y)dy$$

$$= \iint_D \left(\frac{\partial P}{\partial x} + \frac{\partial Q}{\partial y}\right) dx\, dy \tag{4.20}$$

We refer to this as the normal form of Green's theorem. Since the integrand on the right-hand of (4.20) is the divergence of **F**, we can write (4.20) as

$$\oint_C \mathbf{F}\cdot\mathbf{n}\, ds = \iint_D \operatorname{div} \mathbf{F}\, dx\, dy \tag{4.21}$$

In particular, when $\mathbf{F} = \nabla u$, this yields the integral formula

$$\oint_C \frac{\partial u}{\partial n} ds = \iint_D \nabla^2 u\, dx\, dy \tag{4.22}$$

which relates the integral of the Laplacian of u in a domain with the line integral of the normal derivative of u on the boundary of the domain. Other important integral formulas that are derived from (4.21) are asked for in the Exercises. These formulas have important applications for many physical problems involving the Laplace operator ∇^2. Both formulas (4.19) and (4.21) can be extended to vector fields in three dimensions as we shall see in later sections.

4.4 EXERCISES

In each of Problems 1 through 6, evaluate the line integral by using Green's theorem.

1. $\oint_C (x^2 - y^2)\, dx + 2xy\, dy$, where C is the square bounded by the lines $x = 0$, $x = 2$, $y = 0$, $y = 2$.

2. $\oint_C 2xy^2\,dx + x^2y\,dy$, where C is the ellipse $4x^2 + y^2 = 4$.

3. $\oint_C (x\cos x - e^y)\,dx - (y^2 + xe^y)\,dy$, where C is any piecewise smooth simple closed curve.

4. $\oint_C -x^2y\,dx + xy^2\,dy$, where C consists of the upper semicircle $y = \sqrt{(a^2 - x^2)}$ $(-a \le x \le a)$ and the diameter on the x-axis.

5. $\oint_C (\cos x - 3y)\,dx + (x^2 + 2e^y)\,dy$, where C is the circle of radius 2 and center at (1, 2).

6. $\oint_C (3x^3 - y^3)\,dx + (x^3 + 2y^3)\,dy$, where C is the unit circle with center at the origin.

7. Verify Green's theorem for the line integral

$$\oint_C 2xy^3\,dx + 4x^2y^2\,dy$$

where C is the boundary of the region in the first quadrant bounded by $x = 1$, $y = x^3$ and the x-axis.

8. Show that

$$\oint_C (\cos x + xy)\,dx + (x^2 + e^y)dy$$

is zero on any circle C with center on the y-axis. Is the integral independent of path?

9. Let $u(x, y) = x^2 + 2xy - y^2$ be defined in a simply connected domain bounded by a smooth simple closed curve C. Show that

$$\oint_C \frac{\partial u}{\partial n} ds = 0$$

10. Let $u(x, y) = x^2 - 2xy + y^2$. Evaluate $\int_C (\partial u/\partial n)ds$ where C is the unit circle with center at the origin.

11. Evaluate the integral $\int_C (\partial u/\partial n)ds$, where $u(x, y) = x^3 - y^3$ and C is the ellipse $(x/a)^2 + (y/b)^2 = 1$.

12. Evaluate the integral $\int_C u(\partial u/\partial n)ds$, where $u(x, y) = (x + y)^2$ and C is the unit circle with center at the origin.

13. Let

$$F(x, y) = \frac{x\mathbf{i} + (y + 1)\mathbf{j}}{x^2 + (y + 1)^2} \qquad (x, y) \neq (0, -1)$$

Evaluate the integral $\int_C \mathbf{F} \cdot d\mathbf{R}$ around the circle of radius 2 and center at the origin.

14. Let

$$P(x, y) = \frac{-(y - 1)}{x^2 + (y - 1)^2} - \frac{y}{(x + 1)^2 + y^2}$$

and

$$Q(x, y) = \frac{x}{x^2 + (y - 1)^2} + \frac{x + 1}{(x + 1)^2 + y^2}$$

where $(x, y) \neq (0, 1), (-1, 0)$. Evaluate $\int_C Pdx + Qdy$ where C is the circle $x^2 + y^2 = 6$.

15. Derive the following integral formulas from (4.21):

(a) $$\oint_C v \frac{\partial u}{\partial n} ds = \iint_D (v \nabla^2 u + \nabla v \cdot \nabla u) dx\, dy$$

(b) $$\oint_C \left(v \frac{\partial u}{\partial n} - u \frac{\partial v}{\partial n} \right) ds = \iint_D (v \nabla^2 u - u \nabla^2 v) dx\, dy$$

16. Derive the integral formula

$$\frac{1}{2} \oint_C \left(v \frac{\partial u}{\partial x} - u \frac{\partial v}{\partial x} \right) dx + \left(u \frac{\partial v}{\partial y} - v \frac{\partial u}{\partial y} \right) dy$$

$$= \iint_D \left(u \frac{\partial^2 v}{\partial x \partial y} - v \frac{\partial^2 u}{\partial x \partial y} \right) dx\, dy$$

from (4.16).

4.6 PARAMETRIC REPRESENTATION OF SURFACES

A surface S may be defined as a set of points (x, y, z) satisfying an equation of the form

$$f(x, y, z) = 0 \tag{4.23}$$

where f is a continuous function in some domain D. If we regard f as a scalar field, then the surface S is just a level surface of f. We call (4.23) an implicit representation of the surface. If it is possible to solve for one of the variables in terms of the other two, say, z in terms of x and y, then the surface may also be represented by

$$z = z(x,y) \tag{4.24}$$

We call this an explicit representation of the surface. For example, the unit sphere is defined implicitly by the equation $x^2 + y^2 + z^2 - 1 = 0$. If we solve for z, it can also be represented explicitly by the two equations

$$z = -\sqrt{1 - x^2 - y^2} \quad \text{and} \quad z = \sqrt{1 - x^2 - y^2}$$

The first equation describes the lower hemisphere, while the second describes the upper hemisphere.

For many theoretical considerations, however, it is more convenient to represent a surface by yet another method, namely, parametric representation. This method is the counterpart of the parametric representation of a space curve by means of three equations involving a single parameter. Parametrically, then, a surface S is represented by equations of the form

$$x = x(u, v), \quad y = y(u, v), \quad z = z(u, v) \tag{4.25}$$

where u and v are two parameters which range over some domain D* in the uv-plane. The functions on the right-hand side of (4.25) are assumed to be continuous in D*. As (u, v) ranges over D*, the point (x(u,v), y(u,v), z(u,v)) traces the surface (Fig. 4.16). In this sense, the surface may be regarded as the image of the domain D* under the mapping defined by (4.25).

The parametric equations in (4.29) can be combined into a single vector equation

$$\mathbf{R}(u, v) = x(u, v)\mathbf{i} + y(u, v)\mathbf{j} + z(u, v)\mathbf{k} \tag{4.26}$$

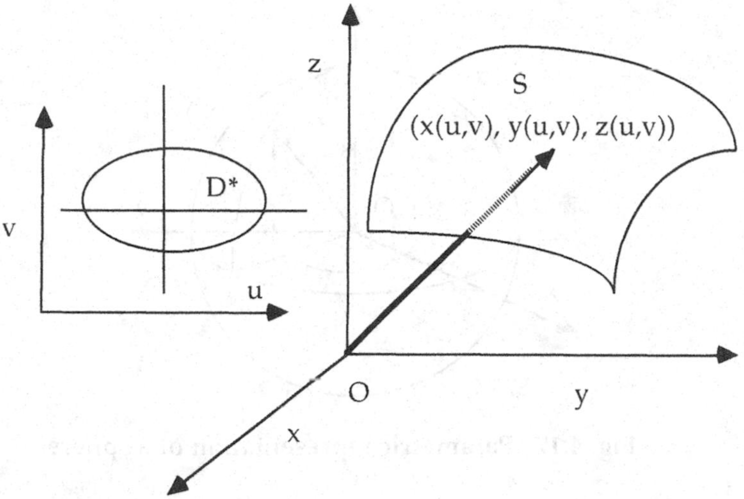

Fig. 4.16 Parametric representation of a surface.

where **R**(u, v) denotes the position vector of the point on the surface corresponding to the point (u, v) in D*. For a surface defined by an equation of the form (4.24), the variables x, y play the role of the parameters u, v, so that its vector equation is given by

$$\mathbf{R}(x, y) = x\mathbf{i} + y\mathbf{j} + z(x, y)\mathbf{k} \tag{4.27}$$

As in the case of space curves, the parametric representation of a surface is not unique. This means that we may replace the parameters u, v in (4.25) by other parameters, say, ξ, η, where ξ, η are related to u, v by the equations

$$\xi = \xi(u, v), \qquad \eta = \eta(u, v)$$

Here it is necessary to require that the Jacobian $\partial(\xi,\eta)/\partial(u,v) \neq 0$ in D* in order to ensure the existence of the inverse transformation

$$u = u(\xi, \eta), \quad v = v(\xi, \eta)$$

This guarantees that the transformation between the (u,v)-plane and the (ξ, η)-plane is one to one.

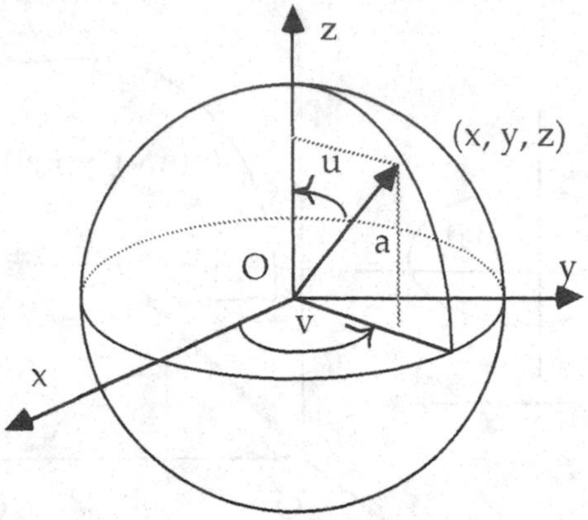

Fig. 4.17 Parametric representation of a sphere.

Example 1. The sphere of radius a and center at the origin can be represented parametrically by the equations

$$x = a \sin u \cos v, \quad y = a \sin u \sin v, \quad z = a \cos u$$

where $0 \le u \le \pi$, and $0 \le v < 2\pi$ (see Fig. 4.17). The upper hemisphere is defined by $0 \le u \le \pi/2$, $0 \le v < 2\pi$, and the lower hemisphere is defined by $\pi/2 \le u \le \pi$, $0 \le v < 2\pi$. The equations $x = a\cos v$, $y = a \sin v$, $z = 0$, when $u = \pi/2$, represents the equator of the sphere.

Example 2. The paraboloid $z = x^2 + y^2$ can be represented para- metrically by the equations

$$x = u, \quad y = v, \quad z = u^2 + v^2$$

where (u, v) ranges over the whole uv-plane. An alternative representation is given by

$$x = \sqrt{u} \cos v, \quad y = \sqrt{u} \sin v, \quad z = u$$

where $u \ge 0$ and $0 \le v < 2\pi$. For a fixed value of z, the equations describe a circle on the surface, and for a fixed value of v the equations represent a parabola, which is the intersection of the plane $y = x \tan v$ and the paraboloid (Fig. 4.18).

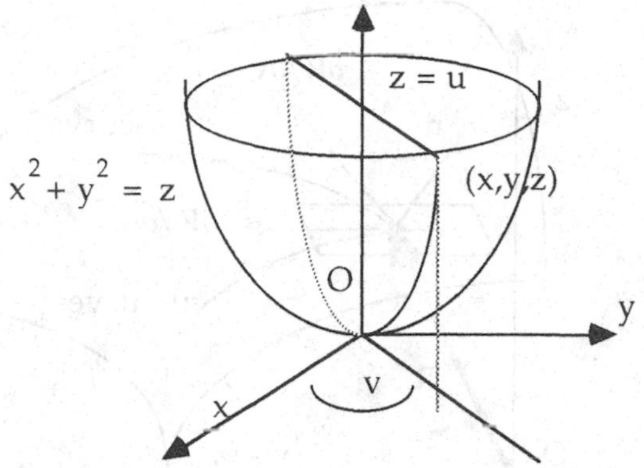

Fig. 4. 18 Parametric representation of a paraboloid.

Tangent Plane and Normal Vector on a Surface

Let S be a surface represented parametrically by

$$x = x(u,v), \quad y = y(u,v), \quad z = z(u,v)$$

where the functions x(u,v), y(u,v), z(u,v) are continuous and have continuous first order derivatives in a domain D* of the uv-plane. Let us consider the position vector

$$\mathbf{R}(u, v) = x(u, v)\mathbf{i} + y(u, v)\mathbf{j} + z(u, v)\mathbf{k}$$

of a point on S. If we set $v = v_0$, then $\mathbf{R}(u,v_0) = x(u,v_0)\mathbf{i} + y(u,v_0)\mathbf{j} + z(u,v_0)\mathbf{k}$ represents a curve on the surface with u as the parameter. Its tangent vector is given by $\partial\mathbf{R}(u,v_0)/\partial u$. Similarly, if we set $u = u_0$, the equation $\mathbf{R}(u_0,v) = x(u_0,v)\mathbf{i} + y(u_0,v)\mathbf{j} + z(u_0,v)\mathbf{k}$ represents another curve on the surface with its tangent vector given by $\partial\mathbf{R}(u_0, v)/\partial v$ (see Fig. 4.19). The two tangent vectors to the curves at the point $\mathbf{R}(u_0,v_0)$ determine a plane, which is the tangent plane to the surface at $\mathbf{R}(u_0,v_0)$. A unit vector normal to the tangent plane is given by

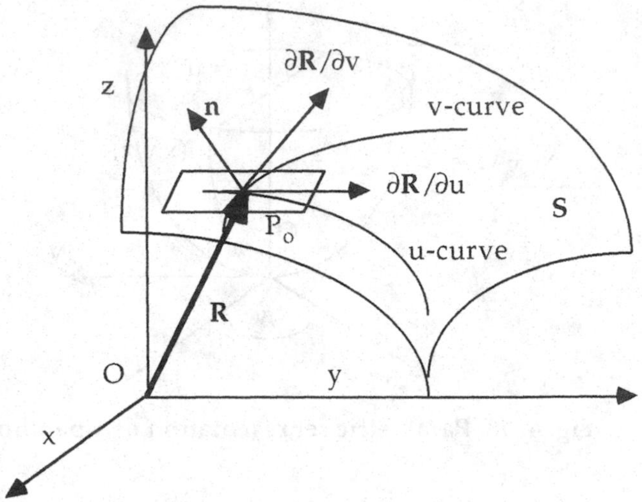

Fig. 4.19 The unit normal vector n and the tangent plane.

$$\mathbf{n} = \frac{\mathbf{R}_u(u_o, v_o) \times \mathbf{R}_v(u_o, v_o)}{|\mathbf{R}_u(u_o, v_o) \times \mathbf{R}_v(u_o, v_o)|}$$

It is worth noting that if the surface is defined by $z = z(x, y)$ so that its vector equation is given by

$$\mathbf{R}(x, y) = x\mathbf{i} + y\mathbf{j} + z(x, y)\mathbf{k}$$

then

$$\frac{\partial \mathbf{R}}{\partial x}(x, y) \times \frac{\partial \mathbf{R}}{\partial y}(x, y) = (\mathbf{i} + z_x\mathbf{k}) \times (\mathbf{j} + z_y\mathbf{k}) = -z_x\mathbf{i} - z_y\mathbf{j} + \mathbf{k}$$

This is the gradient of $f(x, y, z)$, where $f(x, y, z) = z - z(x, y)$, which we know is orthogonal to the surface $f(x, y, z) = 0$.

Example 3. Find a normal vector to the surface represented by the equation

$$\mathbf{R}(u, v) = u(\cos v\,\mathbf{i} + \sin v\,\mathbf{j}) + (1 - u^2)\mathbf{k}, \quad u \geq 0, \ 0 \leq v < 2\pi$$

Solution: First, we determine $\partial \mathbf{R}/\partial u$ and $\partial \mathbf{R}/\partial v$:

$$\frac{\partial \mathbf{R}}{\partial u} = \cos v\ \mathbf{i} + \sin v\ \mathbf{j} - 2u\ \mathbf{k}, \qquad \frac{\partial \mathbf{R}}{\partial v} = u(-\sin v\ \mathbf{i} + \cos v\ \mathbf{j})$$

Hence a normal vector to the surface is given by

$$\mathbf{n} = \frac{\partial \mathbf{R}}{\partial u} \times \frac{\partial \mathbf{R}}{\partial v} = 2u^2(\cos v\ \mathbf{i} + \sin v\ \mathbf{j}) + u\mathbf{k}$$

Smooth, Piecewise Smooth and Orientable Surfaces

Suppose that S is a surface represented by an equation of the form (4.26). We say that S is smooth if $\partial \mathbf{R}/\partial u$ and $\partial \mathbf{R}/\partial v$ are continuous and the normal vector $\partial \mathbf{R}/\partial u \times \partial \mathbf{R}/\partial v$ is not zero at any point on S. Geometrically, this means that at each point on the surface there is a normal vector which changes continuously as it moves over the surface - thus a smooth surface has no corners or edges. If a surface is not smooth but consists of a finite number of surfaces each of which is smooth, then the surface is said to be piecewise smooth. For example, the surface defined by a sphere is smooth and the faces of a cube constitute a piecewise smooth surface.

Let S be a smooth surface. Then at any point on S we can choose a unit normal vector **n**. Obviously, there are two choices and they are opposite in direction. If **n** can be chosen at each point on S in such a manner that it varies continuously as it moves about the surface, then S is said to be *orientable* or *two-sided*. The side of S on which **n** emerges is called the "positive" side of the surface. If S is a closed surface, then by convention **n** is chosen to point outward. Practically all of the quadric surfaces encountered in calculus are orientable surfaces.

It may appear surprising that not all surfaces are orientable. A well known example of a non-orientable surface is the so-called Mobius strip, which has only one side (Fig. 4.20). This surface may be formed by taking a long rectangular strip of paper, giving it one twist, and then joining the ends together. It is readily seen in this case that when a normal vector **n** at a point P on the surface is moved continuously around the strip, its direction reverses when it returns to the point P. Thus, the vector **n** cannot be chosen so as to vary continuously over the entire surface. In other words, the surface is "one-sided" or nonorientable.

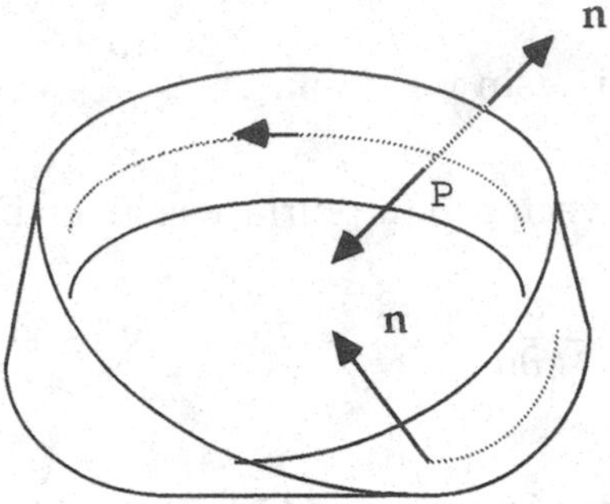

Fig. 4.20 Mobius strip.

If a smooth surface S is bounded by a simple closed curve C we may associate with the orientation of S a positive direction on C. To achieve this, imagine that an object is moving along the boundary C on the positive side of the surface. The direction in which the object must move on C in order that the surface will at all time lie to its left is designated as the positive direction on C (Fig. 4.21). Using this convention, a piecewise smooth surface may also be oriented by choosing the normal vector on each smooth portion of S in such a way that along a common boundary B of two portions of S, the posi-

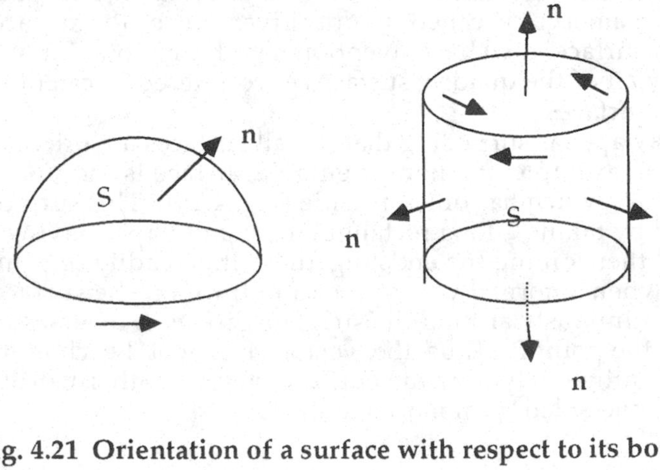

Fig. 4.21 Orientation of a surface with respect to its boundary.

tive direction of B relative to one portion is opposite to the positive direction of B relative to the other portion (see Fig. 4.21). In the sequel we will consider only piecewise smooth orientable surfaces.

4.5 EXERCISES

In each of Problems 1 through 7, describe the surface represented by the given parametric equations, and determine the coordinate curves on the surface. Eliminate the parameters u, v and obtain an implicit reprentation of the surface.

1. $x = a\cos v,\ y = a\sin v,\ z = u \quad (0 \le v < 2\pi,\ u > 0)$
2. $x = u\cos v,\ y = u\sin v,\ z = u^2 \quad (u \ge 0,\ 0 \le v < 2\pi)$
3. $x = u\cos v,\ y = u\sin v,\ z = au \quad (a > 0,\ u \ge 0,\ 0 \le v < 2\pi)$
4. $x = a\sin u\cos v,\ y = b\sin u\sin v,\ z = c\cos u \quad (0 \le u < \pi, 0 \le v \le 2\pi)$
5. $x = a(u + v),\ y = b(u - v),\ z = uv \quad (-\infty < u < \infty,\ -\infty < v < \infty)$
6. $x = a\sin u\cosh v,\ y = b\cos u\cosh v,\ z = c\sinh v \quad (0 \le u < 2\pi,\ -\infty < v < \infty)$
7. $x = au\cosh v,\ y = bu\sinh v,\ z = u \quad (-\infty < u < \infty,\ -\infty < v < \infty)$

In each of Problems 8 through 14, find a parametric representation of the given surface.

8. The plane $ax + by + cz + d = 0$.
9. The parabolic cylinder $z = x^2$.
10. The elliptic paraboloid
$$\frac{x^2}{a^2} + \frac{y^2}{b^2} = 1 - z$$
11. The ellipsoid
$$\frac{x^2}{a^2} + \frac{y^2}{b^2} + \frac{z^2}{c^2} = 1$$
12. The hyperboloid of one sheet
$$\frac{x^2}{a^2} + \frac{y^2}{b^2} - \frac{z^2}{c^2} = 1$$

13. The hyperboloid of two sheets

$$\frac{x^2}{a^2} - \frac{y^2}{b^2} - \frac{z^2}{c^2} = 1$$

14. The hyperbolic paraboloid

$$\frac{x^2}{a^2} - \frac{y^2}{b^2} = \frac{z}{c}$$

In each of Problems 15 through 19, determine a unit normal vector to the given surface, and find an equation of the tangent plane to the surface at the given point.

15. $\mathbf{R}(u, v) = u \cos v\, \mathbf{i} + u \sin v\, \mathbf{j} + u^2\, \mathbf{k}$, $(1, -1, 2)$.
16. $\mathbf{R}(u, v) = u \cos v\, \mathbf{i} + u \sin v\, \mathbf{j} + u\mathbf{k}$, $(\sqrt{3}, 1, 2)$.
17. $\mathbf{R}(u, v) = \sqrt{2}(\sin u \cos v)\mathbf{i} + 2\sqrt{2}(\sin u \sin v)\mathbf{j} + (\sqrt{3} \cos u)\mathbf{k}$, $(1/2, 1, 3/2)$.
18. $\mathbf{R}(u, v) = 2(u + v)\mathbf{i} + (u - v)\mathbf{j} + uv\mathbf{k}$, $(2, -3, -2)$.
19. $\mathbf{R}(u, v) = 2(\sin u \cosh v)\mathbf{i} + 3(\cos u \cosh v)\mathbf{j} + (\sinh v)\mathbf{k}$, $(1, 3\sqrt{3}/2, 0)$.
20. Let S be a smooth surface represented by $\mathbf{R}(u, v)$, $(u, v) \in D^*$, and let Γ be a smooth curve in D^* described by $u = f(t)$, $v = g(t)$ $(a \le t \le b)$. Then $\mathbf{R}^*(t) = \mathbf{R}(f(t), g(t))$ represents a smooth curve C on S. Show that the tangent vector $d\mathbf{R}^*(t)/dt$ to the curve C is a linear combination of the tangent vectors $\partial\mathbf{R}/\partial u$ and $\partial\mathbf{R}/\partial v$ to the coordinate curves on S, and thus deduce that $d\mathbf{R}^*(t)/dt$ is orthogonal to $\partial\mathbf{R}/\partial u \times \partial\mathbf{R}/\partial v$.

4.7 SURFACE AREA

We know that the area of a plane region can be determined by calculating a double integral. In this section we derive a double integral formula for calculating the area of a surface. The idea of area of a surface will later be used in the study of surface integrals of scalar and vector fields. So let S be a smooth surface represented by the equation

$$\mathbf{R}(u, v) = x(u, v)\mathbf{i} + y(u, v)\mathbf{j} + z(u, v)\mathbf{k} \tag{4.28}$$

where (u, v) ranges over a domain D^* in the uv-plane. To facilitate the derivation of an expression for an element of area on the surface, we regard S as the image of the domain D^* under the transformation

(4.28), and consider the image of a rectangular region ABCD in D^* with vertices at (u, v), $(u+\Delta u, v)$, $(u+\Delta u, v+\Delta v)$, $(u, v+\Delta v)$, see Fig. 4.22. Under the transformation (4.28) the sides of the rectangle are mapped onto coordinate curves on S so that the rectangle itself is mapped onto a curvilinear parallelogram, denoted by ΔS, with the corresponding vertices A', B', C', D'. We assert that the area of ΔS is approximately equal to the area of a parallelogram whose sides are

$$\left|\frac{\partial \mathbf{R}}{\partial u}\right| \Delta u \qquad \text{and} \qquad \left|\frac{\partial \mathbf{R}}{\partial v}\right| \Delta v$$

To show this, we imagine the parameter u as representing time when v is held fixed. Then $\partial \mathbf{R}/\partial u$ is a velocity vector, and the term $|\partial \mathbf{R}/\partial u|\Delta u$ represents the distance traveled by the point A' along the u-coordinate curve during the time interval Δu. Now, when Δu is small, $|\partial \mathbf{R}/\partial u|\Delta u$ is approximately equal to the length of the side A'B' of ΔS. Similarly, for small Δv, $|\partial \mathbf{R}/\partial v|\Delta v$ is approximately equal to the length of the side A'D'. Therefore, the area of ΔS is approximately equal to the area of the parallelogram whose sides are equal to the magnitude of the vectors $(\partial \mathbf{R}/\partial u)\Delta u$ and $(\partial \mathbf{R}/\partial v)\Delta v$, for small Δu and Δv.

We know that the magnitude of the cross product of two vectors represents the area of the parallelogram formed by the vectors. Therefore, the area of the parallelogram formed by $(\partial \mathbf{R}/\partial u)\Delta u$ and $(\partial \mathbf{R}/\partial v)\Delta v$ is equal to

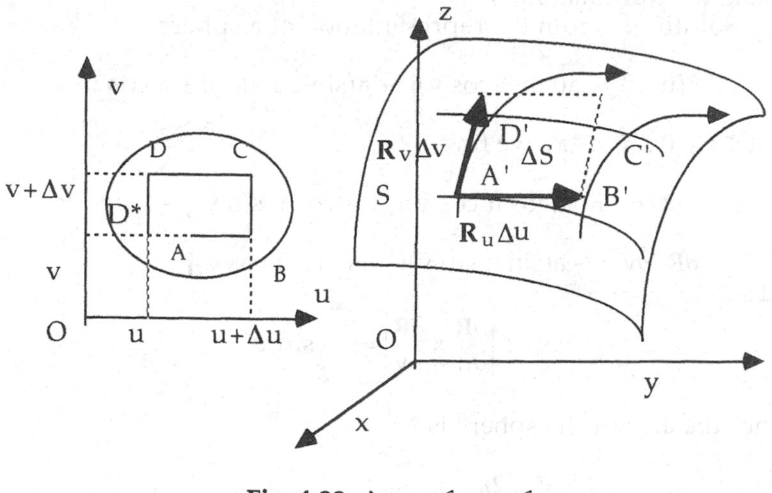

Fig. 4.22 Area of a surface.

$$\left| \frac{\partial \mathbf{R}}{\partial u} \Delta u \times \frac{\partial \mathbf{R}}{\partial v} \Delta v \right| = \left| \frac{\partial \mathbf{R}}{\partial u} \times \frac{\partial \mathbf{R}}{\partial v} \right| \Delta u \, \Delta v$$

and so we have

$$\text{area of } \Delta S = \left| \frac{\partial \mathbf{R}}{\partial u} \times \frac{\partial \mathbf{R}}{\partial v} \right| \Delta u \Delta v$$

Thus we define the area of a surface as follows:

Definition 1. Let S be a smooth surface represented by $\mathbf{R} = \mathbf{R}(u, v)$, where $\mathbf{R}(u, v)$ is continuously differentiable in D*. The area of the surface S is given by the double integral

$$A(\text{area of surface S}) = \iint_{D^*} \left| \frac{\partial \mathbf{R}}{\partial u} \times \frac{\partial \mathbf{R}}{\partial v} \right| du dv \qquad (4.29)$$

The formula

$$dS = \left| \frac{\partial \mathbf{R}}{\partial u} \times \frac{\partial \mathbf{R}}{\partial v} \right| du dv \qquad (4.30)$$

is called the element of surface area on S.

Example 1. Verify the surface area $4\pi a^2$ of a sphere of radius a using the formula (4.29).

Solution: From the representation of a sphere

$$\mathbf{R}(u, v) = a(\sin u \cos v)\mathbf{i} + a(\sin u \sin v)\mathbf{j} + a \cos u\mathbf{k},$$

$0 \leq u \leq \pi$, $0 \leq v < 2\pi$, we find

$$\partial \mathbf{R}/\partial u = a(\cos u \cos v)\mathbf{i} + a(\cos u \sin v)\mathbf{j} - a \sin u\mathbf{k}$$

$$\partial \mathbf{R}/\partial v = - a(\sin u \sin v)\mathbf{i} + a(\sin u \cos v)\mathbf{j}$$

and so

$$\left| \frac{\partial \mathbf{R}}{\partial u} \times \frac{\partial \mathbf{R}}{\partial v} \right| = a^2 \sin u$$

Hence the area of the sphere is

$$\text{Area} = \int_0^{2\pi} \int_0^{\pi} a^2 \sin u \, du \, dv = 4\pi a^2$$

Example 2. Find the surface area of the circular cone

$$\mathbf{R}(u, v) = u \cos v\, \mathbf{i} + u \sin v\, \mathbf{j} + u\, \mathbf{k} \quad (0 \leq u \leq a\,, 0 \leq v < 2\pi)$$

Solution: We have

$$\partial\mathbf{R}/\partial u = \cos v\, \mathbf{i} + \sin v\, \mathbf{j} + \mathbf{k}, \quad \partial\mathbf{R}/\partial v = -u \sin v\, \mathbf{i} + u \cos v\, \mathbf{j}$$

and so

$$\frac{\partial\mathbf{R}}{\partial u} \times \frac{\partial\mathbf{R}}{\partial v} = -u \cos v\, \mathbf{i} - u \sin v\, \mathbf{j} + u\, \mathbf{k}$$

Hence

$$\left|\frac{\partial\mathbf{R}}{\partial u} \times \frac{\partial\mathbf{R}}{\partial v}\right| = \sqrt{2}\, u$$

so that the area is given by

$$A = \int_0^{2\pi}\int_0^{a} \sqrt{2}\, u\, du\, dv = \sqrt{2}\pi a^2$$

Alternate Formulas for Surface Area

If the surface is represented by an equation of the form $z = z(x, y)$, where (x, y) ranges over a domain D on the xy-plane, we may use x and y as the parameters and write

$$\mathbf{R}(x, y) = x\mathbf{i} + y\mathbf{j} + z(x, y)\mathbf{k}$$

Then

$$\partial\mathbf{R}/\partial x = \mathbf{i} + (\partial z/\partial x\, \mathbf{k}, \qquad \partial\mathbf{R}/\partial y = \mathbf{j} + (\partial z/\partial y)\mathbf{k}$$

so that

$$\left|\frac{\partial\mathbf{R}}{\partial x} \times \frac{\partial\mathbf{R}}{\partial y}\right| = \left|-\frac{\partial z}{\partial x}\,\mathbf{i} - \frac{\partial z}{\partial y}\,\mathbf{j} + \mathbf{k}\right|$$

$$= \sqrt{1 + \left(\frac{\partial z}{\partial x}\right)^2 + \left(\frac{\partial z}{\partial y}\right)^2}$$

Hence an alternative formula for the surface area is

$$A = \int\int_D \sqrt{1 + \left(\frac{\partial z}{\partial x}\right)^2 + \left(\frac{\partial z}{\partial y}\right)^2}\, dx\, dy \tag{4.31}$$

so that

$$dS = \sqrt{1 + \left(\frac{\partial z}{\partial x}\right)^2 + \left(\frac{\partial z}{\partial y}\right)^2}\, dx\, dy \tag{4.32}$$

Notice that the integral (4.31) is to be integrated over the domain D which is the projection of S onto the xy-plane.

The formula (4.32) has an interesting and important application. Recall that a unit normal vector on the surface is given by

$$\mathbf{n} = \frac{-z_x\mathbf{i} - z_y\mathbf{j} + \mathbf{k}}{\sqrt{z_x^2 + z_y^2 + 1}}$$

If γ denotes the angle between **n** and **k**, then

$$\cos\gamma = \mathbf{n}\cdot\mathbf{k} = 1/\sqrt{z_x^2 + z_y^2 + 1}$$

Hence, by (4.32), we have

$$dS = \sec\gamma\, dx\, dy \quad \text{or} \quad \cos\gamma\, dS = dx\, dy = dA \tag{4.33}$$

and so (4.31) can be written as

$$A = \int\int_D \sec\gamma\, dx\, dy \tag{4.34}$$

Equation (4.33) says that the projection ($\cos\gamma\ dS$) of the elemen of surface area dS on the xy-plane is precisely the element of area ($dA = dx\ dy$) on that plane (Fig. 4.23). Thus if S lies on a plane on which $\cos\gamma > 0$, then, by (4.33), the area is given by

$$\text{area of S} = (\sec\gamma\,)(\text{area of D})$$

where D is the projection of S on the xy-plane. This result is sometimes called the "area cosine principle."

If the surface S can also be represented by an equation of the form $y = y(x, z)$, $(x, z) \in D'$, or $x = x(y, z)$, $(y, z) \in D''$, then its surface

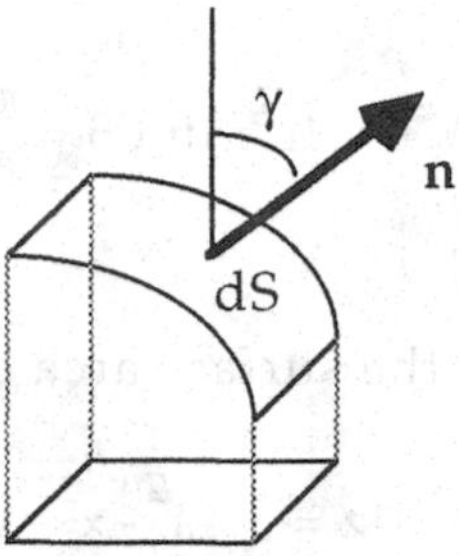

$dA = \cos\gamma \, dS$

Fig. 4.23 Area cosine principle.

area can also be calculated from the formula

$$A = \int\int_{D'} \sec\beta \, dx \, dz \qquad (4.35)$$

or

$$A = \int\int_{D''} \sec\alpha \, dy \, dz \qquad (4.36)$$

Notice that D' and D" are the projections of the surface S on the xz- and yz-planes, respectively.

Example 3. Find the surface area of the paraboloid $z = x^2 + y^2$ that lies inside the cylinder $x^2 + y^2 = a^2$.

Solution: The projection on the xy-plane of the surface in question is defined by $D = \{ (x, y) \mid x^2 + y^2 \leq a^2 \}$. From the equation of the paraboloid, we find

$$z_x = 2x, \qquad z_y = 2y$$

Hence, by (4.34), we have

$$A = \int\int_D \sqrt{1 + 4(x^2 + y^2)} \, dx \, dy$$

Introducing polar coordinates, we obtain

$$A = \int_0^{2\pi}\int_0^a \sqrt{1 + 4r^2}\, r\, dr\, d\theta = \frac{\pi}{6}[(1 + 4a^2)^{3/2} - 1]$$

Example 4. Find the surface area of the portion of the hemisphere

$$z = \sqrt{4a^2 - x^2 - y^2}$$

that is cut off by the cylinder $(x - a)^2 + y^2 = a^2$.

Solution: From the equation of the hemisphere, we find $z_x = -x/z$, $z_y = -y/z$. Hence

$$A = \iint_D \sqrt{1 + (-x/z)^2 + (-y/z)^2}\, dx\, dy$$

$$= 2a \iint_D \frac{dx\, dy}{\sqrt{4a^2 - x^2 - y^2}}$$

where $D = \{(x, y) \mid (x - a)^2 + y^2 \le a^2\}$. In polar coordinates, we observe that the circle bounding the domain D is described by the equation

$$r = 2a \cos\theta \quad (0 \le \theta \le \pi).$$

Thus we obtain

$$A = 2a \int_0^{\pi}\int_0^{2a\cos\theta} \frac{r\, dr\, d\theta}{\sqrt{4a^2 - r^2}} = 2a \int_0^{\pi} \left[-\sqrt{4a^2 - r^2}\right]_0^{2a\cos\theta} d\theta$$

$$= 2a \int_0^{\pi} [-2a \sin\theta + 2a]\, d\theta = -8a^2 + 4\pi a^2$$

4.6 EXERCISES

1. Using the area cosine principle, find the area of the triangle with vertices at (1, 0, 0), (0, 1, 0), (0, 0, 1).
2. Find the area of the portion of the plane $x + y + z = a$ that lies

inside the cylinder $x^2 + y^2 = a^2$.

3. Find the area of the portion of the plane $y = z$ that lies inside the cylinder $x^2 + y^2 - 2ay = 0$.
4. Find the surface area of the surface described by

$$\mathbf{R}(u, v) = (u \cos v)\mathbf{i} + (u \sin v)\mathbf{j} + (1 - u^2)\mathbf{k} \quad (u \geq 0,\ 0 \leq v < 2\pi)$$

above the xy-plane.

5. Find the surface area of the portion of the surface

$$\mathbf{R}(u, v) = u^2\,\mathbf{i} + v^2\,\mathbf{j} + \sqrt{2}uv\mathbf{k}$$

corresponding to $0 \leq u \leq \sqrt{a}$, $0 \leq v < \sqrt{b}$.

6. Find the surface area of the portion of the sphere $x^2 + y^2 + z^2 = a^2$ lying inside the cylinder $x^2 + y^2 = ay$, where $a > 0$.
7. Find the surface area of the portion of the cylinder $x^2 + z^2 = a^2$ that is cut off by the cylinder $y^2 + z^2 = a^2$.
8. Find the surface area of the surface described by the equation $\mathbf{R}(u, v) = u \cos v\ \mathbf{i} + u \sin v\ \mathbf{j} + v\ \mathbf{k}$, where $0 \leq u \leq a$, $0 \leq v \leq 2\pi$.
9. Find the surface area of the portion of the sphere $x^2 + y^2 + z^2 = a^2$ that is cut off by the circular cone $x^2 + y^2 = z^2$.
10. Find the surface area of the portion of the cone $x^2 + y^2 = z^2$ $(z \geq 0)$ that is cut off by the plane $2z = y + 1$.
11. Find the surface area of the portion of the cone $x^2 + y^2 = z^2$ lying inside the cylinder $x^2 + y^2 = 2ay$ $(a > 0)$.
12. Find the surface area of S: $\mathbf{R}(u, v) = u\mathbf{i} + v\mathbf{j} + uv\mathbf{k}$ that lies inside the cylinder $x^2 + y^2 = a^2$.
13. Find the surface area of the portion of the cylinder $x^2 + z^2 = a^2$ in the first octant between $y = 0$ and $y = x$.
14. Given the sphere $x^2 + y^2 + z^2 = 4a^2$ and the paraboloid $x^2 + y^2 = a(z + a)$, where $a > 0$. (a) Find the surface area of the sphere that lies inside the paraboloid. (b) Find the surface area of the paraboloid that lies inside the sphere.
15. Find the area of the surface defined by

$$x = r \cos\theta,\ y = 2r \cos\theta,\ z = \theta \quad (0 \leq r \leq 1,\ 0 \leq \theta < 2\pi).$$

16. Find the area of the portion of the surface $z = xy$ that is inside the cylinder $x^2 + y^2 = 1$.
17. Find the area of the portion of the surface $z = 2(x^{3/2} + y^{3/2})/3$ that lies above the domain bounded by the lines $x = 0$, $y = 1$ and $x = 3y$.
18. A torus S can be represented by the parametric equations

$$x = (R + \cos\phi) \cos\theta,\ y = (R + \cos\phi) \sin\theta,\ z = \sin\theta$$

$(0 \le \phi < 2\pi,\ 0 \le \theta < 2\pi)$, $R > 1$ (R is fixed). Show that the surface area of the torus is given by $4\pi^2 R$. NOTE : The torus has an inner radius of R - 1 and an outer radius of R + 1. For fixed θ, the equations describe a cross section of the torus with ϕ serving as the polar angle.

4.8 SURFACE INTEGRALS

We now study the integration of a scalar or a vector field on a surface. Such an integral is called a surface integral. A surface integral is a natural generalization of the concept of a double integral on a plane domain. In fact, just as we calculate line integrals by reducing them to ordinary single integrals, we also calculate surface integrals by transforming them into double integrals. We first define the surface integral of a scalar field.

Surface Integral of a Scalar Field

Definition 1. Let f be a scalar field defined and continuous in a domain D. Let S be a smooth surface in D represented by $\mathbf{R}(u, v) = x(u,v)\mathbf{i} + y(u,v)\mathbf{j} + z(u,v)\mathbf{k}$, where $\mathbf{R}(u, v)$ is continuously differentiable in a domain D* of the uv-plane. The surface integral of f on S, denoted by $\iint_S f\, dS$, is the double integral

$$\iint_S f(x, y, z)dS$$

$$= \iint_{D^*} f[x(u,v), y(u,v), z(u,v)]\,|\mathbf{R}_u \times \mathbf{R}_v|\,du\,dv \tag{4.37}$$

Notice that when f = 1, the integral (4.37) gives the surface area of S as defined in Sec. 4.7.

If S is piecewise smooth, we define the surface integral of f on S as the sum of the integrals over the pieces of smooth surfaces comprising S.

For a surface that is represented by an equation of the form $z = z(x, y)$, where (x, y) ranges over the projection D of the surface on the xy-plane, the integral (4.37) can be written as

$$\iint_S f(x, y, z)dS = \iint_D f(x, y, z(x,y)) \sec\gamma\, dx\, dy \tag{4.38}$$

where

$$\sec\gamma = \sqrt{1 + z_x^2 + z_y^2}$$

Similarly, if the surface can be represented by the equation $x = x(y, z)$ or $y = y(x, z)$, the integral (4.37) may also be written as

$$\int\int_S f(x, y, z)dS = \int\int_{D'} f(x(y,z), y, z) \sec\alpha \, dy\, dz \qquad (4.39)$$

or

$$\int\int_S f(x, y, z)dS = \int\int_{D''} f(x, y(x,z), z) \sec\beta \, dx\, dz \qquad (4.40)$$

where

$$\sec\alpha = \sqrt{1 + x_y^2 + x_z^2}\ , \qquad \sec\beta = \sqrt{1 + y_x^2 + y_z^2}$$

D' and D" are the projections of the surface on the yz- and xz-coordinate planes, respectively.

Surface integrals of scalar fields occur in many physical problems. For example, suppose we have a thin sheet of material in the shape of a surface S whose density at each point (x, y, z) is given by $\rho(x, y, z)$. Then the mass of the material is given by the surface integral

$$M = \int\int_S \rho(x, y, z)dS$$

Further, the center of mass (x_c, y_c, z_c) of the material are given by the following integrals:

$$x_c = \frac{1}{M}\int\int_S x\rho(x, y, z)dS$$

$$y_c = \frac{1}{M}\int\int_S y\rho(x, y, z)dS$$

$$z_c = \frac{1}{M}\int\int_S z\rho(x, y, z)dS$$

where M is the mass of the material. Finally, the moment of inertia of the material about a line L is given by the surface integral

$$I_L = \int\int_S \delta^2(x, y, z)\rho(x, y, z)dS$$

where $\delta(x,y,z)$ denotes the distance of each point (x, y, z) from the line L.

Example 1. Find the surface integral of $f(x,y,z) = xy + z$ on the upper half of a sphere of radius a.

Solution: We represent the surface by

$$\mathbf{R}(u, v) = a(\sin u \cos v\, \mathbf{i} + \sin u \sin v\, \mathbf{j} + \cos u\, \mathbf{k}),$$

$0 \le u \le \pi/2, 0 \le v < 2\pi$. Then

$$\partial\mathbf{R}/\partial u = a(\cos u \cos v\, \mathbf{i} + \cos u \sin v\, \mathbf{j} - \sin u\, \mathbf{k})$$

and

$$\partial\mathbf{R}/\partial v = a(- \sin u \sin v\, \mathbf{i} + \sin u \cos v\, \mathbf{j})$$

so that

$$|\partial\mathbf{R}/\partial u \times \partial\mathbf{R}/\partial v| = a^2 \sin u$$

Thus

$$\int_0^{2\pi}\int_0^{\pi/2} [(a \sin u \cos v)(a \sin u \sin v) + a \cos u]a^2 \sin u\, du\, dv$$

$$= \int_0^{2\pi}\int_0^{\pi/2} (a^4 \sin^3 u \cos v \sin v + a^3 \cos u \sin u)\, du\, dv$$

$$= \pi a^3$$

Example 2. Evaluate the surface integral of $f(x, y) = x^2 + y^2$ on the upper half of the unit sphere using (4.38).

Solution: We note that the projection D of the surface on the xy-plane is the disk $x^2 + y^2 \le 1$. Now from the equation of the unit sphere, $\phi(x, y, z) = x^2 + y^2 + z^2 = 1$, we find the outward unit normal vector

$$\mathbf{n} = \text{grad}\, \phi / |\text{grad}\, \phi| = x\,\mathbf{i} + y\,\mathbf{j} + z\,\mathbf{k}$$

Thus, $\sec \gamma = 1/z = 1/(1 - x^2 - y^2)^{1/2}$ and so, by (4.38), we have

$$\iint_D (x^2+y^2)\frac{dx\,dy}{\sqrt{1-x^2-y^2}} = \int_0^{2\pi}\int_0^1 \frac{r^3\,dr\,d\theta}{\sqrt{1-r^2}} = \frac{4\pi}{3}$$

Example 3. The density of a hemispherical shell of radius a is given by $\rho = z$. Find the moment of inertia about the z-axis.

Solution: The outward unit normal vector on the hemisphere is given by $\mathbf{n} = (x\mathbf{i} + y\mathbf{j} + z\mathbf{k})/a$. Thus the moment of inertia about the z-axis is equal to

$$I_z = \iint_S z(x^2+y^2)dS = \iint_D z(x^2+y^2)\sec\gamma\,dx\,dy$$

where D is the disk $x^2 + y^2 \le a^2$. Since $\sec\gamma = a/z$ on the surface, we find, using polar coordinates,

$$I_z = a\int_0^{2\pi}\int_0^a r^3\,dr\,d\theta = \frac{\pi a^5}{2}$$

Surface Integral of a Vector Field

We now consider the definition of an integral of a vector field on a surface. To motivate the definition, we consider the calculation of the amount of fluid that flows across a surface S submerged in a fluid flowing in a domain D. Let $\mathbf{V}(x, y, z)$ denote the velocity and $\rho(x, y, z)$ the density of the fluid. Then the rate of flow (mass per unit area per unit time) of the fluid is given by $\mathbf{F} = \rho\mathbf{V}$. This is called the flux density of the fluid. If S is a plane surface and **F** is constant in D, then clearly the mass of fluid flowing across the plane surface is given by $(\mathbf{F}\cdot\mathbf{n})A$, where A denotes the area of the surface and **n** is the unit normal vector on the side of S from which the fluid emerges (Fig. 4.24).

To calculate the amount of fluid flowing across an arbitrary smooth surface S, we subdivide the surface into n smaller surfaces S_i and calculate the amount of fluid flowing across each of the surfaces S_i $(1 \le i \le n)$. So let ΔS_i denote the surface area of S_i, $\mathbf{n}_i$ the unit normal vector on S_i, and $\mathbf{F}_i$ the flux density on S_i. When n is sufficiently large so that each S_i is very small, $\mathbf{n}_i$ and $\mathbf{F}_i$ are nearly con-

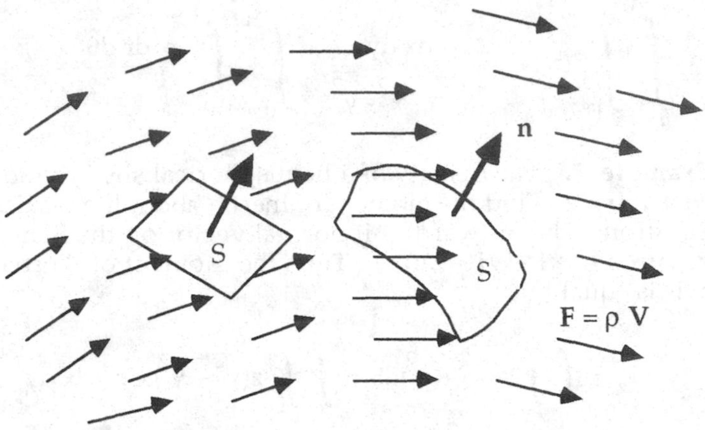

Fig. 4.24 Mass of fluid flowing across a surface S.

stant on S_i. Hence the mass of fluid flowing across S_i is approximately equal to $(\mathbf{F}_i \cdot \mathbf{n}_i)\Delta S_i$, and so the amount of fluid flowing across the entire surface S is approximately equal to the sum $\Sigma_{i=1}^{n} (\mathbf{F}_i \cdot \mathbf{n}_i)\Delta S_i$. The limit of this sum as n increases leads to the double integral

$$\iint_S \mathbf{F} \cdot \mathbf{n} \, dS$$

which is precisely the surface integral of the scalar field $\mathbf{F} \cdot \mathbf{n}$ on S. We call this integral the *surface integral of* **F** *on the surface* S.

Notice that the scalar field $\mathbf{F} \cdot \mathbf{n}$ represents the normal component of the fluid flow on the surface. Thus if the surface is tangential to the flow or the flow is tangential to the surface, then $\mathbf{F} \cdot \mathbf{n} = 0$, which means that there is no fluid flowing across the surface.

Now if the surface S is represented by $\mathbf{R} = \mathbf{R}(u, v)$, we know that the unit normal vector on S is given by

$$\mathbf{n} = \frac{\mathbf{R}_u \times \mathbf{R}_v}{|\mathbf{R}_u \times \mathbf{R}_v|}$$

so that

$$\mathbf{F} \cdot \mathbf{n} = \mathbf{F} \cdot \frac{\partial \mathbf{R}/\partial u \times \partial \mathbf{R}/\partial v}{|\partial \mathbf{R}/\partial u \times \partial \mathbf{R}/\partial v|}$$

Therefore, we define the surface integral of **F** on S as follows:

Definition 2. Let **F** be a vector field defined and continuous in a domain D, and let S be a smooth orientable surface in D represented by $\mathbf{R} = \mathbf{R}(u, v)$, where $\mathbf{R}(u, v)$ is continuously differentiable in D*. The surface integral of **F** on S, denoted by $\iint_S \mathbf{F}\cdot\mathbf{n}\,dS$, is the double integral

$$\iint_S \mathbf{F}\cdot\mathbf{n}\,dS = \iint_{D^*} \mathbf{F}\cdot\frac{\partial \mathbf{R}}{\partial u} \times \frac{\partial \mathbf{R}}{\partial u}\,du\,dv \tag{4.41}$$

If S is piecewise smooth, we define the surface integral (4.41) as the sum of the integrals over the smooth surfaces comprising the surface S. When S is a closed surface, it is customary to take **n** to be the outward unit normal vector.

Other Notations of Surface Integrals

If $\mathbf{F} = P\mathbf{i} + Q\mathbf{j} + R\mathbf{k}$ and we express the unit normal vector **n** on S in terms of its direction cosines, $\mathbf{n} = \cos\alpha\,\mathbf{i} + \cos\beta\,\mathbf{j} + \cos\gamma\,\mathbf{k}$, then the integral (4.41) can be written in the form

$$\begin{aligned}\iint_S \mathbf{F}\cdot\mathbf{n}\,dS &= \iint_S (P\cos\alpha + Q\cos\beta + R\cos\gamma)\,dS \\ &= \iint_S (P\,dy\,dz + Q\,dz\,dx + R\,dx\,dy)\,dS\end{aligned} \tag{4.42}$$

This corresponds to the notation $\int_C P dx + Q dy + R dz$ for the line integral $\int_C \mathbf{F}\cdot d\mathbf{R}$. Further, if the surface is represented by $z = z(x, y)$, $(x, y) \varepsilon D$, so that $dS = \sec\gamma\,dx\,dy$, then the integral (4.41) can also be written in the form

$$\iint_S \mathbf{F}\cdot\mathbf{n}\,dS = \iint_D \mathbf{F}\cdot\mathbf{n}\sec\gamma\,dx\,dy \tag{4.43}$$

Corresponding formulas can also be obtained using the relation $dS = \sec\beta\,dy\,dz$ or $dS = \sec\alpha\,dz\,dx$, whenever the surface S can be represented by $y = y(x, z)$ or $x = x(y, z)$.

Example 4. Calculate the surface integral of the vector field $\mathbf{F}(x, y, z) = x\mathbf{i} + y\mathbf{j} + 3z\mathbf{k}$ on the surface S: $\mathbf{R}(u, v) = a(\cos u\,\mathbf{i} + \sin u\,\mathbf{j}) + v\mathbf{k}$, $0 \le u \le \pi$, $0 \le v \le \pi$.

Solution: On the surface we have

$$\frac{\partial \mathbf{R}}{\partial u} \times \frac{\partial \mathbf{R}}{\partial v} = a(\cos u\,\mathbf{i} + \sin u\,\mathbf{j})$$

and $\mathbf{F} = a(\cos u\,\mathbf{i} + \sin u\,\mathbf{j}) + 3v\mathbf{k}$. Hence

$$\mathbf{F}\cdot\frac{\partial \mathbf{R}}{\partial u} \times \frac{\partial \mathbf{R}}{\partial u} = (a \cos u)^2 + (a \sin u)^2 = a^2$$

so that

$$\int\!\!\int_S \mathbf{F}\cdot\mathbf{n}\, dS = \int_0^{\pi}\!\!\int_0^{\pi} \mathbf{F}\cdot\frac{\partial \mathbf{R}}{\partial u} \times \frac{\partial \mathbf{R}}{\partial v}\, du\, dv = a^2\pi^2 \tag{4.43}$$

Alternatively, we note that the surface can also be represented by the equation

$$y = \sqrt{a^2 - x^2} \qquad (-a \le x \le a,\ 0 \le z \le \pi)$$

The unit normal vector on the surface is given by $\mathbf{n} = (x\,\mathbf{i} + y\,\mathbf{j})/a$, so that

$$\mathbf{F}\cdot\mathbf{n} = (x^2 + y^2)/a$$

The projection of the surface on the xz-coordinate plane is the rectangular domain D: $-a \le x \le a$, $0 \le z \le \pi$. Using the relation $dS = \sec\beta\, dx\, dz = (a/y)\, dx\, dz$, we thus obtain

$$\int\!\!\int_S \mathbf{F}\cdot\mathbf{n}\, dS = \frac{1}{a}\int\!\!\int_D (x^2 + y^2)\sec\beta\, dx\, dz$$

$$= \int_0^{\pi}\!\!\int_{-a}^{a} \frac{a^2\, dx\, dz}{\sqrt{a^2 - x^2}} = \pi a^2 \arcsin (x/a)^4 \Big]_{-a}^{a} = \pi^2 a^2$$

Example 5. Calculate the surface integral of $\mathbf{F}(x, y, z) = x\mathbf{i} + y\mathbf{j} + z\mathbf{k}$ on the hemisphere S: $z = \sqrt{a^2 - x^2 - y^2}$, $(x^2 + y^2 \le a^2)$, where the unit normal vector $\mathbf{n}$ points away from the origin.

Solution: The unit normal vector on the surface is given by $\mathbf{n} = (x\mathbf{i} + y\mathbf{j} + z\mathbf{k})/a$, so that $\mathbf{F}\cdot\mathbf{n} = (x^2 + y^2 + z^2)/a$. Hence, on the surface,

$\mathbf{F}\cdot\mathbf{n} = a$, and so we have

$$\iint_S \mathbf{F}\cdot\mathbf{n}dS = \iint_S a dS = 2\pi a^3$$

where we made use of the fact that the surface area of the hemisphere is $2\pi a^2$.

Example 6. Evaluate the integral $\iint_S \mathbf{F}\cdot\mathbf{n}\, dS$, where $\mathbf{F}(x, y, z) = x^2\mathbf{i} + y^2\mathbf{j} + z^2\mathbf{k}$ and S consists of the faces of the unit cube $0 \le x \le 1$, $0 \le y \le 1$, $0 \le z \le 1$.

Solution: We calculate the surface integrals on the faces S_1, S_2 and S_3 where $x = 1$, $y = 1$, and $z = 1$, and on which the unit normal vectors are $\mathbf{i}$, $\mathbf{j}$, $\mathbf{k}$, respectively. We find

$$\iint_{S_1} \mathbf{F}\cdot\mathbf{n}\, dS = \iint_{S_1} (\mathbf{i} + y^2\mathbf{j} + z^2\mathbf{k})\cdot\mathbf{i}\, dS = \iint_{S_1} dS = 1$$

$$\iint_{S_2} \mathbf{F}\cdot\mathbf{n}\, dS = \iint_{S_2} (x^2\mathbf{i} + \mathbf{j} + z^2\mathbf{k})\cdot\mathbf{j}\, dS = \iint_{S_2} dS = 1$$

$$\iint_{S_3} \mathbf{F}\cdot\mathbf{n}\, dS = \iint_{S_3} (x^2\mathbf{i} + y^2\mathbf{j} + \mathbf{k})\cdot\mathbf{k}\, dS = \iint_{S_3} dS = 1$$

since the area of each of the faces is equal to 1. The surface integrals on the corresponding opposite faces, where $x = 0$, $y = 0$, and $z = 0$, are zero since $\mathbf{F}\cdot\mathbf{n} = 0$ on each of those faces. Therefore, the surface integral is equal to

$$\iint_S \mathbf{F}\cdot\mathbf{n}\, dS = \iint_{S_1} \mathbf{F}\cdot\mathbf{i}\, dS + \iint_{S_2} \mathbf{F}\cdot\mathbf{j}\, dS + \iint_{S_3} \mathbf{F}\cdot\mathbf{k}\, dS = 3$$

Heat Flux Across a Surface

Surface integrals of vector fields also arise in the study of heat conduction in solid bodies. Let $U(x, y, z)$ denote the temperature at a point (x, y, z) of a heat conducting body. According to the law of thermodynamics, heat energy flows in the direction of decreasing temperature (that is, from points of higher temperature to points of lower temperature) at a rate proportional to the temperature gradient ∇U. Since ∇U points in the direction where U increases, it follows that the amount of heat energy flowing across a unit surface area is given by $\mathbf{F} = -k\nabla U$, where k denotes the thermal conductivity of the body. This is known as the heat flux. Therefore the total flux across the surface S of the body is given by $\iint_S \mathbf{F}\cdot \mathbf{n}\, dS$.

Example 7. Suppose that the temperature at each point inside of a ball of radius a is given by $U(x, y, z) = a^2 - x^2 - y^2 - z^2$. What is the heat flux across the surface of the ball, assuming $k = 1$?

Solution: The outward unit normal vector on the ball is $\mathbf{n} = (x\mathbf{i} + y\mathbf{j} + z\mathbf{k})/a$, and $\mathbf{F} = -\nabla U = 2(x\mathbf{i} + y\mathbf{j} + z\mathbf{k})$. Hence $\mathbf{F}\cdot\mathbf{n} = 2(x^2 + y^2 + z^2)/a$ which reduces to $\mathbf{F}\cdot\mathbf{n} = 2a$ on the surface. Therefore the heat flux across the surface of the ball is equal to

$$\iint_S \mathbf{F}\cdot\mathbf{n}\,dS = \iint_S 2a\,dS = 8\pi a^3$$

4.7 EXERCISES

In each of Problems 1 through 5, evaluate the integral $\iint_S f\, dS$ for the given scalar field f and surface S.

1. $f(x, y, z) = xyz$, where S is the triangle with vertices (2, 0, 0), (0, 2, 0), (0, 1, 1).
2. $f(x, y, z) = x^2 + z^2$, where S is the lateral surface of the cylinder $x^2 + y^2 = a^2$ $(0 \le z \le b)$.
3. $f(x, y, z) = z^2$, where S is the upper hemisphere of radius a.
4. $f(x, y, z) = xz$, where S is the paraboloid $x^2 + y^2 = 1 - z$ above the xy-plane.
5. $f(x, y, z) = x + y + z$, where S is the sphere of radius a. (Use the symmetry of the problem.)
6. Let S be the triangle with vertices (1, 0, 0), (0, 1, 0), (0, 0, 1). Argue by symmetry that

$$M = \iint_S x\,dS = \iint_S y\,dS = \iint_S z\,dS$$

Thus, with very little computation, show that

$$M = \frac{1}{3}\text{ (area of triangle)} = \frac{\sqrt{3}}{6}$$

7. Let S be the sphere of radius a. As in Problem 6, show that

$$M = \int\int_S x^2\,dS = \int\int_S y^2\,dS = \int\int_S z^2\,dS$$

and verify that I = $(1/3)a^2$ (area of sphere) = $4\pi a^4/3$.

8. A homogeneous cylindrical shell has radius a and altitude h. Find the moment of inertia about its axis.
9. A material in the shape of a thin paraboloidal shell $x^2 + y^2 = z$ $(0 \le z \le 1)$ has density given by $f(x, y, z) = 1 + z$. Find the total mass of the shell.
10. A surface made of certain metal is in the shape of a thin hemisphere $z = (a^2 - x^2 - y^2)^{1/2}$, $(x^2 + y^2 \le a^2)$. If its density is given by $f(x, y, z) = z$, find the moment of inertia about the z-axis.
11. A certain material is in the shape of a thin cylindrical shell $x^2 + y^2 = 4$ $(0 \le z \le 2)$, If the density is given by $f(x, y, z) = |x| + |y|$, find the mass of the material.
12. Find the average temperature of the material of Problem 9 if the temperature at each point (x, y, z) is given by $T(x, y, z) = z$.

In each of Problems 13 through 19, compute the integral $\iint_S \mathbf{F}\cdot\mathbf{n}\,dS$ for the given vector field **F** and surface S, where **n** is the normal vector on S with positive z-component.

13. $\mathbf{F}(x, y, z) = x\mathbf{i} + y\mathbf{j} + z\mathbf{k}$, where S is the triangle with vertices at (1, 0, 0), (0, 1, 0), (0, 0, 1).
14. $\mathbf{F}(x, y, z) = y\mathbf{i} + z\mathbf{j} + x\mathbf{k}$, where S is the surface of Problem 1.
15. $\mathbf{F}(x, y, z) = x\mathbf{i} + y\mathbf{j} + z\mathbf{k}$, where S is the paraboloid $x^2 + y^2 = 1 - z$ above the xy-plane.
16. $\mathbf{F}(x, y, z) = x^2\mathbf{i} + y\mathbf{j} + z\mathbf{k}$, where S is the cylindrical surface $x^2 + z^2 = a^2$ $(0 \le y \le \pi,\ z \ge 0)$.
17. $\mathbf{F}(x, y, z) = y\mathbf{i} - z\mathbf{j} + x^2\mathbf{k}$, where S is the cylinder $z^2 = 4y$ in the first octant bounded by $x = 1$ and $y = 4$.
18. $\mathbf{F}(x, y, z) = x\mathbf{i} + y\mathbf{j} - 2z\mathbf{k}$, where S is the upper hemisphere of radius a.
19. $\mathbf{F}(x, y, z) = z\mathbf{i} + y\mathbf{j} + x^2\mathbf{k}$, where S is the portion of the paraboloid $y^2 + z^2 = x$ in the first octant bounded by $x = 1$.
20. Compute $\iint_S \mathbf{F}\cdot\mathbf{n}\,dS$, where $\mathbf{F}(x, y, z) = x^2\mathbf{i} + y^2\mathbf{j} + z^2\mathbf{k}$ and S is the faces of the cube $x = \pm 1$, $y = \pm 1$, $z = \pm 1$. The vector **n** is outward.

21. Repeat Problem 20 when $\mathbf{F}(x, y, z) = x\mathbf{i} + y\mathbf{j} + z\mathbf{k}$.
22. Compute $\iint_S \mathbf{F} \cdot \mathbf{n}\, dS$, where $\mathbf{F}(x, y, z) = x\mathbf{i} - y\mathbf{j} + z^2\mathbf{k}$, S is the entire surface of the cylinder $x^2 + y^2 = 4$ $(0 \le z \le 2)$, and $\mathbf{n}$ is the outward unit normal vector.
23. Compute $\iint_S \mathbf{F} \cdot \mathbf{n}\, dS$, where $\mathbf{F}(x, y, z) = z\mathbf{i} + (x + y)\mathbf{j} - x\mathbf{k}$, S is the surface in the first octant consisting of $x^2 + z^2 = 9$, $x = 0$, $y = 0$, $z = 0$, $y = 4$ and $\mathbf{n}$ points away from the origin.
24. Let the temperature at each point (x, y, z) in space be given by $U(x, y, z) = x^2 + y^2$. Find the heat flux across the upper hemishpere of radius a.
25. Find the heat flux across the surface $x^2 + y^2 = 1 - z$ above the xy-plane, if the temperature is given by $U(x, y, z) = y + z$.
26. The flux density of a fluid flow is given by $\mathbf{F} = - y\mathbf{i} + x\mathbf{j} + z\mathbf{k}$. Find the total flux across the sphere of radius R.
27. The flux density of a fluid flow is given by $\mathbf{F} = x\mathbf{i} + y\mathbf{j}$. Find the rate of fluid flowing across the surcface of the paraboloid $x^2 + y^2 = z$ $(0 \le z \le 4)$.
28. Evaluate $\iint_S \operatorname{curl} \mathbf{F} \cdot \mathbf{n}\, dS$, where $\mathbf{F} = yz\mathbf{i} + y^2\mathbf{j} + z^2\mathbf{k}$, S is the lower hemisphere of radius 2, and $\mathbf{n}$ is outward.
29. Evaluate $\iint_S \operatorname{curl} \mathbf{F} \cdot \mathbf{n}\, dS$, where $\mathbf{F} = y\mathbf{i} - x\mathbf{j} + xyz\mathbf{k}$, S is the surface $x^2 + y^2 + 2z^2 = 1$ $(z \ge 0)$.
30. Let S be a smooth surface defined by $\mathbf{R} = \mathbf{R}(u, v)$, $(u, v) \in D^*$. Suppose s, t are new parameters related to u, v by the equation $s = s(u, v)$, $t = t(u, v)$ such that for each $(u, v) \in D^*$ there corresponds one and only one $(s, t) \in D^{**}$ and vice versa. Set $\mathbf{R}^*(s, t) = \mathbf{R}(u, v)$, when u and v are expressed in terms of s and t. Show that $\mathbf{R}_u \times \mathbf{R}_v$ and $\mathbf{R}^*_s \times \mathbf{R}^*_t$ have the same direction if $\partial(s, t)/\partial(u, v) > 0$, and opposite direction if $\partial(s, t)/\partial(u, v) < 0$. Thus verify that if $\partial(s, t)/\partial(u, v) > 0$, then

$$\iint_{D^*} \mathbf{F} \cdot \mathbf{R}_u \times \mathbf{R}_v\, du\, dv = \iint_{D^{**}} \mathbf{F} \cdot \mathbf{R}_s \times \mathbf{R}_t\, ds\, dt$$

31. Show that the surface integral $\iint_S f\, dS$ is independent of the parametric representation of the surface.
32. Let $\mathbf{F} = P\mathbf{i} + Q\mathbf{j} + R\mathbf{k}$ and suppose it is represented by $z = z(x, y)$, $(x, y) \in D$, where D is the projection of S on the xy-plane. When $\mathbf{n}$ has positive z-component, show that

$$\iint_S \mathbf{F} \cdot \mathbf{n}\, dS = \iint_D \left(- P\frac{\partial z}{\partial x} - Q\frac{\partial z}{\partial y} + R \right) dx\, dy$$

Obtain the corresponding formulas when S has the representation $y = y(x, z)$ and $x = x(y, z)$.

4.9 THE DIVERGENCE (GAUSS') THEOREM

We recall that Green's theorem expresses a relationship between a line integral on a simple closed curve in the plane and a double integral taken over the plane domain bounded by the curve. The extension of Green's theorem to three-dimensional space leads to the important divergence theorem or Gauss' theorem. The theorem relates a surface integral on a closed surface to a volume integral (or triple integral) over the domain bounded by the surface. We state the theorem as follows.

THEOREM 1. Let D be a domain in the xyz-space bounded by a piecewise smooth closed surface S, and let **n** denote the outward unit normal vector on S. If **F** is continuously differentiable in a domain containing D, then

$$\iiint_D \operatorname{div} \mathbf{F}\, dx\, dy\, dz = \iint_S \mathbf{F}\cdot\mathbf{n}\, dS \tag{4.44}$$

Proof: Let $\mathbf{F}(x, y, z) = P(x, y, z)\,\mathbf{i} + Q(x, y, z)\,\mathbf{j} + R(x, y, z)\,\mathbf{k}$ and let $\mathbf{n} = \cos\alpha\,\mathbf{i} + \cos\beta\,\mathbf{j} + \cos\gamma\,\mathbf{k}$. Then (4.44) can be written as

$$\iiint_D \left(\frac{\partial P}{\partial x} + \frac{\partial Q}{\partial y} + \frac{\partial R}{\partial z}\right) dx\, dy\, dz$$

$$= \iint_S (P\cos\alpha + Q\cos\beta + R\cos\gamma)\, dS \tag{4.45}$$

Thus the theorem will be proved if we can show that

$$\begin{aligned}
\iiint_D \frac{\partial P}{\partial x}\, dx\, dy\, dz &= \iint_S P\cos\alpha\, dS \\
\iiint_D \frac{\partial Q}{\partial y}\, dx\, dy\, dz &= \iint_S Q\cos\beta\, dS \\
\iiint_D \frac{\partial R}{\partial z}\, dx\, dy\, dz &= \iint_S R\cos\gamma\, dS
\end{aligned} \tag{4.46}$$

As in the proof of Green's theorem, we establish these identities in the special case when the domain D has a non-degenerate projection on the xy-, yz-, and xz-coordinate planes so that corresponding to each projection, the bounding surface S can be split into two surfaces S_1 and S_2. We show such a domain D in Fig. 4.25 whose projection on the yz-coodinate plane is denoted by T. We establish the first identity in (4.46) for this domain. Let S_1 and S_2 be the two parts of the surface and suppose they are represented by the equations

$$S_1: \quad x = f(y, z), \qquad\qquad S_2: \quad x = g(y, z),$$

where $f(y, z) \leq g(y, z)$, $(y, z) \varepsilon T$. By writing the triple integral of $\partial P/\partial x$ in iterated form and then integrating with respect to x first, we find

$$\iiint_D \frac{\partial P}{\partial x}\, dx\, dy\, dz = \iint_T \left[\int_{f(y,z)}^{g(y,z)} \frac{\partial P}{\partial x} dx \right] dy\, dz$$

$$= \iint_T [P(g(y,z), y, z) - P(f(y,z), y, z)]\, dy\, dz \tag{4.47}$$

We will show that the integral on the right-hand side of (4.47) is equivalent to the surface integral $\iint_S P\cos\alpha\, dS$. First we note that

$$\iint_S P\cos\alpha\, dS = \iint_{S_1} P\cos\alpha\, dS + \iint_{S_2} P\cos\alpha\, dS \tag{4.48}$$

Now on S_2 we have $\cos\alpha > 0$, so that $\cos\alpha\, dS = dy\, dz$; hence

$$\iint_{S_2} P\cos\alpha\, dS = \iint_T P[g(y,z), y, z)]\, dy\, dz$$

On S_1 we have $\cos\alpha < 0$, so that $\cos\alpha\, dS = -\, dy\, dz$, and so

$$\iint_{S_1} P\cos\alpha\, dS = -\iint_T P[f(y,z), y, z)]\, dy\, dz$$

Substituting these in (4.48), it follows from (4.47) that

$$\iiint_D \frac{\partial P}{\partial x} dx\, dy\, dz = \iint_S P\cos\alpha\, dS$$

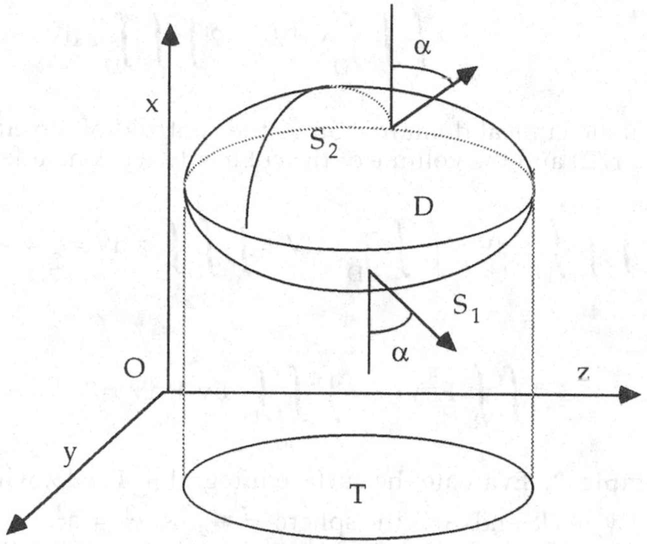

Fig . 4.25 The Divergence Theorem

The other two identities can be established in exactly the same way by considering the projection of the domain on the xz- and on the xy-coordinate planes.

We have seen the usefulness of the Green's theorem in calculating certain line integrals along closed curves. Likewise the divergence theorem can be used to evaluate certain surface integrals over closed surfaces. We illustrate this in the following examples.

Example 1. Evaluate the surface integral $\iint_S \mathbf{F}\cdot \mathbf{n} dS$, where $\mathbf{F}(x, y, z) = x^2\,\mathbf{i} + y^2\,\mathbf{j} + z^2\,\mathbf{k}$ and S consists of the sides of the unit cube $0 \le x \le 1, 0 \le y \le 1, 0 \le z \le 1$.

Solution: We note that div $\mathbf{F} = 2(x + y + z)$. Hence by the divergence theorem, we have

$$\int\int_S \mathbf{F}\cdot \mathbf{n}\, dS = 2\int\int\int_D (x + y + z)dV = 2\int\int\int_D x\, dV$$

$$+ 2\iiint_D y\, dV + 2\iiint_D z\, dV$$

where D is the cubical domain. Since the centroid of the cube is at (1/2, 1/2, 1/2) and the volume of the cube is 1, we conclude that

$$\iiint_D x\, dV = \iiint_D y\, dV = \iiint_D z\, dV = \frac{1}{2} V = \frac{1}{2}$$

Therefore

$$\iint_S \mathbf{F} \cdot \mathbf{n}\, dS = \iiint_D \text{div}\, \mathbf{F}\, dV = 3$$

Example 2. Evaluate the surface integral $\iint_S \mathbf{F} \cdot \mathbf{n} dS$, where $\mathbf{F}(x, y, z) = x\mathbf{i} + y\mathbf{j} + z\mathbf{k}$ and S is the sphere $x^2 + y^2 + z^2 = a^2$.

Solution: Since div $\mathbf{F} = 3$, we have by the divergence theorem,

$$\iint_S \mathbf{F} \cdot \mathbf{n}\, dS = \iiint_D \text{div}\, \mathbf{F}\, dV = 3\iiint_D dV = 4\pi a^3$$

where we noted that the volume of the sphere is $4\pi a^3/3$.

Example 3. Evaluate $\iint_S \mathbf{F} \cdot \mathbf{n} dS$, where $\mathbf{F}(x, y, z) = \sin y\, \mathbf{i} + e^x \mathbf{j} + z^2 \mathbf{k}$ and S consists of the plane $z = 0$ and the upper hemishpere

$$z = \sqrt{a^2 - x^2 - y^2}, \quad x^2 + y^2 \le a^2$$

Solution: By the divergence theorem, we have

$$\iint_S \mathbf{F} \cdot \mathbf{n}\, dS = \iiint_D \text{div}\, \mathbf{F}\, dV = 2\iiint_D z\, dV$$

To calculate the triple integral, we introduce cylindrical coordinates to obtain

$$\iiint_D z\, dx\, dy\, dz = \int_0^{2\pi} \int_0^a \int_0^{\sqrt{a^2 - r^2}} z\, dz\, r\, dr\, d\theta$$

$$= \frac{1}{2}\int_0^{2\pi}\int_0^a (a^2 - r^2) r\, dr\, d\theta = \frac{\pi a^4}{4}$$

Therefore,

$$\iint_S \mathbf{F}\cdot\mathbf{n}\, dS = \frac{\pi a^4}{2}$$

4.10 APPLICATIONS OF THE DIVERGENCE THEOREM

The divergence theorem is used in many physical problems to derive mathematical equations satisfied by the physical entities considered. Before we show some of these derivations, let us see how the theorem provides a natural definition of divergence that is independent of the choice of coordinate system. Let D_r denote the domain bounded by a sphere S_r of radius r and center at P, and let **F** be a vector field that is continuous including its first derivatives in D_r. By the mean value theorem for integrals, we have

$$\iiint_{D_r} \text{div}\, \mathbf{F}\, dV = V(r)\, \text{div}\, \mathbf{F}(Q) \tag{4.49}$$

where V(r) denote the volume of the sphere and Q is a point in D_r. Notice that we purposely did not use dxdydz for the element of volume since we are not necessarily dealing with rectangular cartesian coordinates. Now by the divergence theorem, we have

$$\text{div}\, \mathbf{F}(Q) = \frac{1}{V(r)}\iint_{S_r} \mathbf{F}\cdot\mathbf{n}\, dS$$

As r goes to zero, it is clear that V(r) also tends to zero and the point Q approaches the center P of the sphere. Thus, in taking the limit as r tends to zero, we obtain

$$\text{div}\, \mathbf{F}(Q) = \lim_{r \to 0} \frac{1}{V(r)}\iint_{S_r} \mathbf{F}\cdot\mathbf{n}\, dS \tag{4.50}$$

This formula is sometimes used to define the divergence of a vector field **F**, where S_r is taken to be an arbitrary piecewise smooth and closed surface and the limit is taken as the volume shrinks to the point P. Such a definition clearly does not depend on the coordinate system used, which shows that the concept of a divergence is an intrinsic property of a vector field. On the other hand, if we introduce a rectangular cartesian coordinate and choose S to be a cube with center at (x, y, z), it can be shown that the limit (4.50) leads to the familiar expression

$$\text{div } \mathbf{F} = \frac{\partial P}{\partial x} + \frac{\partial Q}{\partial y} + \frac{\partial R}{\partial z}$$

where P, Q, and R are the components of **F** with respect to the coordinate system.

The Continuity Equation

We now consider some applications of the divergence theorem in fluid dynamics. In Sec. 4.8 we saw that for the flux density of a fluid flow, $\mathbf{F} = \rho\mathbf{V}$, the integral $\iint_S \mathbf{F}\cdot\mathbf{n}\, dS$ represents the total flux across the surface S, that is, the total mass of fluid flowing through the surface (in the normal direction) per unit time. Hence, by (4.50), we can say that the divergence of **F** at a point P is simply the rate of outward flux per unit volume. Equivalently, we can also say that it is the rate of change of mass at P per unit volume. Hence, if div $\mathbf{F} > 0$, then fluid is flowing into the region (presence of a source), and if div $\mathbf{F} < 0$, fluid is being drained out from the region (presence of a sink).

Now consider a volume D of the fluid enclosed by a smooth surface S (Fig. 4.26). The total mass of fluid enclosed by S is given by the volume integral

$$m = \iiint_D \rho\, dV$$

The mass changes at the rate given by

$$\frac{\partial m}{\partial t} = \iiint_D \frac{\partial \rho}{\partial t}\, dV$$

Assuming that there is no point in the fluid where mass is being introduced (a source) or taken away (a sink), the rate at which the mass is changing in D must be equal to the rate at which it is flowing out through the surface S. Therefore,

$$\iint_S \rho\mathbf{V}\cdot\mathbf{n}\, dS = -\iiint_D \frac{\partial \rho}{\partial t}\, dV \tag{4.51}$$

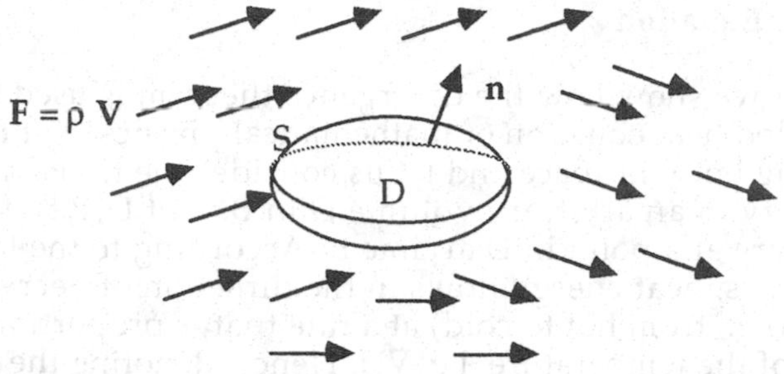

Fig. 4.26 Mass of fluid in a domain D enclosed by a surface S.

where the negative sign indicates that the density is decreasing. Applying the divergence theorem on the left-hand side of (4.51), we obtain

$$\int\int\int_D \text{div}\,(\rho \mathbf{V})\,dV = -\int\int\int_D \frac{\partial \rho}{\partial t}\,dV$$

or

$$\int\int\int_D \left[\text{div}\,(\rho \mathbf{V}) + \frac{\partial \rho}{\partial t}\right] dV = 0 \qquad (4.52)$$

We assume that $\rho \mathbf{V}$ is continuously differentiable so that the integrand in (4.52) is continuous. Then, since the integral holds for an arbitrary volume D, we conclude that the integrand must vanish identically, that is,

$$\text{div}\,(\rho \mathbf{V}) + \frac{\partial \rho}{\partial t} = 0 \qquad (4.53)$$

Indeed, if the integrand were positive (or negative) at some point in D, then because it is continuous it would remain positive (or negaive) in a neighborhood of that point. Hence by choosing that neighborhood for D in (4.52), we would have a non-zero integral, thus contradicting (4.52). In fluid dynamics, equation (4.53) is known as the *continuity equation.* It is actually a statement about conservation of mass.

The Heat Equation

Next we show how the divergence theorem is used to derive the so-called heat equation of mathematical physics. Let B be a heat conducting body in space and let us consider the transfer or flow of heat energy in an arbitrary volume D in B. Let U(P,t) denote the temperature at a point in D at time t. According to the law of thermodynamics, heat energy flows in the direction of decreasing temperature (i.e., from hot to cold) at a rate that is proportional to the gradient of the temperature, i.e. ∇U. Hence, denoting the velocity of heat flow by $\mathbf{v}$ and the thermal conductivity of the body by K (assumed constant), we have $\mathbf{v} = -K\nabla U$, where the negative sign indicates that $\mathbf{v}$ is opposite in direction to ∇U. Now the amount of heat energy flowing across the boundary S of D is given by

$$\int\int_S \mathbf{v}\cdot\mathbf{n}\,dS = -K\int\int_S \nabla U\cdot\mathbf{n}\,dS$$

By the divergence theorem, this is equal to

$$\int\int_S \mathbf{v}\cdot\mathbf{n}\,dS = -K\int\int_S \nabla U\cdot\mathbf{n}\,dS = -K\int\int\int_D \operatorname{div}(\nabla U)dV$$

$$= -K\int\int\int_D \nabla^2 U\,dV \tag{4.54}$$

where $\nabla^2 U = \operatorname{div}(\operatorname{grad} U)$ is the Laplacian of U. By the law of conservation of energy, the total heat energy flowing out of the volume D across its boundary must be equal to the rate at which the heat energy inside the body decreases. Since the total heat energy inside the body D is given by the volume integral

$$H = \int\int\int_D c\rho U\,dV$$

where c is the specific heat and ρ is the density of the body, both assumed to be constant, it follows that

$$-\frac{\partial H}{\partial t} = -\int\int\int_D c\rho\frac{\partial U}{\partial t}\,dV \tag{4.55}$$

Hence, equating (4.54) and (4.55), we obtain

$$\iiint_D \left[c\rho \frac{\partial U}{\partial t} - K\nabla^2 U \right] dV = 0 \tag{4.56}$$

from which we again conclude that

$$\frac{\partial U}{\partial t} - K\nabla^2 U = 0 \tag{4.57}$$

The constant $\kappa = K/c\rho$ in (4.57) is called the coefficient of diffusivity. Equation (4.57) is known as the heat equation. It is one of the principal partial differential equations of mathematics physics.

If the temperature of the body is in equilibrium (steady-stsate), then U is independent of time and so $\partial U/\partial t = 0$. In such a case equation (4.57) becomes

$$\nabla^2 U = 0 \tag{4.58}$$

This is called Laplace's equation, which is another important partial differential equations of mathematical physics.

Gauss' s Law

We recall that the electrostatic field at a point (x, y, z) induced by the presence of a point charge q at the origin is given by

$$\mathbf{E}(x, y, z) = \frac{q}{r^3}\mathbf{R}, \quad r = |\mathbf{R}|$$

By direct calculation, we see that div $\mathbf{E} = 0$ at all points $(x, y, z) \neq (0, 0, 0)$. Hence for any closed surface S not enclosing the origin, the divergence theorem shows that $\iiint_D \text{div}\, \mathbf{E}\, dV = \iint_S \mathbf{E} \cdot \mathbf{n}\, dS = 0$. Thus the net electric flux across any surface S which does not enclose the origin is zero. When S is a closed surface which encloses the origin, the divergence theorem cannot be applied since inside the domain enclosed by S, **E** is not defined at the origin. In this case, we will show that the net flux across the surface S is $4\pi q$, that is,

$$\iint_S \mathbf{E} \cdot \mathbf{n}\, dS = 4\pi q$$

In order to apply the divergence theorem in the domain bounded by the surface S, we enclose the origin in a small sphere Σ of radius ε such that Σ lies inside S (Fig. 4.27). Then applying the divergence theorem in the domain D* that is bounded by S and Σ, we obtain

$$\iiint_{D^*} \operatorname{div} \mathbf{E}\, dV = \iint_S \mathbf{E}\cdot\mathbf{n}\, dS + \iint_\Sigma \mathbf{E}\cdot\mathbf{n}\, dS = 0$$

Hence

$$\iint_S \mathbf{E}\cdot\mathbf{n}\, dS = -\iint_\Sigma \mathbf{E}\cdot\mathbf{n}\, dS$$

The surface integral on the sphere Σ can be calculated. On Σ we note that the unit normal vector **n** is directed toward the origin and is given by $\mathbf{n} = -\mathbf{R}/\varepsilon$, where **R** is the position vector of points on Σ. Thus on Σ we have

$$\mathbf{E}\cdot\mathbf{n} = -\frac{q}{\varepsilon^4}\mathbf{R}\cdot\mathbf{R} = -\frac{q}{\varepsilon^2}$$

and so

$$\iint_S \mathbf{E}\cdot\mathbf{n}\, dS = \iint_\Sigma \frac{q}{\varepsilon^2}\, dS = \frac{q}{\varepsilon^2}(4\pi\varepsilon^2) = 4\pi q$$

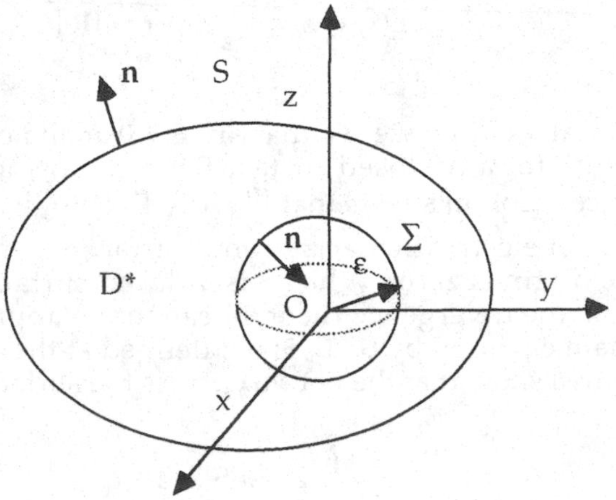

Fig. 4. 27 Gauss's law.

Therefore, the total electric flux across a closed surface S is equal to 4π times the charge q if the charge lies inside S, and it is zero if the charge lies outside of S. This result is known as *Gauss's law.*

4.8 EXERCISES

In each of Problems 1 through 6, use the divergence theorem to evaluate the surface integral $\iint_S \mathbf{F}\cdot \mathbf{n}\, dS$ for the given vector field **F** and surface S, and verify the result by integrating the surface integral directly.

1. $\mathbf{F}(x, y, z) = xy\mathbf{i} + z^2\mathbf{j} + 2yz\mathbf{k}$, S is the surface of the cube $0 \le x, y, z \le 1$.
2. $\mathbf{F}(x, y, z) = y^2\mathbf{i} + yz\mathbf{j} + xz\mathbf{k}$, S is the tetrahedron bounded by $x = 0$, $y = 0$, $z = 0$, and $x + y + z = 1$.
3. $\mathbf{F}(x, y, z) = xz\mathbf{i} + yx\mathbf{j} + zy\mathbf{k}$, S is the surface $x^2 + y^2 = 4$, $z = 0$, $y + z = 2$.
4. $\mathbf{F}(x, y, z) = x\mathbf{i} + y\mathbf{j} + (z - 1)\mathbf{k}$, S is the surface formed by $x^2 + y^2 = (z - 2)^2$, $z = 0$, and $z = 1$.
5. $\mathbf{F}(x, y, z) = xy\mathbf{i} + y^2\mathbf{j} + yz\mathbf{k}$, S consists of an upper hemisphere of radius 2 and the bottom plane $z = 0$.
6. $\mathbf{F}(x, y, z) = x^3\mathbf{i} + x^2y\mathbf{j} + x^2z\mathbf{k}$, S is formed by the cylinder $x^2 + y^2 = 1$ and the planes $z = 0$, $z = 4$.
7. Let **F** be a vector field that is continuously differentiable and satisfies div $\mathbf{F} = 0$ in a domain D, except at a point P. Let S be any closed, piecewise smooth surface enclosing P and let S' be a small sphere with center at P and lying entirely within S. Show that

$$\iint_S \mathbf{F}\cdot\mathbf{n}\, dS = \iint_{S'} \mathbf{F}\cdot\mathbf{n}\, dS$$

8. Let $\mathbf{F}(x, y, z) = (x\mathbf{i} + y\mathbf{j} + z\mathbf{k})/r^n$, where $r = (x^2 + y^2 + z^2)^{1/2}$ and n is a positive integer.
   (a) Show that div $\mathbf{F} = -(n - 3)/r^n$.
   (b) Evaluate the integral $\iint_S \mathbf{F}\cdot\mathbf{n}\, dS$ for $n = 1$, $n = 2$ when S is the sphere $x^2 + y^2 + z^2 = a^2$. Can you use the divergence theorem? Explain.
9. Let **F** be the vector field given in Problem 8 when $n = 3$, so that div $\mathbf{F} = 0$. Let S be the surface of the sphere $x^2 + y^2 + z^2 = a^2$.

(a) Show that

$$\iint_S \mathbf{F}\cdot\mathbf{n}\, dS = 4\pi$$

Does this contradict the divergence theorem? Explain.
(b) If S* is any piecewise, smooth closed surface enclosing the origin, show that

$$\iint_{S^*} \mathbf{F}\cdot\mathbf{n}\, dS = 4\pi$$

(c) What is the value of $\iint_S \mathbf{F}\cdot\mathbf{n}\, dS$ if S does not enclose the origin?

10. If u is a harmonic function, show that div (grad u) = $|\text{grad } u|^2$. Hence, evaluate the surface integral $\iint_S u(\partial u/\partial n)dS$ on the sphere $x^2 + y^2 + z^2 = a^2$ when (a) $u = x + y + z$, (b) $u = x^2 + y^2 + z^2$. (Note that $\partial u/\partial n = \nabla u\cdot \mathbf{n}$.)
11. Evaluate $\iint_S (\partial u/\partial n)dS$ over the sphere $x^2 + y^2 + z^2 = a^2$ when $u(x, y, z) = x^2 + y^2 - z$.
12. Evaluate $\iint_S (\partial u/\partial n)dS$ over the cube $|x| = 1$, $|y| = 1$, $|z| = 1$, when $u(x, y, z) = x^3 + y^2 + z$.
13. Let u and v be twice continuously differentiable functions in a domain D bounded by a piecewise smooth closed surface S. Derive the following formulas from the divergence theorem.

(a) $$\iint_S u\frac{\partial v}{\partial n}\, dS = \iiint_D (\nabla u\cdot\nabla v + u\nabla^2 v)\, dx\, dy\, dz$$

(b) $$\iint_S \left(u\frac{\partial v}{\partial n} - v\frac{\partial u}{\partial n}\right) dS = \iiint_D (u\nabla^2 v - v\nabla^2 u)\, dx\, dy\, dz$$

14. If u is a harmonic function, show that

(a) $$\iint_S \frac{\partial u}{\partial n}\, dS = 0$$

(b) $$\iint_S u\frac{\partial u}{\partial n}\, dS = \iiint_D |\nabla u|^2 dx\, dy\, dz$$

where S is a piecewise smooth surface bounding the domain D.

15. Let u be a harmonic function in a domain D that contains the origin and is bounded by a smooth closed surface S. Let v = 1/r, where $r = (x^2 + y^2 + z^2)^{1/2}$. Show that

$$u(0,0,0) = \frac{1}{4\pi}\int\int_S \left[u\frac{\partial}{\partial n}\left(\frac{1}{r}\right) - \frac{1}{r}\frac{\partial u}{\partial n}\right] dS$$

16. Evaluate the surface integral

$$\int\int_S \left[u\frac{\partial}{\partial n}\left(\frac{1}{r}\right) - \frac{1}{r}\frac{\partial u}{\partial n}\right] dS$$

when $u(x, y, z) = x^2 - y^2 + z + 1$ and S: $x^2 + y^2 + z^2 = 4$.

17. Show that

$$\int\int_S r^2\mathbf{R}\cdot\mathbf{n}\, dS = 5\int\int\int_D r^2 dV$$

for any domain D that is bounded by a smooth surface S.

18. Let $\mathbf{R} = x\,\mathbf{i} + y\,\mathbf{j} + z\,\mathbf{k}$, $r = |\mathbf{R}|$ and n a positive integer. Derive from the divergence theorem the following integrals:

(a) $$\int\int_S r^n\mathbf{R}\cdot\mathbf{n}\, dS = \int\int\int_D (n+3)r^n dV$$

(b) $$\int\int_S r^n\,\mathbf{n}\, dS = \int\int\int_D nr^{n-2}\mathbf{R}\, dV$$

Hint: For part (b) set $\mathbf{F} = \mathbf{A}\,r^n$, where $\mathbf{A}$ is an arbitrary constant vector, and note that $\iint_S \mathbf{A}r^n \cdot \mathbf{n}\, dS = \mathbf{A}\cdot \iint_S r^n\mathbf{n}\, dS$ and $\iiint_D \text{div}(\mathbf{A}r^n)dV = \mathbf{A}\cdot \iiint_D \nabla r^n\, dV$.

19. Let f and **G** be continuously differentiable scalar and vector fields in the whole space, respectively. Using the divergence theorem, prove that

(a) $$\int\int_S (\mathbf{n}\times\mathbf{G})\, dS = \int\int\int_D (\nabla\times\mathbf{G})\, dV$$

$$\text{(b)} \quad \iint_S f\,\mathbf{n}\,dS = \iiint_D \nabla f\,dV$$

Hint: In part (a), set $\mathbf{F} = \mathbf{A} \times \mathbf{G}$, where $\mathbf{A}$ is an arbitrary constant vector and note that

$$\iint_S \mathbf{A} \times \mathbf{G} \cdot \mathbf{n}\,dS = \mathbf{A} \cdot \iint_S (\mathbf{G} \times \mathbf{n})\,dS.$$

In part (b), set $\mathbf{F} = f\,\mathbf{A}$.

4.11 STOKES' THEOREM

We recall that Green's theorem in the plane can be written in the form

$$\iint_D (\mathbf{curl\ F}) \cdot \mathbf{k}\,dx\,dy = \oint_C \mathbf{F} \cdot d\mathbf{R}$$

where $\mathbf{F}(x, y) = P(x, y)\mathbf{i} + Q(x, y)\mathbf{j}$ and D is a plane domain bounded by a simple closed curve C. The extension of Green's theorem to three dimensional vector fields leads to what is known as Stokes' theorem. We establish this theorem in this section and consider some of its applications.

Theorem 1. (Stokes' theorem) Let S be a piecewise smooth orientable surface bounded by a piecewise smooth simple closed curve C. If $\mathbf{F}(x, y, z) = P(x, y, z)\mathbf{i} + Q(x, y, z)\mathbf{j} + R(x, y, z)\mathbf{k}$ is continuously differentiable in a domain containing S and C, then

$$\iint_D (\mathbf{curl\ F}) \cdot \mathbf{n}\,dx\,dy = \oint_C \mathbf{F} \cdot d\mathbf{R} \tag{4.59}$$

where the integration along C is in the positive direction relative to the unit normal vector $\mathbf{n}$ on S (Fig. 4.28).

In terms of the components, equation (4.59) can be written as

$$\iint_S \left[\left(\frac{\partial R}{\partial y} - \frac{\partial Q}{\partial z}\right)\cos\alpha + \left(\frac{\partial P}{\partial z} - \frac{\partial R}{\partial x}\right)\cos\beta + \left(\frac{\partial Q}{\partial x} - \frac{\partial P}{\partial y}\right)\cos\gamma\right] dS$$

$$= \oint_C P\,dx + Q\,dy + R\,dz \tag{4.60}$$

where $\mathbf{n} = \cos\alpha\,\mathbf{i} + \cos\beta\,\mathbf{j} + \cos\gamma\,\mathbf{k}$ is the unit normal vector on S. Hence, to prove the theorem it suffices to show that

$$\int\int_S \left(\frac{\partial P}{\partial z}\cos\beta - \frac{\partial P}{\partial y}\cos\gamma\right) dS = \oint_C P\,dx$$

$$\int\int_S \left(\frac{\partial Q}{\partial x}\cos\gamma - \frac{\partial Q}{\partial z}\cos\alpha\right) dS = \oint_C Q\,dy \tag{4.61}$$

$$\int\int_S \left(\frac{\partial R}{\partial y}\cos\alpha - \frac{\partial R}{\partial x}\cos\beta\right) dS = \oint_C R\,dz$$

since the sum of these integrals yields (4.60). We prove only the first equality in (4.61) since the other two can be established in exactly the same way.

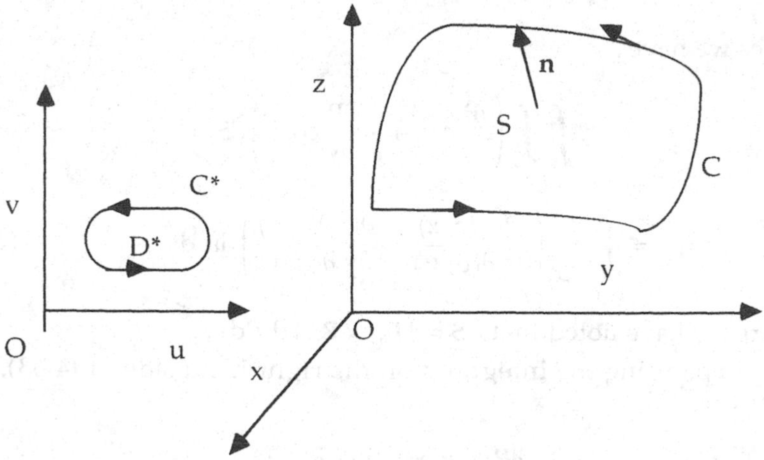

Fig. 4.28 Stokes's theorem

Let the surface S be represented parametrically by the vector equation

$$\mathbf{R}(u, v) = x(u, v)\mathbf{i} + y(u, v)\mathbf{j} + z(u, v)\mathbf{k} \tag{4.62}$$

where the function x(u,v), y(u,v), z(u,v) are assumed to be twice continuously differentiable in a domain D*of the uv-plane (see Fig. 4.28). We assume that the orientation of the boundary C of the surface corresponds to the positive direction of the boundary C* of D* through equation (4.62), and that the positive unit normal vector **n** on the surface relative to the orientation of the boundary C is given by

$$\mathbf{n} = \frac{\mathbf{R}_u \times \mathbf{R}_v}{|\mathbf{R}_u \times \mathbf{R}_v|}$$

Since

$$\frac{\partial \mathbf{R}}{\partial u} \times \frac{\partial \mathbf{R}}{\partial v} = \frac{\partial(y, z)}{\partial(u, v)}\mathbf{i} + \frac{\partial(z, x)}{\partial(u, v)}\mathbf{j} + \frac{\partial(x, y)}{\partial(u, v)}\mathbf{k}$$

it follows that

$$\cos\beta = \frac{\partial(z, x)}{\partial(u, v)} / |\mathbf{R}_u \times \mathbf{R}_v|, \quad \cos\gamma = \frac{\partial(x, y)}{\partial(u, v)} / |\mathbf{R}_u \times \mathbf{R}_v|$$

Hence we have

$$\int\int_S \left(\frac{\partial P}{\partial z}\cos\beta - \frac{\partial P}{\partial y}\cos\gamma\right) dS$$

$$= \int\int_{D^*} \left(\frac{\partial P}{\partial z}\frac{\partial(z, x)}{\partial(u, v)} - \frac{\partial P}{\partial y}\frac{\partial(x, y)}{\partial(u, v)}\right) du\, dv \tag{4.63}$$

where we have noted that $dS = |\mathbf{R}_u \times \mathbf{R}_v|\, du\, dv$.

Expanding the integrand on the right-hand side of (4.63), we obtain

$$\frac{\partial P}{\partial z}\frac{\partial(z, x)}{\partial(u, v)} - \frac{\partial P}{\partial y}\frac{\partial(x, y)}{\partial(u, v)}$$

$$= \frac{\partial P}{\partial z}\left(\frac{\partial z}{\partial u}\frac{\partial x}{\partial v} - \frac{\partial z}{\partial v}\frac{\partial x}{\partial u}\right) - \frac{\partial P}{\partial y}\left(\frac{\partial x}{\partial u}\frac{\partial y}{\partial v} - \frac{\partial x}{\partial v}\frac{\partial y}{\partial u}\right) \tag{4.64}$$

$$= \left(\frac{\partial P}{\partial y}\frac{\partial y}{\partial u} + \frac{\partial P}{\partial z}\frac{\partial z}{\partial u}\right)\frac{\partial x}{\partial v} - \left(\frac{\partial P}{\partial y}\frac{\partial y}{\partial v} + \frac{\partial P}{\partial z}\frac{\partial z}{\partial v}\right)\frac{\partial x}{\partial u}$$

If we add and subtract the term $(\partial P/\partial x)(\partial x/\partial u)(\partial x/\partial v)$ on the right-hand side of (4.64), and note that

$$\frac{\partial P}{\partial u} = \frac{\partial P}{\partial x}\frac{\partial x}{\partial u} + \frac{\partial P}{\partial y}\frac{\partial y}{\partial u} + \frac{\partial P}{\partial z}\frac{\partial z}{\partial u}$$

and

$$\frac{\partial P}{\partial v} = \frac{\partial P}{\partial x}\frac{\partial x}{\partial v} + \frac{\partial P}{\partial y}\frac{\partial y}{\partial v} + \frac{\partial P}{\partial z}\frac{\partial z}{\partial v}$$

we see that

$$\frac{\partial P}{\partial z}\frac{\partial(z, x)}{\partial(u, v)} - \frac{\partial P}{\partial y}\frac{\partial(x, y)}{\partial(u, v)} = \frac{\partial P}{\partial u}\frac{\partial x}{\partial v} - \frac{\partial P}{\partial v}\frac{\partial x}{\partial u}$$

$$= \frac{\partial}{\partial u}\left(P\frac{\partial x}{\partial v}\right) - \frac{\partial}{\partial v}\left(P\frac{\partial x}{\partial u}\right)$$

Therefore, by Green's theorem, the right-hand side of (4.63) gives

$$\int\int_{D^*}\left[\frac{\partial}{\partial u}\left(P\frac{\partial x}{\partial v}\right) - \frac{\partial}{\partial v}\left(P\frac{\partial x}{\partial u}\right)\right] du\, dv$$

$$= \oint P\left(\frac{\partial x}{\partial u} du + \frac{\partial x}{\partial v} dv\right) \tag{4.65}$$

Now as (u, v) ranges on C*, the point [x(u, v), y(u, v), z(u, v)] ranges on C. Moreover, since x= x(u, v), we have

$$dx = \frac{\partial x}{\partial u} du + \frac{\partial x}{\partial v} dv$$

Hence we obtain

$$\oint_C P\left(\frac{\partial x}{\partial u}\,du + \frac{\partial x}{\partial v}\,dv\right) = \oint_C P\,dx \tag{4.66}$$

which proves the first identity in (4.61).

We recall that $\int_C \mathbf{F}\cdot d\mathbf{R}$ is the line integral along C of the tangential component of **F** on C, while $\iint_S \mathbf{F}\cdot \mathbf{n}\, dS$ is the surface integral over S of the normal component of **F** on S. Thus Stokes' theorem says that the integral of the normal component of **curl F** on an orienentable surface S is equal to the line integral of the tangential component of **F** along the boundary of S. In fluid dynamics when **F** is a velocity field, $\int_C \mathbf{F}\cdot d\mathbf{R}$ represents the circulation of the fluid around the closed curve C.

Example 1. Verify Stokes' theorem for $\mathbf{F}(x, y, z) = y^2\mathbf{i} + xy\mathbf{j} + xz\mathbf{k}$, where S: $z = a^2 - x^2 - y^2$, $z \geq 0$.

Solution: We choose the unit normal vector on S to be one with positive z-component so that the orientation of the boundary C: $x^2 + y^2 = a^2$ $(z = 0)$ of S is counterclockwise. Then we have

$$\mathbf{n} = \frac{2x\,\mathbf{i} + 2y\,\mathbf{j} + \mathbf{k}}{\sqrt{4(x^2 + y^2) + 1}}$$

so that $dS = \sec\gamma\, dx\, dy = [1 + 4(x^2 + y^2)]^{1/2}\, dx\, dy$. Since $\mathbf{curl\ F} = -z\mathbf{j} - y\mathbf{k}$, we have

$$\iint_S (\mathbf{curl\ F})\cdot \mathbf{n}\, dS = -\iint_D [2y(a^2 - x^2 - y^2) + y]dx\, dy$$

where D is the circular domain of radius a. Introducing polar coordinates, we obtain

$$\int_0^{2\pi}\int_0^a [2r\sin\theta\,(a^2 - r^2) + r\sin\theta]\, r\, dr\, d\theta = 0$$

On the other hand, evaluating the line integral of **F** on the boundary $x^2 + y^2 = a^2$ $(z = 0)$, we find

$$\oint_C y^2 dx + xy\, dy = a^3\int_0^{2\pi} (-\sin^3\theta + \cos^2\theta\,\sin\theta)\, d\theta$$

$$= a^3 \int_0^{2\pi} (-\sin\theta + 2\cos^2\theta \sin\theta)\, d\theta = 0$$

thus verifying the theorem.

Example 2. By Stokes' theorem, evaluate $\iint_S (\mathbf{curl\ F}) \cdot \mathbf{n} dS$, where

$$\mathbf{F}(x, y, z) = (1 - z)y\mathbf{i} + z\, e^x \mathbf{j} + (x \sin z)\mathbf{k}$$

and

$$S:\ z = (a^2 - x^2 - y^2)^{1/2}.$$

Solution: Taking the unit normal vector **n** on S to be outward, we find

$$\int\int_S (\mathbf{curl\ F}) \cdot \mathbf{n}\, dS = \oint_C \mathbf{F} \cdot d\mathbf{R}$$

$$= \oint_C (1 - z)y\, dx + ze^x\, dy + x \sin z\, dz$$

where $C: x^2 + y^2 = a^2$ is the boundary of S oriented counterclockwise. Since $z = 0$ on C, the line integral on C reduces to

$$\oint_C \mathbf{F} \cdot d\mathbf{R} = \oint_C y\, dx = -a^2 \int_0^{2\pi} \sin^2\theta\, d\theta$$

$$= -a^2 \int_0^{2\pi} (1 - \cos 2\theta)\, d\theta / 2 = -\pi a^2$$

4.12 SOME APPLICATIONS OF STOKES' THEOREM

In Theorem 2 of Sec. 4.4 we proved that if **curl F** = **0** in a simply connected domain D, then the line integral of **F** is independent of path in D. This fact now readily follows from Stokes' theorem. Indeed, if **curl F** = **0** in a simply connected domain D, then for any piecewise smooth simple closed curve C in D, application of Stokes' theorem gives

$$\oint_C \mathbf{F} \cdot d\mathbf{R} = \int\int_S (\mathbf{curl\ F}) \cdot \mathbf{n}\, dS = 0$$

Here S is any piecewise smooth surface bounded by C. Thus the line integral of **F** vanishes on every closed curve in D, and this implies independence of path for $\int \mathbf{F}\cdot d\mathbf{R}$ in D.

We have seen that the divergence theorem can be used to give an alternative definition of the divergence of a vector field. Likewise, Stokes' theorem provides us an alternative definition of the curl of a vector field **F**. To see this, let S_r denote the circular disk bounded by the circle C_r of radius r and center at a point P. By the mean value theorem for integrals, there exists a point Q in S_r such that

$$A(r)\, \mathbf{curl}\, \mathbf{F}(Q)\cdot \mathbf{n} = \iint_{S_r} (\mathbf{curl}\, \mathbf{F})\cdot \mathbf{n}\, dS$$

Here **n** is a unit normal vector on S_r and A(r) is the area of S_r (see Fig. 4.29). By Stokes' theorem, we obtain

$$\mathbf{curl}\, \mathbf{F}(Q)\cdot \mathbf{n} = \frac{1}{A(r)} \oint_{C_r} \mathbf{F}\cdot d\mathbf{R}$$

Now as we let r tend to zero, the point Q approaches the point P, and so in the limit we obtain

$$\mathbf{curl}\, \mathbf{F}(Q)\cdot \mathbf{n} = \lim_{r \to 0} \frac{1}{A(r)} \oint_{C_r} \mathbf{F}\cdot d\mathbf{R} \qquad (4.67)$$

This defines the component of **curl F** at the point P on the surface S_r in the direction of the normal vector **n** regardless of the coor-

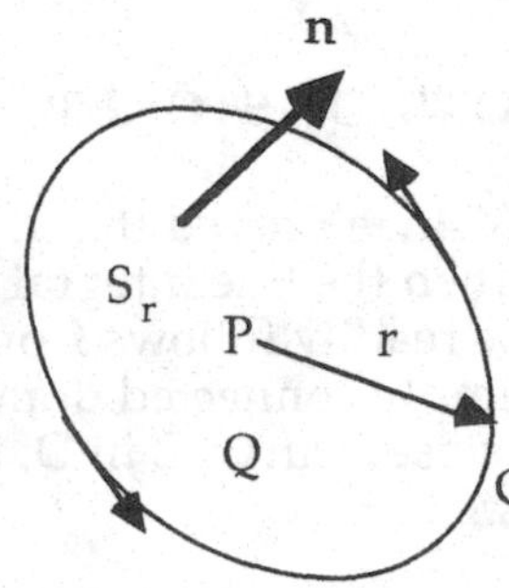

Fig. 4.29 Curl of a vector field.

dinate system used. However, if we let **n** be the unit vectors **i**, **j**, **k** successively, then (4.67) will give the three components of **curl F** in the rectangular cartesian coordinate system.

As in the discussion of divergence, let us consider the flux density **F** = ρ**V** of a fluid flow with velocity field **V** and density ρ. We observe that $\int_C \mathbf{F}\cdot d\mathbf{R}$ is the line integral of the tangential component of **F** on C. Its value is called the circulation of **F** around C. Thus equation (4.67) says that the component of **curl F** along the unit vector **n** is equal to the circulation per unit area of **F** around the boundary of a surface that is perpendicular to **n**. If the circulation is zero at every point, then **curl F** = **0**, which means that the flow is irrotational. Therefore, we call a vector field **F** for which **curl F** = **0** an irrotational vector field.

It follows from (4.67) that the value of **curl F**(P)· **n** will be maximum when **n** is in the same direction as **curl F**. This means that the circulation motion of the fluid is greatest on a surface that is perpendicular to **curl F**. In fluid dynamics, **curl F** is called the vorticity vector.

Example 1. Consider a fluid with constant density ρ that rotates around the z-axis with angular speed ω. Then the angular velocity **w** is represented by **w** = ω**k** and the velocity is given by **V** = **w** x **R** = ω(- yi + xj). Thus the flux density is **F** = ρω(- yi + xj), for which we find

$$\mathbf{curl\ F} = 2\rho\omega\mathbf{k}$$

and, therefore

$$\mathbf{curl\ F}\cdot\mathbf{k} = 2\rho\omega$$

On the other hand, the circulation of **F** around the circle $x^2 + y^2 = r^2$ is given by

$$\oint_{C_r} \mathbf{F}\cdot d\mathbf{R} = \oint_{C_r} \rho\omega\,(-y\,dx + x\,dy) = \int_0^{2\pi} \rho\omega r^2\,d\theta = 2\pi\rho\omega r^2$$

Dividing this by the area of the circle, we find

$$2\pi\rho\omega\, r^2/\pi\, r^2 = \mathbf{curl\ F}\cdot\mathbf{k} = 2\rho\omega$$

in agreement with (4.67).

Lastly, we consider the application of Stokes' theorem in the study of electromagnetism. A basic law of electromagnetic theory is that if **E** and **H** represent the electric and magnetic fields at time t, then **curl E** = $-\partial\mathbf{H}/\partial t$. Hence, by Stokes' theorem,

$$\oint_C \mathbf{E}\cdot d\mathbf{R} = \iint_S \text{curl } \mathbf{E}\cdot \mathbf{n}\, dS = -\iint_S \frac{\partial \mathbf{H}}{\partial t}\cdot \mathbf{n}\, dS$$

$$= -\frac{\partial}{\partial t}\iint_S \mathbf{H}\cdot \mathbf{n}\, dS \tag{4.68}$$

where C is a closed circuit and S is any smooth surface bounded by C. This equation is known as *Faraday's law*. The integral $\int_C \mathbf{E}\cdot d\mathbf{R}$ represents the electromotive force (voltage) around the circuit C, while $\iint_S \mathbf{H}\cdot\mathbf{n}\, dS$ represents the magnetic flux across the surface S. Thus Faraday's law says that the electromotive force around a closed circuit is equal to the negative of the rate of change of the magnetic flux through the circuit.

It is interesting to note that the curl can also be defined as

$$\text{curl } \mathbf{F}\,(P) = \lim_{r\to 0} \frac{1}{V(r)} \iint_{S_r} (\mathbf{n} \times \mathbf{F})\, dS \tag{4.69}$$

where V(r) denotes the volume of the sphere S_r of radius *r*. This formula can be established by the use of the divergence theorem. Note that the integrand in (4.69) is vector-valued and so the integral must be considered componentwise.

4.9 EXERCISES

In each of Problems 1 through 4, evaluate the surface integral $\iint_S$ **curl F· n** dS by using Stokes' theorem, and verify the result by integrating the integral directly. The unit normal vector **n** has a non-negative z-component.

1. **F**(x, y, z) = xz**i** - y**j** + xy**k**, where S is the triangle with vertices (1, 0, 0), (0, 1, 0), (0, 0, 1).
2. **F**(x, y, z) = y**i** + z**j** + x**k**, where S is the portion of the plane y + z = 2 intercepted by the cylinder $x^2 + y^2 = 4$.
3. **F**(x, y, z) = yz**i** - xz**j** + **k**, where S is the portion of the paraboloid $x^2 + y^2 = z$ for $0 \le z \le 1$.

4. $\mathbf{F}(x, y, z) = yz\mathbf{i} - xz\mathbf{j} + xy\mathbf{k}$, where S is the hemisphere $z = (a^2 - x^2 - y^2)^{1/2}$
5. Evaluate $\int_C y\,dx + z\,dy + dz$, where C is the intersection of the cylinder $x^2 + y^2 = 2ax$ and the plane $x = z$ oriented counterclockwise as viewed from a point high on the positive z-axis.
6. Evaluate $\int_C (y^2 + z^2)dx + (x^2 + z^2)dy + (x^2 + y^2)dz$, where C is the triangle with vertices (1, 1, 0), (0, 1, 1), (1, 0, 1) oriented counterclockwise as viewed from the point (1, 1, 1).
7. Evaluate $\int_C y^2\,dx + xy\,dy + xz\,dz$, where C is the intersection of the hemisphere $x^2 + y^2 + z^2 = 4x$ $(z \geq 0)$ and the cylinder $x^2 + y^2 = 2x$ oriented counterclockwise as viewed from a point high on the positive z-axis.
8. Evaluate $\int_C xz\,dx + yz\,dy + y^2dz$, where C is the intersection of the surface $z = xy$ and the cylinder $x^2 + y^2 = a^2$ oriented counterclockwise as viewed from the point (0, 0, 2a).
9. Evaluate $\int_C (bz - cy)\,dx + (cx - az)\,dy + (ay - bx)\,dz$, where C is the intersection of the cylinder $x^2 + y^2 = R^2$ and the plane $ax + by + cz = d$ oriented counterclockwise as viewed from a point high on the positive z-axis.
10. Show that

$$\oint_C u\nabla v \cdot d\mathbf{R} = \int\int_D (\nabla u \times \nabla v)\cdot \mathbf{k}\,dx\,dy$$

where u and v are differentiable functions of x, y and D is a simply connected domain bounded by a smooth closed curve C.

11. By Stokes' theorem, show that

$$\oint_C f\,d\mathbf{R} = \int\int_S \mathbf{n} \times \nabla f\,dS$$

where C is a smooth simple closed curve bounding a smooth surface S. Hint: Let $\mathbf{F} = f\mathbf{A}$, where $\mathbf{A}$ is an arbitrary constant vector.

12. Show that

$$\oint_C \frac{d\mathbf{R}}{r} = \mathbf{i}\oint_C \frac{dx}{r} + \mathbf{j}\oint_C \frac{dy}{r} = \int\int_D \frac{\mathbf{R} \times \mathbf{k}}{r^3}\,dx\,dy$$

for any simple closed curve C that does not enclose the origin on the xy-plane.

13. This problem shows how the Stokes' theorem may be extended to a surface with a finite number of holes. Let

$$\mathbf{F}(x, y, z) = \frac{-y}{x^2 + y^2}\,\mathbf{i} + \frac{x}{x^2 + y^2}\,\mathbf{j} + z\,\mathbf{k}$$

and let S be the portion of the hemisphere $z = (a^2 - x^2 - y^2)^{1/2}$ without the piece cut off by the cylinder $x^2 + y^2 = \varepsilon^2$, where ε is sufficiently small ($\varepsilon < a$).

(a) Show that **curl F** = **0** for all points (x, y, z) except those on the z-axis.

(b) Show that $\iint_S \mathbf{curl\ F}\cdot\mathbf{n}dS = 0$, and thus deduce that $\int_C \mathbf{F}\cdot d\mathbf{R} = \int_\Gamma \mathbf{F}\cdot d\mathbf{R}$ where C: $x^2 + y^2 = a^2$ and Γ is the intersection of the hemisphere and the cylinder.

(c) Evaluate the integral $\int_C \mathbf{F}\cdot d\mathbf{R}$.

5

TENSORS IN RECTANGULAR CARTESIAN COORDINATE SYSTEMS

5.1 INTRODUCTION

Up to this stage, we have been concerned with vector analysis. Among other things, we have studied the algebra of vectors, the various field properties - gradient, divergence, and curl - together with their representations in orthogonal curvilinear coordinate systems, and the various integral theorems involving the divergence and the curl. Having learned this material, a student might now assume that all quantities in physics and applied mathematics can be described in terms of scalars and vectors. Unfortunately, this is not the case. There are many physical quantities of more complex in nature that cannot be described or represented by scalars or vectors. Examples are the stress at a point of a solid body due to internal forces, the deformation of an arbitrary element of volume of an elastic body, and even the simple concept of moments of inertia. These quantities can be described and represented adequately only by the more sophisticated mathematical entities called *tensors*. As we shall see later, scalars and vectors are actually special cases of tensors. In this and the last chapters, we shall study the principal properties of tensors and consider some of their elementary applications.

The concept of a tensor, like that of a vector, is an invariant concept. By this we mean that it is an entity that does not depend on any frame of reference or coordinate system. However, just as a vector can be represented by its so-called components when referred to a particular coordinate system, a tensor can also be so represented. The components of a tensor may be constants or functions. Whether a given set of constants and/or functions really represent a tensor depends on the manner by which the members of the set transform under a change of coordinate system. The situation is identical with that of vectors. Relative to a given rectangular cartesian coordinate system, we know that a vector is uniquely determined by its components. With respect to a different coordinate system, the same vector will have different components, and the new components are related to the old components in a definite manner depending on the transformation of the coordinate systems. Thus the key property of a tensor is the manner by which its components transform under a change of coordinate system.

Tensors that are represented in rectangular cartesian coordinate system are commonly called cartesian tensors. In the study of such tensors, one is interested in the way the components transform from one rectangular cartesian coordinate system to another. Therefore, the kinds of coordinate transformations involved in cartesian tensors are transformations of rectangular coordinate systems. On the other hand, tensors that are represented in general curvilinear coordinate systems are called general tensors or simply tensors. Clearly, since cartesian coordinate systems are special cases of general coordinate systems, cartesian tensors are special cases of general tensors.

In this book, we shall present separate discussion of cartesian and general tensors. Our reason for doing this is twofold. First, a student who is interested only in learning cartesian tensors can do so without having to sieve out these results from a mass of information pertaining to general tensors. Second, most students find the transition from vectors to general tensors rather abrupt and thus encounter considerable difficulty. By first studying cartesian tensors, it is hoped that the student will have acquired sufficient understanding, appreciation and familiarity with some aspects of tensors so that by the time the student studies general tensors, he or she will find the transition quite natural.

5.2 NOTATION AND SUMMATION CONVENTION

It is sometimes said that tensor analysis is a study of (tensorial) notations. There is a certain degree of truth in this statement. Because tensors are complex entities that can have a large number of components, it is necessary, indeed imperative, to introduce certain notations and convention to facilitate the representation and manipulation of tensors. This section is designed to introduce the student to these notations and convention.

Instead of the usual notation x, y, z for rectangular cartesian coordinates, we now adopt the notation x_1, x_2, x_3 or briefly x_i, with the understanding that the subscript or index i assumes the values 1, 2, 3. Likewise, we denote the orthonormal basis $\mathbf{i}, \mathbf{j}, \mathbf{k}$, by $\mathbf{i}_1, \mathbf{i}_2, \mathbf{i}_3$ respectively, or simply by $\mathbf{i}_i$ for brevity. With this notation we can now write the position vector of a point with coordinates x_i in the form

$$\mathbf{R} = \sum_{i=1}^{3} x_i \mathbf{i}_i \tag{5.1}$$

using the summation sign Σ. Likewise the analytic representation of a vector **A** with components a_i can be written as

$$\mathbf{A} = \sum_{i=1}^{3} a_i \mathbf{i}_i \tag{5.2}$$

To achieve further simplification in our notation, we now drop the summation sign and adopt the so-called Einstein summation convention. The convention stipulates that whenever an index or a subscript is repeated in a term, the term must be summed with respect to that index for all admissible values of the index. In our case, since we are dealing only with three dimensional space, the admissible values of an index are 1, 2, 3. Thus, for example, the summations (5.1) and (5.2) will be written simply as

$$\mathbf{R} = x_i \mathbf{i}_i$$

and

$$\mathbf{A} = a_i \mathbf{i}_i$$

where, in both cases, summation from 1 to 3 with respect to the repeated index i is implied. Other examples are given below.

Example 1. The term a_{ii} implies the sum $a_{11} + a_{22} + a_{33}$.

Example 2. The term $a_i b_i$ implies the sum $a_1 b_1 + a_2 b_2 + a_3 b_3$.

Example 3. The expression $a_i b_j - a_j b_i$ does not imply any summation because none of the indices is repeated in the same term.

The expression stands for the nine quantities

$$c_{ij} = a_i b_j - a_j b_i \qquad (i, j = 1, 2, 3)$$

Example 4. The term $a_{ij}b_{jk}$ implies summation with respect to the repeated index j from 1 to 3. Thus,

$$a_{ij}b_{jk} = a_{i1}\, b_{1k} + a_{i2}\, b_{2k} + a_{i3}\, b_{3k} = c_{ik} \quad (i, k = 1, 2, 3)$$

Example 5. Let $\phi\,(x_1\,, x_2\,, x_3\,)$ be a scalar field. Then

$$\begin{aligned} \text{grad}\ \phi(x_1, x_2, x_3) &= \frac{\partial\phi}{\partial x_j}\,\mathbf{i}_j \\ &= \frac{\partial\phi}{\partial x_1}\,\mathbf{i}_1 + \frac{\partial\phi}{\partial x_2}\,\mathbf{i}_2 + \frac{\partial\phi}{\partial x_3}\,\mathbf{i}_3 \end{aligned}$$

Example 6. Let $\phi\,(x_1\,, x_2\,, x_3\,)$ be a scalar field, where $x_i = x_i(t)$, $i = 1, 2, 3$. Then

$$\begin{aligned} \frac{d\phi(x_i)}{dt} &= \frac{\partial\phi}{\partial x_i}\frac{dx_i}{dt} \\ &= \frac{\partial\phi}{\partial x_1}\frac{dx_1}{dt} + \frac{\partial\phi}{\partial x_2}\frac{dx_2}{dt} + \frac{\partial\phi}{\partial x_3}\frac{dx_3}{dt} \end{aligned}$$

5.1 EXERCISES

1. Write out in full each of the following, where δ_{ij} is the Kronecker delta.
   (a) $\delta_{ij}\, a_j$, (b) $\delta_{ij}\, a_i\, b_j$, (c) $a_{ij}\, x_i\, x_j$, (d) $a_{ik}\, b_{kj} = \delta_{ij}$,
   (e) $s = a_{ij}\, b_{ij}$,
   (f) $g_{ij} = (\partial x_i/\partial u_k)(\partial x_j/\partial u_k)$
2. Let ϕ be a differentiable function of x_i $(i = 1, 2, 3)$. If $x_i = f_i(t)$, where $f_i(t)$ has continuous derivative in (a, b), find $d\phi/dt$ and express it without using the summation symbol.

3. Let ϕ be a twice continuously differentiable function of x_i in a domain D, and suppose $x_i = g_i(u_1, u_2, u_3)$ where g_i is twice continuously differentiable function in D* for i = 1, 2, 3. Find $\partial\phi/\partial u_i$ and $\partial^2\phi/\partial u_i \partial u_j$ using the summation convention.
4. Let ϕ be a function of x_i and suppose $x_i = \alpha_{ij} x^*_j$ (i = 1, 2, 3). Find $\partial\phi/\partial x_i$ for i = 1, 2, 3.

5.3 TRANSFORMATIONS OF RECTANGULAR CARTESIAN COORDINATE SYSTEMS

In this section we study transformations of rectangular cartesian coordinate systems. Let x_i (referred to as the old system) and x^*_i (referred to as the new system) be two rectangular cartesian coordinate systems with the corresponding base vectors $\mathbf{i}_i$ and $\mathbf{i}^*_i$ ($1 \le i \le 3$). For convenience we assume that the two coordinate systems are both right-handed and that they have a common origin (see Fig.5.1). Consider a point P whose coordinates with respect to the old and the new coordinate systems are x_i and x^*_i, repectively. Then $\mathbf{R} = x_i\,\mathbf{i}_i$ is the position vector of the point with respect to the old system, and $\mathbf{R}^* = x^*_i\,\mathbf{i}^*_i$ the position vector with respect to the new system. Since

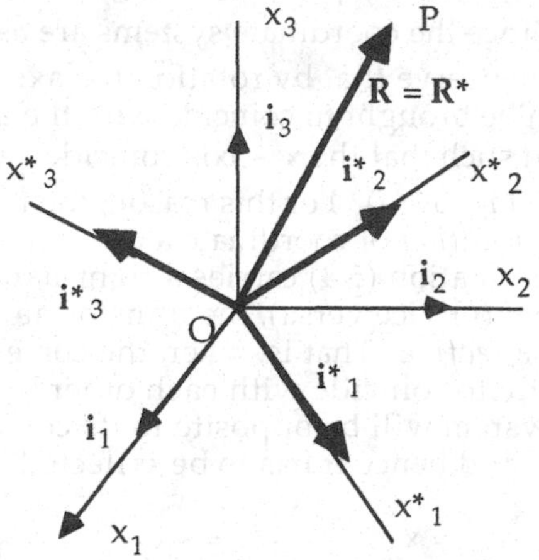

Fig. 5.1 Transformation of cartesian coordinate systems.

$\mathbf{R} = \mathbf{R}^*$ (see Fig. 5.1), we have

$$x^*_i \, \mathbf{i}^*_i = x_j \, \mathbf{i}_j \tag{5.3}$$

Notice that we have used different letters for the indices to indicate their being independent of each other. Now to obtain the relationship between the coordinates x_i and x^*_i, we may solve (5.3) for x^*_i in terms of x_i or solve for x_i in terms of x^*_i. To solve for x^*_i in terms of x_i, we take the dot product of both sides of equation (5.3) with $\mathbf{i}^*_k$ $(1 \leq k \leq 3)$. Since

$$\mathbf{i}^*_i \cdot \mathbf{i}^*_k = \delta_{ik} = \begin{Bmatrix} 1, & i = k \\ 0, & i \neq k \end{Bmatrix}$$

we find

$$x^*_i \, \delta_{ik} = (\mathbf{i}_j \cdot \mathbf{i}^*_k) x_j \qquad (1 \leq k \leq 3)$$

or

$$x^*_i = \alpha_{ij} \, x_j \qquad (i = 1, 2, 3) \tag{5.4}$$

where we define

$$\alpha_{ij} = \mathbf{i}^*_i \cdot \mathbf{i}_j \qquad (i, j = 1, 2, 3) \tag{5.5}$$

The symbol δ_{ij}, which has the value 1 whenever the subscripts are identical and zero value otherwise, is called the Kronecker delta.

Equation (5.4) defines the transformation from the rectangular cartesian coordinate system x_i to the rectangular cartesian coordinate system x^*_i. Since the coordinate systems are assumed to be both right-handed, we observe that by rotation the axes of the new coordinate system can be brought to coincide with the axes of the old coordinate system such that the x_i - axis coincides with the x^*_i - axis for $i = 1, 2, 3$. See Fig. 5.2(a). For this reason, the transformation (5.4) is said to define a *rotation* of coordinate axes.

If the transformation (5.4) carries a right-handed system into a left-handed system (or vice versa), the transformation will consist of a rotation and a *reflection*. That is, when the corresponding coordinate axes are made to coincide with each other by rotation, one of the axes in one system will be opposite to its corresponding axis in the other system, and hence it has to be reflected. For example, the transformation

$$x^*_1 = x_2, \qquad x^*_2 = -x_1, \qquad x^*_3 = -x_3$$

defines a left-handed coordinate system x^*_i (assuming x_i to be right-

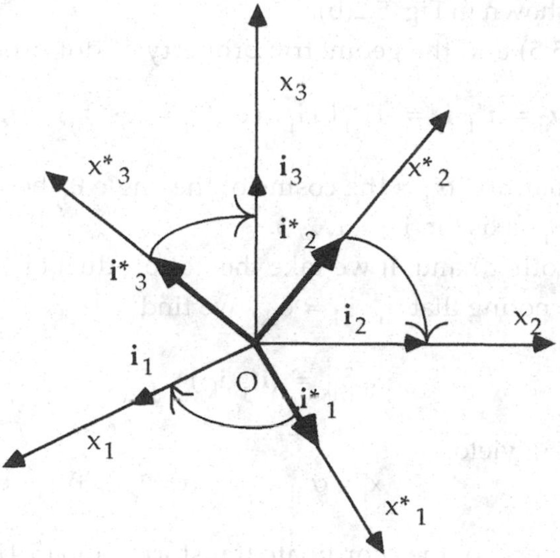

(a) Rotation of axes

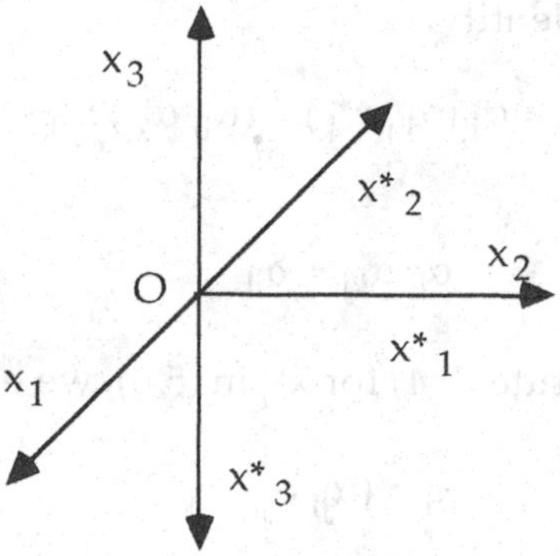

(b) Rotation about x_3 - axis followed by reflection.

Fig. 5.2 Rotation and reflection of coordinate axes.

handed). The corresponding axes can be made to coincide with each other by rotation about the x_3-axis followed by reflection of the x_3- or

x^*_3 - axis, as shown in Fig. 5.2(b).

From (5.5) and the geometric property of dot product, it follows that

$$\alpha_{ij} = \mathbf{i}^*_i \cdot \mathbf{i}_j = |\mathbf{i}^*_i| \, |\mathbf{i}_j| \cos\theta_{ij} = \cos\theta_{ij} \qquad (i, j = 1, 2, 3)$$

so that the quantity α_{ij} is the cosine of the angle θ_{ij} between the x^*_i - axis and the x_j - axis for $i, j = 1, 2, 3$.

On the other hand, if we take the dot product of $\mathbf{i}_k$ with both sides of (5.3), noting that $\mathbf{i}_i \cdot \mathbf{i}_k = \delta_{ik}$, we find

$$x_i \, \delta_{ik} = (\mathbf{i}^*_j \cdot \mathbf{i}_k) \, x^*_j$$

which, by (5.5), yields

$$x_i = \alpha_{ji} \, x^*_j \qquad (i = 1, 2, 3) \tag{5.6}$$

This is the inverse of the coordinate transformation (5.4).

The relationship between the quantities α_{ij} ($= \mathbf{i}^*_i \cdot \mathbf{i}_j$) in (5.4) and α_{ji} ($= \mathbf{i}^*_j \cdot \mathbf{i}_i$) in (5.6) is easily seen when we substitute (5.6) for x_i in (5.4) to obtain the identity

$$x^*_i = \alpha_{ij}(\alpha_{kj} \, x^*_k) = (\alpha_{ij} \, \alpha_{kj}) \, x^*_k$$

which implies that

$$\alpha_{ij} \, \alpha_{kj} = \delta_{ik} \tag{5.7}$$

Similarly, if we substitute (5.4) for x^*_i in (5.6), we obtain

$$x_i = (\alpha_{ji} \, \alpha_{jk}) \, x_k$$

which implies

$$\alpha_{ji} \, \alpha_{jk} = \delta_{ik} \tag{5.8}$$

Matrix Representations of Coordinate Transformations

We can understand better the meaning and significance of equations (5.4), (5.6), (5.7) and (5.8) if we express them in matrix form. First we observe that the coefficients of the transformation (5.4) can be represented as a 3 x 3 matrix

$$A = (\alpha_{ij}) = \begin{bmatrix} \alpha_{11} & \alpha_{12} & \alpha_{13} \\ \alpha_{21} & \alpha_{22} & \alpha_{23} \\ \alpha_{31} & \alpha_{32} & \alpha_{33} \end{bmatrix} \tag{5.9}$$

called the *transformation matrix* . Notice that the first subscript of an element α_{ij} in (5.9) indicates the row and the second subscript the column of the matrix in which the element is located. For example, the subscripts of the element α_{23} indicate second row and third column, the position in which the element is located. It is convenient to regard the rows in (5.9) as row vectors and the columns as column vectors. Thus, by (5.7) we see that the rows of the matrix (5.9) forms an orthonormal set of row vectors. Similarly, by (5.8) the columns form an orthonormal set of column vectors.

Likewise, the matrix corresponding to the inverse transformation (5.6) is given in (5.10). Notice that the rows and columns of this matrix are precisely the columns and rows of the matrix (5.9). The matrix (5.10) is called the transpose of (5.9), and is denoted by A^T. In general, the transpose of a matrix is the matrix obtained when the rows are arranged as columns or the columns arranged as rows. Thus

$$A^T = (\alpha_{ji}) = \begin{bmatrix} \alpha_{11} & \alpha_{21} & \alpha_{31} \\ \alpha_{12} & \alpha_{22} & \alpha_{32} \\ \alpha_{13} & \alpha_{23} & \alpha_{33} \end{bmatrix} \tag{5.10}$$

To be able to represent the transformations (5.4), (5.6) in matrix form, we need to introduce the concept of matrix multiplication. Let $B = (b_{ij})$ and $C = (c_{ij})$ be two 3 x 3 matrices. The product BC of the two matrices is defined as the 3 x 3 matrix $D = (d_{ij})$, whose elements are given by .

$$d_{ij} = b_{i1} c_{1j} + b_{i2} c_{2j} + b_{i3} c_{3j} = b_{ik} c_{kj} \quad (i, j = 1, 2, 3)$$

Notice that d_{ij} is the element in the i-th row and j-th column of the matrix D. It is simply the scalar product of the i-th row vector of B and the j-th column vector of C. Thus the i-th row of D $(1 \leq i \leq 3)$

consists of the scalar products of the i-th row vector of B and the column vectors of C.

For example, if

$$B = \begin{bmatrix} 1 & 0 & 2 \\ 2 & 1 & -1 \\ 0 & 3 & 0 \end{bmatrix} \quad \text{and} \quad C = \begin{bmatrix} 0 & 2 & 1 \\ 1 & 1 & 2 \\ 3 & 1 & -1 \end{bmatrix}$$

then

$$BC = \begin{bmatrix} 6 & 4 & -1 \\ -2 & 4 & 5 \\ 3 & 3 & 6 \end{bmatrix} \quad \text{and} \quad CB = \begin{bmatrix} 4 & 5 & -2 \\ 3 & 7 & 1 \\ 5 & -2 & 5 \end{bmatrix}$$

Notice that $BC \neq CB$, which shows that matrix multiplication is not commutative. Also, if X is a matrix consisting only of one column (a column matrix) with elements x_1, x_2, x_3, then

$$BX = \begin{bmatrix} 1 & 0 & 2 \\ 2 & 1 & -1 \\ 0 & 3 & 0 \end{bmatrix} \begin{bmatrix} x_1 \\ x_2 \\ x_3 \end{bmatrix} = \begin{bmatrix} x_1 + 2x_3 \\ 2x_1 + x_2 - x_3 \\ 3x_2 \end{bmatrix}$$

It follows from the definition that as long as the row vectors of B and the column vectors of C belong to the same vector space, the product BC is always defined regardless of the number of rows in and the number of columns in C. Indeed, if the matrix B has m rows and the matrix C has n columns, then the product BC will have m rows and n columns - an m x n matrix.

Using the definition, it can be easily shown that matrix multiplication obeys the familiar associative and distributive laws of ordnary multiplication of numbers provided the products are defined. Thus, if A, B, and C are 3 x 3 matrices, then

$$\begin{aligned} &(a) \quad A(BC) = (AB)C \\ &(b) \quad A(B + C) = AB + AC \\ &(c) \quad (B + C)A = BA + CA \end{aligned} \tag{5.11}$$

Now let X denote the column matrix whose elements are x_1, x_2, x_3, and similarly for X*. Then the transformation (5.4) can be written in the matrix form

$$X^* = AX \tag{5.12}$$

Notice that the matrix multiplication AX entails taking the scalar product of the row vectors of A with the column vector X, resulting in a column vector X*. Similarly, the inverse transformation (5.6) can be written in the form

$$X = A^T X^* \tag{5.13}$$

where A^T is the transpose of A. If we substitute (5.12) for X* in (5.13), or if we substitute (5.13) for X in (5.12), we obtain the result

$$A^T A = A\,A^T = I \tag{5.14}$$

where I is the identity matrix

$$I = \begin{bmatrix} 1 & 0 & 0 \\ 0 & 1 & 0 \\ 0 & 0 & 1 \end{bmatrix}$$

Equation (5.14) expresses the properties (5.7) and (5.8) in matrix form. It means that A^T is the inverse of the matrix A, or A is the inverse of the matrix A^T. If we denote the inverse of the matrix A by A^{-1}, then we have

$$A^{-1} = A^T \quad \text{and} \quad (A^T)^{-1} = A$$

A matrix whose transpose is its own inverse is called an *orthogonal* matrix. It follows that the matrix of a transformation that represents pure rotation is an orthogonal matrix.

From the above results we see that the rows of the matrix (5.9) are precisely the components of the unit vectors $\mathbf{i}^*_i$ (i = 1, 2, 3) with respect to the base vectors $\mathbf{i}_i$ (i = 1, 2, 3), that is,

$$\mathbf{i}^*_i = \alpha_{ij}\,\mathbf{i}_j \quad (i = 1, 2, 3) \tag{5.15}$$

Likewise, the columns of A are the components of the unit vectors $\mathbf{i}_i$ (i = 1, 2, 3) with respect to the base vectors $\mathbf{i}^*_i$ (i = 1, 2, 3), that is,

$$\mathbf{i}_i = \alpha_{ji}\,\mathbf{i}^*_j \quad (i = 1, 2, 3) \tag{5.16}$$

Notice that equations (5.15) and (5.16) are analogous to equations

(5.4) and (5.6) which define the coordinate transformations.

Finally, since we assume that both coordinate systems are right-handed, it follows from the definition of scalar triple product of vectors that

$$\mathbf{i}^*_1 \cdot \mathbf{i}^*_2 \times \mathbf{i}^*_3 = \det A = 1$$

and

$$\mathbf{i}_1 \cdot \mathbf{i}_2 \times \mathbf{i}_3 = \det A^T = 1$$

On the other hand, if the transformation (5.12) defines a left-handed coordinate system x^*_i, then

$$\mathbf{i}^*_1 \cdot \mathbf{i}^*_2 \times \mathbf{i}^*_3 = \det A = -1$$

We summarize our results as a theorem.

Theorem 1. Let (5.12) represents a transformation of rectangular cartesian coordinates. Then A is an orthogonal matrix such that det A = 1 if the transformation preserves orientation of the axes (pure rotation), or det A = -1 if the transformation reverses the orientation (rotation and/or reflection).

Example 1. The unit base vectors $\mathbf{i}^*_i$ of a new coordinate system x^*_i are given by

$$\mathbf{i}^*_1 = \frac{\mathbf{i}_2 + \mathbf{i}_3}{\sqrt{2}}, \quad \mathbf{i}^*_2 = \frac{\mathbf{i}_1 - \mathbf{i}_2 + \mathbf{i}_3}{\sqrt{3}}, \quad \mathbf{i}^*_3 = \frac{2\mathbf{i}_1 + \mathbf{i}_2 - \mathbf{i}_3}{\sqrt{6}}$$

Find the equations (5.4) and (5.6) of the coordinate transformation.

Solution: In view of the resemblance of (5.4) and (5.15), we immediately deduce the transformation

$$x^*_1 = \frac{x_2 + x_3}{\sqrt{2}}, \quad x^*_2 = \frac{x_1 - x_2 + x_3}{\sqrt{3}}, \quad x^*_3 = \frac{2x_1 + x_2 - x_3}{\sqrt{6}}$$

Thus the matrix of the transformation is given by

$$A = \begin{bmatrix} 0 & 1/\sqrt{2} & 1/\sqrt{2} \\ 1/\sqrt{3} & -1/\sqrt{3} & 1/\sqrt{3} \\ 2/\sqrt{6} & 1/\sqrt{6} & -1/\sqrt{6} \end{bmatrix}$$

Since $\mathbf{i}^*_1 \cdot \mathbf{i}^*_2 \times \mathbf{i}^*_3 = \det A = 1$, the new coordinate system is right-handed. The inverse transformation is given by

$$x_1 = \frac{x^*_2}{\sqrt{3}} + \frac{2x^*_3}{\sqrt{6}}, \quad x_2 = \frac{x^*_1}{\sqrt{2}} - \frac{x^*_2}{\sqrt{3}} + \frac{x^*_3}{\sqrt{6}}, \quad x_3 = \frac{x^*_1}{\sqrt{2}} + \frac{x^*_2}{\sqrt{3}} - \frac{x^*_3}{\sqrt{6}}.$$

5.2 EXERCISES

In the following problems, the x_i coordinate system is assumed to be right-handed.

1. What is the equation of the transformation in which each point (x_1, x_2, x_3) assumes the new coordinates $(x_3, -x_1, x_2)$? Is the new coordinate system right-handed or left-handed?
2. The new coordinate system x^*_i is obtained by rotating the x_i coordinate system about the x_3-axis through an angle θ (counterclockwise). What is the equation of the transformation?
3. The new coordinate system x^*_i is obtained by rotating the x_i-axes so that the positive x_1, x_2, x_3 axes are in the same direction as the positive x^*_3, x^*_1, x^*_2 axes, respectively. Find the equation of the transformation and determine whether the x^*_i coordinate system is right-handed or left-handed.
4. The new base vectors of a x^*_i coordinate system are given by

$$\mathbf{i}^*_1 = (\mathbf{i}_1 + \mathbf{i}_2 + \mathbf{i}_3)/\sqrt{3}, \quad \mathbf{i}^*_2 = (2\mathbf{i}_1 - \mathbf{i}_2 - \mathbf{i}_3)/\sqrt{6}$$

$$\mathbf{i}^*_3 = (\mathbf{i}_2 - \mathbf{i}_3)/\sqrt{2}$$

Find the equations of the coordinate transformation and determine whether the new coordinate system is right-handed or left-handed.

5. Repeat Problem 4 when

$$\mathbf{i}^*_1 = (\mathbf{i}_1 + \mathbf{i}_3)/\sqrt{2}, \quad \mathbf{i}^*_2 = (-\mathbf{i}_1 + \mathbf{i}_2 + \mathbf{i}_3)/\sqrt{3},$$

$$\mathbf{i}^*_3 = (\mathbf{i}_1 + 2\,\mathbf{i}_2 - \mathbf{i}_3)/\sqrt{6}$$

6. Show that the transformation $x^*_i = \alpha_{ij} x_j$, where

$$(\alpha_{ij}) = \begin{bmatrix} 2/3 & 2/3 & 1/3 \\ -2/3 & 1/3 & 2/3 \\ 1/3 & -2/3 & 2/3 \end{bmatrix}$$

represents a rotation of the coordinate axes. Is the new coordinate system right-handed or left-handed?

7. Suppose that the x^*_i coordinate system is also right-handed. Then it is true that $\mathbf{i}^*_i = \mathbf{i}^*_j \times \mathbf{i}^*_k$, where i, j, k are cyclic permutations of 1, 2, 3. By using the relations (5.15) and (5.16), obtain the relations

$$\alpha_{ir} = \alpha_{jp}\,\alpha_{kq} - \alpha_{jq}\,\alpha_{kp}$$

where i, j, k and p, q, r are cyclic permutations of 1, 2, 3.

8. Obtain the corresponding relations as in Problem 7 if the x^* coordinate system is left-handed.

5.4 TRANSFORMATION LAW FOR VECTORS

As a first step toward the definition of cartesian tensors, let us examine how the components of a vector transform under a change of coordinate system. To avoid unnecessary repetition of terminologies, let us agree that whenever we speak of a coordinate transformation in this chapter, we mean a transformation from one rectangular cartesian coordinate system into another rectangular cartesian coordinate system. Thus, the equations of our coordinate transformations are precisely those derived in Sec. 5.3, namely,

$$x^*_i = \alpha_{ij}\, x_j \qquad \text{or} \qquad x_i = \alpha_{ji}\, x^*_j$$

where $\alpha_{ij} = \mathbf{i}^*_i \cdot \mathbf{i}_j$ (i, j = 1, 2, 3) satisfy the relation

$$\alpha_{ik}\,\alpha_{jk} = \alpha_{ki}\,\alpha_{kj} = \delta_{ij}$$

Now consider a vector **A** and suppose that its components with respect to the coordinate systems x_i and x^*_i are a_i and a^*_i, respectively. Then we have

$$\mathbf{A} = a^*_i \, \mathbf{i}^*_i = a_j \, \mathbf{i}_j \tag{5.17}$$

To determine the relationship between the components a_j and a^*_i, we take the dot product of $\mathbf{i}^*_k$ with both sides of (5.17) to obtain

$$a^*_i \, \delta_{ik} = (\mathbf{i}^*_k \cdot \mathbf{i}_j)a_j$$

which leads to

$$a^*_i = \alpha_{ij} a_j = \frac{\partial x^*_i}{\partial x_j} a_j \tag{5.18}$$

On the other hand, if we take the dot product of $\mathbf{i}_k$ with both sides of (5.17), we obtain

$$a_i = \alpha_{ji} a_j = \frac{\partial x_i}{\partial x^*_j} a^*_j \tag{5.19}$$

Equation (5.18) or its inverse (5.19) demonstrates how the components of a vector transform when we change coordinate system. We call this the transformation law for vectors. Thus, in terms of the transformation law (5.18) or (5.19), we may redefine a vector as follows:

Definition 1. A vector is a quantity consisting of three components a_i (i = 1, 2, 3) which transform under a change of coordinates according to the law (5.18) or (5.19), where a_i and a^*_i are the components of the vector with respect to the coordinate systems x_i and x^*_i, respectively.

This definition casts vectors in the setting of cartesian tensors. In fact, Definition 1 is precisely the definition of a cartesian tensor of order 1 (or rank 1). In the context of this definition, a scalar becomes a quantity that remains invariant (does not change in numerical value) under any coordinate transformation. In tensor language, scalars are tensors of order (or rank) zero.

Example 1. Show that the Euclidean distance between two points in space is invariant (a scalar) with respect to any coordinate transformation.

Solution: The Euclidean distance between any two points in space with the coordinates x_i and y_i is given by

$$d^2 = (x_i - y_i)(x_i - y_i) = (x_i - y_i)^2 \quad \text{(summation on i)}$$

Under a tranformation of coordinates, the new coordinates of the points are given by

$$x^*_i = \alpha_{ij}\, x_j \quad \text{and} \quad y^*_i = \alpha_{ik}\, y_k$$

Hence, in the new coordinate system, we have

$$\begin{aligned} d^{*2} &= (x^*_i - y^*_i)(x^*_i - y^*_i) \\ &= \alpha_{ij}\,(x_j - y_j)\,\alpha_{ik}\,(x_k - y_k) \\ &= \alpha_{ij}\,\alpha_{ik}\,(x_j - y_j)(x_k - y_k) \end{aligned}$$

Since

$$\alpha_{ij}\,\alpha_{ik} = \delta_{jk} = \begin{cases} 1 \text{ when } j = k \\ 0 \text{ when } j \neq k \end{cases}$$

it follows that

$$d^{*2} = \delta_{jk}(x_j - y_j)(x_k - y_k) = (x_j - y_j)(x_j - y_j) = d^2$$

Example 2. Show that the dot product of two vectors is invariant under any coordinate transformation.

Solution: Let $\mathbf{A} = a_i\, \mathbf{i}_i$ and $\mathbf{B} = b_j\, \mathbf{i}_j$ be two vectors. Then

$$\mathbf{A} \cdot \mathbf{B} = (a_i b_j)\mathbf{i}_i \cdot \mathbf{i}_j = \delta_{ij}\, a_i b_j = a_i b_i$$

For this to be numerically the same in any other coordinate system, we need to show that $a_i b_i = a^*_j b^*_j$, where a^*_j and b^*_j are the components of the vectors in the new coordinate system x^*_i defined by the transformation $x^*_i = \alpha_{ij}\, x_j$. Now, according to the transformation law for vectors, we have

$$a^*_i = \alpha_{ip}\, a_p, \qquad b^*_i = \alpha_{iq}\, b_q$$

Hence,

$$a^*_i b^*_i = \alpha_{ip}\alpha_{iq}\, a_p b_q = \delta_{pq}\, a_p b_q = a_p b_p$$

Example 3. Given the transformation of coordinates $x^*_i = \alpha_{ij} x_j$, where

$$(\alpha_{ij}) = \begin{bmatrix} \sqrt{2}/2 & \sqrt{2}/2 & 0 \\ \sqrt{3}/3 & -\sqrt{3}/3 & \sqrt{3}/3 \\ -\sqrt{6}/6 & \sqrt{6}/6 & \sqrt{6}/3 \end{bmatrix}$$

If a vector **A** has the components [1, 2, -1] with respect to the x_i-coordinate system, find its components in the x^*_i - system.

Solution: According to the transformation law for vectors, we have

$$a^*_1 = \alpha_{1j} a_j = \frac{\sqrt{2}}{2} + 2\left(\frac{\sqrt{2}}{2}\right) = \frac{3\sqrt{2}}{2}$$

$$a^*_2 = \alpha_{2j} a_j = \frac{\sqrt{3}}{3} - 2\left(\frac{\sqrt{3}}{3}\right) - \frac{\sqrt{3}}{3} = -\frac{2\sqrt{3}}{3}$$

$$a^*_3 = \alpha_{3j} a_j = -\frac{\sqrt{6}}{6} + 2\left(\frac{\sqrt{6}}{6}\right) - \frac{\sqrt{6}}{3} = -\frac{\sqrt{6}}{6}$$

Notice that by definition the elements in the rows of the matrix (α_{ij}) are the components of the unit vectors $\mathbf{i}^*_i$ (i = 1, 2, 3). Thus the components a^*_i are just the dot product of the vector **A** with the base vectors $\mathbf{i}^*_i$ (i = 1, 2, 3).

Example 4. Let a_i and b_i be the components of two nonzero vectors **A** and **B**, respectively, in a rectangular cartesian coordinate system x_i, and set $\mathbf{A} \times \mathbf{B} = c_i \mathbf{i}_i$, where $c_i = a_j b_k - a_k b_j$, i, j, k being the cyclic permutations of 1, 2, 3. Let a^*_i, b^*_i, and c^*_i be the respective components of the vectors **A**, **B**, and **A** x **B** in the new coordinate system x^*_i defined by $x^*_i = \alpha_{ij} x_j$, where

$$(\alpha_{ij}) = \begin{bmatrix} 0 & 0 & 1 \\ -1 & 0 & 0 \\ 0 & 1 & 0 \end{bmatrix}$$

Show that $c^*_i = -(a^*_j b^*_k - a^*_k b^*_j)$, where i, j, k are cyclic permutations of 1, 2, 3.

Solution: Under the given coordinate transformation, the new components of the vectors **A** , **B** , **A** x **B** are given by

$$a^*_1 = \alpha_{1j}\, a_j = a_3, \quad a^*_2 = \alpha_{2j}\, a_j = -a_1, \quad a^*_3 = \alpha_{3j}\, a_j = a_2$$
$$b^*_1 = \alpha_{1j}\, b_j = b_3, \quad b^*_2 = \alpha_{2j}\, b_j = -b_1, \quad b^*_3 = \alpha_{3j}\, b_j = b_2$$
$$c^*_1 = \alpha_{1j}\, c_j = c_3 = a_1 b_2 - a_2 b_1, \quad c^*_2 = \alpha_{2j}\, c_j = -c_1 = -(a_2 b_3 - a_3 b_2)$$
$$c^*_3 = \alpha_{3j}\, c_j = c_2 = a_3\, b_1 - a_1\, b_3$$

Thus we see that

$$a^*_2 b^*_3 - a^*_3 b^*_2 = -a_1 b_2 + a_2 b_1 = -c^*_1$$
$$a^*_3 b^*_1 - a^*_1 b^*_3 = a_2 b_3 - a_3 b_2 = -c^*_2$$
$$a^*_1 b^*_2 - a^*_2 b^*_1 = -a_3 b_1 + a_1 b_3 = -c^*_3$$

This shows that the transform of the components $c_i = a_j b_k - a_k b_j$ differ from $a^*_j b^*_k - a^*_k b^*_j$ by the factor (-1).

Notice that in Example 4, det $(\alpha_{ij}) = -1$ which means that the new coordinate system x^*_i is left-handed (as we assume that the coordinate system x_i is right-handed). This example illustrates the fact that, in general, the components of the cross product of two vectors do not transform according to (5.18) or (5.19). Therefore, the cross product of two vectors is not a vector in the sense of Definition 1. It is called a pseudovector or an axial vector.

5.3 EXERCISES

1. Consider the trajectory of a moving particle $x_i = x_i(t)$ $(t \geq 0)$, where t denotes time. Define

$$v_i(t) = \lim_{\Delta t \to 0} \frac{x_i\,(t + \Delta t\,) - x_i\,(t)}{\Delta t} \qquad (i = 1, 2, 3)$$

Show that v_i (i = 1, 2, 3) are components of a vector (the velocity vector).

2. Let a transformation from a coordinate system x_i to a coordinate system x^*_i be defined by $x^*_i = \alpha_{ij} x_j$, where

$$(\alpha_{ij}) = \begin{bmatrix} \sqrt{3}/2 & 1/2 & 0 \\ -\sqrt{3}/4 & 3/4 & 1/2 \\ 1/4 & -\sqrt{3}/4 & \sqrt{3}/2 \end{bmatrix}$$

(a) If $\mathbf{A} = -\mathbf{i}_1 + 2\mathbf{i}_2 - 2\mathbf{i}_3$, find its components in the x^*_i coordinate system.
(b) If $\mathbf{B}^* = 2\mathbf{i}^*_1 + 2\mathbf{i}^*_2 - 3\mathbf{i}^*_3$, find its components in the x_i coordinate system.
(c) Find the scalar product of the two vectors **A** and **B***. (Does it matter which components you used in taking the scalar product?)
(d) Determine the vector product of **A** and **B*** in both coordinate systems and show that the components are related by the transformation law $c^*_i = \alpha_{ij} c_j$, where c_i denote the components in the x_i-system and c^*_i the components in the x^*_i-system.

3. Consider a transformation defined by $x^*_i = \alpha_{ij} x_j$, where

$$(\alpha_{ij}) = \begin{bmatrix} \sqrt{3}/3 & \sqrt{3}/3 & -\sqrt{3}/3 \\ \sqrt{2}/2 & -\sqrt{2}/2 & 0 \\ \sqrt{6}/6 & \sqrt{6}/6 & 2\sqrt{6}/6 \end{bmatrix}$$

(a) Show that the transformation changes a right-handed coordinate system into a left-handed coordinate system x^*_i (or vice versa).
(b) Find the components of the vector $\mathbf{A} = 6\mathbf{i}_1 + 12\mathbf{i}_2 + 6\mathbf{i}_3$ in the x^*_i - coordinate system.
(c) Find the components of the vector $\mathbf{B}^* = \sqrt{3}\,\mathbf{i}^*_1 + \sqrt{2}\,\mathbf{i}^*_2 + \sqrt{6}\,\mathbf{i}^*_3$ in the x_i - coordinate system.
(d) Using the components c_i of $\mathbf{A} \times \mathbf{B}$ in the x_i-coordinate system, determine the components of $\mathbf{A} \times \mathbf{B}$ in the x^*_i - coordinate system by the transformation law $c^*_i = \alpha_{ij} c_j$.

4. Let a_i and b_i denote the components of the vectors **A** and **B**, respectively.
(a) Show that the components of the vector product **A** x **B** can be written as $c_i = a_j b_k - a_k b_j$, where i, j, k are cyclic permutations of 1, 2, 3 (that is, ijk = 123, 231, 312).
(b) Under a coordinate transformation $x^*_i = \alpha_{ij} x_j$, show that

$$c^*_i = a^*_j b^*_k - a^*_k b^*_j = \alpha_{jp}\alpha_{kq}(a_p b_q - a_q b_p) \quad (p, q = 1, 2, 3)$$

or

$$c^*_i = (\alpha_{jp}\alpha_{kq} - \alpha_{jq}\alpha_{kp})(a_p b_q - a_q b_p) \quad (p, q = 1, 2\,; 2, 3\,; 3, 1)$$

(c) From the fact that $\mathbf{i}^*_j = \alpha_{jp}\mathbf{i}_p$ and $\mathbf{i}^*_k = \alpha_{kq}\mathbf{i}_q$, show that

$$\mathbf{i}^*_j \times \mathbf{i}^*_k = \alpha_{jp}\alpha_{kq}(\mathbf{i}_p \times \mathbf{i}_q) \quad (p, q = 1, 2, 3)$$

$$= (\alpha_{jp}\alpha_{kq} - \alpha_{jq}\alpha_{kp})(\mathbf{i}_p \times \mathbf{i}_q) \quad (p, q = 1, 2; 2, 3; 3, 1)$$

(d) Thus if both coordinate systems have the same orientation (both right-handed or left-handed), show that

$$\mathbf{i}^*_i = (\alpha_{jp}\alpha_{kq} - \alpha_{jq}\alpha_{kp})\,\mathbf{i}_r$$

where p, q, r are also cyclic permutations of 1, 2, 3. On the other hand, the coordinate systems have opposite orientation, show that

$$\mathbf{i}^*_i = -(\alpha_{jp}\alpha_{kq} - \alpha_{jq}\alpha_{kp})\mathbf{i}_r$$

(e) Since $\mathbf{i}^*_i = \alpha_{ir}\mathbf{i}_r$, deduce that $\alpha_{jp}\alpha_{kq} - \alpha_{jq}\alpha_{kp} = \pm\alpha_{ir}$, where the plus or minus sign prevails according as the transformation preserves or reverses the orientation of the coordinate systems.
(f) Finally, from (b) show that $c^*_i = \Delta\,\alpha_{ir}\,c_r$, where $\Delta = \det(\alpha_{ij}) = \pm 1$. [A quantity consisting of three components which transform according to the law given in (f) is called a pseudotensor of order one or axial vector.]

5.5 CARTESIAN TENSORS

In the preceding section we introduced the definition of cartesian tensors of order 0 and 1, which are essentially scalars and vecors. We now consider cartesian tensors of higher order. The order of a tensor has to do with the number of components of the tensor. In three-dimensional space, a tensor is said to be of order or rank r if it has 3^r components. Thus scalars and vectors are of order 0 and 1 since they have $3^0 = 1$ and $3^1 = 3$ components, respectively.

To motivate the definition of tensors of higher order, we consider a simple physical example of a quantity that is described by a cartesian tensor of order 2. Let a particle of mass m be located at the point (x_1, x_2, x_3). The so-called moment of inertia of the particle is a set of nine numbers defined by

$$I_{ij} = m(\delta_{ij}\, x_k x_k - x_i x_j) \qquad (i, j = 1, 2, 3) \tag{5.20}$$

When i = j, this yields

$$I_{11} = m[(x_2)^2 + (x_3)^2], \quad I_{22} = m[(x_3)^2 + (x_1)^2]$$

$$I_{33} = m[(x_1)^2 + (x_2)^2]$$

These are called the moments of inertia of the particle about the x_1-, x_2-, x_3- axes, respectively, and are the types of moments of inertia usually studied in a standard calculus course. When $i \neq j$, (5.20) gives

$$I_{12} = -m\, x_1 x_2 = I_{21}, \qquad I_{23} = -m\, x_2 x_3 = I_{32}$$

$$I_{31} = -m\, x_3 x_1 = I_{13}$$

These are called the products of inertia.

Now let us examine what happens to the moment of inertia of the particle when we introduce a different coordinate system. Let $x^*_i = \alpha_{ij}\, x_j$ define a new coordinate system and let (x^*_1, x^*_2, x^*_3) be the new coordinates of the particle. Then, by defintion, the moment of inertia of the particle in the new coordinate system x^*_i is given by

$$I^*_{ij} = m(\delta^*_{ij}\, x^*_k x^*_k - x^*_i x^*_j) \quad (i, j = 1, 2, 3) \tag{5.21}$$

To relate these quantities with the I_{ij} (i, j = 1, 2, 3), we first observe that under the coordinate transformation, we have

$$x^*_k x^*_k = x_r x_r$$

(invariance of Euclidean distance), and

$$\delta^*_{ij} = \alpha_{ip}\,\alpha_{jq}\,\delta_{pq}$$

that is, $\delta^*_{ij} = \delta_{ij}$ (because the Kronecker delta has the same value in any coordinate system). Substituting these identities in (5.21), we obtain

$$\begin{aligned} I^*_{ij} &= m[\alpha_{ip}\,\alpha_{jq}\,\delta_{pq}\,x_r\,x_r - \alpha_{ip}\,\alpha_{jq}\,x_p\,x_q] \\ &= \alpha_{ip}\,\alpha_{jq}\,m(\delta_{pq}\,x_r\,x_r - x_p\,x_q) \qquad (5.22) \\ &= \alpha_{ip}\,\alpha_{jq}\,I_{pq} \end{aligned}$$

Equation (5.22) describes the manner by which the moment of inertia of a particle transforms under a change of coordinate system. Quantities that transform according to the formula (5.22) are called cartesian tensors of order two. This is the classical definition of tensors of order 2 in terms of components relative to various coordinate systems. Thus the quantities I_{ij} given in (5.20) define a second order tensor. It is called the moment of inertia tensor, and the quantities I_{ij} are the components of this tensor. The components may be viewed as constituting a representation of the tensor with respect to a given coordinate system.

We now formally define a second order tensor.

Definition 2. (Tensor of Order Two) A tensor of order two is a quantity consisting of 3^2 components a_{ij} (i, j = 1, 2, 3), which transform under a change of coordinate system $x^*_i = \alpha_{ij}\,x_j$ according to the law

$$a^*_{ij} = \alpha_{ip}\,\alpha_{jq}\,a_{pq} \qquad (i, j = 1, 2, 3) \qquad (5.23)$$

where a_{ij} and a^*_{ij} are the components of the tensor with respect to the coordinate systems x_i and x^*_i, respectively.

The transformation law (5.23) can be inverted if we multiply both sides of the equation by $\alpha_{ir}\,\alpha_{js}$ and sum with respect to i and j. We obtain

$$a_{rs} = \alpha_{ir}\,\alpha_{js}\,a^*_{ij} \qquad (5.24)$$

in which we used the relation $\alpha_{ij}\,\alpha_{ik} = \delta_{jk}$.

It is convenient that a second order tensor can be written in a matrix form

$$(a_{ij}) = \begin{bmatrix} a_{11} & a_{12} & a_{13} \\ a_{21} & a_{22} & a_{23} \\ a_{31} & a_{32} & a_{33} \end{bmatrix}$$

Because of this, certain algebraic operations involving vectors and tensors of order two can be calculated by matrix algebra and the results expressed in matrix form. As a matter of fact, if we set

$$A = (a_{ij}), \qquad A^* = (a^*_{ij}), \qquad \Lambda = (\alpha_{ij})$$

then the product $b_{iq} = \alpha_{ip}a_{pq}$ represents the element in the i-th row and q-th column of the matrix product $\Lambda A = B$, while the product $b_{iq}\alpha_{jq} = \alpha_{ip}\alpha_{jq}a_{pq}$ represents the element in the i-th row and j-th column of the matrix product $B\Lambda^T = \Lambda A \Lambda^T$. Therefore, the transformation law (5.23) can be written in the matrix form

$$A^* = \Lambda\, A\, \Lambda^T \qquad (5.25)$$

Since $\Lambda^{-1} = \Lambda^T$, the inverse transformation law (5.24) has the matrix form

$$A = \Lambda^T A^* \Lambda \qquad (5.26)$$

It should be noted that matrix notation fails for tensors of higher order.

Example 1. A trivial example of a second order tensor is the Kronecker delta δ_{ij} which, by definition, is independent of any coordinate system, that is, $\delta_{ij} = \delta^*_{ij}$. Hence, we can write

$$\delta^*_{ij} = \alpha_{ip}\,\alpha_{jq}\,\delta_{pq}$$

since $\alpha_{ip}\alpha_{jp} = \delta_{ij}$. It is clear that the matrix representation of the Kronecker delta is the identity matrix, that is, $(\delta_{ij}) = I$.

Example 2. Let a_i and b_j be the components of two vectors. The nine quantities $c_{ij} = a_i b_j$ (i, j = 1, 2, 3) form the components of a second order tensor, known as the outer product of the two vectors.

To prove that (c_{ij}) is indeed a second order tensor, we need to show that it transforms according to the law (5.23). Now under a change of coordinates, we know that the components a_i, b_j transform according to the laws:

$$a^*_i = \alpha_{ip}\, a_p, \qquad b^*_j = \alpha_{jq}\, b_q$$

Hence

$$c^*_{ij} = a^*_i\, b^*_j = \alpha_{ip}\, \alpha_{jq}\, a_p\, b_p = \alpha_{ip}\, \alpha_{jq}\, c_{pq}$$

in accordance with (5.23).

Example 3. Given the second order tensor

$$(a_{ij}) = \begin{bmatrix} 0 & -1 & 3 \\ 1 & 0 & 2 \\ -3 & -2 & 0 \end{bmatrix}$$

Find the components of this tensor in the coordinate system x^*_i defined by $x^*_i = \alpha_{ij}\, x_j$, where

$$(\alpha_{ij}) = \begin{pmatrix} 0 & 0 & 1 \\ -1 & 0 & 0 \\ 0 & 1 & 0 \end{pmatrix}$$

Solution: The new components are given by

$$a^*_{ij} = \alpha_{ip}\, \alpha_{jq}\, a_{pq} \quad (i, j = 1, 2, 3)$$

Thus

$$\begin{aligned} a^*_{11} &= \alpha_{1p}\, \alpha_{1q}\, a_{pq} = \alpha_{13}\, \alpha_{13}\, a_{33} = 0 \\ a^*_{12} &= \alpha_{1p}\, \alpha_{2q}\, a_{pq} = \alpha_{13}\, \alpha_{21}\, a_{31} = 3 \\ a^*_{13} &= \alpha_{1p}\, \alpha_{3q}\, a_{pq} = \alpha_{13}\, \alpha_{32}\, a_{32} = -2 \end{aligned}$$

$$a^*_{21} = \alpha_{2p}\,\alpha_{1q}\,a_{pq} = \alpha_{21}\,\alpha_{13}\,a_{13} = -3$$
$$a^*_{22} = \alpha_{2p}\,\alpha_{2q}\,a_{pq} = \alpha_{21}\,\alpha_{21}\,a_{11} = 0$$
$$a^*_{23} = \alpha_{2p}\,\alpha_{3q}\,a_{pq} = \alpha_{21}\,\alpha_{32}\,a_{12} = 1$$
$$a^*_{31} = \alpha_{3p}\,\alpha_{3q}\,a_{pq} = \alpha_{32}\,\alpha_{13}\,a_{23} = 2$$
$$a^*_{32} = \alpha_{3p}\,\alpha_{2q}\,a_{pq} = \alpha_{32}\,\alpha_{21}\,a_{21} = -1$$
$$a^*_{33} = \alpha_{3p}\,\alpha_{3q}\,a_{pq} = \alpha_{32}\,\alpha_{32}\,a_{22} = 0$$

Hence

$$(a^*_{ij}) = \begin{bmatrix} 0 & 3 & -2 \\ -3 & 0 & 1 \\ 2 & -1 & 0 \end{bmatrix}$$

The student should check this result by using the formula (5.25).

Example 4. Let a body of unit mass be located at a point P: (1, 1, 0). By (5.20) the components of the moment of inertia tensor are

$$I_{11} = 1, \qquad I_{22} = 1, \qquad I_{33} = 2$$
$$I_{12} = I_{21} = -1, \quad I_{23} = I_{32} = 0, \quad I_{13} = I_{31} = 0$$

or, in matrix form

$$(I_{ij}) = \begin{bmatrix} 1 & -1 & 0 \\ -1 & 1 & 0 \\ 0 & 0 & 2 \end{bmatrix}$$

Now suppose we rotate the coordinate system about the x_3-axis so that the x_1-axis passes through the point P. Let us designate this axis as x^*_1 and the others as x^*_2 and x^*_3 (see Fig. 5.3). Obviously, $\mathbf{i}^*_1 = (\mathbf{i}_1 + \mathbf{i}_2)/\sqrt{2}$, $\mathbf{i}^*_2 = (-\mathbf{i}_1 + \mathbf{i}_2)/\sqrt{2}$, $\mathbf{i}^*_3 = \mathbf{i}_3$, so that the transformation effecting the rotation is defined by the matrix

$$(\alpha_{ij}) = \begin{bmatrix} \sqrt{2}/2 & \sqrt{2}/2 & 0 \\ -\sqrt{2}/2 & \sqrt{2}/2 & 0 \\ 0 & 0 & 1 \end{bmatrix}$$

(Remember $\alpha_{ij} = \mathbf{i}^*_i \cdot \mathbf{i}_j$.) Thus, with respect to the x^*_i - coordinate

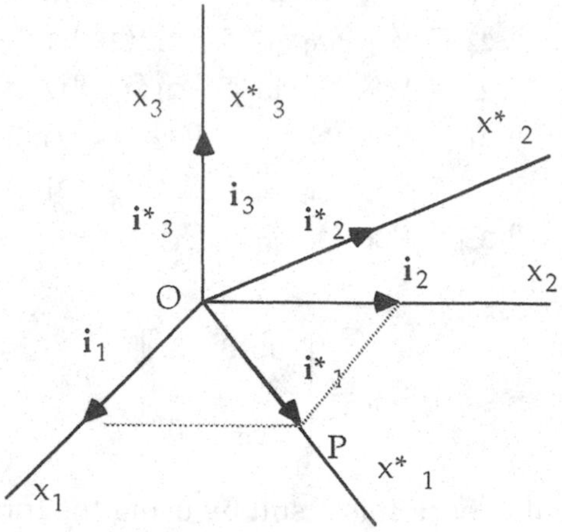

Fig. 5.3 Principal axes of a tensor.

system, the moment of inertia tensor becomes

$$(I^*_{ij}) = \begin{bmatrix} 0 & 0 & 0 \\ 0 & 2 & 0 \\ 0 & 0 & 2 \end{bmatrix}$$

as is readily verified by using (5.25). Notice that in the x^*_i- coordinate system, the point P has the coordinates ($\sqrt{2}$, 0, 0).

Therefore, in the x^*_i-coordinate system all the product inertias are zero, and the moments of inertia about the the x^*_1 -, x^*_2- and x^*_3 - axes are $I^*_{11} = 0$, $I^*_{22} = 2$ and $I^*_{33} = 2$, respectively. The axes x^*_i with respect to which $I_{ij} = 0$ for all $i \neq j$ are called the *principal axes* of the tensor. This topic is discussed further in Sec. 5.8.

Example 5. Suppose that a certain quantity has components defined by $a_{11} = x_1x_2$, $a_{12} = x_2^2$, $a_{21} = x_1^2$, $a_{22} = x_1x_2$ with respect to a rectangular cartesian coordinate system x_1, x_2 on a plane. If the quantity were a (plane) tensor of order two, its components in any other rectangular cartesian coordinate system x^*_i would be given by

$$a^*_{11} = x^*_1 x^*_2, \quad a^*_{12} = x^{*2}_2, \quad a^*_{21} = x^{*2}_1, \quad a^*_{22} = x^*_1 x^*_2$$

However, under a rotation of axes,

$$\begin{aligned} x^*_1 &= x_1 \cos\theta + x_2 \sin\theta \\ x^*_2 &= -x_1 \sin\theta + x_2 \cos\theta \end{aligned}$$

we see that

$$\begin{aligned} a^*_{11} = \alpha_{1i}\alpha_{1j} a_{ij} &= x_1 x_1 \cos^2\theta + x_2^2 \sin\theta\cos\theta \\ &\quad + x_1^2 \sin\theta\cos\theta + x_1 x_2 \sin^2\theta \\ &= x_1 x_2 + (x_1^2 + x_2^2)\sin\theta\cos\theta \end{aligned}$$

while, on the other hand,

$$\begin{aligned} x^*_1 x^*_2 &= (x_1 \cos\theta + x_2 \sin\theta)(-x_1 \sin\theta + x_2 \cos\theta) \\ &= x_1 x_2 (\cos^2\theta - \sin^2\theta) + (x_2^2 - x_1^2)\sin\theta\cos\theta \end{aligned}$$

Thus, $a^*_{11} \neq x^*_1 x^*_2$, and, therefore, the quantity is not a tensor. In other words, the quantity given here cannot represent any physical entity.

The definition of a cartesian tensor of order r (r > 2) is a straight-forward extension of Definition 2.

Definition 3. A cartesian tensor of order r (r > 2) is a quantity consisting of 3^r components $a_{i_1 i_2 \ldots i_r}$ which transform under a change of coordinates according to the law

$$a^*_{i_1 i_2 \ldots i_r} = \alpha_{i_1 j_1}\alpha_{i_2 j_2} \ldots \alpha_{i_r j_r} a_{j_1 j_2 \ldots j_r} \tag{5.27}$$

$$(i_k, j_k = 1, 2, 3; k = 1, 2, \ldots, r)$$

where $a^*_{i_1 i_2 \ldots i_r}$ and $a_{j_1 j_2 \ldots j_r}$ are the components of the tensor with respect to the x^*_i and x_i coordinate systems, respectively.

Thus, for example, a tensor of order 3 will have 27 components a_{ijk} (i, j, k = 1, 2, 3) which transform according to the law

$$a^*_{ijk} = \alpha_{ip}\,\alpha_{jq}\,\alpha_{kr}\,a_{pqr}$$

under a change of coordinates $x^*_i = \alpha_{ij}\,x_j$, $i = 1, 2, 3$.

Example 6. Let a_i, b_i, and c_i $(i = 1, 2, 3)$ be the components of three nonzero vectors. Then the 27 quantities

$$d_{ijk} = a_i\,b_j\,c_k \quad (i, j, k = 1, 2, 3)$$

form the components of a third order tensor known as the outer product of the vectors. To see this, we first note that under a change of coordinates, we have

$$a^*_i = \alpha_{ip}a_p, \quad b^*_j = \alpha_{jq}b_q, \quad c^*_k = \alpha_{kr}c_r$$

Hence

$$d^*_{ijk} = a^*_i b^*_j c^*_k = \alpha_{ip}\alpha_{jq}\alpha_{kr}\,a_p b_q c_r = \alpha_{ip}\alpha_{jq}\alpha_{kr}\,d_{pqr}$$

which shows that the quantities d_{ijk} indeed transform as components of a third order tensor.

5.6 STRESS TENSOR

We consider another simple example of a physical quantity that is described by a second order tensor. When a solid body is under the influence of a system of forces and remains in equilibrium, there is a tendency for the shape of the body to be distorted or deformed. A solid body in such a state is said to be under strain. The resulting internal forces in a body that is in a state of strain give rise to a stress at each point of the body. The stress at a point is measured as force per unit area and depends not only on the force but also on the orientation of the surface relative to the force. To illustrate this, consider a uniform thin cylindrical rod of cross-sectional area a that is being stretched by a force **F** (Fig. 5.4). Consider a plane section whose unit normal vector **n** makes an angle θ with the axis of the rod. By the area cosine principle, the area of this plane section is equal to $A = a \sec \theta$. Thus the stress vector across the plane section is equal to

$$\mathbf{S} = \mathbf{F}/A = (\mathbf{F}/a)\cos\theta \tag{5.28}$$

It is clear from (5.28) that the stress vector depends on **F** as well as on the orientation of the plane, i.e., **n** or θ. In fact, we see that the magnitude of **S** is maximum when $\theta = 0$, and it is zero when $\theta = \pi/2$.

The stress vector **S** on the plane area can be resolved into two

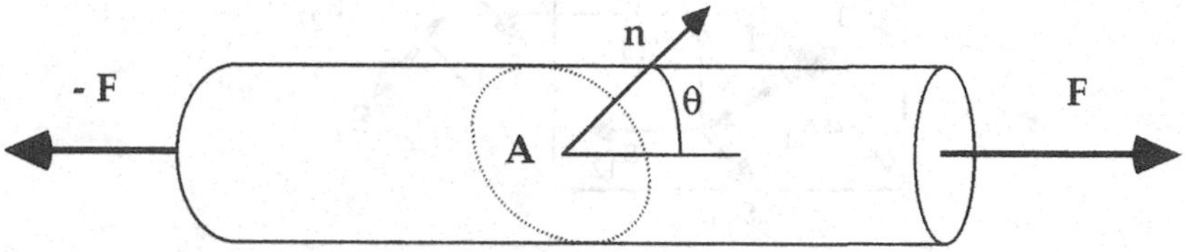

Fig. 5.4 Stress in a thin rod.

components, one perpendicular to the plane section (along the unit normal vector **n**) denoted by s_n, and the other parallel to the plane section (perpendicular to **n**), denoted by s_t. The component s_n along **n** is called the tensile or normal stress, while the component s_t tangential to the plane section is called the shearing stress.

To examine the stress at a point P in a solid body under strain, we set up a coordinate system x_1, x_2, x_3 and consider an arbitrary element of volume (a small rectangular parallelepiped) of the body with P as one of its vertices (see Fig. 5.5). Denote the length of the edges of the element by $\Delta x_1, \Delta x_2, \Delta x_3$ as shown in Fig. 5.5. Clearly there will be stresses on each face of the element. Let $\mathbf{S}_i$ denote the stress vector on the face that is perpendicular to the vector $\mathbf{i}_i$, and let $\mathbf{S}'_i$ denote the stress vector on the opposite face (perpendicular to $\mathbf{i}_i$). Since the body is in equilibrium, the sum of the forces (stress times area) acting on the element must be zero. Thus we have

$$(\mathbf{S}_1 + \mathbf{S}'_1)\Delta x_2 \Delta x_3 + (\mathbf{S}_2 + \mathbf{S}'_2)\Delta x_1 \Delta x_3 + (\mathbf{S}_3 + \mathbf{S}'_3)\Delta x_2 \Delta x_1 = 0 \qquad (5.29)$$

Since the quantities $\Delta x_1, \Delta x_2, \Delta x_3$ are arbitrary, the equation implies that $\mathbf{S}'_1 = -\mathbf{S}_1$, $\mathbf{S}'_2 = -\mathbf{S}_2$, $\mathbf{S}'_3 = -\mathbf{S}_3$. Hence the stress vectors on opposite faces of the element are equal in magnitude but opposite in direction. Moreover, since (5.29) holds for any element of the body

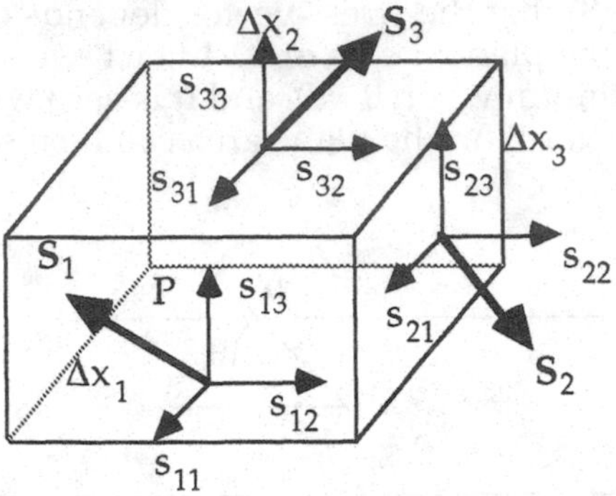

Fig. 5.5 Stress tensor.

at P, we conclude that the stress vectors $\mathbf{S}_i$ $(1 \le i \le 3)$ represent the stress at P acting on surface areas that are normal to the unit vectors $\mathbf{i}_i$ $(1 \le i \le 3)$, respectively.

Now each of the stress vectors $\mathbf{S}_i$ $(1 \le i \le 3)$ can be represented by its components s_{i1}, s_{i2}, s_{i3}, so that $\mathbf{S}_i = s_{ij}\mathbf{i}_j$ $(i = 1, 2, 3)$. The quantities s_{11}, s_{22}, s_{33} are the respective normal components of the stress vectors $\mathbf{S}_1, \mathbf{S}_2, \mathbf{S}_3$, and they are called the tensile or normal stresses; the quantities s_{ij} for $i \ne j$ are called the shearing stresses. Thus, at each point, the stress is represented by the nine quantities s_{ij} $(i, j = 1, 2, 3)$, which can be written in the matrix form

$$(s_{ij}) = \begin{bmatrix} s_{11} & s_{12} & s_{13} \\ s_{21} & s_{22} & s_{23} \\ s_{31} & s_{32} & s_{33} \end{bmatrix} \tag{5.30}$$

However, the shearing stresses s_{ij}, $i \ne j$, are not altogether independent of one another. In fact, since the element of volume $\Delta x_1, \Delta x_2, \Delta x_3$ is also in equilibrium with respect to rotation, the sum of the moments of force or torques about any axis must be zero. So if we consider the net torque about the edge through P parallel to the x_3-axis (see Fig. 5.5), we obtain

$$(s_{12}\,\Delta x_2\,\Delta x_3)\Delta x_1 \;-\; (s_{21}\,\Delta x_1\,\Delta x_3)\Delta x_2 = 0$$

The stresses s_{11}, s_{22}, s_{31}, s_{32} are balanced by equal and opposite stresses on the opposite faces of the element, and s_{13}, s_{23}, s_{33} have zero moment arm as they are directed along the x_3-axis. Thus we obtain,

$$s_{12} = s_{21}$$

In a similar manner, by considering the net torque about the edges parallel to the x_1 - and x_2 - axes, we find

$$s_{23} = s_{32}\,, \qquad s_{31} = s_{13}$$

Therefore, $s_{ij} = s_{ji}$ (i, j = 1, 2, 3) which means that the matrix (5.30) is symmetric.

Next we show that the quantities s_{ij} do transform according to the transformation law

$$s^*_{ij} = \alpha_{ip}\,\alpha_{jq}\,s_{pq} \tag{5.31}$$

where s^*_{ij} (i, j = 1, 2, 3) represent the stress with respect to a new coordinate system x^*_i. First, we show that the components of the stress vector $\mathbf{S}_n$ across a surface with unit normal vector $\mathbf{n} = \alpha_{ni}\mathbf{i}_i$ are given by $s_{ni} = \mathbf{S}_i \cdot \mathbf{n} = s_{ij}\alpha_{nj}$ (i = 1, 2, 3), so that $\mathbf{S}_n = s_{ni}\mathbf{i}_i$. We consider an infinitesimal tetrahedron with P as a vertex and with the slant face perpendicular to $\mathbf{n}$ as shown in Fig. 5.6. Let ΔA denote the area of the slant face. Then, by the area cosine principle, the areas of the sides with normal vectors $\mathbf{i}_1$, $\mathbf{i}_2$, and $\mathbf{i}_3$ are given by $\alpha_{n1}\Delta A$, $\alpha_{n2}\,\Delta A$, and $\alpha_{n3}\,\Delta A$, respectively. Now since the element is in equilibrium, the force on the slant face must be equal to the sum of the forces on the three faces of the tetrahedron. Thus, we have

$$\mathbf{S}_n\,\Delta A \;=\; \mathbf{S}_1\,\alpha_{n1}\,\Delta A + \mathbf{S}_2\,\alpha_{n2}\,\Delta A + \mathbf{S}_3\,\alpha_{n3}\,\Delta A$$

Equating the respective components, we obtain

$$\begin{aligned} s_{nj} &= s_{1j}\,\alpha_{n1} + s_{2j}\,\alpha_{n2} + s_{3j}\,\alpha_{n3} \\ &= \alpha_{np}\,s_{pj} = \mathbf{n}\cdot\mathbf{S}_j \qquad (j = 1, 2, 3) \end{aligned} \tag{5.32a}$$

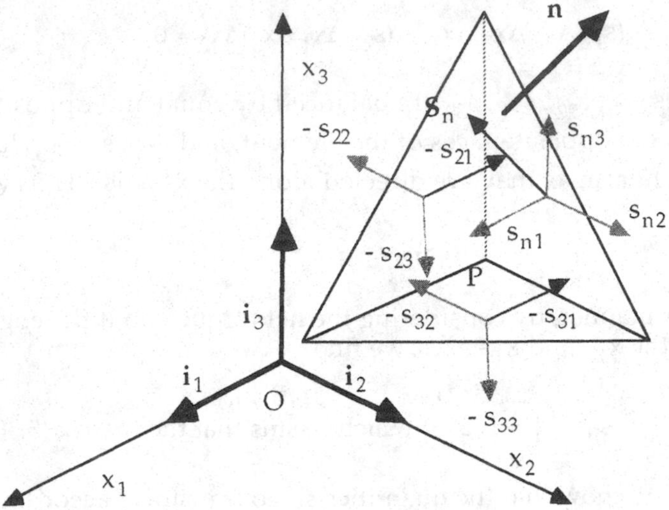

Fig. 5.6 Stress vector across an arbitrary surface area.

Since $s_{ij} = s_{ji}$, we may write this in matrix form as

$$\begin{bmatrix} s_{n1} \\ s_{n2} \\ s_{n3} \end{bmatrix} = \begin{bmatrix} s_{11} & s_{12} & s_{13} \\ s_{21} & s_{22} & s_{23} \\ s_{31} & s_{32} & s_{33} \end{bmatrix} \begin{bmatrix} \alpha_{n1} \\ \alpha_{n2} \\ \alpha_{n3} \end{bmatrix} \qquad (5.32b)$$

Now let $\mathbf{i}^*_i$ be the base vectors of a new coordinate system x^*_i and let $\mathbf{S}^*_i$ denote the stress vector across a surface that is perpendicular to $\mathbf{i}^*_i$, i = 1, 2, 3. Then , by (5.32a), we have

$$\mathbf{S}^*_i = \alpha_{ip}\, s_{pq}\, \mathbf{i}_q$$

On the other hand, if s^*_{ij} denote the components of $\mathbf{S}^*_i$ with respect to the new coordinate system x^*_i , then , $\mathbf{S}^*_i = s^*_{ij}\, \mathbf{i}^*_j$. Thus we have

$$s^*_{ij}\, \mathbf{i}^*_j = \alpha_{ip}\, s_{pq}\, \mathbf{i}_q \qquad (5.33)$$

We solve for s^*_{ij} by taking the dot product of both sides of (5.33) with $\mathbf{i}^*_k$, noting that $\mathbf{i}^*_j \cdot \mathbf{i}^*_k = \delta_{jk}$. Since $\alpha_{jq} = \mathbf{i}^*_j \cdot \mathbf{i}_q$, we finally obtain

$$s^*_{ij} = \alpha_{ip}\, \alpha_{jq}\, s_{pq} \qquad (5.34)$$

Thus the quantities s_{ij} transform as components of a second order tensor. This tensor is known as the stress tensor.

Example 1. The stress tensor at a point P is given by

$$(s_{ij}) = \begin{bmatrix} 2 & -1 & 1 \\ -1 & 2 & -1 \\ 1 & -1 & 2 \end{bmatrix}$$

What is the stress vector on a surface normal to the unit vector

$$\mathbf{n} = \frac{2}{3}\,\mathbf{i}_1 + \frac{1}{3}\,\mathbf{i}_2 - \frac{2}{3}\,\mathbf{i}_3$$

at the point? What is the normal stress and the shearing stress on such a surface?

Solution: According to (5.32a), the components of the stress vector $\mathbf{S}_n$ on the surface normal to the vector $\mathbf{n}$ are given by

$$s_{n1} = \mathbf{S}_1 \cdot \mathbf{n} = 2\left(\frac{2}{3}\right) + (-1)\left(\frac{1}{3}\right) + \left(\frac{-2}{3}\right) = \frac{1}{3}$$

$$s_{n2} = \mathbf{S}_2 \cdot \mathbf{n} = (-1)\left(\frac{2}{3}\right) + 2\left(\frac{1}{3}\right) + (-1)\left(\frac{-2}{3}\right) = \frac{2}{3}$$

$$s_{n3} = \mathbf{S}_3 \cdot \mathbf{n} = \left(\frac{2}{3}\right) + (-1)\left(\frac{1}{3}\right) + 2\left(\frac{-2}{3}\right) = -1$$

The normal stress, that is, the stress along the direction of $\mathbf{n}$, is given by

$$s_n = \mathbf{S}_n \cdot \mathbf{n} = \frac{2}{3}\, s_{n1} + \frac{1}{3}\, s_{n2} - \frac{2}{3}\, s_{n3} = \frac{10}{9}$$

Thus the shearing stress is

$$s_t = \sqrt{|\mathbf{S}_n|^2 - s_n^2} = \sqrt{\frac{14}{9} - \frac{100}{81}} = \frac{\sqrt{26}}{9}$$

Example 2. For stress on a plane, the components of a stress tensor are given by

$$(s_{ij}) = \begin{bmatrix} s_{11} & s_{12} \\ s_{21} & s_{22} \end{bmatrix}$$

where $s_{12} = s_{21}$. Then, by (5.32a), the components s_{n1} and s_{n2} of the stress vector across a plane normal to $\mathbf{n} = \cos\theta\, \mathbf{i}_1 + \sin\theta\, \mathbf{i}_2$ are given by

$$s_{n1} = s_{11} \cos\theta + s_{21} \sin\theta$$
$$s_{n2} = s_{12} \cos\theta + s_{22} \sin\theta$$

Thus the normal stress on such a plane is equal to

$$\begin{aligned} s_n = \mathbf{S}_n \cdot \mathbf{n} &= s_{n1} \cos\theta + s_{n2} \sin\theta \\ &= s_{11} \cos^2\theta + 2\, s_{12} \sin\theta \cos\theta + s_{22} \sin^2\theta \end{aligned}$$

and the shearing stress is given by

$$s_t = \sqrt{|\mathbf{S}_n|^2 - s_n^2}$$

For example, if

$$(s_{ij}) = \begin{bmatrix} 2 & -1 \\ -1 & 2 \end{bmatrix}$$

then the components of the stress across a plane normal to the unit vector $\mathbf{n} = (\sqrt{3}/2)\, \mathbf{i}_1 + (1/2)\, \mathbf{i}_2$ are

$$s^*_1 = 2(\sqrt{3}/2) + (-1)(1/2) = \sqrt{3} - 1/2$$
$$s^*_2 = -\sqrt{3}/2 + 2(1/2) = -\sqrt{3}/2 + 1$$

Thus the normal stress across such a plane is

$$s^*_n = \left(\sqrt{3} - \frac{1}{2}\right)\frac{\sqrt{3}}{2} + \left(-\frac{\sqrt{3}}{2} + 1\right)\frac{1}{2} = 2 - \frac{\sqrt{3}}{2}$$

and the shearing stress is given by

$$s^*_t = \sqrt{s^{*2}_1 + s^{*2}_2 - s^{*2}_n}$$

$$= \sqrt{(\sqrt{3} - 1/2)^2 + (1 - \sqrt{3}/2)^2 - (2 - \sqrt{3}/2)^2} = \frac{1}{2}$$

Example 3. Consider a stress tensor given by

$$(s_{ij}) = \begin{bmatrix} 0 & 1 & 1 \\ 1 & 2 & 1 \\ 1 & 1 & 0 \end{bmatrix}$$

If we change the coordinate system through the transformation $x^*_i = \alpha_{ij} x_j$, where

$$(\alpha_{ij}) = \begin{bmatrix} \sqrt{2}/2 & 0 & -\sqrt{2}/2 \\ \sqrt{3}/3 & -\sqrt{3}/3 & \sqrt{3}/3 \\ \sqrt{6}/6 & \sqrt{6}/3 & \sqrt{6}/6 \end{bmatrix}$$

then in the x^*_i-coordinate system the stress tensor becomes

$$(s^*_{ij}) = \begin{bmatrix} -1 & 0 & 0 \\ 0 & 0 & 0 \\ 0 & 0 & 3 \end{bmatrix}$$

This can, of course, be verified by using the transformation law

$$s^*_{ij} = \alpha_{ip}\,\alpha_{jq}\,s_{pq}$$

For instance, we find

$$s^*_{11} = \alpha_{11}\,\alpha_{13}\,s_{13} + \alpha_{13}\,\alpha_{11}\,s_{31}$$

$$= -\frac{1}{2} - \frac{1}{2} = -1$$

since $\alpha_{12} = 0$, $s_{11} = 0$, and $s_{33} = 0$. Thus in the x^*_i coordinate system, the stress consists only of normal stresses as all shearing stresses are zero.

5.4 EXERCISES

1. The equation of a quadric surface with center at the origin of a coordinate system x_i is given by $a_{ij} x_i x_j = 1$. If the equation of the surface under rotation of axes is $a^*_{ij} x^*_i x^*_j = 1$, show that the cofficients a_{ij} form components of a second order tensor.
2. Verify the result of Problem 1 for the conic

$$a_{11}x^2 + 2a_{12}\,x\,y + a_{22}\,y^2 = 1, \quad a_{12} = a_{21}$$

under the rotation of axes

$$x^* = x \cos\theta + y \sin\theta$$
$$y^* = -x \sin\theta + y \cos\theta$$

3. A body of mass M is located at the point (2, -2, 3). Find the components of its moment of inertia tensor.
4. The moment of inertia of a body of mass m about an axis in the direction of a unit vector $\mathbf{u} = u_i\,\mathbf{i}_i$ is given by

$$I_u = I_{ij}\,u_i u_j$$

where I_{ij} is the moment of inertia defined in Sec. 5.5. Show that $I_u = m[(x_k x_k - (x_i u_i)^2]$. What is the moment of inertia of the body in Problem 3 about the line $x_1 = x_2$ $(x_3 = 0)$?
5. The components of the moment of inertia of a certain body with respect to a coordinate system x_i are given by

$$(I_{ij}) = \begin{bmatrix} a & 0 & 0 \\ 0 & b & 0 \\ 0 & 0 & c \end{bmatrix}$$

Find the moment of inertia of the body about a line in the direction of the vector $\mathbf{A} = \mathbf{i}_1 - \mathbf{i}_2 + \mathbf{i}_3$.
6. Find the moment of inertia of a body of mass M located at (2, 1, 3) about an axis in the direction of the vector $\mathbf{A} = 2\mathbf{i}_1 - \mathbf{i}_2 + 2\mathbf{i}_3$.

7. The moment of inertia tensor of a certain body is given by

$$\left(I_{ij}\right)=\begin{bmatrix} 0 & 2 & 1 \\ 2 & 0 & 1 \\ 1 & 1 & 3 \end{bmatrix}$$

Show that in the coordinate system x^*_i defined by $x^*_i = \alpha_{ij}x_j$, where

$$\left(\alpha_{ij}\right)=\begin{bmatrix} 1/\sqrt{2} & -1/\sqrt{2} & 0 \\ 1/\sqrt{3} & 1/\sqrt{3} & -1/\sqrt{3} \\ 1/\sqrt{6} & 1/\sqrt{6} & 2/\sqrt{6} \end{bmatrix}$$

$I^*_{11} = -2$, $I^*_{22} = 1$, $I^*_{33} = 4$.

8. A stress tensor has components given by

$$\left(s_{ij}\right)=\begin{bmatrix} 3 & 0 & -1 \\ 0 & 2 & 4 \\ -1 & 4 & 0 \end{bmatrix}$$

Find I^*_{11}, I^*_{22}, and I^*_{33} in the coordinate system x^*_i defined by $x^*_i = \alpha_{ij}\, x_j$, where

$$(\alpha_{ij}) = \begin{bmatrix} \sqrt{3}/2 & 1/2 & 0 \\ -\sqrt{3}/4 & 3/4 & 1/2 \\ 1/4 & -\sqrt{3}/4 & \sqrt{3}/2 \end{bmatrix}$$

9. Verify that the quantities

$$(a_{ij}) = \begin{bmatrix} x_2^2 & -x_1 x_2 \\ -x_1 x_2 & x_1^2 \end{bmatrix}$$

are components of a second order plane tensor.

10. Show that the quantities

$$(a_{ij}) = \begin{bmatrix} x_1 x_2 & x_2^2 \\ x_1^2 & -x_1 x_2 \end{bmatrix}$$

are not components of a tensor.

11. The stress tensor at a point has components given by

$$(s_{ij}) = \begin{bmatrix} 2 & -1 & 2 \\ -1 & 3 & 0 \\ 2 & 0 & -1 \end{bmatrix}$$

Find the stress vector across an area normal to the vector $\mathbf{A} = \mathbf{i}_1 - \mathbf{i}_2 + \mathbf{i}_3$. What are the normal and the shearing stresses across such an area?

12. Find the normal and shearing stresses across an area normal to $\mathbf{A} = 2\mathbf{i}_1 + \mathbf{i}_2 - 2\mathbf{i}_3$ at a point where the stress tensor is defined by

$$(s_{ij}) = \begin{bmatrix} 1 & 2 & 0 \\ 2 & -1 & 2 \\ 0 & 2 & 3 \end{bmatrix}$$

13. The stress tensor at a point on the $x_1 x_2$ -plane has components given by

$$(s_{ij}) = \begin{vmatrix} 2 & 3 \\ 3 & -2 \end{vmatrix}$$

Find the normal and the shearing stresses across a plane whose normal is $\mathbf{A} = \mathbf{i}_1 + 2\,\mathbf{i}_2$.

14. The components of a stress tensor are given by

$$(s_{ij}) = \begin{bmatrix} s_{11} & 0 & 0 \\ 0 & s_{22} & 0 \\ 0 & 0 & s_{33} \end{bmatrix}$$

What are the components of the tensor in the coordinate system x^*_i where the positive x^*_2-axis coincides with the positive x_3-axis, and the positive x^*_1-, x^*_3-axes lie on the x_1x_2 plane and make angles of $\pi/3$ and $-\pi/6$, respectively, with the positive x_1-axis.

15. Let a_1, a_2, a_3 be the components of a vector, and define

$$(t_{ij}) = \begin{bmatrix} 0 & -a_3 & a_2 \\ a_3 & 0 & -a_1 \\ -a_2 & a_1 & 0 \end{bmatrix}$$

Show that

$$(t^*_{ij}) = \begin{bmatrix} 0 & -a^*_3 & a^*_2 \\ a^*_3 & 0 & -a^*_1 \\ -a^*_2 & a^*_1 & 0 \end{bmatrix}$$

under any rotation of axes $x^*_i = \alpha_{ij} x_j$ such that $\det(\alpha_{ij}) = 1$.

16. Let $A = (a_{ij})$ be a matrix whose elements a_{ij} are components of a second order tensor, and let B be a column matrix whose elements b_i are components of a vector. Show that the elements of the matrix product AB are components of a vector.
17. Write down the transformation law and its inverse for a fourth order tensor.
18. Show that if the corresponding components of two tensors of order *r* are equal in one particular rectangular cartesian coordinate system, then they are equal in all rectangular cartesian coordinate systems.

5.7 ALGEBRA OF CARTESIAN TENSORS

Tensors possess much the same algebraic properties as vectors with respect to various operations that combine tensors to produce other tensors. These operations are defined as follows.

Definition 1. The sum of two tensors of order r is a tensor of the same order whose components consist of the sum of the corresponding components of the two tensors.

For example, if a_{ij} and b_{km} are the components of two second order tensors, then their sum is a tensor of second order whose components are given by $a_{ij} + b_{ij}$. It is easy to see that $c_{ij} = a_{ij} + b_{ij}$ transform according to the law

$$c^*_{ij} = \alpha_{ip}\,\alpha_{jq}\,c_{pq}$$

where $c^*_{ij} = a^*_{ij} + b^*_{ij}$. In fact, since a_{ij} and b_{ij} are components of tensors, we know that

$$a^*_{ij} = \alpha_{ip}\,\alpha_{jq}\,a_{pq}\,, \qquad b^*_{ij} = \alpha_{ip}\,\alpha_{jq}\,b_{pq}$$

Hence

$$c^*_{ij} = a^*_{ij} + b^*_{ij} = \alpha_{ip}\,\alpha_{jq}(a_{pq} + b_{pq}) = \alpha_{ip}\,\alpha_{jq}\,c_{pq}$$

Definition 2 (Outer Product.) The outer product (or simply product) of two tensors of order r and s is a tensor of order $r + s$ whose components consist of the products of the various components of the tensors.

For example, let a_{ij} and b_{kmn} be the components of a second order and a third order tensor. Then the outer product of these two tensors is a tensor of order five whose components consist of $c_{ijkmn} = a_{ij}\,b_{kmn}$ (i, j, k, m, n = 1, 2, 3). The tensorial character of this outer product follows from those of a_{ij} and b_{kmn}. In fact, since $a^*_{ij} = \alpha_{ip}\alpha_{jq}a_{pq}$ and $b^*_{kmn} = \alpha_{kr}\alpha_{ms}\alpha_{nt}b_{rst}$, we have

$$\begin{aligned} c^*_{ijkmn} &= a^*_{ij}\,b^*_{kmn} = \alpha_{ip}\alpha_{jq}\,a_{pq}\,\alpha_{kr}\alpha_{ms}\alpha_{nt}\,b_{rst} \\ &= \alpha_{ip}\alpha_{jq}\alpha_{kr}\alpha_{ms}\alpha_{nt}\,a_{pq}b_{rst} \\ &= \alpha_{ip}\alpha_{jq}\alpha_{kr}\alpha_{ms}\alpha_{nt}\,c_{pqrst} \end{aligned}$$

Definition 3 (Contraction.) The operation of setting two of the indices in the components of a tensor of order $r \geq 2$ equal and then summing with respect to the repeated index is called contraction.

For example, consider the components a_{ij} of a second order tensor. The contraction of this tensor yields the scalar

$$a_{ii} = a_{11} + a_{22} + a_{33}$$

which is just the trace of the matrix (a_{ij}). For a third order tensor with components a_{ijk}, there are three possible contractions, namely,

$$\begin{aligned} a_{iik} &= a_{11k} + a_{22k} + a_{33k} \quad (k = 1, 2, 3) \\ a_{iji} &= a_{1j1} + a_{2j2} + a_{3j3} \quad (j = 1, 2, 3) \\ a_{ijj} &= a_{i11} + a_{i22} + a_{i33} \quad (i = 1, 2, 3) \end{aligned}$$

It can be shown (see below) that each of these contractions results in a vector. In fact, the contraction of a tensor reduces the order of the tensor by 2. We state this as a theorem.

Theorem 1 (Reduction of order by contraction.) The contraction of a tensor of order $r \geq 2$ results in a tensor of order $r - 2$.

We illustrate the proof of Theorem 1 for a third order tensor a_{ijk}, where the contraction is with respect to the indices i and j. First we note that under a coordinate transformation, we have

$$a^*_{ijk} = \alpha_{ip}\,\alpha_{jq}\,\alpha_{kr}\,a_{pqr}$$

Setting i = j (summation is implied), we find

$$\begin{aligned} a^*_{iik} &= \alpha_{ip}\alpha_{iq}\alpha_{kr}\,a_{pqr} \\ &= \delta_{pq}\alpha_{kr}\,a_{pqr} \\ &= \alpha_{kr}\,a_{ppr} \end{aligned}$$

since $\alpha_{ip}\alpha_{iq} = \delta_{pq}$. Hence if we set $c^*_k = a^*_{iik}$ (sum on i) and $c_r = a_{ppr}$ (sum on p), then we have

$$c^*_k = \alpha_{kr}\,c_r$$

which shows the tensor character of c_r of order one (a vector).

It is evident that contraction of a tensor can be repeated with respect to the various indices until the tensor reduces to a vector or a scalar depending on whether the original order of the tensor is odd or even.

Definition 4 (Inner Product.) An inner product of two tensors is a contraction of the outer product with respect to two indices, each belonging to a component of the tensors.

For example, the components of the outer product of two vectors $\mathbf{A} = a_i \mathbf{i}_i$ and $\mathbf{B} = b_i \mathbf{i}_i$ are given by $c_{ij} = a_i b_j$. By contraction, we obtain the inner product

$$c_{ii} = a_i b_i = a_1 b_1 + a_2 b_2 + a_3 b_3$$

which is the dot or scalar product of the two vectors. On the other hand, for two second order tensors with components a_{ij} and b_{km}, there are four possible inner products, namely,

$$c_{ijim} = a_{ij} b_{im}, \quad c_{ijki} = a_{ij} b_{ki} \quad \text{(sum on i)}$$

$$c_{ijjm} = a_{ij}\, b_{jm}, \quad c_{ijkj} = a_{ij}\, b_{kj} \quad \text{(sum on j)}$$

By Theorem 1, each of these inner products gives rise to a second order tensor.

Quotient Rule

Thus far we have seen that the components of a tensor transform according to definite laws under rotation of the coordinate axes. Now suppose we are given a set of 3^r quantities

$$X_{i_1 i_2 \ldots i_r} \; (i_1, i_2, \ldots, i_r = 1, 2, 3) \tag{5.35}$$

How do we know that the quantities are components of a tensor without actually going to the trouble of checking how they transform under a change of coordinates? It turns out that there is a rule, known as the *quotient rule,* which enables us to determine whether or not the given quantities do form a tensor. We state the rule as a theorem and demonstrate its proof for $r = 2$.

Theorem 2. If the outer or inner product of the set of quantities (5.35) and an arbitrary tensor of any order yields a tensor of appropriate order, then the quantities are components of a tensor of order r.

Let the given quantities be X_{ij} and suppose that the outer product $X_{ij}\, a_k$ with an arbitrary vector a_k is a tensor of order 3. Set $b_{ijk} = X_{ij}\, a_k$ and let X^*_{ij} represent the quantities in a new coordinate system . Since b_{ijk} are components of a third order tensor, we have

$$\begin{aligned} b^*_{ijk} = X^*_{ij}\, a^*_k &= \alpha_{ip}\,\alpha_{jq}\,\alpha_{kr}\,b_{pqr} \\ &= \alpha_{ip}\,\alpha_{jq}\,\alpha_{kr}\,X_{pq} a_r \\ &= \alpha_{ip}\,\alpha_{jq}\,X_{pq}\,\alpha_{kr}\,a_r \\ &= \alpha_{ip}\,\alpha_{jq}\,X_{pq}\,a^*_k \end{aligned}$$

which implies

$$X^*_{ij} = \alpha_{ip}\,\alpha_{jq}\,X_{pq}$$

Thus the quantities X_{ij} are components of a second order tensor.

Example 1. Let the components of a second order tensor be given by

$$(t_{ij}) = \begin{bmatrix} 2 & -1 & 0 \\ 1 & 2 & 2 \\ 0 & -3 & 1 \end{bmatrix}$$

and let the components a_i of a vector be 1, -2, 2. Find the inner product $t_{ij}a_j$ and $a_i t_{ij}$.

Solution: Set $b_i = t_{ij}\, a_j$. Then we have

$$\begin{aligned} b_1 &= t_{1j}\, a_j = 2(1) + (-1)(-2) + 0\,(2) = 4 \\ b_2 &= t_{2j}\, a_j = 1(1) + 2(-2) + 2(2) = 1 \\ b_3 &= t_{3j}\, a_j = 0(1) + (-3)(-2) + 1(2) = 8 \end{aligned}$$

Next, let $c_j = a_i\, t_{ij}$. Then

$$\begin{aligned} c_1 &= a_i\, t_{i1} = 1(2) + (-2)(1) + 2(0) = 0 \\ c_2 &= a_i\, t_{i2} = 1(-1) + (-2)2 + 2(-3) = -11 \\ c_3 &= a_i\, t_{i3} = 1(0) + (-2)2 + 2(1) = -2 \end{aligned}$$

Example 2. Let s_{ij} denote the components of a stress tensor at a point. Then the stress vector $\mathbf{S}^*_n$ across an area with normal vector $\mathbf{n} = n_i\, \mathbf{i}_i$ is the inner product of the stress tensor and the vector **n**, that is, $s^*_{nj} = s_{ij}\, n_i = s_{ji}\, n_i$, since $s_{ij} = s_{ji}$.

Symmetric and Anti-Symmetric Tensors

Many tensors that arise in physical problems possess certain simplifying properties. For example, we have seen that the components s_{ij} of a stress tensor satisfy the condition $s_{ij} = s_{ji}$ for i, j = 1, 2, 3. We call such a tensor *symmetric tensor*. A second order tensor whose components satisfy the condition $a_{ij} = -a_{ji}$ is called an *anti-symmetric tensor.* It is clear that for such a tensor, $a_{ii} = 0$ (no sum on i) for i = 1, 2, 3. Thus a second order anti-symmetric tensor has the matrix representation

$$\begin{bmatrix} 0 & a_{12} & a_{13} \\ -a_{12} & 0 & a_{23} \\ -a_{13} & -a_{23} & 0 \end{bmatrix}$$

Such a matrix is called a skew-symmetric matrix.

The definition of symmetric or anti-symmetric tensor can be extended as well to tensors of higher order. For example, a third order tensor with components a_{ijk} is said to be *symmetric* or *anti-symmetric* with respect to the first two indices according as $a_{ijk} = a_{jik}$ or $a_{ijk} = -a_{jik}$. It is said to be symmetric or anti-symmetric with respect to the last two indices if $a_{ijk} = a_{ikj}$ or $a_{ijk} = -a_{ikj}$.

It is important to remember that if a tensor is symmetric or anti-symmetric in one coordinate system, it remains symmetric or anti-symmetric in all other coordinate systems. This can, of course, be verified through the transformation law governing the tensor. For example, suppose $a_{ij} = a_{ji}$ in a coordinate system x_i. Under a transformation of coordinates $x^*_i = \alpha_{ij} x_j$, we see that

$$a^*_{ij} = \alpha_{ip}\alpha_{jq} a_{pq} = \alpha_{ip}\alpha_{jq} a_{qp} = \alpha_{jq}\alpha_{ip} a_{qp} = a^*_{ji}$$

Thus the tensor is also symmetric in the x^*_i coordinate system.

We observe that every tensor of order $r \geq 2$ can be expressed as a sum of a symmetric and an anti-symmetric tensors of the same order. For example, for a third order tensor with components a_{ijk}, we set

$$s_{ijk} = (a_{ijk} + a_{jik})/2, \quad t_{ijk} = (a_{ijk} - a_{jik})/2$$

Then, clearly, $a_{ijk} = s_{ijk} + t_{ijk}$ and $s_{ijk} = s_{jik}$, $t_{ijk} = - t_{jik}$, so that s_{ijk} and t_{ijk} are components of a symmetric and an anti-symmetric tensors of third order, respectively.

5.5 EXERCISES

1. Let the components of a second order tensor be given by

$$(t_{ij}) = \begin{bmatrix} 3 & 1 & -2 \\ 1 & 0 & 3 \\ 2 & -1 & 4 \end{bmatrix}$$

and let $\mathbf{A} = 2\mathbf{i}_1 - 3\mathbf{i}_2 + 2\mathbf{i}_3$, $\mathbf{B} = -\mathbf{i}_1 + 2\mathbf{i}_2 - 2\mathbf{i}_3$. Find the inner product (a) $t_{ij}a_j$, (b) $t_{ij}a_i$, (c) $t_{ij}a_ib_j$, where a_i, b_j denote the components of **A** and **B**, respectively.

2. Given the components of second order tensors

$$(a_{ij}) = \begin{bmatrix} 2 & -1 & 0 \\ 1 & 3 & -2 \\ -2 & 1 & 4 \end{bmatrix}, \quad (b_{ij}) = \begin{bmatrix} 2 & 0 & 1 \\ 3 & -4 & 2 \\ 2 & 3 & 2 \end{bmatrix}$$

find the inner product (a) $a_{ij}b_{jk}$, (b) $a_{ik}b_{jk}$, (c) $a_{ij}b_{ik}$, (d) $a_{ij}b_{ij}$.

3. Let a_1, a_2, a_3 be given three numbers.

   (a) If a_ib_i is a scalar for an arbitrary vector b_i, show that a_i $(1 \le i \le 3)$ must be components of a vector.

   (b) If a_ib_j are components of a second order tensor for any vector b_i, show that a_i $(1 \le i \le 3)$ must be components of a vector.

4. Let X_{ijk} (i, j, k = 1, 2, 3) be given quantities having the property that for arbitrary vectors **A**, **B**, **C** with components a_i, b_i, c_i, the inner product $X_{ijk}\,a_ib_jc_k$ is a scalar. Show that X_{ijk} must be components of a third order tensor.

5. Let T denote a second order tensor whose components are given by

$$(a_{ij}) = \begin{bmatrix} 2 & 4 & -2 \\ 0 & 3 & 2 \\ 4 & 2 & 1 \end{bmatrix}$$

Find the respective components s_{ij} and t_{ij} of the symmetric and the anti-symmetric tensors whose sum is the tensor T.

6. Let a_{ij} (i, j = 1, 2, 3) denote the components of an anti-symmetric tensor. Show that for any numbers b_1, b_2, b_3, it is true that $a_{ij} b_i b_j = 0$.
7. Let a_{ij} and b_{ij} (i, j = 1, 2, 3) denote the components of a symmetric and an anti-symmetric tensors, respectively. Show that $a_{ij}b_{ij} = 0$.
8. Let a_{ijk} (i, j, k = 1, 2, 3) denote the components of a third order tensor. If the tensor is symmetric with respect to the indices j, k, show that it is symmetric with respect to the same indices in any other coordinate system.
9. Let X_{ij} (i, j =1, 2, 3) be given quantities. If $X_{ij}\ a_i a_j$ is always a scalar for components a_i of an arbitrary vector, show that $(X_{ij} + X_{ji})$ are components of a second order tensor. Hence, if $X_{ij} = X_{ji}$, deduce that X_{ij} are components of a symmetric tensor of order two.
10. Let X_{ijk} (i, j, k = 1, 2, 3) be given quantities. If the inner product $X_{ijk}b_{jk} = c_i$ $(1 \le i \le 3)$ forms components of a vector, where b_{ij} are components of an arbitrary symmetric tensor, show that $(X_{ijk} + X_{ikj})$ form components of a tensor. Hence, if $X_{ijk} = X_{ikj}$, deduce that X_{ijk} are components of a tensor.
11. Referring to Problem 10, if the quantities $X_{ijk}b_{jk} = c_i$ $(1 \le i \le 3)$ form components of a vector, where b_{ij} are components of an arbitrary anti-symmetric tensor, show that $(X_{ijk} - X_{ikj})$ are components of a tensor. Hence, if $X_{ijk} = - X_{ikj}$, deduce that X_{ijk} are components of a tensor.

5.8 PRINCIPAL AXES OF SECOND ORDER TENSORS

In Example 4 of Sec. 5.5, we saw that by introducing an appropriate coordinate system, the components I_{ij} $(i \ne j)$ of the moment of inertia tensor become zero, thus leaving only the moments of inertia I_{11}, I_{22}, I_{33} with respect to the coordinate axes. Likewise, in Example 3, Sec. 5.6, the shearing stress components s_{ij} $(i \ne j)$ vanish under an appropriate coordinate transformation so that the components of the tensor consisted only of normal stresses along the coordinate axes. The coordinate axes with respect to which the components of a second order tensor vanish are called the *principal axes* of the tensor. In this section, we study the problem of determining the principal axes of a given second order tensor.

Let t_{ij} (i, j = 1, 2, 3) denote the components of a second order tensor T, and let a_i (i = 1, 2, 3) denote the components of a vector **A**. We know that the inner product of T and **A** (with respect to the second index of t_{ij}) is a vector with components given by $b_i = t_{ij}a_i$ (i = 1, 2, 3). It turns out that the principal axes of the tensor T are determined by those vectors **A** whose inner product with the tensor are parallel to the vectors themselves, that is,

$$t_{ij}\, a_j = \lambda\, a_i \quad (i = 1, 2, 3) \tag{5.36}$$

where λ is a scalar. Such vectors (nonzero), if they exist, are called the eigenvectors of the tensor, and their directions are called the principal directions. The principal axes of the tensor are simply the axes determined by the principal directions. The values of the scalar λ that satisfy (5.36) are called the eigenvalues of the tensor. As it turns out, the eigenvalues are precisely the components t^*_{11}, t^*_{22}, t^*_{33} of the tensor with respect to the principal axes.

Therefore, to determine the principal axes of the tensor T, we must solve the system of equations (5.36) for the eigenvectors corresponding to each eigenvalue. Now the system of equations defined by (5.36) consists of

$$\begin{aligned}
(t_{11} - \lambda)a_1 + t_{12}\, a_2 + t_{13}\, a_3 &= 0 \\
t_{21}\, a_1 + (t_{22} - \lambda)a_2 + t_{23}\, a_3 &= 0 \\
t_{31}\, a_1 + t_{32}\, a_2 + (t_{33} - \lambda)a_3 &= 0
\end{aligned} \tag{5.37}$$

for the components a_1, a_2, a_3 of an eigenvector. Since we are interested in nonzero vectors, it follows from linear algebra that the determinant of the coefficients of the system (5.37) must vanish, that is,

$$\Delta(\lambda) = \begin{vmatrix} t_{11} - \lambda & t_{12} & t_{13} \\ t_{21} & t_{22} - \lambda & t_{23} \\ t_{31} & t_{32} & t_{33} - \lambda \end{vmatrix} = 0 \tag{5.38}$$

This determinant leads to a cubic equation in λ, known as the characteristic equation of the tensor. Thus the eigenvalues of the tensor are simply the zeros or roots of the characteristic equation. For each eigenvalue λ, the system (5.37) can be solved for the components a_1, a_2, a_3 of the corresponding eigenvector, which is non-zero. Clearly, if **A** is an eigenvector, so is c**A** for any constant c. Thus an eigenvector of a tensor is unique up to a constant factor. In linear algebra the problem of solving for λ and the quantities a_1, a_2, a_3 from the system (5.37) is known as an *eigenvalue problem*.

Eigenvalue Problems for Symmetric Tensors.

Since most second order tensors encountered in practice are symmetric, we shall confine our study of the eigenvalue problem (5.37) to symmetric tensors. For a symmetric tensor, we first show that the eigenvalues are all real. Indeed, suppose λ is an eigenvalue which might be complex, and let a_i $(1 \leq i \leq 3)$ denote the components of the corresponding eigenvector **A**. Then

$$t_{ij}\, a_j = \lambda\, a_i \quad (i = 1, 2, 3)$$

Multiplying this equation by the conjugate a^*_i of a_i and summing with respect to i, we obtain

$$a^*_i\, t_{ij}\, a_j = \lambda\, a^*_i\, a_i \tag{5.39}$$

If we take the complex conjugate of both sides of this equation, noting that the quantities t_{ij} are real, we find

$$a_i t_{ij}\, a^*_j = \lambda^*\, a_i\, a^*_i \tag{5.40}$$

Since the tensor is symmetric, we can show that the left-hand sides of (5.39) and (5.40) are identical. Indeed, since $t_{ij} = t_{ji}$, we have

$$a^*_i\, t_{ij}\, a_j = a^*_j\, t_{ji}\, a_i = a_i\, t_{ji}\, a^*_j = a_i\, t_{ij} a^*_j$$

Therefore,

$$\lambda\, a^*_i\, a_i = \lambda^*\, a_i\, a^*_i \quad \text{or} \quad (\lambda - \lambda^*)\, a_i\, a^*_i = 0$$

Since $a^*_i\, a_i > 0$, this implies $\lambda - \lambda^* = 0$, which means that λ is a real number.

Another important characteristic of a symmetric tensor is that its eigenvectors corresponding to distinct eigenvalues are orthogonal. This lends the principal axes of a symmetric tensor a natural axes for a new rectangular cartesian coordinate system. To prove the orthogonality property, let a_i and b_i denote the components of two eigenvectors **A** and **B** corresponding to the two distinct eigenvalues λ and μ , respectively. Then

$$t_{ij}\, a_j = \lambda\, a_i\,, \qquad t_{ij}\, b_j = \mu\, b_i$$

If we multiply the first equation by b_i, the second equation by a_i, and then sum with respect to i, we obtain

$$b_i\, t_{ij}\, a_j = \lambda\, a_i b_i\,, \qquad a_i\, t_{ij}\, b_j = \mu\, a_i b_i$$

Since $t_{ij} = t_{ji}$, by re-arranging the factors and changing indices, we find

$$\lambda\, a_i b_i = b_i t_{ij} a_j = a_j t_{ij} b_i = a_j t_{ji}\, b_i = a_i t_{ij}\, b_j = \mu\, a_i b_i$$

therefore,

$$(\lambda - \mu)\, a_i b_i = 0$$

Since $\lambda \neq \mu$, we conclude that $a_i b_i = \mathbf{A} \cdot \mathbf{B} = 0$, which says that the eigenvectors are orthogonal.

With symmetric tensors, even when the eigenvalues are not all distinct, we can still determine three eigenvectors that are mutually orthogonal. We shall not prove this fact here but shall simply illustrate it by an example (see Example 2).

Finally, we show that when we introduce a new coordinate system x^*_i with axes along the eigenvectors (ordered so as to be right-handed), the new components t^*_{ij} of the tensor satisfy the relation $t^*_{ij} = 0$ for $i \neq j$ with t^*_{11} , t^*_{22} , t^*_{33} assuming the values of the three eigenvalues λ_1 , λ_2 , λ_3. So let $n^{(i)}{}_k$ (k = 1, 2, 3) denote the components of the normalized eigenvector $\mathbf{i}^*_i = \mathbf{A}_i / |\mathbf{A}_i|$ (i = 1, 2, 3) that corresponds to the eigenvalue λ_i (i = 1, 2, 3), that is, $\mathbf{i}^*_i = n^{(i)}{}_k \mathbf{i}_k$. We assume that the normalized eigenvectors $\mathbf{i}^*_i$ are so labelled that they form a right-handed triple. Then the transformation from the x_i coordinate system to the new x^*_i-coordinate system with base vectors $\mathbf{i}^*_i$ is given by

$$x^*_i = \alpha_{ij}\, x_j\,, \quad \text{where } \alpha_{ij} = \mathbf{i}^*_i \cdot \mathbf{i}_j = n^{(i)}_j$$

Thus the row elements in the transformation matrix $\Lambda = (\alpha_{ij})$ are the components of the normalized eigenvectors. With respect to the new coordinate system x^*_i, we now show that the components of the tensor t_{ij} assume the following values:

$$t^*_{11} = \lambda_1\,, \quad t^*_{22} = \lambda_2\,, \quad t^*_{33} = \lambda_3\,, \quad t^*_{ij} = 0 \quad \text{for } i \neq j \tag{5.41}$$

By the transformation law, we have

$$t^*_{ij} = \alpha_{ip}\, \alpha_{jq}\, t_{pq} = n^{(i)}_p\, n^{(j)}_q\, t_{pq} \tag{5.42}$$

From (5.36) we have

$$t_{pq}\, n^{(j)}_q = \lambda_j\, n^{(j)}_p \quad \text{(no sum on j)}$$

so that (5.42) becomes

$$t^*_{ij} = n^{(i)}_p\, n^{(j)}_q\, t_{pq} = \lambda_j\, n^{(i)}_p\, n^{(j)}_p = \lambda_j\, \delta_{ij}$$

This yields the relations given in (5.41).

Example 1. Consider the symmetric tensor

$$(t_{ij}) = \begin{bmatrix} 1 & 1 & 0 \\ 1 & 2 & 1 \\ 0 & 1 & 1 \end{bmatrix}$$

The characteristic equation is given by

$$\begin{vmatrix} 1-\lambda & 1 & 0 \\ 1 & 2-\lambda & 1 \\ 0 & 1 & 1-\lambda \end{vmatrix} = \lambda(1-\lambda)(\lambda-3) = 0$$

and so the eigenvalues are $\lambda = 0, 1, 3$. For $\lambda = 0$, the system of equations (5.37) becomes

$$a_1 + a_2 = 0\,,\quad a_1 + 2a_2 + a_3 = 0\,,\quad a_2 + a_3 = 0$$

from which we find

$$a_1 = -a_2\,,\quad a_3 = -a_2$$

Choosing $a_2 = -1$, we obtain the eigenvector $\mathbf{A}_1 = \mathbf{i}_1 - \mathbf{i}_2 + \mathbf{i}_3$.

Next, when $\lambda = 1$, the system of equations (5.37) reduces to

$$a_2 = 0\,,\quad a_1 + a_2 + a_3 = 0$$

By choosing $a_1 = 1$, we obtain the corresponding eigenvector $\mathbf{A}_2 = \mathbf{i}_1 - \mathbf{i}_3$. Finally, when $\lambda = 3$, the system of equations becomes

$$-2a_1 + a_2 = 0,\quad a_1 - a_2 + a_3 = 0,\quad a_2 - 2a_3 = 0$$

from which we find

$$a_2 = 2a_1\,,\quad a_1 = a_3$$

Setting $a_1 = 1$, we obtain the corresponding eigenvector $\mathbf{A}_3 = \mathbf{i}_1 + 2\mathbf{i}_2 + \mathbf{i}_3$.

It is easily verified that the three eigenvectors are mutually orthogonal. Moreover, since the scalar triple product ($\mathbf{A}_1\,\mathbf{A}_2\,\mathbf{A}_3$) is positive, it follows that the eigenvectors form a right-handed triple. (Note that in case the eigenvectors do not form a right-handed triple, we need only change the order of any two of them and re-label the vectors accordingly.) Thus the principal axes of the tensor are along the direction of the eigenvectors. The transformation from the rectangular cartesian coordinate x_i to the new coordinate system x^*_i determined by the principal axes is defined by $x^*_i = \alpha_{ij}\,x_j$, where $\alpha_{ij} = \mathbf{i}^*_i \cdot \mathbf{i}_j$. Thus, taking the dot products of the base vectors, the transformation is represented by the matrix

$$(\alpha_{ij}) = \begin{bmatrix} 1/\sqrt{3} & -1/\sqrt{3} & 1/\sqrt{3} \\ 1/\sqrt{2} & 0 & -1/\sqrt{3} \\ 1/\sqrt{6} & 2/\sqrt{6} & 1/\sqrt{6} \end{bmatrix}$$

Now, with respect to the principal axes, the components of the given tensor are represented by the matrix

$$(t^*_{ij}) = \begin{bmatrix} 0 & 0 & 0 \\ 0 & 1 & 0 \\ 0 & 0 & 3 \end{bmatrix}$$

It is instructive for the student to verify the elements t^*_{ij} either from the transformation law $t^*_{ij} = \alpha_{ip}\alpha_{jq}t_{pq}$ or from the matrix product

$$(t^*_{ij}) = (\alpha_{ip})\,(t_{pq})\,(\alpha_{jq})^T .$$

Example 2. Consider the symmetric tensor

$$(t_{ij}) = \begin{bmatrix} 3 & 0 & 0 \\ 0 & 4 & \sqrt{3} \\ 0 & \sqrt{3} & 6 \end{bmatrix}$$

Its characteristic equation is given by

$$\begin{vmatrix} 3-\lambda & 0 & 0 \\ 0 & 4-\lambda & \sqrt{3} \\ 0 & \sqrt{3} & 6-\lambda \end{vmatrix} = (3-\lambda)(\lambda-7)(\lambda-3) = 0$$

and so the eigenvalues are $\lambda = 3, 3, 7$ with $\lambda = 3$ being a root of multiplicity 2. Corrresponding to $\lambda = 3$, the system of equations (5.37) gives

$$a_2 + \sqrt{3}\,a_3 = 0, \qquad \sqrt{3}\,a_2 + 3a_3 = 0$$

which are identical. Hence $a_2 = -\sqrt{3}\,a_3$, so that a_1 and a_3 are arbitrary constants. By choosing these constants judiciously, we can obtain two orthogonal eigenvectors corresponding to the eigenvalue $\lambda = 3$. In fact, if we choose $a_1 = 1$, $a_2 = 0$, $a_3 = 0$, we obtain $\mathbf{A}_1 = \mathbf{i}_1$. By choosing $a_1 = 0$, $a_2 = \sqrt{3}$, $a_3 = -1$, we obtain a second eigenvector $\mathbf{A}_2 = \sqrt{3}\,\mathbf{i}_2 - \mathbf{i}_3$, which is orthogonal to $\mathbf{A}_1$.

Finally, when $\lambda = 7$, we solve the system of equations

$$-4a_1 = 0, \quad -3a_2 + \sqrt{3}a_3 = 0, \quad \sqrt{3}a_2 - a_3 = 0$$

for which a solution is given by $a_1 = 0$, $a_2 = 1$, $a_3 = \sqrt{3}$. Thus $\mathbf{A}_3 = \mathbf{i}_2 + \sqrt{3}\mathbf{i}_3$ is an eigenvector corresponding to $\lambda = 7$. Clearly this vector is orthogonal to both $\mathbf{A}_1$ and $\mathbf{A}_2$. These three eigenvectors are the principal axes of the given tensor. With respect to these principal axes, the tensor assumes the values

$$t^*_{11} = t^*_{22} = 3, \quad t_{33} = 7, \quad t^*_{ij} = 0 \quad \text{for } i \neq j$$

These values can be readily verified from the transformation law $t^*_{ij} = \alpha_{ip}\,\alpha_{jq}\,t_{pq}$, where $\alpha_{ij} = \mathbf{i}^*_i \cdot \mathbf{i}_j$ and $\mathbf{i}^*_i = \mathbf{A}_i / |\mathbf{A}_i|$. In fact, we find

$$(\alpha_{ij}) = \begin{bmatrix} 1 & 0 & 0 \\ 0 & \sqrt{3}/2 & -1/2 \\ 0 & 1/2 & \sqrt{3}/2 \end{bmatrix}$$

5.6 EXERCISES

1. The components of a moment of inertia tensor are given by

$$(I_{ij}) = \begin{bmatrix} 1 & 1 & 1 \\ 1 & 2 & 0 \\ 1 & 0 & 2 \end{bmatrix}$$

(a) Find the eigenvalues and the principal axes of the tensor.
(b) Express the tensor with respect to its principal axes.
(c) Write the equation of the transformation relating the coordinates x_i with that determined by the principal axes.
(d) Verify the results in (b) by using the transformation law for second order tensor.

2. Repeat Problem 1 for the moment of inertia tensor

$$(I_{ij}) = \begin{bmatrix} 2 & 0 & 0 \\ 0 & 3 & 2 \\ 0 & 2 & 3 \end{bmatrix}$$

3. The components of a stress tensor are given by

$$(s_{ij}) = \begin{bmatrix} 0 & 2 & 1 \\ 2 & 0 & 1 \\ 1 & 1 & 3 \end{bmatrix}$$

Determine the coordinate system x^*_i in which the shearing stresses vanish. What are the components of the stress tensor in this new coordinate system?

4. Repeat Problem 3 for the stress tensor

$$(s_{ij}) = \begin{bmatrix} 1 & -4 & 2 \\ -4 & 1 & -2 \\ 2 & -2 & -2 \end{bmatrix}$$

5. Find the normal stresses of the stress tensor

$$(s_{ij}) = \begin{bmatrix} 0 & 0 & a \\ 0 & 0 & b \\ a & b & c \end{bmatrix}$$

with respect to its principal axes.

6. The coefficients a_{ij} in the equation $a_{ij}x_ix_j = 1$ ($a_{ij} = a_{ji}$) of a quadric surface form the components of a second order symmetric tensor (see Problem 1, Exercise 5.4). Show that with respect to the principal axes of the tensor, the equation of the quadric surface becomes

$$\lambda_1 x_1^{*2} + \lambda_2 x_2^{*2} + \lambda_3 x_3^{*2} = 1$$

(In this case, the principal axes are also called the principal axes of the quadric surface.)

7. Express the equation

$$2\,x_1^2 + 2\,x_2^2 + 2\,x_3^2 - 2\,x_1x_2 - 2\,x_2x_3 + 2\,x_1x_3 = 1$$

of a quadric surface with respect to its principal axes. Can you determine what type of a surface it represents?

8. A quadric surface is defined by the equation

$$x_1^2 + 5\,x_2^2 + 3\,x_3^2 + 8\,x_2x_3 - 8\,x_1x_3 = 1$$

Find the equation of the surface with respect to its principal axes. (The surface is an elliptic hyperboloid of one sheet.)

9. Show that the quadric surface defined by

$$x_1^2 + x_2^2 + 5\,x_3^2 + 2\,x_1x_3 + 4\,x_2x_3 = 1$$

is an elliptic cylinder.

5.9 DIFFERENTIATION OF CARTESIAN TENSOR FIELDS

We recall that a scalar or a vector field consists of points of a given domain together with the corresponding scalar or vector at each of those points. Similarly, if at each point of a domain there is defined a tensor, then we have a tensor field. For example, an elastic body that is under the influence of an external force gives rise to a second order tensor field since at each of its interior points there is a stress tensor. Thus a tensor field, say of order two, is characterized by a set of tensor components $a_{ij}(x_1, x_2, x_3)$ or briefly $a_{ij}(x_k)$, which depend on the coordinates x_i $(1 \le i \le 3)$. Sometimes the components of a tensor field may also depend on a parameter that is independent of the space coordinates x_i. For example, the components may depend on a parameter *t* which represents time, thus leading to what is called an unsteady tensor field.

We observe that a tensor field of zero order is simply a scalar field. Now if $\phi(x_k)$ denotes a scalar field, then under a transformation of coordinates $x^*_i = \alpha_{ij}\, x_j$ (or $x_j = \alpha_{ij}\, x^*_i$), we have

$$\phi(x_j) = \phi(\alpha_{ij}\, x^*_i) = \phi^*(x^*_i) \tag{5.43}$$

On the other hand, a tensor field of order one is simply a vector field. If $a_i(x_p)$ denote the components of a vector field, then under a coordinate transformation, we have

$$\alpha_{ip}\, a_p(x_q) = \alpha_{ip}\, a_p(\alpha_{jq}\, x^*_j) = a^*_i\,(x^*_j) \tag{5.44}$$

Transformation laws for tensors of higher order can be written in a similar manner.

It is evident that all the algebraic operations we have discussed before pertaining to (constant) tensors also hold for tensor fields.

Tensor Gradients

Henceforth, we shall assume that the components of a tensor field can be differentiated as often as needed and that the derivatives are continuous in the domain of definition. Let us first examine how the partial derivatives $\partial\phi(x_i)/\partial x_j$ of a scalar field transform under a change of coordinates. From (5.43) and by the chain rule, we find

$$\frac{\partial \phi^*}{\partial x^*_i} = \frac{\partial \phi}{\partial x_j}\frac{\partial x_j}{\partial x^*_i} = \alpha_{ij}\frac{\partial \phi}{\partial x_j} \tag{5.45}$$

since $\partial x_j/\partial x^*_i = \alpha_{ij}$, in view of the inverse transformation $x_j = \alpha_{ij}x^*_i$. Hence, if we set $a^*_i = \partial\phi^*(x^*_j)/\partial x^*_i$ and $a_i = \partial\phi(x_j)/\partial x_i$, (5.45) becomes

$$a^*_i = \alpha_{ij}\, a_j \tag{5.46}$$

This shows that the partial derivatives $\partial\phi(x_j)/\partial x_i$ of a scalar field transform as components of a vector. Indeed, this vector field is grad $\phi(x_j)$, which is the gradient of $\phi(x_j)$. Thus grad $\phi(x_j)$ is a tensor field of order one. Further, since $\mathbf{i}_j = \alpha_{ij}\,\mathbf{i}^*_i$, it follows from (5.45) that

$$\frac{\partial \phi^*}{\partial x^*_i}\mathbf{i}^*_i = \alpha_{ij}\frac{\partial \phi}{\partial x_j}\mathbf{i}^*_i = \frac{\partial \phi}{\partial x_j}\mathbf{i}_j \tag{5.47}$$

This shows that the representation of grad $\phi(x_j)$ is invariant with respect to transformation of coordinates.

Next, we consider the derivatives of a vector field and examine how they transform under a change of coordinates. The components of a vector field transform according to (5.44). Hence, by the chain rule, we find

$$\frac{\partial a^*_i}{\partial x^*_j} = \alpha_{ip}\frac{\partial a_p}{\partial x_q}\frac{\partial x_q}{\partial x^*_j} = \alpha_{ip}\alpha_{jq}\frac{\partial a_p}{\partial x_q} \tag{5.48}$$

Setting $t^*_{ij} = \partial a^*_i/\partial x^*_j$ and $t_{pq} = \partial a_p/\partial x_q$, we conclude from (5.48) that the partial derivatives t_{ij} form components of a second order tensor field. Similarly, if $a_{ij}(x_k)$ denote the components of a second order tensor field, the partial derivatives $\partial a_{ij}/\partial x_k = t_{ijk}$ become the components of a third order tensor field. In general, it can be shown by induction that the partial derivatives of a tensor field of order r results in a tensor field of order $r + 1$. We call the partial derivatives of a tensor field of arbitrary order the *tensor gradient*. Thus the tensor gradient of a vector field is a second order tensor. We state this result as a theorem.

Theorem 1. The tensor gradient of a cartesian tensor field of order r is a tensor field of order $r + 1$.

Caution: It should be noted that Theorem 1 is true only for cartesian tensor fields because for such tensors all the admissible coordinate transformations are of the form $x^*_i = \alpha_{ij} x_j$, where the coefficients matrix (α_{ij}) is an orthogonal constant matrix. The transformation equations are linear and they represent rotations of the coordinate axes.

It is clear from Theorem 1 that differentiation of a tensor field may be repeated as often as permissible, and it will always yield a tensor field of one order higher.

Divergence of a Tensor Field

Now consider the components of the tensor gradient of a vector field $\partial a_i / \partial x_j$. By contraction we obtain

$$\frac{\partial a_i}{\partial x_i} = \frac{\partial a_1}{\partial x_1} + \frac{\partial a_2}{\partial x_2} + \frac{\partial a_3}{\partial x_3}$$

which is precisely the divergence of the vector field. Further, from the transformation formula (5.48), we find

$$\frac{\partial a^*_i}{\partial x^*_i} = \alpha_{ip} \alpha_{iq} \frac{\partial a_p}{\partial x_q} = \delta_{pq} \frac{\partial a_p}{\partial x_q} = \frac{\partial a_p}{\partial x_p} \tag{5.49}$$

Equation (5.49) shows that the divergence of a vector field is invariant under transformation of coordinates or, equivalently, that the divergence of a vector field is a scalar.

Likewise, contraction with respect to the indices j, k in the tensor gradient of a second order tensor $\partial a_{ij} / \partial x_k$ leads to the components $b_i = \partial a_{ij} / \partial x_j$ of a vector field. In fact, from the transformation law

$$\frac{\partial a^*_{ij}}{\partial x^*_k} = \alpha_{ip} \alpha_{jq} \alpha_{kr} \frac{\partial a_{pq}}{\partial x_r}$$

we obtain

$$b^*_i = \frac{\partial a^*_{ij}}{\partial x^*_j} = \alpha_{ip} \delta_{qr} \frac{\partial a_{pq}}{\partial x_r} = \alpha_{ip} \frac{\partial a_{pq}}{\partial x_q} = \alpha_{ip} b_p$$

which shows that the quantities b_i indeed form components of a vector field. Similarly, the contraction with respect to the indices i, k yields components of a vector field $b_j = \partial a_{ij}/\partial x_i$. The quantities $\partial a_{ij}/\partial x_i$ and $\partial a_{ij}/\partial x_j$ are called the components of the divergence of the tensor field a_{ij} with respect to the indices i and j, respectively. In the general case, we define the divergence of a tensor field with respect to any one of its indices as the contraction of the corresponding tensor gradient with respect to that index. It follows from Theorem 1 above and Theorem 1 of Sec. 5.7 that the divergence of a tensor field of order r yields a tensor field of order $r - 1$.

Example 1. Let $a_1 = x_2x_3$, $a_2 = x_1^2x_3$, $a_3 = x_2x_3^2$ be the components of a vector field in a domain D. Then the components of the gradient of this vector field are given by

$$\left(\frac{\partial a_i}{\partial x_j}\right) = \begin{bmatrix} \partial a_1/\partial x_1 & \partial a_1/\partial x_2 & \partial a_1/\partial x_3 \\ \partial a_2/\partial x_1 & \partial a_2/\partial x_2 & \partial a_2/\partial x_3 \\ \partial a_3/\partial x_1 & \partial a_3/\partial x_2 & \partial a_3/\partial x_3 \end{bmatrix}$$

$$= \begin{bmatrix} 0 & x_3 & x_2 \\ 2x_1x_3 & 0 & x_1^2 \\ 0 & x_3^2 & 2x_2x_3 \end{bmatrix}$$

The divergence is given by

$$\frac{\partial a_i}{\partial x_i} = 2x_2x_3$$

Example 2. Let the components of a second order tensor field be given by

$$(a_{ij}) = \begin{bmatrix} 0 & x_1x_2 & x_3^2 \\ x_1^2 & x_2x_3 & 0 \\ x_3x_1 & 0 & x_3x_2 \end{bmatrix}$$

Then the components of the divergence with respect to the index i are given by the formula

$$b_j = \frac{\partial a_{ij}}{\partial x_i} = \frac{\partial a_{1j}}{\partial x_1} + \frac{\partial a_{2j}}{\partial x_2} + \frac{\partial a_{3j}}{\partial x_3} \qquad (j = 1, 2, 3)$$

which yields

$$b_1 = x_1, \quad b_2 = x_2 + x_3, \quad b_3 = x_2$$

These are components of a vector field.

The divergence with respect to the second index j is given by

$$b_i = \frac{\partial a_{ij}}{\partial x_j} = \frac{\partial a_{i1}}{\partial x_1} + \frac{\partial a_{i2}}{\partial x_2} + \frac{\partial a_{i3}}{\partial x_3} \qquad (i = 1, 2, 3)$$

which gives

$$b_1 = x_1 + 2x_3, \quad b_2 = 2x_1 + x_3, \quad b_3 = x_1 + x_2$$

Notice that the two divergences are not the same.

5.10 STRAIN TENSOR

As a physical example of an application of the discussion in the preceding section, we present here the concept of strain which arises in the mechanics of continuous media. When a solid body is acted on by external forces which are in equilibrium, its configuration exhibits certain deformation. The deformation is usually described in terms of the change in the relative distance between two infinitely close points before and after deformation. This gives rise to what is called a strain tensor. To describe this, suppose P and Q are two points which are very close to each other in a body subjected to external forces in equilibrium. Let the position vectors of these points with respect to a rectangular cartesian coordinate system be denoted by $\mathbf{R}$ and $\mathbf{R} + \Delta\mathbf{R}$ (Fig. 5.7) before deformation. Suppose that after deformation the points P and Q have moved to points P' and Q', respectively. Let $\mathbf{U}(\mathbf{R})$ denote the displacement of P to P' and $\mathbf{U}(\mathbf{R} + \Delta\mathbf{R})$ the displacement of Q to Q'. Notice that the displacement vector $\mathbf{U}$ depends on the position of the point since different points in the body will in general experience different displacements. Then the position vectors of P' and Q' are given by

$$OP' = \mathbf{R} + \mathbf{U}(\mathbf{R}) \quad \text{and} \quad OQ' = (\mathbf{R} + \Delta\mathbf{R}) + \mathbf{U}(\mathbf{R} + \Delta\mathbf{R})$$

so that the relative position vector between P' and Q' is

$$\begin{aligned}\Delta\mathbf{R}' &= OQ' - OP' = (\mathbf{R} + \Delta\mathbf{R}) + \mathbf{U}(\mathbf{R} + \Delta\mathbf{R}) - (\mathbf{R} + \mathbf{U}(\mathbf{R})) \\ &= \Delta\mathbf{R} + \mathbf{U}(\mathbf{R} + \Delta\mathbf{R}) - \mathbf{U}(\mathbf{R})\end{aligned}$$

Thus the change in the relative position vectors between P, Q and P', Q' is represented by

$$\Delta\mathbf{U} = \Delta\mathbf{R}' - \Delta\mathbf{R} = \mathbf{U}(\mathbf{R} + \Delta\mathbf{R}) - \mathbf{U}(\mathbf{R}) \tag{5.52}$$

Now let $\mathbf{R} = x_i\mathbf{i}_i$, $\Delta\mathbf{R} = \Delta x_i\, \mathbf{i}_i$, $\Delta\mathbf{R}' = \Delta x'_i\, \mathbf{i}_i$, and $\mathbf{U}(\mathbf{R}) = u_i(x_j)\mathbf{i}_i$, where u_i are continuously differentiable. Let us consider the i-th component of (5.52), $\Delta u_i = \Delta x'_i - \Delta x_i = u_i(x_j + \Delta x_j) - u_i(x_j)$. Since the points P, Q are sufficiently close, the quantities Δx_i are very small. Hence, by the mean value theorem of differential calculus, we can

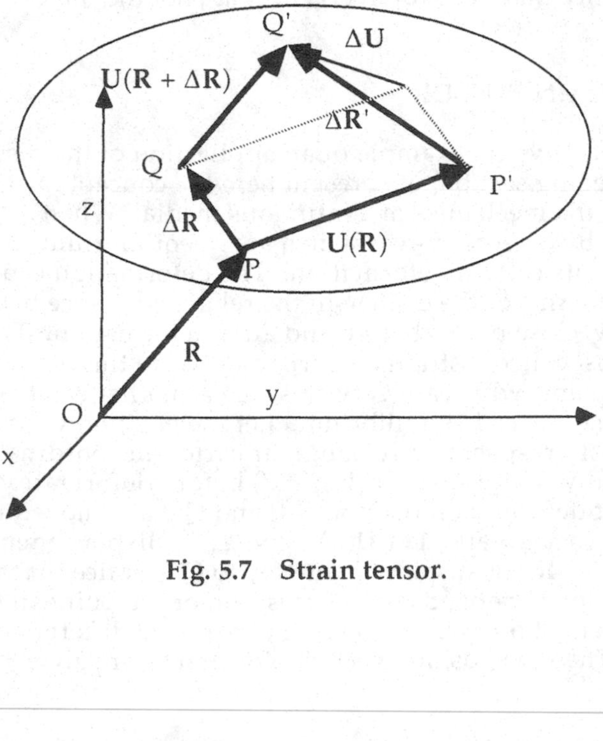

Fig. 5.7 Strain tensor.

write

$$\Delta u_i = u_i(x_j + \Delta x_j) - u_i(x_j) = \frac{\partial u_i}{\partial x_j}\Delta x_j \quad \text{(sum on j)} \tag{5.53}$$

Thus the components of the change in the relative position vectors of the points before and after deformation are linear combinations of the Δx_j ($j = 1, 2, 3$). In differential form, this can be written as

$$d\mathbf{U} = \begin{bmatrix} du_1 \\ du_2 \\ du_3 \end{bmatrix} = \begin{bmatrix} \frac{\partial u_1}{\partial x_1} & \frac{\partial u_1}{\partial x_2} & \frac{\partial u_1}{\partial x_3} \\ \frac{\partial u_2}{\partial x_1} & \frac{\partial u_2}{\partial x_2} & \frac{\partial u_2}{\partial x_3} \\ \frac{\partial u_3}{\partial x_1} & \frac{\partial u_3}{\partial x_2} & \frac{\partial u_3}{\partial x_3} \end{bmatrix} \begin{bmatrix} dx_1 \\ dx_2 \\ dx_3 \end{bmatrix} = (T_{ij})\, d\mathbf{R}$$

The quantities

$$T_{ij} = \frac{\partial u_i}{\partial x_j} \qquad (i, j = 1, 2, 3) \tag{5.54}$$

are components of the tensor gradient of the displacement vector **U**. By Theorem 1, this tensor gradient is a second order tensor field and it is called the *displacement tensor*. We shall see below that the symmetric part of the displacement tensor corresponds to the *strain tensor* .

We begin by noting that the strain at the point P in the direction of $\Delta\mathbf{R}$ is defined as the ratio

$$S = \frac{|\Delta\mathbf{R}'| - |\Delta\mathbf{R}|}{|\Delta\mathbf{R}|} \tag{5.55}$$

If we assmue that $|\Delta\mathbf{U}|$ is much smaller than either $|\Delta\mathbf{R}|$ or $|\Delta\mathbf{R}'|$, then the change in the distance $|\Delta\mathbf{R}'| - |\Delta\mathbf{R}|$ may be approximated by the projection of $\Delta\mathbf{U}$ on $\Delta\mathbf{R}$ (see Fig. 5.8), that is

$$|\Delta\mathbf{R}'| - |\Delta\mathbf{R}| = \Delta\mathbf{U}\cdot\frac{\Delta\mathbf{R}}{|\Delta\mathbf{R}|}$$

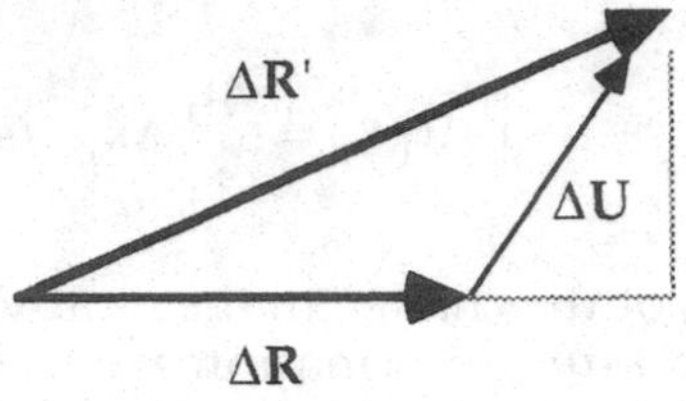

Fig. 5.8 Projection of ΔU on ΔR.

Hence (5.55) can be written as

$$S = \frac{\Delta \mathbf{U} \cdot \Delta \mathbf{R}}{|\Delta \mathbf{R}|^2} = \frac{1}{|\Delta \mathbf{R}|^2} \Delta u_i \Delta x_i$$

In view of (5.53), this becomes

$$S = \frac{1}{|\Delta \mathbf{R}|^2} \Delta u_i \Delta x_i = \frac{1}{|\Delta \mathbf{R}|^2} \frac{\partial u_i}{\partial x_j} \Delta x_i \Delta x_j$$

$$= \frac{\partial u_i}{\partial x_j} n_i n_j \tag{5.56}$$

where $n_i = \Delta x_i / |\Delta \mathbf{R}|$ $(i = 1, 2, 3)$ are the direction cosines of **ΔR**. Thus the strain is a quadratic function of the direction cosines of **ΔR** with coefficients which are the components of the displacement tensor (5.54).

In matrix form, the strain (5.56) can be written as

$$S = [n_1, n_2, n_3] \begin{bmatrix} S_{11} & S_{12} & S_{13} \\ S_{21} & S_{22} & S_{23} \\ S_{31} & S_{32} & S_{33} \end{bmatrix} \begin{bmatrix} n_1 \\ n_2 \\ n_3 \end{bmatrix}$$

where

$$S_{ij} = \frac{1}{2}\left(\frac{\partial u_i}{\partial x_j} + \frac{\partial u_j}{\partial x_i}\right) = S_{ji} \quad (i, j = 1, 2, 3) \tag{5.57}$$

The quantities S_{ij} are the components of a second order tensor field called the *strain tensor*. It is clear that (S_{ij}) is just the symmetric part of the displacement tensor when (5.54) is written as

$$(T_{ij}) = (S_{ij}) + (K_{ij})$$

where the components of the skew-symmetric part (K_{ij}) are given by

$$K_{ij} = \frac{1}{2}\left(\frac{\partial u_i}{\partial x_j} - \frac{\partial u_j}{\partial x_i}\right) \quad (i, j = 1, 2, 3) \tag{5.58}$$

The skew-symmetric tensor field (K_{ij}) contributes to the deformation of the body in a different way. To see what kind of deformation it causes, let us associate a vector $\mathbf{w} = \omega_i \mathbf{i}_i$ with the three independent components K_{12}, K_{23}, K_{31} of the skew-symmetric tensor by defining

$$\omega_k = \frac{1}{2}\left(\frac{\partial u_j}{\partial x_i} - \frac{\partial u_i}{\partial x_j}\right) = K_{ji} \tag{5.59}$$

where i, j, k are cyclic permutations of 1, 2, 3. Then, from (5.51), we have

$$\Delta u_i = S_{ij}\,\Delta x_j + K_{ij}\,\Delta x_j = S_{ij}\,\Delta x_j - \omega_k\,\Delta x_j \tag{5.60}$$

Now the term $-\omega_k\,\Delta x_j$ in (5.60) is precisely the i-th component of the cross product $\mathbf{w} \times \Delta\mathbf{R}$. This product represents the displacement due to an infinitesimal rotation represented by $\mathbf{w}$ (see Fig. 5.9). We therefore conclude that the term $K_{ij}\,\Delta x_j$ in (5.60) corresponds to a deformation due to an infinitesimal rigid-body rotation about an axis passing through P and parallel to $\mathbf{w}$.

From the foregoing discussion, it follows that if the displacement tensor is zero, that is, $\partial u_i/\partial x_j = 0$ for i, j = 1, 2, 3, then (5.52) implies $\Delta\mathbf{R} = \Delta\mathbf{R}'$, which means that the deformation consists only of translation. Geometrically, this means that the configuration PP'Q'Q in Fig. 5.7 is a parallelogram. On the other hand, if the strain tensor (S_{ij}) is zero, then from (5.60) we obtain

$$\Delta x'_i = = \Delta x_i - \omega_k\,\Delta x_j$$

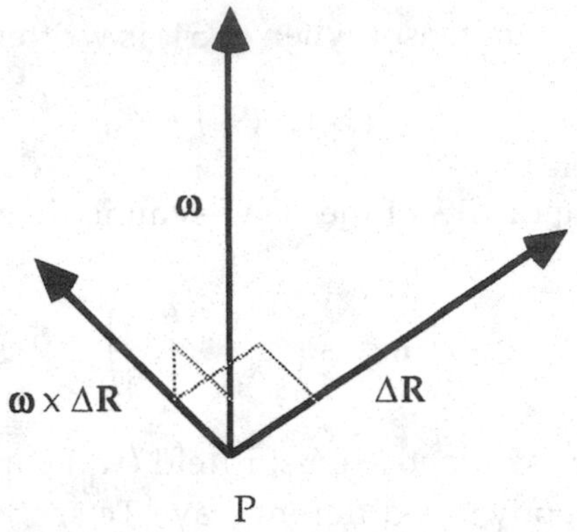

Fig. 5.9 Deformation due to an infinitesimal rotation.

which shows that the deformation consists of a translation and a rotation.

It is interesting to note that by definition (5.59), we have

$$\mathbf{w} = \omega_i \, \mathbf{i}_i = (1/2)\, \mathbf{curl\ U}$$

where **U** is the displacement vector of the point P. Hence, if **curl U** = **0**, then there is no rotation and so the deformation is purely due to translation and strain.

5.7 EXERCISES

1. Verify that the quantities

$$(a_{ij}) = \begin{bmatrix} x_2^2 & -x_1x_2 \\ -x_1x_2 & x_1^2 \end{bmatrix}$$

are components of a second order plane tensor.

2. Show that the quantities

$$(a_{ij}) = \begin{bmatrix} x_1x_2 & x_2^2 \\ x_1^2 & -x_1x_2 \end{bmatrix}$$

are not components of a tensor.

3. Show that the curl of a vector field is a pseudotensor. (Hint: Let a_i denote the components of a vector field **F** and write **curl F** = $P_i\mathbf{i}_i$, where $P_i = \partial a_k/\partial x_j - \partial a_j/\partial x_k$, and i, j, k are cyclic permutations of 1, 2, 3 . Show that

$$\frac{\partial a^*_k}{\partial x_j} - \frac{\partial a^*_j}{\partial x_k} = \alpha_{jr}\alpha_{ks}\left(\frac{\partial a_s}{\partial x_r} - \frac{\partial a_r}{\partial x_s}\right)$$

r, s = 1, 2, 3 and then proceed as in Problem 4, Exercise 5.3.)

4. Find the tensor gradient of each of the following:

(a) $\mathbf{F}(x_1, x_2, x_3) = x_1 \sin x_3\, \mathbf{i}_1 - x_2 \exp(x_1)\, \mathbf{i}_2 + x_2 x_3\, \mathbf{i}_3$

(b) $\mathbf{F}(x_1, x_2, x_3) = (x_1 - x_2^2)\mathbf{i}_1 + x_3 \ln(1 + x_1^2)\, \mathbf{i}_2 - x_1^2 \cos x_3\, \mathbf{i}_3$

5. Find all the divergences of each of the following:

(a) $$(s_{ij}) = \begin{bmatrix} x_1^2 - x_2^2 & x_2x_3 & x_1x_3 \\ x_2x_1 & x_2^2 - x_3^2 & x_2x_3 \\ x_1x_3 & x_2x_3 & x_3^2 - x_1^2 \end{bmatrix}$$

(b) $$(t_{ij}) = \begin{bmatrix} x_2^2 x_3 & x_1 - x_2 & x_1x_2 \\ x_2 + x_3 & x_3^2 x_1 & x_1^2 \\ x_2^2 & x_2 - x_3 & x_1^2 x_2 \end{bmatrix}$$

6. Let a_{ij} denote the components of a second order tensor field. Prove that $\partial a_{ij}/\partial x_k$ are components of a third order tensor field.

7. Let ϕ be a scalar field and a_i components of a vector field. Prove that $\partial(\phi a_i)/\partial x_j$ are components of a second order tensor field.
8. Find the symmetric and the skew-symmetric part of the tensor gradient of

$$\mathbf{F}(x_1, x_2, x_3) = (x_1 x_2 - x_3^2)\, \mathbf{i}_1 + x_2 \exp(x_3)\, \mathbf{i}_2 + \sin(x_1 x_3)\, \mathbf{i}_3$$

9. Let a_{ijk} denote the components of a third order tensor field. Write the symmetric and the skew-symmetric part of the divergence of the tensor with respect to the last index.
10. Let ω_k be as defined in (5.59). Show that $\partial\omega_k/\partial x_k = 0$ (summation with respect to k).
11. The most general linear relationship between the stress tensor (s_{ij}) and the strain tensor (S_{ij}) is given by

$$s_{ij} = \lambda_{ijkm} S_{km}$$

where λ_{ijkm} are components of a fourth order tensor called the elastic tensor. From the symmetric property of the stress and strain tensors, show that $\lambda_{ijkm} = \lambda_{jikm}$ and $\lambda_{ijkm} = \lambda_{ijmk}$. If, in addition, $\lambda_{ijkm} = \lambda_{kmij}$, show that the elastic tensor has 21 components.

6

TENSORS IN GENERAL COORDINATES

In this chapter we shall discuss tensors in general curvilinear coordinate systems. Such tensors are called general tensors or simply tensors. Obviously, cartesian tensors are special cases of general tensors since cartesian coordinate systems are particular types of coordinate systems. As a preview of what prevails in the theory of general tensors and to make the transition from cartesian to general tensors somewhat smooth, we shall first discuss tensors in oblique cartesian coordinate systems. Oblique cartesian coordinate systems are systems in which the coordinate axes are straight lines that are not necessarily perpendicular to each other.

6.1 OBLIQUE CARTESIAN COORDINATES

Suppose we change from a rectangular cartesian basis defined by the unit vectors $\mathbf{i}_i$ to a new basis defined by the vectors $\mathbf{e}_i$ ($1 \le i \le 3$), where the new basis is not necessarily orthonormal. Then we can construct an oblique cartesian coordinate system by taking three straight lines in the direction of the base vectors $\mathbf{e}_i$ and intersecting at a common point O, called the origin (Fig. 6.1). The straight lines are the oblique coordinate axes. For reasons that will become evident later, we designate the coordinate axis along the vector $\mathbf{e}_i$ by x^i ($1 \le i \le 3$), using superscript. We reserve the notation x_i for rectangular cartesian coordinate system. Now let P be a point in space and let $\mathbf{R}$ denote its position vector. Then, the oblique coordinates of P with respect to the basis $\mathbf{e}_i$ are the ordered triple of numbers x^1, x^2, x^3 such that

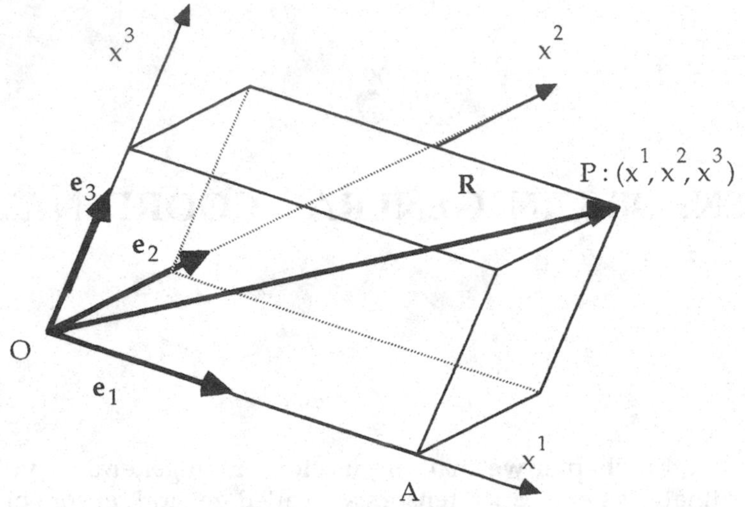

Fig. 6. 1 Oblique cartesian coordinate system.

$$\mathbf{R} = x^1\,\mathbf{e}_1 + x^2\,\mathbf{e}_2 + x^3\,\mathbf{e}_3 \tag{6.1}$$

The coordinate system is said to be right-handed or left-handed according as the base vectors $\mathbf{e}_1$, $\mathbf{e}_2$, $\mathbf{e}_3$, in that order, form a right-handed or left-handed triple.

It should be pointed out that the coordinates of a point in an oblique coordinate system depend on the base vectors $\mathbf{e}_i$. In other words, if we choose a different set of base vectors along the same coordinate axes, the coordinates of the point will change. Moreover, unlike rectangular coordinates, the oblique coordinates (x^1, x^2, x^3) of a point do not necessarily indicate distances measured along the coordinate axes. For instance, referring to Fig. 6.1, we observe that $OA = x^1\mathbf{e}_1$, so that $|x^1| = |OA|/|\mathbf{e}_1|$. Hence, unless $\mathbf{e}_1$ is a unit vector, $|x^1|$ does not represent the length of the line segment OA.

We pause here to say a few words about subscripts and superscripts with respect to the summation convention. The summation convention now applies to repeated indices of which one is a subscript and the other a superscript. Using this convention, the sum

(6.1) can be written simply as $\mathbf{R} = x^i \mathbf{e}_i$. (A reason for using superscript to denote the coordinates of points in oblique coordinate sys tems is now apparent.) As another example, the expression

$$d = g_{ij}\, x^i x^j$$

implies two summations, one with respect to the index i and the other with respect to the index j. The summation yields

$$d = g_{11}(x^1)^2 + g_{12}x^1x^2 + g_{13}x^1x^3 + g_{21}x^2x^1 + g_{22}(x^2)^2 + g_{23}x^2x^3 + g_{31}x^3x^1 + g_{32}x^3x^2 + g_{33}(x^3)^2$$

Notice that we have used parentheses to distinguish between superscripts and exponentiation.

For repeated indices that both appear as subscripts or superscripts, the summation convention applies provided one of the indices appears in a numerator term and the other in a denominator term. For example,

$$du^i = \frac{\partial u^i}{\partial x^j} dx^j = \frac{\partial u^i}{\partial x^1} dx^1 + \frac{\partial u^i}{\partial x^2} dx^2 + \frac{\partial u^i}{\partial x^3} dx^3$$

where the summation is with respect to the repeated index j, which appears both times as a superscript in a term in the numerator and a term in the denominator.

Example 1. Let (3, 4) be the rectangular cartesian coordinates of a point on a plane with unit vectors $\mathbf{i}_1, \mathbf{i}_2$. Find the coordinates of the point in an oblique coordinate system determined by the basis $\mathbf{e}_1 = 3\mathbf{i}_1$, $\mathbf{e}_2 = \mathbf{i}_1 + 2\mathbf{i}_2$ (Fig. 6.2).

Solution: The position vector of the given point with respect to the basis $\mathbf{i}_1, \mathbf{i}_2$ is $\mathbf{R} = 3\mathbf{i}_1 + 4\mathbf{i}_2$. Let (x^1, x^2) be the oblique coordinates of the point. Then, by definition,

$$x^1\mathbf{e}_1 + x^2\mathbf{e}_2 = \mathbf{R} = 3\mathbf{i}_1 + 4\mathbf{i}_2$$

Substituting for the basis $\mathbf{e}_1, \mathbf{e}_2$ and collecting similar terms, we find

$$(3x^1 + x^2)\mathbf{i}_1 + 2x^2\mathbf{i}_2 = 3\mathbf{i}_1 + 4\mathbf{i}_2$$

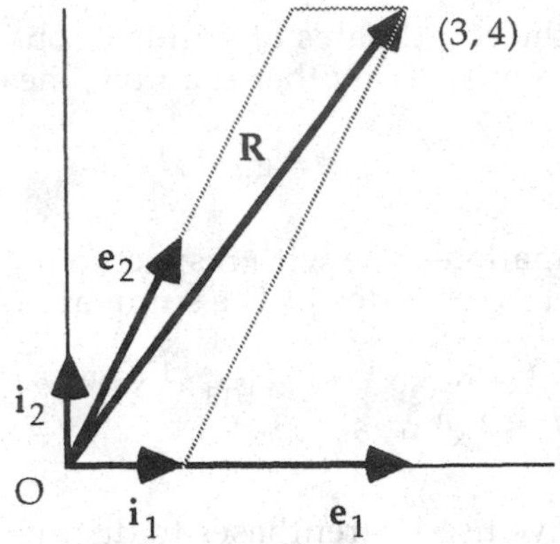

Fig. 6.2 Oblique coordinates of a point on a plane.

This implies that

$$3x^1 + x^2 = 3 \quad \text{and} \quad 2x^2 = 4$$

and so $x^1 = 1/3$ and $x^2 = 2$. Thus in terms of the basis $\mathbf{e}_1$, $\mathbf{e}_2$, the position vector of the point is given by $\mathbf{R} = (1/3)\mathbf{e}_1 + 2\mathbf{e}_2$.

Example 2. Let (2, 2, -1) be the coordinates of a point with respect to a rectangular cartesian coordinate system. Find its coordinates in an oblique coordinate system defined by the basis

$$\mathbf{e}_1 = \mathbf{i}_1 - \mathbf{i}_2, \qquad \mathbf{e}_2 = \mathbf{i}_1 - 2\mathbf{i}_2, \qquad \mathbf{e}_1 = \mathbf{i}_2 + \mathbf{i}_3$$

Solution: Let (x^1, x^2, x^3) be the oblique coordinates of the point. Then, by definition,

$$x^1\mathbf{e}_1 + x^2\mathbf{e}_2 + x^3\mathbf{e}_3 = 2\mathbf{i}_1 + 2\mathbf{i}_2 - \mathbf{i}_3$$

Substituting for the base vectors $\mathbf{e}_1$, $\mathbf{e}_2$, $\mathbf{e}_3$, and equating the corresponding coefficients of $\mathbf{i}_1$, $\mathbf{i}_2$, $\mathbf{i}_3$, we find

$$x^1 + x^2 = 2, \qquad -x^1 - 2x^2 + x^3 = 2, \qquad x^3 = -1$$

which yields $x^1 = 7$, $x^2 = -5$, $x^3 = -1$.

6.2 RECIPROCAL BASIS; TRANSFORMATIONS OF OBLIQUE COORDINATE SYSTEMS

Let x^i denote an oblique coordinate system determined by a basis $\mathbf{e}_i$ and suppose x^{*i} is another oblique coordinate system determined by another basis $\mathbf{e}^*_i$. We assume that the two systems have a common origin (see Fig. 6.3). We wish to obtain the equations that relate these two oblique coordinate systems. For this purpose, we need to introduce the notion of a *reciprocal basis* of a given basis.

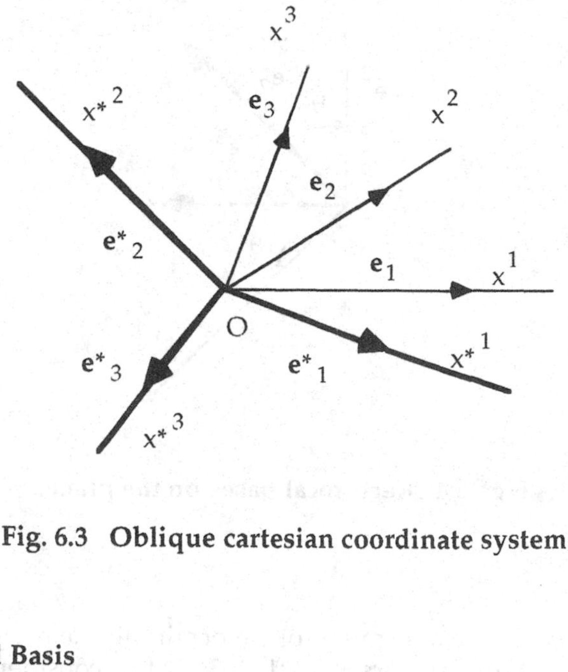

Fig. 6.3 Oblique cartesian coordinate systems.

Reciprocal Basis

Definition 1. The reciprocal basis of a given basis $\mathbf{e}_i$ is a set of vectors $\mathbf{e}^i$ that satisfies the condition

$$\mathbf{e}_i \cdot \mathbf{e}^j = \delta_i^j = \begin{Bmatrix} 1, & i = j \\ 0, & i \neq j \end{Bmatrix} \tag{6.2}$$

(Notice that the Kronecker delta is now denoted by δ_i^j.)

From the definition we see that for distinct values of i, j, k, the vector $\mathbf{e}^i$ is orthogonal to the vectors $\mathbf{e}_j$ and $\mathbf{e}_k$, and conversely, the vector $\mathbf{e}_i$ is orthogonal to the vectors $\mathbf{e}^j$ and $\mathbf{e}^k$. Further, by the property of dot product, it follows that

$$\mathbf{e}_i \cdot \mathbf{e}^i = |\mathbf{e}_i|\,|\mathbf{e}^i| \cos \theta_i = 1$$

This means that the angle θ_i between the vectors $\mathbf{e}_i$ and $\mathbf{e}^i$ is an acute angle. In the two-dimensional case, the situation is illustrated in Fig. 6.4.

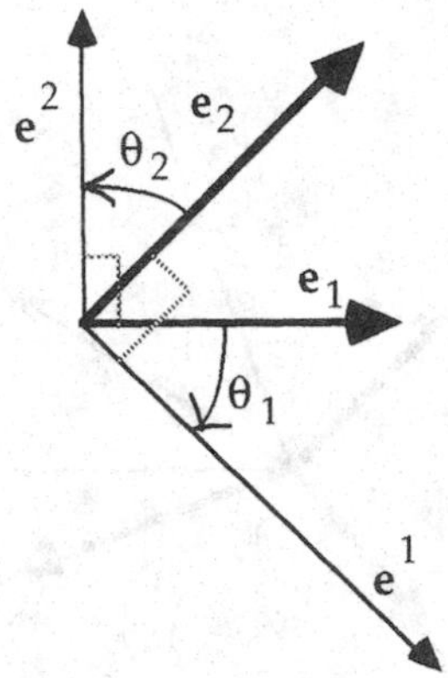

Fig. 6.4 Reciprocal bases on the plane.

We now derive a formula for determining the reciprocal basis of a given set of base vectors $\mathbf{e}_i$, i = 1, 2, 3. Let us consider first the vector $\mathbf{e}^1$. Since $\mathbf{e}^1$ is orthogonal to both the vectors $\mathbf{e}_2$ and $\mathbf{e}_3$, it must be parallel to the cross product $\mathbf{e}_2 \times \mathbf{e}_3$. Hence, there is a scalar m such that

$$\mathbf{e}^1 = m(\mathbf{e}_2 \times \mathbf{e}_3)$$

Taking the dot product with $\mathbf{e}_1$, we get

$$1 = \mathbf{e}_1 \cdot \mathbf{e}^1 = m(\mathbf{e}_1 \cdot \mathbf{e}_2 \times \mathbf{e}_3)$$

Hence we find

$$m = \frac{1}{\mathbf{e}_1 \cdot \mathbf{e}_2 \times \mathbf{e}_3}$$

and so we have

$$\mathbf{e}^1 = \frac{\mathbf{e}_2 \times \mathbf{e}_3}{\mathbf{e}_1 \cdot \mathbf{e}_2 \times \mathbf{e}_3}$$

By following the same procedure, we also find

$$\mathbf{e}^2 = \frac{\mathbf{e}_3 \times \mathbf{e}_1}{\mathbf{e}_2 \cdot \mathbf{e}_3 \times \mathbf{e}_1}, \qquad \mathbf{e}^3 = \frac{\mathbf{e}_1 \times \mathbf{e}_2}{\mathbf{e}_3 \cdot \mathbf{e}_1 \times \mathbf{e}_2}$$

These results can be represented simply by the formula

$$\mathbf{e}^i = \frac{\mathbf{e}_j \times \mathbf{e}_k}{V} \tag{6.3}$$

where i, j, k are cyclic permutations of 1, 2, 3, and

$$V = \mathbf{e}_1 \cdot \mathbf{e}_2 \times \mathbf{e}_3 = \mathbf{e}_2 \cdot \mathbf{e}_3 \times \mathbf{e}_1 = \mathbf{e}_3 \cdot \mathbf{e}_1 \times \mathbf{e}_2 \tag{6.4}$$

Notice that the quantity V represents in absolute value the volume of the parallelepiped formed by the vectors $\mathbf{e}_1$, $\mathbf{e}_2$, $\mathbf{e}_3$. V is positive or negative according as the vectors $\mathbf{e}_1$, $\mathbf{e}_2$, $\mathbf{e}_3$ form a right-handed or left-handed triples. In the sequel, we assume that $V > 0$ so that our coordinate system is right-handed.

It is clear that the reciprocal basis given by (6.3) is uniquely determined. Moreover, if we determine the reciprocal basis of the vectors $\mathbf{e}^i$, we are led back to the basis $\mathbf{e}_i$ (i = 1, 2, 3). In fact, it can be readily verified that

$$\mathbf{e}_i = \frac{\mathbf{e}^j \times \mathbf{e}^k}{V'}, \quad V' = \mathbf{e}^1 \cdot \mathbf{e}^2 \times \mathbf{e}^3 \tag{6.5}$$

where i, j, k assume cyclic permutations of 1, 2, 3, and $VV' = 1$. It follows that when a basis consists of orthonormal vectors, the reciprocal basis is the basis itself. For example, the reciprocal basis of $\mathbf{i}_1, \mathbf{i}_2, \mathbf{i}_3$ is again $\mathbf{i}_1, \mathbf{i}_2, \mathbf{i}_3$.

Example 1. Find the reciprocal basis of the base vectors

$$\mathbf{e}_1 = \mathbf{i}_1 + \mathbf{i}_2, \qquad \mathbf{e}_2 = -\mathbf{i}_1 + 2\mathbf{i}_2 + \mathbf{i}_3, \qquad \mathbf{e}_3 = 2\mathbf{i}_1 - \mathbf{i}_2 + \mathbf{i}_3.$$

Solution: First we find

$$V = \mathbf{e}_1 \cdot \mathbf{e}_2 \times \mathbf{e}_3 = \begin{vmatrix} 1 & 1 & 0 \\ -1 & 2 & 1 \\ 2 & -1 & 1 \end{vmatrix} = 6$$

Hence, by (6.3), we have

$$\mathbf{e}^1 = \frac{\mathbf{e}_2 \times \mathbf{e}_3}{V} = \frac{1}{2}(\mathbf{i}_1 + \mathbf{i}_2 - \mathbf{i}_3)$$

$$\mathbf{e}^2 = \frac{\mathbf{e}_3 \times \mathbf{e}_1}{V} = \frac{1}{6}(-\mathbf{i}_1 + \mathbf{i}_2 + 3\mathbf{i}_3)$$

$$\mathbf{e}^3 = \frac{\mathbf{e}_1 \times \mathbf{e}_2}{V} = \frac{1}{6}(\mathbf{i}_1 - \mathbf{i}_2 + 3\mathbf{i}_3)$$

Example 2. Find the reciprocal basis of the base vectors

$$\mathbf{e}_1 = 2\mathbf{i}_1 + \mathbf{i}_2, \qquad \mathbf{e}_2 = 2\mathbf{i}_1 + 3\mathbf{i}_2$$

Solution: Since we are dealing with plane vectors, the formula (6.3) can not be applied. However, the definition of a reciprocal basis remains valid. So let us assume that

$$\mathbf{e}^1 = a\mathbf{i}_1 + b\mathbf{i}_2$$

Then, applying the conditions $\mathbf{e}^1 \cdot \mathbf{e}_1 = 1$ and $\mathbf{e}^1 \cdot \mathbf{e}_2 = 0$, we obtain

$$2a + b = 1, \quad 2a + 3b = 0$$

from which we find $a = 3/4$, $b = -1/2$. Therefore,

$$\mathbf{e}^1 = \frac{3}{4}\mathbf{i}_1 - \frac{1}{2}\mathbf{i}_2$$

Similarly, by assuming $\mathbf{e}^2 = a\mathbf{i}_1 + b\mathbf{i}_2$ and applying the conditions $\mathbf{e}^2 \cdot \mathbf{e}_1 = 0$, $\mathbf{e}^2 \cdot \mathbf{e}_2 = 1$, we find

$$\mathbf{e}^2 = -\frac{1}{4}\mathbf{i}_1 + \frac{1}{2}\mathbf{i}_2$$

These are the reciprocal base vectors corresponding to the given basis.

Equations of Transformations

We now return to the problem of finding the relationship between the two oblique coordinate systems x^i and x^{*i} defined by the respective bases $\mathbf{e}_i$ and $\mathbf{e}^*_i$. Consider a point in space and denote its coordinates with respect to the two systems by (x^1, x^2, x^3) and (x^{*1}, x^{*2}, x^{*3}). Since we assume a common origin, the position vector of the point with respect to the two bases is given by

$$\mathbf{R} = x^{*i}\,\mathbf{e}^*_i = x^j\,\mathbf{e}_j \tag{6.6}$$

(Remember the summation convention.) Let $\mathbf{e}^{*i}$ denote the reciprocal basis corresponding to the basis $\mathbf{e}^*_i$. Taking the dot product of both sides of (6.6) with $\mathbf{e}^{*i}$, noting that $\mathbf{e}^*_i \cdot \mathbf{e}^{*j} = \delta_i{}^j$, we find

$$x^{*i} = (\mathbf{e}^{*i} \cdot \mathbf{e}_j)\,x^j = \alpha^{*i}{}_j\,x^j \qquad (i = 1, 2, 3) \tag{6.7}$$

where we have set

$$\alpha^{*i}{}_j = \mathbf{e}^{*i} \cdot \mathbf{e}_j \qquad (1, j = 1, 2, 3) \tag{6.8}$$

Equation (6.7) with (6.8) defines the transformation from the oblique coordinate system x^i to the oblique coordinate system x^{*i} as determined by the bases $\mathbf{e}_i$ and $\mathbf{e}^*_i$.

On the other hand, by taking the dot product of both sides of (6.6) with the reciprocal basis $\mathbf{e}^i$, we find

$$x^i = (\mathbf{e}^i \cdot \mathbf{e}^*_j)\,x^{*j} = \alpha_*{}^i{}_j\,x^{*j} \qquad (i = 1, 2, 3) \tag{6.9}$$

where

$$\alpha_*{}^i{}_j = \mathbf{e}^i \cdot \mathbf{e}^*_j \qquad (1, j = 1, 2, 3) \tag{6.10}$$

Equation (6.9) with (6.10) is the inverse of the transformation (6.7).

Thus the coordinate transformations from the x^i system to the x^{*i} system and vice versa are given by equations (6.7) and (6.9), respectively. These transformations coincide with those obtained in the preceding chapter when the bases consist of orthonormal vectors.

Example 3. Find the coordinate transformation that relate the oblique coordinate systems x^i and x^{*i} defined by the respective bases

$$\mathbf{e}_1 = \mathbf{i}_1 + 2\mathbf{i}_2, \quad \mathbf{e}_2 = \mathbf{i}_1 + 2\mathbf{i}_2 + \mathbf{i}_3, \quad \mathbf{e}_3 = -\mathbf{i}_2 + \mathbf{i}_3$$

and

$$\mathbf{e}^*_1 = \mathbf{i}_1 + \mathbf{i}_2, \quad \mathbf{e}^*_2 = \mathbf{i}_2 + 2\mathbf{i}_3, \quad \mathbf{e}^*_3 = 2\mathbf{i}_1 + \mathbf{i}_2 - \mathbf{i}_3$$

Solution: First, we find the reciprocal bases corresponding to the given bases. By (6.3), since $V = \mathbf{e}_1 \cdot \mathbf{e}_2 \times \mathbf{e}_3 = 1$, we find

$$\mathbf{e}^1 = 3\mathbf{i}_1 - \mathbf{i}_2 - \mathbf{i}_3, \quad \mathbf{e}^2 = -2\mathbf{i}_1 + \mathbf{i}_2 + \mathbf{i}_3, \quad \mathbf{e}^3 = 2\mathbf{i}_1 - \mathbf{i}_2$$

Similarly, since $V' = \mathbf{e}^*_1 \cdot \mathbf{e}^*_2 \times \mathbf{e}^*_3 = 1$, we find

$$\mathbf{e}^{*1} = -3\mathbf{i}_1 + 4\mathbf{i}_2 - 2\mathbf{i}_3, \quad \mathbf{e}^{*2} = \mathbf{i}_1 - \mathbf{i}_2 + \mathbf{i}_3, \quad \mathbf{e}^{*3} = 2\mathbf{i}_1 - 2\mathbf{i}_2 + \mathbf{i}_3$$

Hence, by (6.8), we have

$$\begin{aligned}
&\alpha^{*1}{}_1 = \mathbf{e}^{*1} \cdot \mathbf{e}_1 = 5, && \alpha^{*1}{}_2 = \mathbf{e}^{*1} \cdot \mathbf{e}_2 = 3, && \alpha^{*1}{}_3 = \mathbf{e}^{*1} \cdot \mathbf{e}_3 = -6 \\
&\alpha^{*2}{}_1 = \mathbf{e}^{*2} \cdot \mathbf{e}_1 = -1, && \alpha^{*2}{}_2 = \mathbf{e}^{*2} \cdot \mathbf{e}_2 = 0, && \alpha^{*2}{}_3 = \mathbf{e}^{*2} \cdot \mathbf{e}_3 = 2 \\
&\alpha^{*3}{}_1 = \mathbf{e}^{*3} \cdot \mathbf{e}_1 = -2, && \alpha^{*3}{}_2 = \mathbf{e}^{*3} \cdot \mathbf{e}_2 = -1, && \alpha^{*3}{}_3 = \mathbf{e}^{*3} \cdot \mathbf{e}_3 = 3
\end{aligned}$$

Therefore, the coordinate transformation from the x^i - to the x^{*i} - coordinate system is defined by

$$\begin{aligned}
x^{*1} &= 5x^1 + 3x^2 - 6x^3 \\
x^{*2} &= -x^1 \qquad\quad + 2x^3 \\
x^{*3} &= -2x^1 - x^2 + 3x^3
\end{aligned}$$

The inverse transformation is given by

$$x^1 = 2x^{*1} - 3x^{*2} + 6x^{*3}$$
$$x^2 = -x^{*1} + 3x^{*2} - 4x^{*3}$$
$$x^3 = x^{*1} - x^{*2} + 3x^{*3}$$

which can be verified by calculating the coefficients $\alpha^{i}{}_{*j}$ from (6.10).

Properties of Transformation Matrix

The linear transformation (6.7) can be written in the matrix form

$$X^* = \Lambda^* X \tag{6.11}$$

where the coefficient matrix Λ^* is given by

$$\Lambda^* = (\alpha^{*i}{}_j) = \begin{bmatrix} \alpha^{*1}{}_1 & \alpha^{*1}{}_2 & \alpha^{*1}{}_3 \\ \alpha^{*2}{}_1 & \alpha^{*2}{}_2 & \alpha^{*2}{}_3 \\ \alpha^{*3}{}_1 & \alpha^{*3}{}_2 & \alpha^{*3}{}_3 \end{bmatrix} \tag{6.12}$$

$X^* = [x^{*1}, x^{*2}, x^{*3}]^T$ and $X = [x^1, x^2, x^3]^T$. Notice that the superscripts and the subscripts denote rows and columns, respectively. Now from (6.8) we readily deduce the relations

$$\mathbf{e}^{*i} = \alpha^{*i}{}_j \mathbf{e}^j, \qquad \mathbf{e}_i = \alpha^{*j}{}_i \mathbf{e}^*{}_j \tag{6.13}$$

These imply that the elements in the i-th row of (6.12) are simply the components of the reciprocal base vector $\mathbf{e}^{*i}$ (i = 1, 2, 3) with respect to the basis $\mathbf{e}^i$. Likewise the elements in the j-th column are the components of the base vector $\mathbf{e}_j$ (j = 1, 2, 3) with respect to the basis $\mathbf{e}^*{}_i$.

Similarly, the inverse transformation (6.9) can be written in matrix form as

$$X = \Lambda_* X^* \tag{6.14}$$

where

$$\Lambda_* = (\alpha_*{}^i{}_j) = \begin{bmatrix} \alpha_*{}^1{}_1 & \alpha_*{}^1{}_2 & \alpha_*{}^1{}_3 \\ \alpha_*{}^2{}_1 & \alpha_*{}^2{}_2 & \alpha_*{}^2{}_3 \\ \alpha_*{}^3{}_1 & \alpha_*{}^3{}_2 & \alpha_*{}^3{}_3 \end{bmatrix} \tag{6.15}$$

Here we observe that the elements in the i-th row of the matrix (6.15) are components of the vector $\mathbf{e}^i$ with respect to the reciprocal basis $\mathbf{e}^{*i}$ and the elements in the j-th column are components of the vector $\mathbf{e}^*_j$ with respect to the basis $\mathbf{e}_i$. That is,

$$\mathbf{e}^i = \alpha_*{}^i{}_j\, \mathbf{e}^{*j}, \qquad \mathbf{e}^*_i = \alpha_*{}^j{}_i\, \mathbf{e}_j \tag{6.16}$$

Now from (6.13), (6.16) and the definition of reciprocal basis, we see that

$$\delta^i{}_j = \mathbf{e}^i \cdot \mathbf{e}_j = (\alpha^i{}_{*p}\mathbf{e}^{*p})\cdot(\alpha^{*q}{}_j\mathbf{e}^*_q)$$

$$= \alpha^i{}_{*p}\, \alpha^{*q}{}_j\, \mathbf{e}^{*p}\cdot\mathbf{e}^*_q = \alpha^i{}_{*p}\, \alpha^{*q}{}_j\, \delta_*{}^p{}_q$$

$$= \alpha^i{}_{*p}\, \alpha^{*p}{}_j$$

Since the right-hand side of the above equation is the scalar product of the i-th row of Λ^* and the j-th column of Λ_*, and $\delta^i{}_j$ is the Kronecker delta function, we conclude that the product $\Lambda^*\Lambda_*$ is the identity matrix. Thus, $\Lambda^*\Lambda_* = I$. Likewise, we have

$$\mathbf{e}^{*i}\cdot \mathbf{e}^*_j = \alpha^{*i}{}_p\, \alpha^q{}_{*j}\, \mathbf{e}^p\cdot \mathbf{e}_q = \alpha^{*i}{}_p\, \alpha^p{}_{*j}$$

which implies that $\Lambda_*\Lambda^* = I$. Therefore, the matrices defined in (6.12) and (6.15) are inverses of each other, that is,

$$\Lambda^{*-1} = \Lambda_*, \qquad \Lambda_*^{-1} = \Lambda^*$$

Thus, in Example 3, the matrix of the transformation is given by

$$\Lambda^* = \begin{bmatrix} 5 & 3 & -6 \\ -1 & 0 & 2 \\ -2 & -1 & 3 \end{bmatrix}$$

and the matrix of the inverse transformation is

$$\Lambda_* = \begin{bmatrix} 2 & -3 & 6 \\ -1 & 3 & -4 \\ 1 & -1 & 3 \end{bmatrix}$$

It is readily verified that

$$\Lambda^*\Lambda_* = \Lambda_*\Lambda^* = I$$

6.1 EXERCISES

1. Let (a_1, a_2, a_3) and (b^1, b^2, b^3) denote the ordered triples (-2, 1, 3) and (1, 2, 3), respectively. Write down the elements represented by $a_i b^j$. What is the term $a_i b^i$?
2. Let a^i_j denote the element in the i-th row and j-th column of a matrix (a^i_j). What does $a_i^{\ i}$ represent?
3. Let u^j $(1 \le j \le 3)$ be some functions of the variables x^1, x^2, x^3 that are differentiable. Write down the expressions
$$a^i = \frac{\partial u^i}{\partial x^j} b^j, \qquad a_i = \frac{\partial u^j}{\partial x^i} b_j$$
4. If $u^i = a^i_j x^j$ $(1 \le i \le 3)$, where a^i_j are all constants, show that
$$a^i_j = \frac{\partial u^i}{\partial x^j}$$
5. If $a^i = (\partial u^i/\partial x^j)b^j$ and a^i and b^j are functions of the x^i's, find $\partial a^i/\partial x^k$.
6. Let **R** be a twice continuously differentiable function of x^i $(1 \le i \le 3)$, and set $\mathbf{A}_i = \partial \mathbf{R}/\partial x^i$. Show that $\partial \mathbf{A}_i/\partial x^j = \partial \mathbf{A}_j/\partial x^i$ for all i, j.

In each of Problems 7 through 11, the coordinates of a point with respect to a rectangular cartesian coordinate system are given. Determine the coordinates of the point in the oblique cartesian coordinate system defined by the given basis.

7. P: (1, 2) ; $\mathbf{e}_1 = 2\mathbf{i}_1 - \mathbf{i}_2$, $\mathbf{e}_2 = 3\mathbf{i}_1 + \mathbf{i}_2$.
8. P: (-2, 2) ; $\mathbf{e}_1 = \mathbf{i}_1 + \mathbf{i}_2$, $\mathbf{e}_2 = -\mathbf{i}_1 + 2\mathbf{i}_2$.
9. P: (2, -1, 2) ; $\mathbf{e}_1 = \mathbf{i}_1 + \mathbf{i}_2$, $\mathbf{e}_2 = \mathbf{i}_2 + \mathbf{i}_3$, $\mathbf{e}_3 = \mathbf{i}_1 + \mathbf{i}_3$.
10. P: (3, -1, 2) ; $\mathbf{e}_1 = \mathbf{i}_1$, $\mathbf{e}_2 = \mathbf{i}_1 + \mathbf{i}_2$, $\mathbf{e}_3 = \mathbf{i}_1 + \mathbf{i}_2 + \mathbf{i}_3$.
11. P: (2, 3, -2) ; $\mathbf{e}_1 = \mathbf{i}_1 + \mathbf{i}_2$, $\mathbf{e}_2 = \mathbf{i}_1 + 2\mathbf{i}_2 + \mathbf{i}_3$, $\mathbf{e}_3 = 2\mathbf{i}_1 - \mathbf{i}_2 - \mathbf{i}_3$.
12. Derive the formula (6.5) and show that VV' = 1.
13. Show that if a basis is orthonormal, its reciprocal basis coincides with itself.
14. Find the reciprocal basis of $\mathbf{e}_1 = \mathbf{i}_1$, $\mathbf{e}_2 = \mathbf{i}_1 + \mathbf{i}_2$, $\mathbf{e}_3 = \mathbf{i}_1 + \mathbf{i}_2 + \mathbf{i}_3$.
15. Find the basis of which the reciprocal basis is: $\mathbf{e}^1 = \mathbf{i}_2 + \mathbf{i}_3$, $\mathbf{e}^2 = \mathbf{i}_1 + \mathbf{i}_3$, $\mathbf{e}^3 = \mathbf{i}_1 + \mathbf{i}_2$.
16. Find the reciprocal basis of $\mathbf{e}_1 = \mathbf{i}_1 + \mathbf{i}_2$, $\mathbf{e}_2 = \mathbf{i}_1 + 2\mathbf{i}_2 + \mathbf{i}_3$, $\mathbf{e}_3 = 2\mathbf{i}_1 - \mathbf{i}_2 - \mathbf{i}_3$.

In each of Problems 17 through 22, find the equation of the transformation relating the coordinate systems x^i and x^{*i} defined by the given bases, and determine whether the transformation preserves or reverses the orientation of the coordinate axes. Assume a common origin for both system.

17. $\mathbf{e}_1 = \mathbf{i}_1 + \mathbf{i}_2 + \mathbf{i}_3$, $\mathbf{e}_2 = -\mathbf{i}_2$, $\mathbf{e}_3 = -\mathbf{i}_1 + \mathbf{i}_3$; $\mathbf{e}^*_1 = \mathbf{i}_1 + \mathbf{i}_2$, $\mathbf{e}^*_2 = \mathbf{i}_2 + \mathbf{i}_3$, $\mathbf{e}^*_3 = \mathbf{i}_1 + \mathbf{i}_3$.
18. $\mathbf{e}_1 = 2\mathbf{i}_1 + \mathbf{i}_2$, $\mathbf{e}_2 = -\mathbf{i}_1 + \mathbf{i}_2$; $\mathbf{e}^*_1 = \mathbf{i}_1 + 2\mathbf{i}_2$, $\mathbf{e}^*_2 = -2\mathbf{i}_1 + \mathbf{i}_2$.
19. $\mathbf{e}_1 = \mathbf{i}_2 - \mathbf{i}_3$, $\mathbf{e}_2 = -\mathbf{i}_1 + \mathbf{i}_2$, $\mathbf{e}_3 = \mathbf{i}_1 + \mathbf{i}_2 + \mathbf{i}_3$; $\mathbf{e}^*_1 = \mathbf{i}_1$, $\mathbf{e}^*_2 = \mathbf{i}_1 + 2\mathbf{i}_3$, $\mathbf{e}^*_3 = \mathbf{i}_1 + 2\mathbf{i}_2$.
20. $\mathbf{e}_1 = 2\mathbf{i}_1$, $\mathbf{e}_2 = \mathbf{i}_1 + 2\mathbf{i}_2$; $\mathbf{e}^*_1 = 2\mathbf{i}_1 + \mathbf{i}_2$, $\mathbf{e}^*_2 = \mathbf{i}_2$.
21. $\mathbf{e}_1 = 2\mathbf{i}_1 + \mathbf{i}_2$, $\mathbf{e}_2 = \mathbf{i}_1 + 2\mathbf{i}_2 + \mathbf{i}_3$, $\mathbf{e}_3 = 3\mathbf{i}_1 - \mathbf{i}_2 + 2\mathbf{i}_3$; $\mathbf{e}^*_1 = -\mathbf{i}_1 + 2\mathbf{i}_2 + \mathbf{i}_3$, $\mathbf{e}^*_2 = \mathbf{i}_1 + \mathbf{i}_2 + \mathbf{i}_3$, $\mathbf{e}^*_3 = 2\mathbf{i}_1 - 3\mathbf{i}_2 + 2\mathbf{i}_3$.
22. $\mathbf{e}_1 = 2\mathbf{i}_2 - \mathbf{i}_3$, $\mathbf{e}_2 = \mathbf{i}_1 + 3\mathbf{i}_2 - \mathbf{i}_3$, $\mathbf{e}_3 = 2\mathbf{i}_1 - \mathbf{i}_2 + \mathbf{i}_3$; $\mathbf{e}^*_1 = \mathbf{i}_1 + 2\mathbf{i}_2 + \mathbf{i}_3$, $\mathbf{e}^*_2 = -\mathbf{i}_1 + 2\mathbf{i}_2 + \mathbf{i}_3$, $\mathbf{e}^*_3 = \mathbf{i}_1 - 3\mathbf{i}_2 - 2\mathbf{i}_3$.

6.3 TENSORS IN OBLIQUE CARTESIAN COORDINATE SYSTEMS

As an introduction to the study of tensors in general curvilinear coordinate systems, we now discuss tensors in oblique cartesian coordinate systems. As in our discussion of cartesian tensors, we first consider vectors in oblique cartesian coordinate systems and examine how their components transform under a change of the coordinates.

Tensors of Order One - Vectors

So suppose x^i is an oblique cartesian coordinate system determined by a given basis $\mathbf{e}_i$ $(1 \le i \le 3)$. We know that corresponding to the given basis there exists another basis $\mathbf{e}^i$ $(1 \le i \le 3)$, called the reciprocal basis. Hence every vector may be represented in terms of either the given basis $\mathbf{e}_i$ or the reciprocal basis $\mathbf{e}^i$. This means that every nonzero vector has two distinct sets of components. To distinguish these sets of components and to conform with the summation convection, we denote the components of a vector $\mathbf{A} \ne \mathbf{0}$ with respect to the reciprocal basis $\mathbf{e}^i$ by a_i, and the components with respect to the basis $\mathbf{e}_i$ by a^i. Thus we have

$$\mathbf{A} = a_i\, \mathbf{e}^i = a^i\, \mathbf{e}_i \tag{6.17}$$

To determine the components a_i and a^i from (6.17), we take the dot product of $\mathbf{A}$ with $\mathbf{e}_i$ and with $\mathbf{e}^i$, using the defnition of reciprocal basis. We find

$$\mathbf{A} \cdot \mathbf{e}_i = (a_i\, \mathbf{e}^i) \cdot \mathbf{e}_i = a_i \tag{6.18}$$

and

$$\mathbf{A} \cdot \mathbf{e}^i = (a^i\, \mathbf{e}_i\,) \cdot \mathbf{e}^i = a^i \tag{6.19}$$

We call the components a_i and a^i $(1 \le i \le 3)$ the *covariant* and the *contravariant* components of the vector $\mathbf{A}$, respectively.

For example, suppose

$$\mathbf{A} = a_1\mathbf{e}^1 + a_2\mathbf{e}^2 = a^1\mathbf{e}_1 + a^2\mathbf{e}_2$$

$(a_i > 0,\ a^i > 0,\ i = 1, 2.)$ Then, by the parallelogram law of vector addition, the vector forms the diagonal of the parallelogram whose

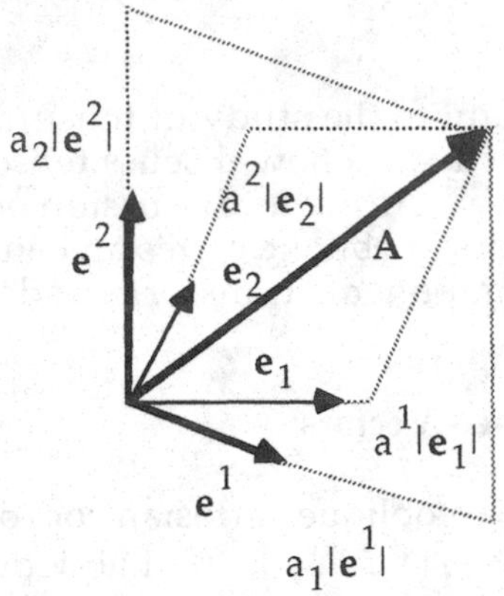

Fig. 6.5 Covariant and contravariant components of a vector.

adjacent sides are represented by the vectors $a^1\mathbf{e}_1$ and $a^2\mathbf{e}_2$, involving the contravariant components a^1, a^2, as shown in Fig. 6.5. It is also the diagonal of the parallelogram whose sides are represented by $a_1\mathbf{e}^1$ and $a_2\mathbf{e}^2$, involving the covariant components.

Actually, the terms covariant and contravariant have to do with the manner in which the components of the vector transform under a change of oblique coordinate systems. To see how the components transform, suppose x^{*i} is another oblique coordinate system determined by a basis $\mathbf{e}^*_i$ $(1 \le i \le 3)$. Let a^*_i and a^{*i} denote the components of the vector $\mathbf{A}$ with respect to the bases $\mathbf{e}^{*i}$ and $\mathbf{e}^*_i$, respectively. Then we have

$$\mathbf{A} = a^*_i\, \mathbf{e}^{*i} = a_j\, \mathbf{e}^j \tag{6.20}$$

and

$$\mathbf{A} = a^{*i}\, \mathbf{e}^*_i = a^j\, \mathbf{e}_j \tag{6.21}$$

Now if we take the dot product of $\mathbf{e}^*_k$ with both sides of (6.20), noting that $\mathbf{e}^{*i} \cdot \mathbf{e}^*_k = \delta^{*i}{}_k$, we find

$$a^*_i = (\mathbf{e}^*_i \cdot \mathbf{e}^j)a_j = \alpha_*{}^j{}_i\, a_j \quad (1 = 1, 2, 3) \tag{6.22}$$

where we noted that $\mathbf{e}^*_i \cdot \mathbf{e}^j = \alpha_*{}^j{}_i$. This is the same form as the relationship between $\mathbf{e}^*_i$ and $\mathbf{e}_i$ as shown in (6.16). Thus the components a_i transform "in the same fashion" as the base vectors $\mathbf{e}_i$ ($1 \leq i \leq 3$), and for this reason they are called *covariant* components. The base vectors $\mathbf{e}_i$ are sometimes also called *covariant* basis. Similarly, by taking the dot product of both sides of (6.21) with $\mathbf{e}^{*k}$, we obtain

$$a^{*i} = (\mathbf{e}^{*i} \cdot \mathbf{e}_j)\, a^j = \alpha^{*i}{}_j\, a^j \quad (1 = 1, 2, 3) \tag{6.23}$$

This is the same form as the relationship in (6.16) between the reciprocal bases $\mathbf{e}^{*i}$ and $\mathbf{e}^i$, and so the components a^i ($1 \leq i \leq 3$) transform "in the opposite fashion" to the base vectors $\mathbf{e}_i$. Accordingly, the components a^i are called *contravariant* components of the vector.

The inverse transformation of (6.22) can be found by taking the dot product of both sides of (6.20) with $\mathbf{e}_j$. We find

$$a_j = \mathbf{e}^{*i} \cdot \mathbf{e}_j\, a^*{}_i = \alpha^{*i}{}_j a^*{}_i \quad (j = 1, 2, 3) \tag{6.24}$$

where $\mathbf{e}^{*i} \cdot \mathbf{e}_j = \alpha^{*i}{}_j$. Likewise, the inverse of the transformation (6.24) can be obtained by taking the dot product of both sides of (6.21) with $\mathbf{e}^j$. It is given by

$$a^j = \alpha_*{}^j{}_i\, a^{*i} \quad (j = 1, 2, 3) \tag{6.25}$$

where $\alpha_*{}^j{}_i = \mathbf{e}^*{}_i \cdot \mathbf{e}^j$.

It should be remembered, however, that the two sets of components represent the same vector referred to two different bases of a given coordinate system.

Definition 1. A *tensor of order one* is a vector with covariant components a_i or contravariant components a^i ($1 \leq i \leq 3$) with respect to an oblique cartesian coordinate system x^i defined by a basis $\mathbf{e}_i$ ($1 \leq i \leq 3$). Under a change of coordinates, the components transform according to the laws (6.22) and (6.23) or (6.24) and (6.25).

Example 1. Consider an oblique coordinate system x^i defined by the basis

$$\mathbf{e}_1 = \mathbf{i}_1 + \mathbf{i}_2, \quad \mathbf{e}_2 = \mathbf{i}_2 - \mathbf{i}_3, \quad \mathbf{e}_3 = \mathbf{i}_1 + 2\mathbf{i}_3$$

Find the covariant and the contravariant components of the vector $\mathbf{A} = 2\mathbf{i}_1 + \mathbf{i}_2 - 3\mathbf{i}_3$.

Solution: By (6.18) the covariant components are given by

$$\begin{aligned} a_1 &= \mathbf{A}\cdot\mathbf{e}_1 = (2\mathbf{i}_1 + \mathbf{i}_2 - 3\mathbf{i}_3)\cdot(\mathbf{i}_1 + \mathbf{i}_2) = 3 \\ a_2 &= \mathbf{A}\cdot\mathbf{e}_2 = (2\mathbf{i}_1 + \mathbf{i}_2 - 3\mathbf{i}_3)\cdot(\mathbf{i}_2 - \mathbf{i}_3) = 4 \\ a_3 &= \mathbf{A}\cdot\mathbf{e}_3 = (2\mathbf{i}_1 + \mathbf{i}_2 - 3\mathbf{i}_3)\cdot(\mathbf{i}_1 + 2\mathbf{i}_3) = -4 \end{aligned}$$

These are the components of the vector with respect to the reciprocal basis $\mathbf{e}^i$ $(1 \le i \le 3)$. Since $\mathbf{e}_1\cdot\mathbf{e}_2 \times \mathbf{e}_3 = 1$, the reciprocal base vectors are given by

$$\begin{aligned} \mathbf{e}^1 &= \mathbf{e}_2 \times \mathbf{e}_3 = (\mathbf{i}_2 - \mathbf{i}_3) \times (\mathbf{i}_1 + 2\mathbf{i}_3) = 2\mathbf{i}_1 - \mathbf{i}_2 - \mathbf{i}_3 \\ \mathbf{e}^2 &= \mathbf{e}_3 \times \mathbf{e}_1 = (\mathbf{i}_1 + 2\mathbf{i}_3) \times (\mathbf{i}_1 + \mathbf{i}_2) = -2\mathbf{i}_1 + 2\mathbf{i}_2 + \mathbf{i}_3 \\ \mathbf{e}^3 &= \mathbf{e}_1 \times \mathbf{e}_2 = (\mathbf{i}_1 + \mathbf{i}_2) \times (\mathbf{i}_2 - \mathbf{i}_3) = -\mathbf{i}_1 + \mathbf{i}_2 + \mathbf{i}_3 \end{aligned}$$

Thus, $\mathbf{A} = 3\mathbf{e}^1 + 4\mathbf{e}^2 - 4\mathbf{e}^1$. Next, by (6.19), the contravariant components of $\mathbf{A}$ are

$$\begin{aligned} a^1 &= \mathbf{A}\cdot\mathbf{e}^1 = (2\mathbf{i}_1 + \mathbf{i}_2 - 3\mathbf{i}_3)\cdot(2\mathbf{i}_1 - \mathbf{i}_2 - \mathbf{i}_3) = 6 \\ a^2 &= \mathbf{A}\cdot\mathbf{e}^2 = (2\mathbf{i}_1 + \mathbf{i}_2 - 3\mathbf{i}_3)\cdot(-2\mathbf{i}_1 + 2\mathbf{i}_2 + \mathbf{i}_3) = -5 \\ a^3 &= \mathbf{A}\cdot\mathbf{e}^3 = (2\mathbf{i}_1 + \mathbf{i}_2 - 3\mathbf{i}_3)\cdot(-\mathbf{i}_1 + \mathbf{i}_2 + \mathbf{i}_3) = -4 \end{aligned}$$

hence, $\mathbf{A} = 6\mathbf{e}_1 - 5\mathbf{e}_2 - 4\mathbf{e}_3$.

In practice, certain vector quantities are naturally represented by vectors whose components transform according to (6.22) or (6.23). Accordingly, the vectors are called covariant vectors or contravariant vectors. However, as we shall see later, the covariant and contravariant components of a vector are related; that is, it is possible to switch from covariant to contravariant components or vice versa. This property also holds for general tensors.

Example 2. Let $\phi(x^i)$ be a scalar field in a rectangular cartesian coordinate system x^i and let $x^{*i} = \alpha^{*i}{}_j x^j$ define an oblique coordinate system x^{*i}. Verify that $\partial\phi/\partial x^i$ $(i = 1, 2, 3)$ are covariant components of the gradient of ϕ.

Solution: We note that in rectangular coordinate system, we have $\mathbf{e}_i = \mathbf{i}_i = \mathbf{e}^i$, and so

$$\text{grad}\,\phi = \sum_{i=1}^{3} \frac{\partial\phi}{\partial x^i}\mathbf{i}_i = \frac{\partial\phi}{\partial x^i}\mathbf{e}^i$$

Now under the given transformation, we find by the chain rule

$$\frac{\partial \phi}{\partial x^i} = \frac{\partial \phi}{\partial x^{*j}} \frac{\partial x^{*j}}{\partial x^i} = \alpha^{*j}{}_i \frac{\partial \phi}{\partial x^{*j}}$$

which is of the form

$$a_i = \alpha^{*j}{}_i \, a^*_j \qquad (i = 1, 2, 3)$$

where

$$a_i = \frac{\partial \phi}{\partial x^i}, \qquad a^*_i = \frac{\partial \phi}{\partial x^{*i}}$$

This is the transformation law (6.24) for the covariant components of a vector. Further, since $\mathbf{e}^{*j} = \alpha^{*j}{}_i \, \mathbf{e}^i$, we see that

$$\text{grad}\,\phi = \frac{\partial \phi}{\partial x^i} \mathbf{e}^i = \frac{\partial \phi}{\partial x^{*j}} \frac{\partial x^{*j}}{\partial x^i} \mathbf{e}^i$$

$$= \frac{\partial \phi}{\partial x^{*j}} \alpha^{*j}{}_i \, \mathbf{e}^i = \frac{\partial \phi}{\partial x^{*j}} \mathbf{e}^{*j}$$

This shows that the natural expression for grad ϕ is in terms of its covariant components. Hence we say that grad ϕ is a covariant vector field.

Example 3. Let $\mathbf{R}(t) = x^i(t)\mathbf{e}_i$, $t \geq 0$ be the position vector of an object moving in space with respect to an oblique coordinate system with basis $\mathbf{e}_i$. Then the velocity of the object is given by

$$\mathbf{v}(t) = \frac{dx^i(t)}{dt} \mathbf{e}_i$$

If we change to another coordinate system defined by $x^{*j} = \alpha^{*j}{}_k \, x^k$ with basis $\mathbf{e}^*_i$, the components $dx^{*i}(t)/dt$ of the velocity vector relative to the x^{*i} coordinate system are related to $dx^i(t)/dt$ by the equation

$$\frac{dx^{*i}(t)}{dt} = \frac{dx^{*i}}{dx^j} \frac{dx^j(t)}{dt} = \alpha^{*i}{}_j \frac{dx^j(t)}{dt}$$

This is the transformation law (6.23) for the contravariant components of a vector. Further, since $\alpha^{*i}{}_j \, \mathbf{e}^*_i = \mathbf{e}_j$, we have

$$\mathbf{v}^*(t) = \frac{dx^{*i}(t)}{dt}\mathbf{e}^*_i = \alpha^{*i}_{\ j}\frac{dx^j(t)}{dt}\mathbf{e}^*_i$$

$$= \alpha^{*i}_{\ j}\mathbf{e}^*_i\frac{dx^j(t)}{dt} = \frac{dx^j(t)}{dt}\mathbf{e}_j = \mathbf{v}(t)$$

Thus the natural expression for velocity is in terms of contravariant components. Hence we say that velocity is a contravariant vector.

Example 4. Let $\mathbf{A} = a^i\mathbf{e}_i = a_i\mathbf{e}^i$ and $\mathbf{B} = b^i\mathbf{e}_i = b_i\mathbf{e}^i$ be two vectors represented by their covariant as well as contravariant components. Then the scalar product of **A** and **B** can be obtained in four different ways, namely

$$\begin{aligned}\mathbf{A}\cdot\mathbf{B} &= (a^i\,\mathbf{e}_i)\cdot(b_j\,\mathbf{e}^j) = a^i\,b_i\\ &= (a_i\,\mathbf{e}^i)\cdot(a^j\,\mathbf{e}_j) = a_i\,b^i\\ &= (a^i\mathbf{e}_i\,)\cdot(b^j\,\mathbf{e}_j) = a^i\,b^j\mathbf{e}_i\cdot\mathbf{e}_j\\ &= (a_i\,\mathbf{e}^i)\cdot(b_j\,\mathbf{e}^j) = a_i\,b_j\mathbf{e}^i\cdot\mathbf{e}^j\end{aligned}$$

It appears that the natural expression for the scalar product of two vectors involves the covariant components of one vector and the contravariant components of the other. In fact, under a coordinate transformaion, we see that

$$a^{*i}b^*_i = (\alpha^{*i}_{\ p}\,a^p)(\alpha\,{*_i}^q\,b_q) = (\alpha^{*i}_{\ p}\,\alpha\,{*_i}^q)\,a^pb_q = a^pb_p$$

since $\alpha^{*i}_{\ p}\alpha\,{*_i}^q = \delta_p^{\ q}$. Likewise, we have

$$a^*_ib^{*i} = (\alpha\,{*_i}^p\,a_p)(\alpha^{*i}_{\ q}\,b^q) = a_pb^p$$

noting that $\alpha{*_i}^p\,\alpha^{*i}_{\ q} = \delta^{*p}_{\ q}$. This shows again the invariant property of the scalar product. The other two expressions involving both covariant or both contravavariant components require the sets of scalar products $\mathbf{e}_i\cdot\mathbf{e}_j$ or $\mathbf{e}^i\cdot\mathbf{e}^j$. These sets define what is called the metric tensor which is discussed in a later section.

Tensors of Higher Order

To define tensors of higher order, we need to introduce a more general kind of notation that is universal for general tensors. First, from a given coordinate transformation $x^{*i} = \alpha^{*i}{}_j\, x^j$ or its inverse $x^i = \alpha_*{}^i{}_j\, x^{*j}$, we observe that

$$\alpha^{*}{}^{i}{}_{j} = \frac{\partial x^{*i}}{\partial x^{j}} \quad \text{and} \quad \alpha_{*}{}^{i}{}_{j} = \frac{\partial x^{i}}{\partial x^{*j}} \tag{6.26}$$

Accordingly, the transformation laws for the covariant (6.22) and the contravariant (6.23) components of a vector (tensor of order one) can be written equivalently as

$$a^{*}{}_{i} = \frac{\partial x^{j}}{\partial x^{*i}}\, a_{j} \quad \text{and} \quad a^{*i} = \frac{\partial x^{*i}}{\partial x^{j}}\, a^{j} \qquad (i = 1, 2, 3)$$

(Recall the summation convention with respect to repeated indices that are both subscripts or superscripts.)

Definition 2. A tensor of order two is a quantity consisting of 3^2 (nine) covariant component a_{ij} or 3^2 contravariant components a^{ij} which transform under a change of coordinates according to the law

$$a^{*}{}_{ij} = \frac{\partial x^{p}}{\partial x^{*i}} \frac{\partial x^{q}}{\partial x^{*j}}\, a_{pq} \qquad (i, j = 1, 2, 3) \tag{6.27}$$

or

$$a^{*ij} = \frac{\partial x^{*i}}{\partial x^{p}} \frac{\partial x^{*j}}{\partial x^{q}}\, a^{pq} \qquad (i, j = 1, 2, 3) \tag{6.28}$$

As we shall see later on, these are the same transformation laws that govern general tensors of order two when x^i and x^{*i} denote general curvilinear coordinates. Notice that in cartesian coordinate systems (rectangular or oblique), the quantities $\partial x^p/\partial x^{*q}$ and $\partial x^{*p}/\partial x^q$ are constants because the coordinate transformations involve only linear equations. In general curvilinear coordinate systems, these quantities are no longer constants. This is a distinct difference between cartesian tensors (in rectangular or oblique coordinate system) and general tensors.

Example 5. The components of a second order symmetric cartesian tensor are given by

$$(a_{ij}) = \begin{bmatrix} 2 & -1 & 0 \\ -1 & 0 & 3 \\ 0 & 3 & 2 \end{bmatrix}$$

Find the covariant and the contravariant components of the tensor in the oblique coordinate system x^i defined by the basis

$$\mathbf{e}^*_1 = \mathbf{i}_1 + \mathbf{i}_2, \quad \mathbf{e}^*_2 = \mathbf{i}_2 - \mathbf{i}_3, \quad \mathbf{e}^*_3 = \mathbf{i}_1 + 2\mathbf{i}_3$$

(See Example 1.)

Solution: We note that for cartesian tensors, $\mathbf{e}_i = \mathbf{e}^i = \mathbf{i}_i$ $(1 \le i \le 3)$ and $a_{ij} = a^{ij}$. Hence, from Example 1, we have

$$\mathbf{e}^{*1} = 2\mathbf{i}_1 - \mathbf{i}_2 - \mathbf{i}_3, \quad \mathbf{e}^{*2} = -2\mathbf{i}_1 + 2\mathbf{i}_2 + \mathbf{i}_3, \quad \mathbf{e}^{*3} = -\mathbf{i}_1 + \mathbf{i}_2 + \mathbf{i}_3$$

so that $\alpha^{*i}{}_j = \mathbf{e}^{*i} \cdot \mathbf{e}_j$ is defined by the matrix

$$\left(\alpha^{*i}{}_j\right) = \begin{bmatrix} 2 & -1 & -1 \\ -2 & 2 & 1 \\ -1 & 1 & 1 \end{bmatrix}$$

The inverse matrix defined by $\alpha_*{}^i{}_j = \mathbf{e}^i \cdot \mathbf{e}^*_j$ is given by

$$\left(\alpha_*{}^i{}_j\right) = \begin{bmatrix} 1 & 0 & 1 \\ 1 & 1 & 0 \\ 0 & -1 & 2 \end{bmatrix}$$

Hence, by the formula $a^*_{ij} = \alpha_*{}_i{}^p \alpha_*{}_j{}^q a_{pq}$, we find

$$a^*_{11} = \alpha_*{}_1{}^p \alpha_*{}_1{}^q a_{pq} = 0, \qquad a^*_{12} = a^*_{21} = \alpha_*{}_1{}^p \alpha_*{}_2{}^q a_{pq} = -4$$

$$a^*_{13} = a^*_{31} = \alpha_*{}_1{}^p \alpha_*{}_3{}^q a_{pq} = 7, \qquad a^*_{22} = \alpha_*{}_2{}^p \alpha_*{}_2{}^q a_{pq} = -4$$

$$a^*_{23} = a^*_{32} = \alpha_*{}_2{}^p \alpha_*{}_3{}^q a_{pq} = 1, \qquad a^*_{33} = \alpha_*{}_3{}^p \alpha_*{}_3{}^q a_{pq} = 10$$

or, in matrix form

$$(a^*_{ij}) = \begin{bmatrix} 0 & -4 & 7 \\ -4 & -4 & 1 \\ 7 & 1 & 10 \end{bmatrix}$$

The contravariant components are given by the formula

$$a^{*ij} = \alpha^{*i}{}_p \, \alpha^{*j}{}_q \, a^{pq} .$$

Thus we find

$$a^{*11} = \alpha^{*1}{}_p \, \alpha^{*1}{}_q \, a^{pq} = 20, \quad a^{*12} = a^{*21} = \alpha^{*1}{}_p \, \alpha^{*2}{}_q \, a^{pq} = -25$$

$$a^{*13} = a^{*31} = \alpha^{*1}{}_p \alpha^{*3}{}_q \, a^{pq} = -15, \quad a^{*22} = \alpha^{*2}{}_p \, \alpha^{*2}{}_q \, a^{pq} = 30$$

$$a^{*23} = a^{*32} = \alpha^{*2}{}_p \alpha^{*3}{}_q \, a^{pq} = 19, \quad a^{*33} = \alpha^{*3}{}_p \, \alpha^{*3}{}_q \, a^{pq} = 12$$

or, in matrix form

$$(a^{*ij}) = \begin{bmatrix} 20 & -25 & -15 \\ -25 & 30 & 19 \\ -15 & 19 & 12 \end{bmatrix}$$

Notice that both (a^{*ij}) and (a_{*ij}) are symmetric matrices; thus we see that the symmetric property of the tensor is invariant under a coordinate transformation.

Example 6. Let a_i and b_i be the covariant components of two vectors, and define $c_{ij} = a_i \, b_j$ (i, j = 1, 2, 3). Show that c_{ij} are components of a second order tensor.

Solution: The components a_i and b_j transform according to the law (6.22). Hence, under a change of coordinate system, we have

$$c^*{}_{ij} = a^*{}_i \, b^*{}_j = \frac{\partial x^p}{\partial x^{*i}} \frac{\partial x^q}{\partial x^{*j}} a_p b_q$$

$$= \frac{\partial x^p}{\partial x^{*i}} \frac{\partial x^q}{\partial x^{*j}} c_{pq}$$

which is in accord with the transformation law (6.27) for the covariant components of a second order tensor.

Definition 3. A *tensor of order r* is a quantity consisting of 3^r covariant components

$$a_{i_1 i_2 \ldots i_r} \quad (i_1, i_2, \ldots, i_r = 1, 2, 3)$$

or 3^r contravariant components

$$a^{i_1 i_2 \ldots i_r} \quad (i_1, i_2, \ldots, i_r = 1, 2, 3)$$

which transform under a change of coordinates according to the law

$$a^*_{i_1 i_2 \ldots i_r} = \frac{\partial x^{j_1}}{\partial x^{*i_1}} \frac{\partial x^{j_2}}{\partial x^{*i_2}} \cdots \cdots \frac{\partial x^{j_r}}{\partial x^{*i_r}} a^*_{j_1 j_2 \ldots j_r} \tag{6.29}$$

or

$$a^{*i_1 i_2 \ldots i_r} = \frac{\partial x^{*i_1}}{\partial x^{j_1}} \frac{\partial x^{*i_2}}{\partial x^{j_2}} \cdots \cdots \frac{\partial x^{*i_r}}{\partial x^{j_r}} a^{*j_1 j_2 \ldots j_r} \tag{6.30}$$

Although we shall not be dealing too much with it, we should mention that there are also so-called *mixed tensors* whose components have covariant as well as contravariant indices. For example, if we take the outer product of two vectors using the covariant components of one vector and the contravariant components of the other, say, $c_i^{\,j} = a_i b^j$, we obtain components of a second order mixed tensor. The components transform according to the law

$$c^{*\,j}_{\;i} = \frac{\partial x^p}{\partial x^{*i}} \frac{\partial x^{*j}}{\partial x^q} c_p^{\,q} \tag{6.31}$$

where $c^{*\,j}_{\;i} = a^*_i b^{*j}$. As another example, the outer product of a second order tensor with components a_{ij} and a vector with components b^k leads to third order mixed tensor whose components are given by $c_{ij}^{\;k} = a_{ij} b^k$. The components of this tensor are governed by the transformation law

$$c^{*\;k}_{\;ij} = \frac{\partial x^p}{\partial x^{*i}} \frac{\partial x^q}{\partial x^{*j}} \frac{\partial x^{*k}}{\partial x^r} c_{pq}^{\;r} \tag{6.32}$$

where $c^{*\;k}_{\;ij} = a^*_{ij} b^{*k}$.

6.4 ALGEBRA OF TENSORS IN OBLIQUE COORDINATES

With slight modifications, tensors in nonrectangular cartesian coordinate systems possess much the same algebraic properties as cartesian tensors. The modifications are necessitated by the difference in the transformation laws governing covariant and contravariant components of tensors. For instance, the sum of two tensors will be defined only if the tensors have the same order and the components have the same character, that is, the number of covariant (subscripts) or contravariant (superscripts) indices must match. For example, if a_{jk} and b_{qr} are the components of two second order tensors both having two covariant indices, then the sum of the two tensors is a tensor whose components are given by $c_{jk} = a_{jk} + b_{jk}$ (j, k = 1, 2, 3). On the other hand, the outer product of tensors are always defined regardless of the order and character of the individual components. The order and character of the product is determined in an obvious manner. For example, the outer product of two tensors with respective components a_{ij} and b^{kmn} is a tensor whose components are given by $c_{ij}{}^{kmn} = a_{ij}b^{kmn}$, which are of the mixed type. The verification of these facts are similar to those given for cartesian tensors in the previous chapter.

One aspect of the algebra of cartesian tensors that needs certain modification is the operation of contraction and, as a result, the operation of inner product. In order for the contraction of two tensors to be valid, the contraction must be performed over a pair of indices, one of which must be a subscript (covariant index) and the other a superscript (contravariant index). Under this condition, Theorem 1 of Sec. 5.7 continues to hold. For example, let us consider the contraction of a third order mixed tensor with the components $a^i{}_{jk}$. Contraction with respect to the first subscript yields $b_j = a^i{}_{ij}$ while contraction with respect to the second subscript gives $c_i = a^j{}_{ij}$. Both of these contractions lead to tensors of the first order or vectors. In fact, from the transformation law

$$a^{*i}{}_{jk} = \frac{\partial x^{*i}}{\partial x^p} \frac{\partial x^q}{\partial x^{*j}} \frac{\partial x^r}{\partial x^{*k}} a^p{}_{qr}$$

and the identity

$$\frac{\partial x^{*i}}{\partial x^p} \frac{\partial x^q}{\partial x^{*i}} = \delta^q_p$$

we see that

$$a^{*i}_{\ ik} = \delta^q_p \frac{\partial x^r}{\partial x^{*k}} a^p_{\ qr} = \frac{\partial x^r}{\partial x^{*k}} a^p_{\ pr}$$

or

$$b^*_{\ k} = \frac{\partial x^r}{\partial x^{*k}} b_r$$

where $b^*_j = a^{*i}_{\ ij}$. This shows that the quantities $b_r = a^p_{\ pr}$ $(r = 1, 2, 3)$ are components of a vector. A similar proof can also be shown for the quantities $c_i = a^j_{\ ij}$.

Therefore, the inner product of two tensors of order r and s is a tensor of order r + s - 2, provided, of course, that the contraction is performed over a pair of indices one of which is a subscript and the other a superscript.

Theorem 2 of Sec. 5.7 (quotient rule) has its counterpart for noncartesian tensors. That is, if

$$T^{j_1 \ldots j_p}_{i_1 \ldots i_q}$$

are quantities which have the property that their outer or inner product with an arbitrary tensor yields a nonzero tensor of an appropriate order, then the quantities are components of a mixed tensor of order k, where k = m + n.

To indicate the method of proof, let us suppose that $T^i_{\ jk}$ are quantities which have the property that for an arbitrary vector with components a^k, the inner product $T^i_{\ jk}a^k = b^i_{\ j}$ results in components $b^i_{\ j}$ of a mixed tensor of order two. Let $T^{*i}_{\ jk}$ denote the value of the quantities $T^i_{\ jk}$ in the coordinate system x^{*i}. For an arbitrary vector a^k, let a^{*k} be the corresponding components in the coordinate system x^{*i}, and set $T^{*i}_{\ jk}\, a^{*k} = b^{*i}_{\ j}$. Since $b^{*i}_{\ j}$ are components of a mixed tensor, they satisfy the transformation law

$$b^{*p}_{\ q} = \frac{\partial x^{*p}}{\partial x^s} \frac{\partial x^{*t}}{\partial x^q} b^s_t$$

Likewise, we have

$$a^k = \frac{\partial x^k}{\partial x^{*r}} a^{*r}$$

Hence

$$T^{*p}_{qr} a^{*r} = b^{*p}_{q} = \frac{\partial x^{*p}}{\partial x^s} \frac{\partial x^t}{\partial x^{*q}} b^s_t$$

$$= \frac{\partial x^{*p}}{\partial x^s} \frac{\partial x^t}{\partial x^{*q}} T^s_{tk} a^k = \frac{\partial x^{*p}}{\partial x^s} \frac{\partial x^t}{\partial x^{*q}} \frac{\partial x^k}{\partial x^{*r}} a^{*r} T^s_{tk}$$

so that

$$(T^{*p}_{qr} - \frac{\partial x^{*p}}{\partial x^s} \frac{\partial x^t}{\partial x^{*q}} \frac{\partial x^k}{\partial x^{*r}} T^s_{tk}) a^{*r} = 0$$

Since this is true for an arbitrary vector, we conclude that the term inside the parentheses must vanish. Thus the quantities $T^i{}_{jk}$ transform as components of a third order mixed tensor.

It should be noted that all the algebraic properties satisfied by tensors in oblique coordinate systems are also satisfied by general tensors. For this reason, we shall not present a separate discussion of the algebra of general tensors.

Example 1. Suppose the quantities T_{ij} have the property that $T_{ij}\, a^i b^j$ is a scalar in any coordinate system, where a^i and b^j are components of two arbitrary vectors. Show that T_{ij} are components of a second order tensor.

Solution: Let T^*_{ij} denote the values of T_{ij} in the coordinate system x^{*i}. For any vectors with components a^i and b^j, we have

$$T^*_{ij} a^{*i} b^{*j} = T_{pq} a^p b^q, \qquad \text{(a scalar)}$$

Since

$$a^p = \frac{\partial x^p}{\partial x^{*i}} a^{*i}, \qquad b^q = \frac{\partial x^q}{\partial x^{*j}} b^{*j}$$

we find

$$T^*_{ij} a^{*i} b^{*j} = T_{pq} \frac{\partial x^p}{\partial x^{*i}} a^{*i} \frac{\partial x^q}{\partial x^{*j}} b^{*j}$$

$$= T_{pq} \frac{\partial x^p}{\partial x^{*i}} \frac{\partial x^q}{\partial x^{*j}} a^{*i} b^{*j}$$

Hence

$$(T^*_{ij} - T_{pq} \frac{\partial x^p}{\partial x^{*i}} \frac{\partial x^q}{\partial x^{*j}}) a^{*i} b^{*j} = 0$$

from which we conclude that the term inside the parentheses must be zero. Thus we obtain

$$T^*_{ij} = \frac{\partial x^p}{\partial x^{*i}} \frac{\partial x^q}{\partial x^{*j}} T_{pq}$$

which shows that the quantities T_{ij} indeed transform as components of a second order tensor.

Example 2. Let a^i and b_i be the contravariant and covariant components of two vectors **A** and **B**, respectively. Then the quantities $c^i_{.j} = a^i b_j$ form as components of a second order mixed tensor, which is the outer product of **A** and **B**. In fact, since

$$a^{*i} = \frac{\partial x^{*i}}{\partial x^p} a^p, \qquad b^*_j = \frac{\partial x^q}{\partial x^{*j}} b_q$$

we see that

$$c^{*i}_{\cdot j} = a^{*i} b^*_j = \frac{\partial x^{*i}}{\partial x^p} \frac{\partial x^q}{\partial x^{*j}} a^p b_q$$

$$= \frac{\partial x^{*i}}{\partial x^p} \frac{\partial x^q}{\partial x^{*j}} c^p_{\cdot q}$$

The contraction of $c^{i}_{.j}$ yields $c^{i}_{.i} = a^{i}b_{i} = \mathbf{A}\cdot\mathbf{B}$, which is the inner product of the vectors.

6.2 EXERCISES

1. Deduce from (6.17), (6.18), (6.19) the expression

$$\mathbf{A} = (\mathbf{A}\cdot \mathbf{e}_i)\mathbf{e}^i = (\mathbf{A}\cdot \mathbf{e}^i)\mathbf{e}_i$$

and then, by replacing $\mathbf{A}$ by $\mathbf{e}^{*i}$ and $\mathbf{e}^{*}_{i}$, obtain the results (6.13) and (6.16).

2. In each of the following cases, find the covariant and contravariant components of the given vector in the oblique coordinate system defined by the given basis $\mathbf{e}_i$ ($i = 1, 2, 3$).

   (a) $\mathbf{A} = \mathbf{i}_1 + 2\mathbf{i}_2$; $\mathbf{e}_1 = -\mathbf{i}_1 + 2\mathbf{i}_2$, $\mathbf{e}_2 = \mathbf{i}_1 + \mathbf{i}_2$.
   (b) $\mathbf{A} = -2\mathbf{i}_1 + 3\mathbf{i}_2$; $\mathbf{e}_1 = \mathbf{i}_1 + 2\mathbf{i}_2$, $\mathbf{e}_2 = -\mathbf{i}_1 + \mathbf{i}_2$.
   (c) $\mathbf{A} = \mathbf{i}_1 + 2\mathbf{i}_2 - 3\mathbf{i}_3$; $\mathbf{e}_1 = \mathbf{i}_1 + \mathbf{i}_2$, $\mathbf{e}_2 = \mathbf{i}_2 + \mathbf{i}_3$, $\mathbf{e}_3 = \mathbf{i}_1 + \mathbf{i}_3$.
   (d) $\mathbf{A} = 3\mathbf{i}_1 - \mathbf{i}_2 + 2\mathbf{i}_3$; $\mathbf{e}_1 = \mathbf{i}_1 - \mathbf{i}_2$, $\mathbf{e}_2 = \mathbf{i}_2 - \mathbf{i}_3$, $\mathbf{e}_3 = \mathbf{i}_1 + \mathbf{i}_3$.

3. A vector field $\mathbf{F}$ has components $F_1 = x_1x_2$, $F_2 = x_3x_2$, $F_3 = x_1x_3$ with respect to a rectangular cartesian coordinate system x_i. Find its components in the oblique coordinate system x^i defined by the basis $\mathbf{e}_i$ given in Problem 2(c).

4. Repeat Problem 3 for the vector field $\mathbf{F} = x_1^2\,\mathbf{i}_1 + x_2^2\,\mathbf{i}_2 + x_3^2\,\mathbf{i}_3$, where the oblique coordinate system is defined by the basis $\mathbf{e}_i$ of Problem 2(d).

5. Find the covariant and contravariant components of the cartesian tensor

$$(a_{ij}) = \begin{bmatrix} 2 & 0 & 3 \\ 0 & -1 & 0 \\ 0 & 2 & 1 \end{bmatrix}$$

in the oblique coordinate system defined by the basis

$$\mathbf{e}_1 = \mathbf{i}_1 + \mathbf{i}_3, \quad \mathbf{e}_2 = \mathbf{i}_1 + 2\mathbf{i}_2 + \mathbf{i}_3, \quad \mathbf{e}_3 = -\mathbf{i}_1 + \mathbf{i}_2 + \mathbf{i}_3.$$

6. Find the covariant and contravariant components of the carte-

sian tensor

$$(t_{ij}) = \begin{bmatrix} x_1 & -x_2 & 0 \\ x_2 & 0 & x_3 \\ 0 & -x_3 & x_1 \end{bmatrix}$$

in the oblique coordinate system defined by the basis

$$\mathbf{e}_1 = -\mathbf{i}_1 + \mathbf{i}_2, \ \mathbf{e}_2 = \mathbf{i}_2 + \mathbf{i}_3, \ \mathbf{e}_3 = \mathbf{i}_1 - \mathbf{i}_3$$

7. Write the transformation law for a third order (a) covariant tensor, (b) contravariant tensor, (c) mixed tensor with two covariant indices.
8. Write the transformation law for a fifth order mixed tensor having two covariant and three contravariant indices.
9. Let $\mathbf{A} = (a_{ij})$, $\mathbf{A}^* = (a^*_{ij})$, $\Lambda_* = (\alpha_*{}^i{}_j)$, $\alpha_*{}^i{}_j = \partial x^i/\partial x^j$, with superscripts indicating rows. Show that the transformation law

$$a^*_{ij} = \frac{\partial x^p}{\partial x^{*i}} \frac{\partial x^q}{\partial x^{*j}} a_{pq}$$

is equivalent to the matrix multiplication $A_* = \Lambda_*{}^T A \, \Lambda_*$.

10 Obtain the corresponding matrix formula for the tranformation law

$$a^{*ij} = \frac{\partial x^{*i}}{\partial x^p} \frac{\partial x^{*j}}{\partial x^q} a^{pq}$$

11. If a_{ij} and b^{kmn} are components of two tensors, show that the quantities $c_{ij}{}^{kmn} = a_{ij}b^{kmn}$ are components of a mixed tensor, covariant of order two and contravariant of order three.
12. If a_{ij} and b^{km} are components of two tensors, show that $c_i{}^k = a_{ij}b^{jk}$ are components of mixed tensor of order two.
13. Suppose that the quantities $X_i{}^{pq}$ have the property that for arbitrary tensor with components $a_j{}^r$, the outer product $b_{ij}{}^{pqr} = X_i{}^{pq}a_j{}^r$ are components of a tensor. Show that $X_i{}^{pq}$ are components of a mixed tensor.

6.5 THE METRIC TENSOR

In Sec. 3.11 we introduce the notion of a metric in connection with the expression for arc length in a curvilinear coordinate system. The metric is defined by the quantities g_{ij} given by

$$g_{ij} = \frac{\partial \mathbf{R}}{\partial u^i} \cdot \frac{\partial \mathbf{R}}{\partial u^j} \qquad (i, j = 1, 2, 3) \tag{6.33}$$

where u^i denotes the curvilinear coordinates. Now in an oblique coordinate system x^i defined by a basis $\mathbf{e}_i$, we have $\mathbf{R} = x^i \, \mathbf{e}_i$ so that

$$\frac{\partial \mathbf{R}}{\partial u^i} = \frac{\partial \mathbf{R}}{\partial x^i} = \mathbf{e}_i \qquad (i, j = 1, 2, 3)$$

Thus, in an oblique coordinate system, the quantities g_{ij} are given by

$$g_{ij} = \mathbf{e}_i \cdot \mathbf{e}_j \qquad (i, j = 1, 2, 3) \tag{6.34}$$

If the base vectors $\mathbf{e}_i$ $(1 \le i \le 3)$ are unit vectors, then $g_{ij} = \cos\theta_{ij}$, where θ_{ij} denotes the angle between the x^i - and x^j - coordinate axes. In particular, if the coordinate system is rectangular, so that the basis is an orthogonormal set, then $g_{ij} = \delta_{ij}$.

The quantities g_{ij} defined in (6.33) form the covariant components of a second order tensor. In fact, from the transformation law (6.16) of the basis and (6.26), we see that

$$g^*_{ij} = \mathbf{e}^*_i \cdot \mathbf{e}^*_j = \frac{\partial x^p}{\partial x^{*i}} \frac{\partial x^q}{\partial x^{*j}} \mathbf{e}_p \cdot \mathbf{e}_q = \frac{\partial x^p}{\partial x^{*i}} \frac{\partial x^q}{\partial x^{*j}} g_{pq} \tag{6.35}$$

Thus the quantities g_{ij} transform as covariant components of a second order tensor. This tensor is called the *metric tensor* of space with respect to the coordinate system x^i. Since $g_{ij} = g_{ji}$, the metric tensor is symmetric.

Using the reciprocal basis $\mathbf{e}^i$, we can likewise define the quantities g^{ij} by the equation

$$g^{ij} = \mathbf{e}^i \cdot \mathbf{e}^j \qquad (i, j = 1, 2, 3) \tag{6.36}$$

From the transformation law of the reciprocal basis, it follows that

$$g^{*ij} = \frac{\partial x^{*i}}{\partial x^p} \frac{\partial x^{*j}}{\partial x^q} g^{pq} \tag{6.37}$$

Thus the quantities g^{ij} are the contravariant components of a second order tensor. This tensor is precisely the metric tensor whose covariant components are the quantities g_{ij}.

We note that the g_{ij} and g^{ij} satisfy the important relation

$$g_{ij}\, g^{jk} = \delta_i^{\,k} \tag{6.38}$$

To see this, we observe that the metric tensor relates the basis $\mathbf{e}_i$ to its reciprocal basis $\mathbf{e}^i$ according to the formula

$$\mathbf{e}_i = g_{ij}\, \mathbf{e}^j \tag{6.39}$$

This follows from the identity $\mathbf{A} = (\mathbf{A}\cdot \mathbf{e}_j)\mathbf{e}^j$ when we replace $\mathbf{A}$ by $\mathbf{e}_i$. Now, if we take the dot product of both sides of (6.39) with $\mathbf{e}^k$, noting that $\mathbf{e}_i\cdot \mathbf{e}^k = \delta_i^{\,k}$, we find

$$\delta_i^{\,k} = g_{ij}\mathbf{e}^j\cdot \mathbf{e}^k = g_{ij}\, g^{jk}$$

which establishes (6.38).

The relation (6.38) implies that the matrix (g^{ij}) is the inverse of the matrix (g_{ij}), and vice versa. This enables us to relate the cofactors of the elements of (g_{ij}) with the elements of (g^{ij}). Indeed, from linear algebra, we recall that if $A = (a_{ij})$ is a nonsingular matrix, its inverse is given by the formula

$$A^{-1} = \frac{\operatorname{adj} A}{|A|}$$

where adj A denotes the adjoint of A. The adjoint of A is defined as the transpose of the matrix (A_{ij}), that is, adj $A = (A_{ij})^T$, where A_{ij} is the cofactor of the element a_{ij}. Now since (g_{ij}) is symmetric, its adjoint matrix is also symmetric because the cofactor of g_{ij} is the same as the cofactor of g_{ji}. Hence, if we denote the cofactor of g_{ij} by G^{ij} and the determinant of (g_{ij}) by g, then from the above formula, we have

$$G^{ij} = g\, g^{ij} = \text{cofactor of } g_{ij} \tag{6.40}$$

This result is required in Sec. 6.11.

Example 1. Let $\mathbf{e}_1 = \mathbf{i}_1 + \mathbf{i}_2$, $\mathbf{e}_2 = \mathbf{i}_2 + \mathbf{i}_3$, $\mathbf{e}_3 = \mathbf{i}_1 + \mathbf{i}_3$. Find the quantities g_{ij} and g^{ij}, and show that $(g_{ij})(g^{ij})$ is the identity matrix.

Solution: By (6.34) we find

$$g_{11} = \mathbf{e}_1 \cdot \mathbf{e}_1 = 2, \qquad g_{12} = g_{21} = \mathbf{e}_1 \cdot \mathbf{e}_1 = 1$$
$$g_{22} = \mathbf{e}_2 \cdot \mathbf{e}_2 = 2, \qquad g_{13} = g_{31} = \mathbf{e}_1 \cdot \mathbf{e}_3 = 1$$
$$g_{33} = \mathbf{e}_3 \cdot \mathbf{e}_3 = 2, \qquad g_{23} = g_{32} = \mathbf{e}_2 \cdot \mathbf{e}_3 = 1$$

Next, we determine the reciprocal basis $\mathbf{e}^i$ (i = 1, 2, 3). Since $\mathbf{e}_1 \cdot \mathbf{e}_2 \times \mathbf{e}_3 = 2$, we find

$$\mathbf{e}^1 = (\mathbf{e}_2 \times \mathbf{e}_3)/2 = (\mathbf{i}_1 + \mathbf{i}_2 - \mathbf{i}_3)/2$$
$$\mathbf{e}^2 = (\mathbf{e}_3 \times \mathbf{e}_1)/2 = (-\,\mathbf{i}_1 + \mathbf{i}_2 + \mathbf{i}_3)/2$$
$$\mathbf{e}^3 = (\mathbf{e}_1 \times \mathbf{e}_2)/2 = (\mathbf{i}_1 - \mathbf{i}_2 + \mathbf{i}_3)/2$$

Therefore, by (6.36), we have

$$g^{11} = \mathbf{e}^1 \cdot \mathbf{e}^1 = 3/4, \qquad g^{12} = g^{21} = \mathbf{e}^1 \cdot \mathbf{e}^2 = -1/4$$
$$g^{22} = \mathbf{e}^2 \cdot \mathbf{e}^2 = 3/4, \qquad g^{13} = g^{31} = \mathbf{e}^1 \cdot \mathbf{e}^3 = -1/4$$
$$g^{33} = \mathbf{e}^3 \cdot \mathbf{e}^3 = 3/4, \qquad g^{23} = g^{32} = \mathbf{e}^2 \cdot \mathbf{e}^3 = -1/4$$

It follows that

$$(g_{ij})(g^{ij}) = \begin{bmatrix} 2 & 1 & 1 \\ 1 & 2 & 1 \\ 1 & 1 & 2 \end{bmatrix} \begin{bmatrix} 3/4 & -1/4 & -1/4 \\ -1/4 & 3/4 & -1/4 \\ -1/4 & -1/4 & 3/4 \end{bmatrix} = \begin{bmatrix} 1 & 0 & 0 \\ 0 & 1 & 0 \\ 0 & 0 & 1 \end{bmatrix}$$

Relationships Between Covariant and Contravariant Components

The covariant and contravariant components of a tensor are related to each other through the metric tensor. Here we derive the relationship for tensors of the first order or vectors. Let us consider a vector $\mathbf{A} = a_j \mathbf{e}^j = a^j \mathbf{e}_j$ represented by its covariant as well as contravariant components in a coordinate system defined by the basis $\mathbf{e}_i$ $(1 \le i \le 3)$. Taking the dot product of the vector with the base vector $\mathbf{e}_i$, noting that $\mathbf{e}^j \cdot \mathbf{e}_k = \delta_k{}^j$, we find

$$a_i = (\mathbf{e}^j \cdot \mathbf{e}_i)a_j = (\mathbf{e}_i \cdot \mathbf{e}_j)a^j = g_{ij}a^j \qquad (i = 1, 2, 3) \tag{6.41}$$

In matrix notation this can be written as

$$\begin{bmatrix} a_1 \\ a_2 \\ a_3 \end{bmatrix} = \begin{bmatrix} g_{11} & g_{12} & g_{13} \\ g_{21} & g_{22} & g_{23} \\ g_{31} & g_{32} & g_{33} \end{bmatrix} \begin{bmatrix} a^1 \\ a^2 \\ a^3 \end{bmatrix}$$

Similarly, by taking the dot product of the vector with the reciprocal base vector $\mathbf{e}^i$, we find

$$a^i = (\mathbf{e}_j \cdot \mathbf{e}^i)a^j = (\mathbf{e}^i \cdot \mathbf{e}^j)a_j = g^{ij}a_j \qquad (i = 1, 2, 3) \tag{6.42}$$

Thus we can shift from covariant to contravariant components or vice versa through the metric tensor. Notice that both (6.41) and (6.42) are inner product of the vector with the metric tensor. In a rectangular cartesian coordinate, we have $g_{ij} = \delta_{ij}$, so that $a^i = a_i$. Hence for cartesian tensors there is no distinction between covariant and contravariant components.

The relationships among the components a_i, a^i, a^*_i, a^{*i} of a vector under a transformation of coordinates are best depicted in the following schematic diagram:

x^i - coordinate system

$$a_i = g_{ij}a^j = \frac{\partial x^{*j}}{\partial x^i} a^*_j \quad \longleftrightarrow \quad a^i = g^{ij}a_j = \frac{\partial x^i}{\partial x^{*j}} a^{*j}$$

$$\updownarrow \qquad\qquad \updownarrow$$

x^{*i} - coordinate system (6.43)

$$a^*_i = g^*_{ij} a^{*j} = \frac{\partial x^j}{\partial x^{*i}} a_j \quad \text{<-------->} \quad a^{*i} = g^{*ij} a^*_j = \frac{\partial x^{*i}}{\partial x^j} a^j$$

Similar relationships also hold between the covariant and contravariant components of tensors of higher order. For example, if a^{ij} are the contravariant components of a second order tensor, then its covariant components are given by

$$a_{ij} = g_{ip}\, g_{jq} a^{pq} \tag{6.44}$$

To see this, let a'_{ij} denote the components of the tensor in rectangular cartesian coordinates x_i. Then we have

$$a_{ij} = \frac{\partial x_p}{\partial x^i} \frac{\partial x_q}{\partial x^j} a'_{pq} \quad \text{(sum on p and q)}$$

Since $a'^{pq} = a'_{pq}$ for cartesian tensors and

$$a'^{pq} = \frac{\partial x_p}{\partial x^k} \frac{\partial x_q}{\partial x^m} a^{km} \quad \text{(sum on k and m)}$$

it follows that

$$a_{ij} = \frac{\partial x_p}{\partial x^i} \frac{\partial x_p}{\partial x^k} \frac{\partial x_q}{\partial x^j} \frac{\partial x_q}{\partial x^m} a^{km} \quad \text{(sum on p and q)} \tag{6.45}$$

Now from (6.16) and (6.26), we have

$$\mathbf{e}_i = \frac{\partial x_p}{\partial x^i} \mathbf{i}_p \quad \text{(sum on p)}$$

so that by (6.34)

$$g_{ij} = \mathbf{e}_i \cdot \mathbf{e}_j = \frac{\partial x_p}{\partial x^i} \frac{\partial x_p}{\partial x^j} \quad \text{(sum on p)}$$

Thus (6.45) becomes

$$a_{ij} = g_{ik} g_{jm} a^{km}$$

which is (6.44). Conversely, if a_{ij} are known, then the contravariant components of the tensor are given by

$$a^{ij} = g^{ip}g^{jq}\, a_{pq} \tag{6.46}$$

Both (6.44) and (6.46) involve taking the inner product of the tensor twice with the metric tensor.

It follows from (6.38) that

$$g^{ip}g^{jq}g_{pq} = g^{ip}\delta_p^{\ j} = g^{ij}$$

so that by (6.46) we conclude that the quantities g^{ij} are the contravariant components of the metric tensor.

The inner product of the metric tensor with tensors of order at least two can lead to mixed tensors. For example, the inner product of the metric tensor with a second order tensor with components a_{ij} and a^{ij} leads to a mixed tensor with components defined by

$$a_i^{\cdot k} = g_{ij}a^{jk} = g^{kj}a_{ij} \tag{6.47}$$

and

$$a^i_{\cdot k} = g^{ij}a_{jk} = g_{kj}a^{ij} \tag{6.48}$$

In general the components $a_i^{\cdot k}$ and $a^i_{\cdot k}$ are not identical (see Problem 8).

From (6.47) and (6.48) we observe that taking the inner product of a tensor with the metric tensor has the effect of lowering or raising the index with respect to which the summation is performed according to whether the covariant components g_{ij} or the contravariant components g^{ij} are involved in the product. The spot from which an index has been lowered or raised is indicated by a dot. As further example, let us consider the inner product of the metric tensor with a fourth-order mixed tensor with components a_{ij}^{pq}. The inner product leads to four mixed tensors of the fifth order with their components given by

$$g_{kp}a_{ij}^{pq} = a_{ijk}^{\cdot\, q}, \qquad g^{jk}a_{ij}^{pq} = a_{i\,\cdot}^{pqk}$$

$$g_{kq}a_{ij}^{pq} = a_{ijk}^{p\,\cdot}, \qquad g^{ik}a_{ij}^{pq} = a_{\cdot\, j}^{pqk}$$

We conclude this section by deriving the relationship between the determinant g of the metric tensor matrix (g_{ij}) and the quantity $V = \mathbf{e}_1 \cdot \mathbf{e}_2 \times \mathbf{e}_3$. By (6.39) we note that $\mathbf{e}_1 = g_{1j}\mathbf{e}^j$, $\mathbf{e}_2 = g_{2j}\mathbf{e}^j$, $\mathbf{e}_3 = g_{3j}\mathbf{e}^j$ so that

$$V = \mathbf{e}_1 \cdot \mathbf{e}_2 \times \mathbf{e}_3 = g_{1i}\, g_{2j}\, g_{3k}\, \mathbf{e}^i \cdot \mathbf{e}^j \times \mathbf{e}^k \tag{6.49}$$

[Note that in (6.49) summations are to be performed with respect to the repeated indices i, j, and k.] Now since $V' = \mathbf{e}^1 \cdot \mathbf{e}^2 \times \mathbf{e}^3$ has the same value under cyclic permutations of 1, 2, 3, and since $\mathbf{e}^i \cdot \mathbf{e}^j \times \mathbf{e}^k = 0$ whenever two of the indices are identical, it follows that

$$\begin{aligned} V &= (g_{11}\, g_{22}\, g_{33} - g_{11}\, g_{23}\, g_{32})\, \mathbf{e}^1 \cdot \mathbf{e}^2 \times \mathbf{e}^3 \\ &\quad + (g_{12}\, g_{23}\, g_{31} - g_{12}\, g_{21}\, g_{33})\, \mathbf{e}^1 \cdot \mathbf{e}^2 \times \mathbf{e}^3 \\ &\quad + (g_{13}\, g_{21}\, g_{32} - g_{13}\, g_{22}\, g_{31})\, \mathbf{e}^1 \cdot \mathbf{e}^2 \times \mathbf{e}^3 \\ &= \begin{vmatrix} g_{11} & g_{12} & g_{13} \\ g_{21} & g_{22} & g_{23} \\ g_{31} & g_{32} & g_{33} \end{vmatrix} (\mathbf{e}^1 \cdot \mathbf{e}^2 \times \mathbf{e}^3) = g V' \end{aligned} \tag{6.50}$$

But $V' = 1/V$; therefore, we have $V^2 = g$ or

$$V = \sqrt{g} \tag{6.51}$$

Example 2. Let $\mathbf{A} = a_i\mathbf{e}^i = a^i\mathbf{e}_i$ and $\mathbf{B} = b_i\mathbf{e}^i = b^i\,\mathbf{e}_i$. In terms of the covariant components, the scalar product of these vectors is given by

$$\mathbf{A} \cdot \mathbf{B} = (a_i\mathbf{e}^i) \cdot (b_j\mathbf{e}^j) = g^{ij} a_i\, b_j$$

which involves the contravariant components of the metric tensor. In terms of the contravariant components, the scalar product is given by

$$\mathbf{A} \cdot \mathbf{B} = (\, a^i\mathbf{e}_i) \cdot (b^j\mathbf{e}_j) = g_{ij} a^i\, b^j$$

involving the covariant components of the metric tensor.

Example 3. Express the vector product of two vectors in terms of their covariant and contravariant components, assuming a right-handed coordinate system.

Solution: First, suppose $\mathbf{A} = a^i\mathbf{e}_i$ and $\mathbf{B} = b^j\mathbf{e}_j$. Then, by definition, we have

$$\mathbf{A} \times \mathbf{B} = a^i b^j(\mathbf{e}_i \times \mathbf{e}_j) = (a^i b^j - a^j b^i)(\mathbf{e}_i \times \mathbf{e}_j)$$

where i, j take on the values 1, 2; 2, 3; 3, 1 only. Since

$$\mathbf{e}^k = (\mathbf{e}_i \times \mathbf{e}_j)/V \qquad (V = \mathbf{e}_1 \cdot \mathbf{e}_2 \times \mathbf{e}_3)$$

it follows that

$$\mathbf{A} \times \mathbf{B} = V(a^i b^j - a^j b^i)\mathbf{e}^k$$

where i, j, k are cyclic permutations of 1, 2, 3. Thus we find

$$\mathbf{A} \times \mathbf{B} = V[(a^2 b^3 - a^3 b^2)\mathbf{e}^1 + (a^3 b^1 - a^1 b^3)\mathbf{e}^2 + (a^1 b^2 - a^2 b^1)\mathbf{e}^3]$$

$$= V\begin{vmatrix} \mathbf{e}^1 & \mathbf{e}^2 & \mathbf{e}^3 \\ a^1 & a^2 & a^3 \\ b^1 & b^2 & b^3 \end{vmatrix}$$

Next, suppose $\mathbf{A} = a_i\mathbf{e}^i$ and $\mathbf{B} = b_j\mathbf{e}^j$. Then, by a similar argument, we find

$$\mathbf{A} \times \mathbf{B} = V^{-1}(a_i b_j - a_j b_i)\mathbf{e}_k$$

$$= \frac{1}{V}\begin{vmatrix} \mathbf{e}_1 & \mathbf{e}_2 & \mathbf{e}_3 \\ a_1 & a_2 & a_3 \\ b_1 & b_2 & b_3 \end{vmatrix}$$

Example 4. The covariant components of a second order tensor with respect to the basis $\mathbf{e}_1 = \mathbf{i}_1 + \mathbf{i}_2$, $\mathbf{e}_2 = \mathbf{i}_2 + \mathbf{i}_3$, $\mathbf{e}_3 = \mathbf{i}_1 + \mathbf{i}_3$ (see Example 1) are given by

$$(a_{ij}) = \begin{bmatrix} 2 & 0 & 1 \\ -1 & 2 & 0 \\ 0 & 2 & -1 \end{bmatrix}$$

Find the components a^{ij}, $a_i{}^{\cdot k}$ and $a^i{}_{\cdot k}$.

Solution: From Example 1, we have

$$(g_{ij}) = \begin{bmatrix} 2 & 1 & 1 \\ 1 & 2 & 1 \\ 1 & 1 & 2 \end{bmatrix}, \qquad (g^{ij}) = \frac{1}{4}\begin{bmatrix} 3 & -1 & -1 \\ -1 & 3 & -1 \\ -1 & -1 & 2 \end{bmatrix}$$

Hence, by (6.46), we find

$$a^{11} = g^{1p}g^{1q}a_{pq} = 21/16\,, \qquad a^{12} = g^{1p}g^{2q}a_{pq} = -23/16$$

$$a^{13} = g^{1p}\,g^{3q}a_{pq} = 9/16\,, \qquad a^{21} = g^{2p}g^{1q}\,a_{pq} = -19/16$$

$$a^{22} = g^{2p}\,g^{2q}a_{pq} = -17/16\,, \qquad a^{23} = g^{2p}g^{3q}\,a_{pq} = 1/16$$

$$a^{31} = g^{3p}g^{1q}a_{pq} = -3/16\,, \qquad a^{32} = g^{3p}g^{2q}\,a_{pq} = 17/16$$

$$a^{33} = g^{3p}g^{3q}a_{pq} = -15/16\,,$$

The student may check these results by performing the matrix multiplication

$$(a^{ij}) = (g^{ij}\,)(a_{ij})(g^{ij}\,)^T$$

Next, by (6.47), we find

$$a_{1}{}^{\cdot 1} = g_{1j}a^{j1} = 5/4\,, \quad a_{1}{}^{\cdot 2} = g_{1j}a^{j2} = -3/4\,, \quad a_{1}{}^{\cdot 3} = g_{1j}a^{j3} = 1/4$$

$$a_{2}{}^{\cdot 1} = g_{2j}a^{j1} = -5/4\,, \quad a_{2}{}^{\cdot 2} = g_{2j}a^{j2} = 7/4\,, \quad a_{2}{}^{\cdot 3} = g_{2j}a^{j3} = -1/4$$

$$a_{3}{}^{\cdot 1} = g_{3j}a^{j1} = -1/4\,, \quad a_{3}{}^{\cdot 2} = g_{3j}a^{j2} = 7/4\,, \quad a_{3}{}^{\cdot 3} = g_{3j}a^{j3} = -5/4$$

and by (6.48), we have

$$a^{1}{}_{.1} = g^{1j}a_{j1} = 7/4, \quad a^{1}{}_{.2} = g^{1j}a_{j2} = -1, \quad a^{1}{}_{.3} = g^{1j}a_{j3} = 1$$

$$a^{2}{}_{.1} = g^{2j}a_{j1} = -5/4, \quad a^{2}{}_{.2} = g^{2j}a_{j2} = 1, \quad a^{2}{}_{.3} = g^{2j}a_{j3} = 0$$

$$a^{3}{}_{.1} = g^{3j}a_{j1} = -1/4, \quad a^{3}{}_{.2} = g^{3j}a_{j2} = 1, \quad a^{3}{}_{.3} = g^{3j}a_{j3} = -1$$

6.3 EXERCISES

1. Let

$$(a_{ij}) = \begin{bmatrix} -1 & 2 & 0 \\ 2 & 0 & 3 \\ 0 & 3 & -2 \end{bmatrix}$$

relative to the basis $\mathbf{e}_1 = \mathbf{i}_2 + \mathbf{i}_3$, $\mathbf{e}_2 = \mathbf{i}_1 + \mathbf{i}_3$, $\mathbf{e}_3 = \mathbf{i}_1 + \mathbf{i}_2 + \mathbf{i}_3$. Find the components a^{ij}, $a^{i}_{.k}$ and $a_{i}^{.k}$ of the tensor.

2. Let

$$(a_{ij}) = \begin{bmatrix} x^1 & 0 & x^3 \\ 0 & x^2 & x^1 \\ x^2 & x^3 & 0 \end{bmatrix}$$

Find the components a^{ij}, $a^{i}_{.k}$ and $a_{i}^{.k}$ of the tensor, relative to the basis

$$\mathbf{e}_1 = \mathbf{i}_2 + \mathbf{i}_3, \qquad \mathbf{e}_2 = \mathbf{i}_1 + \mathbf{i}_2 + 2\mathbf{i}_3, \qquad \mathbf{e}_3 = \mathbf{i}_1 + \mathbf{i}_2 + \mathbf{i}_3.$$

3. Suppose

$$(a^{ij}) = \begin{bmatrix} 0 & x^2 & -x^3 \\ -x^2 & 0 & x^1 \\ x^3 & -x^1 & 0 \end{bmatrix}$$

relative to the basis of Problem 1. Find the components a_{ij}, $a^{i}_{.k}$ and $a_{i}^{.k}$ of the tensor.

4. The metric tensor g_{ij} of an oblique coordinate system determined by a basis $\mathbf{e}_i$ is given by

$$(g_{ij}) = \begin{bmatrix} 2 & 0 & 0 \\ 0 & 1 & 1 \\ 0 & 1 & 3 \end{bmatrix}$$

(a) What is the distance between the two points whose oblique coordinates are given by (2, 1, -3) and (4, -3, -1)?

(b) Find the scalar product of the two vectors $\mathbf{A} = 2\mathbf{e}_1 + \mathbf{e}_2 - 2\mathbf{e}_3$ and $\mathbf{B} = -\mathbf{e}_1 + 3\mathbf{e}_2 + 2\mathbf{e}_3$.

(c) Express the vector product of **A** and **B** in terms of the reciprocal basis $\mathbf{e}^l$.

5. The metric tensor of an oblique coordinate system is given by

$$(g_{ij}) = \begin{bmatrix} 2 & 1 & 0 \\ 1 & 2 & 1 \\ 0 & 1 & 1 \end{bmatrix}$$

with respect to a basis $\mathbf{e}_i$. If $\mathbf{A} = 3\mathbf{e}_1 - \mathbf{e}_2 + 2\mathbf{e}_3$ and $\mathbf{B} = 2\mathbf{e}_1 + 3\mathbf{e}_2 + 4\mathbf{e}_3$, find $\mathbf{A}\cdot\mathbf{B}$ and $\mathbf{A}\times\mathbf{B}$. (Express in terms of the reciprocal basis $\mathbf{e}^i$.)

6. Show that $a^i{}_{.k} = a_i{}^{.k}$ [see (6.47) and (6.48)] if and only if $a_{ji} = a_{ij}$ for all i, j.

7. Write down the formula relating the components a_{ijk} and a^{ijk} of a third order tensor.

8. Show that

$$g_{ir}a^{rjk} = g^{jp}g^{kq}a_{ipq}$$

9. Show that

$$g_{ij}a^{jk} = g^{kj}a_{ij} \text{ and } g^{ij}a_{jk} = g_{kj}a^{ij}$$

10. Verify the result $V = \sqrt{g}$ in (6.51) for the basis $\mathbf{e}_1 = \mathbf{i}_1 + \mathbf{i}_2 - 2\mathbf{i}_3$, $\mathbf{e}_2 = 2\mathbf{i}_1 - \mathbf{i}_2 + \mathbf{i}_3$, $\mathbf{e}_3 = \mathbf{i}_1 - 2\mathbf{i}_2 + \mathbf{i}_3$.

11. Obtain the formula $\mathbf{e}^i = g^{ij}\mathbf{e}_j$.

12. Using the result of Problem 11, show that $V' = \sqrt{G}$, where $V' = \mathbf{e}^1\cdot\mathbf{e}^2\times\mathbf{e}^3$ and G is the determinant of the matrix (g^{ij}). Thus deduce that $gG = 1$.

6.6 TRANSFORMATIONS OF CURVILINEAR COORDINATES

We recall from Sec. 3.8 that a curvilinear coordinates system u^i (notice the shift to the use of superscript) is defined by a set of equations

$$u^i = u^i(x_1, x_2, x_3) \quad (i = 1, 2, 3) \tag{6.52}$$

where the functions $u^i(x_1, x_2, x_3)$ are single-valued and continuously differentiable in a domain D with cartesian coordinates x_1, x_2, x_3. It is assumed that the Jacobian $\partial(u^1, u^2, u^3)/\partial(x_1, x_2, x_3) \neq 0$ at each point in D, so that the inverse of (6.52) exists in a neighborhood of every point in D. Let

$$x_i = x_i(u^1, u^2, u^3) \quad (i = 1, 2, 3) \tag{6.53}$$

denote the inverse transformation of (6.52). Naturally, since the inverse of (6.53) leads back to (6.52), the Jacobian $\partial(x_1, x_2, x_3)/\partial(u^1, u^2, u^3)$ of the transformation (6.53) is also different from zero in a corresponding domain of the coordinates u^i. As a matter of fact, the Jacobian of the transformations (6.52) and (6.53) satisfy the important relation

$$\frac{\partial(u^1, u^2, u^3)}{\partial(x_1, x_2, x_3)} \frac{\partial(x_1, x_2, x_3)}{\partial(u^1, u^2, u^3)} = 1 \tag{6.54}$$

This is easily verified by the chain rule of partial differentiation

$$\frac{\partial u^i}{\partial u^j} = \frac{\partial u^i}{\partial x_k} \frac{\partial x_k}{\partial u^j} = \delta^i_j \tag{6.55}$$

Local and Reciprocal Bases

So suppose we have a curvilinear coordinate system in our space defined by (6.52) and let the inverse transformation be given by (6.53). Then at each point P in space we can introduce a local basis consisting of the tangent vectors to the u^i-coordinate curves passing through P (see Fig. 6.6). The base vectors are given by

$$\mathbf{e}_i = \frac{\partial \mathbf{R}}{\partial u^i} = \sum_{k=1}^{3} \frac{\partial x_k}{\partial u^i} \mathbf{i}_k \quad (i = 1, 2, 3) \tag{6.56}$$

where $\mathbf{R}$ is the position vector of the point P, and the $x_k(u^i)$ are the inverse functions defined in (6.53). Thus the metric tensor at the point P has the components

$$g_{ij} = \mathbf{e}_i \cdot \mathbf{e}_j = \sum_{k=1}^{3} \frac{\partial x_k}{\partial u^i} \frac{\partial x_k}{\partial u^j} \tag{6.57}$$

which are functions of the coordinates of the point P.

Corresponding to the local basis (6.56), we can also define the local reciprocal basis $\mathbf{e}^i$ in exactly the same way as in Sec. 6.2. Hence they are determined by the formula

$$\mathbf{e}^i = \frac{\mathbf{e}_j \times \mathbf{e}_k}{V}, \quad V = \mathbf{e}_1 \cdot \mathbf{e}_2 \times \mathbf{e}_3 \tag{6.58}$$

for i = 1, 2, 3 where i, j, k are cyclic permutations of 1, 2, 3. Alternatively, the local reciprocal base vectors $\mathbf{e}^i$ are more conveniently determined by the formula

$$\mathbf{e}^i = \nabla u^i = \frac{\partial u^i}{\partial x_k} \mathbf{i}_k \quad (i = 1, 2, 3) \tag{6.59}$$

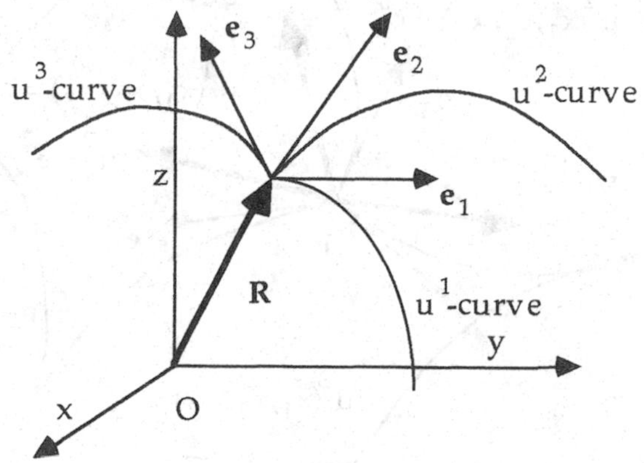

Fig. 6.6 Local base vectors.

where the u^i are the transformation functions defined in (6.52). That (6.59) provides the local reciprocal basis follows readily from (6.55) since

$$\delta^i_j = \frac{\partial u^i}{\partial x_k}\frac{\partial x_k}{\partial u^j} = \nabla u^i \cdot \frac{\partial \mathbf{R}}{\partial u^j} = \nabla u^i \cdot \mathbf{e}_j$$

Now we know that ∇u^i is a normal vector to the coordinate surface $u^i(x_1, x_2, x_3) = c_i$, $i = 1, 2, 3$. Thus we see that while the local basis consists of the tangent vectors to the coordinate curves at the point P, the local reciprocal basis consists of the normal vectors to the coordinate surfaces at P (Fig. 6.7). Therefore, if the curvilinear coordinate system is orthogonal, then for each value of the index i, the vectors $\mathbf{e}_i$ and $\mathbf{e}^i$ are parallel and have the same direction, so that $|\mathbf{e}_i|\,|\mathbf{e}^i| = 1$ for $i = 1, 2, 3$.

It is instructive to note that the Jacobian of the transformation (6.52) can be written as

$$\frac{\partial(u^1, u^2, u^3)}{\partial(x_1, x_2, x_3)} = \nabla u^1 \cdot \nabla u^2 \times \nabla u^3$$

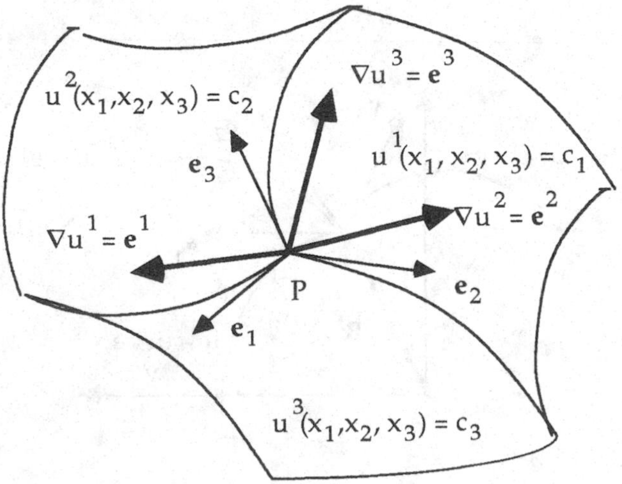

Fig. 6.7 Local and reciprocal base vectors.

$$= \mathbf{e}^1 . \mathbf{e}^2 \times \mathbf{e}^3 = V' \tag{6.60}$$

while the Jacobian of the inverse transformation (6.53) can be written as

$$\frac{\partial(x_1, x_2, x_3)}{\partial(u^1, u^2, u^3)} = \frac{\partial \mathbf{R}}{\partial u^1} \cdot \frac{\partial \mathbf{R}}{\partial u^2} \times \frac{\partial \mathbf{R}}{\partial u^3}$$

$$= \mathbf{e}_1 \cdot \mathbf{e}_2 \times \mathbf{e}_3 = V \tag{6.61}$$

(Cf. Probs. 1 and 2, Exer. 3.7.) In Sec. 6.2 we established that VV' = 1, thus we have again verified (6.55) from another viewpoint.

Example 1. The position vector of a point in the cylindrical coordinate system is given by

$$\mathbf{R} = r \cos\theta\, \mathbf{i} + r \sin\theta\, \mathbf{j} + z\mathbf{k}$$

Let us set $u^1 = r$, $u^2 = \theta$, $u^3 = z$, noting that $\mathbf{i} = \mathbf{i}_1$, $\mathbf{j} = \mathbf{i}_2$, $\mathbf{k} = \mathbf{i}_3$. Then the local basis consists of the vectors

$$\mathbf{e}_1 = \frac{\partial \mathbf{R}}{\partial u^1} = \cos\theta\, \mathbf{i}_1 + \sin\theta\, \mathbf{i}_2$$

$$\mathbf{e}_2 = \frac{\partial \mathbf{R}}{\partial u^2} = -r \sin\theta\, \mathbf{i}_1 + r \cos\theta\, \mathbf{i}_2$$

$$\mathbf{e}_3 = \frac{\partial \mathbf{R}}{\partial u^3} = \mathbf{i}_3$$

so that the metric tensor is given by

$$(g_{ij}) = \mathbf{e}_i \cdot \mathbf{e}_j = \begin{bmatrix} 1 & 0 & 0 \\ 0 & r^2 & 0 \\ 0 & 0 & 1 \end{bmatrix}$$

The reciprocal basis consists of the vectors

$$\mathbf{e}^1 = \nabla r = \frac{x_1}{r}\mathbf{i}_1 + \frac{x_2}{r}\mathbf{i}_2 = \mathbf{e}_1$$

$$\mathbf{e}^2 = \nabla\theta = -\frac{x_2}{r^2}\mathbf{i}_1 + \frac{x_1}{r^2}\mathbf{i}_2 = -\frac{\sin\theta}{r}\mathbf{i}_1 + \frac{\cos\theta}{r}\mathbf{i}_2$$

$$\mathbf{e}^3 = \nabla z = \mathbf{i}_3$$

and so

$$(g^{ij}) = \mathbf{e}^i \cdot \mathbf{e}^j = \begin{bmatrix} 1 & 0 & 0 \\ 0 & 1/r^2 & 0 \\ 0 & 0 & 1 \end{bmatrix}$$

Coordinates Transformation

Now suppose u^{*i} is another curvilinear coordinate system defined by the transformation equations

$$u^{*i} = u^{*i}(x_1, x_2, x_3) \qquad (i = 1, 2, 3) \tag{6.62}$$

where the functions $u^{*i}(x_1, x_2, x_3)$ possess the same properties as those in (6.52). If we substitute (6.53) for x_i in (6.62), we obtain the transformation equations relating the two curvilinear coordinate systems u^i and u^{*i}. Suppose that these equations are given by

$$u^{*i} = u^{*i}(u^1, u^2, u^3) \qquad (i = 1, 2, 3) \tag{6.63}$$

In particular cases, these equations might be relating a rectangular and an oblique coordinate system; a rectangular and a cylindrical or spherical coordinate system; or a cylindrical and a spherical coordinate system (see Prob. 8). Since

$$\frac{\partial u^{*i}}{\partial u^j} = \frac{\partial u^{*i}}{\partial x_k}\frac{\partial x_k}{\partial u^j}$$

it follows from the definition of matrix multiplication that

$$\frac{\partial(u^{*1}, u^{*2}, u^{*3})}{\partial(u^1, u^2, u^3)} = \frac{\partial(u^{*1}, u^{*2}, u^{*3})}{\partial(x_1, x_2, x_3)}\frac{\partial(x_1, x_2, x_3)}{\partial(u^1, u^2, u^3)} \tag{6.64}$$

This is the analogue of the chain rule of partial differentiation for Jacobians. In particular, this shows that the Jacobian of the transformation (6.63) also differs from zero; hence, the inverse transformation equations, say

$$u^i = u^i(u^{*1}, u^{*2}, u^{*3}) \qquad (i = 1, 2, 3) \tag{6.65}$$

exist. Of course, these equations can also be determined by substituting the inverse of (6.62) for x_i in (6.52).

Now let $\mathbf{e}^{*i}$ denote the local basis corresponding to the curvilinear coordinates u^{*i} defined by (6.62). We wish to obtain the relations between the local bases $\mathbf{e}_i$ and $\mathbf{e}^i$ of the two coordinate systems u^i and u^{*i}. From the definition (6.56) and by the chain rule, we find

$$\mathbf{e}^*_i = \frac{\partial \mathbf{R}}{\partial u^{*i}} = \frac{\partial \mathbf{R}}{\partial u^j}\frac{\partial u^j}{\partial u^{*i}} = \frac{\partial u^j}{\partial u^{*i}}\mathbf{e}_j \tag{6.66}$$

where the coefficients $\partial u^j/\partial u^{*i}$ are to be calculated from (6.65). The inverse relation is given by

$$\mathbf{e}_j = \frac{\partial u^{*i}}{\partial u^j}\mathbf{e}^*_i \tag{6.67}$$

Similarly, the relations between the reciprocal bases $\mathbf{e}^i$ and $\mathbf{e}^{*i}$ can be obtained by using the formula (6.59) along with the chain rule. We find

$$\begin{aligned}\mathbf{e}^{*i} &= \nabla u^{*i} = \frac{\partial u^{*i}}{\partial x_k}\mathbf{i}_k \\ &= \frac{\partial u^{*i}}{\partial u^j}\frac{\partial u^j}{\partial x_k}\mathbf{i}_k = \frac{\partial u^{*i}}{\partial u^j}\nabla u^j \\ &= \frac{\partial u^{*i}}{\partial u^j}\mathbf{e}^j\end{aligned} \tag{6.68}$$

The inverse of (6.68) is given by

$$\mathbf{e}^i = \frac{\partial u^i}{\partial u^{*j}}\mathbf{e}^{*j} \tag{6.69}$$

[Compare these equations with (6.13) and (6.16).]

Example 2. Find the relationship between the cylindrical coordinates

$$x_1 = r\cos\theta, \quad x_2 = r\sin\theta, \quad x_3 = z$$

and the elliptic cylindrical coordinates

$$x_1 = a\cosh u\cos v, \quad x_2 = a\sinh u\sin v, \quad x_3 = w$$

Calculate the Jacobian $\partial(r, \theta, z)/\partial(u, v, w)$.

Solution: We note that

$$(x_1)^2 + (x_2)^2 = r^2 = a^2(\cosh^2 u - \sin^2 v)$$

hence

$$r = a[\cosh^2 u - \sin^2 v]^{1/2}$$

Further,

$$\tan\theta = \frac{x_2}{x_1} = \frac{a\sinh u\sin v}{a\cosh u\cos v} = \tanh u\tan v$$

so that

$$\theta = \arctan(\tanh u\tan v)$$

and $z = w$. Since

$$\frac{\partial r}{\partial w} = 0, \quad \frac{\partial \theta}{\partial w} = 0, \quad \frac{\partial z}{\partial u} = 0, \text{ and } \frac{\partial z}{\partial v} = 0$$

it follows that

$$\frac{\partial(r,\theta,z)}{\partial(u,v,w)} = \frac{\partial r}{\partial u}\frac{\partial \theta}{\partial v} - \frac{\partial r}{\partial v}\frac{\partial \theta}{\partial u}$$

$$= \frac{a(\sinh^2 u\sec^2 v + \operatorname{sech}^2 u\sec^2 v)}{\sqrt{\cosh^2 u - \sin^2 v}\,(1 + \tanh^2 u\tan^2 v)}$$

6.4 EXERCISES

In each of Problems 1 through 5, the transformation equations relating the rectangular cartesian coordinates x_i and a curvilinear coordinate system u^i are given. Find the local basis $\mathbf{e}_i$, the local reciprocal basis $\mathbf{e}^i$, and the covariant and contravariant components of the metric tensor.

1. The spherical coordinate system:

$$x_1 = r \sin\phi \cos\theta, \quad x_2 = r \sin\phi \sin\theta, \quad x_3 = r \cos\phi$$

where $u^1 = r$, $u^2 = \phi$, and $u^3 = \theta$.

2. The parabolic cylindrical coordinate system:

$$x_1 = (v^2 - u^2)/2, \quad x_2 = vw, \quad x_3 = z$$

where $u^1 = v$, $u^2 = w$, and $u^3 = z$.

3. The parabolic coordinate system:

$$x_1 = vw \cos\theta, \quad x_2 = vw \sin\theta, \quad x_3 = (v^2 - w^2)/2$$

where $u^1 = v$, $u^2 = w$, and $u^3 = \theta$.

4. The elliptic cylindrical coordinate system:

$$x_1 = a \cosh v \cos w, \quad x_2 = a \sinh v \sin w, \quad x_3 = z$$

where $u^1 = v$, $u^2 = w$, and $u^3 = z$ (a is a constant).

5. The curvilinear coordinate system:

$$x_1 = u^1 u^2, \quad x_2 = [(u^1)^2 + (u^2)^2]/2, \quad x_3 = u^3$$

6. Find the transformation equations relating the cylindrical coordinates r, θ, z and the parabolic cylindrical coordinates v, w, z. Calculate the Jacobian $\partial(r, \theta, z)/\partial(v, w, z)$.
7. Find the transformation equations relating the cylindrical coordinates r, θ, z and the parabolic coordinates v, w, ϕ. Calculate the Jacobian $\partial(r, \theta, z)/\partial(v, w, \phi)$.
8. (a) Find the transformation equations relating the cylindrical coordinates r, θ, z and the spherical coordinates R, ϕ, θ. Also find the inverse transformation.
   (b) Let $u^1 = r$, $u^2 = \theta$, $u^3 = z$ and $u^{*1} = R$, $u^{*2} = \phi$, $u^{*3} = \theta$. Evaluate $\partial u^i/\partial u^{*j}$ and $\partial u^{*i}/\partial u^j$ and represent them in matrix form.
9. Let $\mathbf{e}_i$ denote the local basis in cylindrical coordinate system. Using the results of Prob. 8(b) and (6.66), obtain the local basis

$\mathbf{e}^*_i$ in spherical coordinates. Check your results with that of Problem 1.

6.7 GENERAL TENSORS

We now discuss tensors in general curvilinear coordinate systems. Such tensors are called general tensors or simply tensors. As we will see here, the theory of general tensors is analogous to that of tensors in an oblique coordinate system restricted to a point. For in a general coordinate system, although we no longer have a fixed basis that is valid for the whole space, it is possible as we see in Sec. 6.6 to set up at each point a local coordinate system with the tangent vectors to the coordinate curves as the base vectors and the normal vectors to the coordinate surfaces as the reciprocal base vectors. Relative to these bases the entire edifice of tensor algebra that we have developed for oblique caartesian coordinate systems can be carried over.

First, suppose $\mathbf{A}$ is a vector with covariant and contravariant components a_i and a^i at a point in space with respect to a local basis $\mathbf{e}_i$ of a given curvilinear coordinate system u^i. Let a^*_i and a^{*i} denote the corresponding components of the vector at the same point with respect to another curvilinear coordinate system u^{*i} with basis $\mathbf{e}^*_i$, where $u^{*i} = u^{*i}(u^1, u^2, u^3)$ $(i = 1, 2, 3)$. Then at the point in question, we have $\mathbf{A} = a^*_i \mathbf{e}^{*i} = a_j \mathbf{e}^j$, and so

$$a^*_i = \mathbf{A} \cdot \mathbf{e}^*_i = (a_j\, \mathbf{e}^j) \cdot \mathbf{e}^*_i$$

Substituting the formula (6.66) for $\mathbf{e}^*_i$ and noting that $\mathbf{e}^j \cdot \mathbf{e}_i = \delta^j{}_i$, we finally obtain the transformation law

$$a^*_i = \frac{\partial u^j}{\partial u^{*i}}\, a_j \tag{6.70}$$

$(i = 1, 2, 3)$ for the covariant components. Similarly, from the identity $\mathbf{A} = a^{*i}\mathbf{e}^*_i = a^j \mathbf{e}_j$, we find

$$a^{*i} = \mathbf{A} \cdot \mathbf{e}^{*i} = (a^j\, \mathbf{e}_j) \cdot \mathbf{e}^{*i}$$

which, by (6.68), leads to the transformation law

$$a^{*i} = \frac{\partial u^{*i}}{\partial u^j} a^j \tag{6.71}$$

(i = 1, 2, 3) for the contravariant components of the vector. These are the laws governing the components of a tensor of the first order (a vector).

For tensors of higher order, the transformation laws are precisely those given in Sec. 6.3 with respect to oblique coordinate systems. We need simply to replace the variables x^i and x^{*i} by u^i and u^{*i}, respectively, to indicate general coordinate system. Thus, for instance, a tensor of order 2 consists of 3^2 covariant components a_{ij} or 3^2 contravariant component a^{ij} or mixed components $a_i^{\cdot j}$ and $a^i_{\cdot j}$, which transform under a change of coordinates according to the laws:

$$\begin{aligned}
&\text{(i)} \quad a^*_{ij} = \frac{\partial u^p}{\partial u^{*i}} \frac{\partial u^q}{\partial u^{*j}} a_{pq} \\
&\text{(ii)} \quad a^{*ij} = \frac{\partial u^{*i}}{\partial u^p} \frac{\partial u^{*j}}{\partial u^q} a^{pq} \\
&\text{(iii)} \quad a^{*\cdot j}_{i} = \frac{\partial u^p}{\partial u^{*i}} \frac{\partial u^{*j}}{\partial u^q} a_p^{\cdot q} \\
&\text{(iv)} \quad a^{*i}_{\cdot j} = \frac{\partial u^{*i}}{\partial u^p} \frac{\partial u^q}{\partial u^{*j}} a^p_{\cdot q}
\end{aligned} \tag{6.72}$$

An important example of a second order tensor is the metric tensor (g_{ij}) defined by (6.57). Indeed, from the definition and the formula (6.66), we see that

$$\begin{aligned}
g^*_{ij} = \mathbf{e}^*_i \cdot \mathbf{e}^*_j &= \frac{\partial u^p}{\partial u^{*i}} \frac{\partial u^q}{\partial u^{*j}} \mathbf{e}_p \cdot \mathbf{e}_q \\
&= \frac{\partial u^p}{\partial u^{*i}} \frac{\partial u^q}{\partial u^{*j}} g_{pq}
\end{aligned}$$

so that the quantities g_{ij} transform as covariant components of the metric tensor.

From the results of our discussion in Sec. 6.5, we can verify that at each point of the space the various components a_{ij}, a^{ij}, $a_i{}^{\cdot j}$, and $a^i{}_{\cdot j}$ of a second order tensor are related to one another by the metric tensor at that point.

Example 1. Let ϕ be a scalar field in a rectangular cartesian coordinate x_i. If we change to general coordinates $u^i = u^i(x_1, x_2, x_3)$ $(i = 1, 2, 3)$, the gradient of ϕ becomes

$$\nabla\phi = \frac{\partial\phi}{\partial x_i}\mathbf{i}_i = \frac{\partial\phi}{\partial u^j}\frac{\partial u^j}{\partial x_i}\mathbf{i}_i = \frac{\partial\phi}{\partial u^j}\mathbf{e}^j$$

since $(\partial u^j/\partial x_i)\mathbf{i}_i = \nabla u^j = \mathbf{e}^j$. This means that the natural expression for $\nabla\phi$ is in terms of the local reciprocal basis $\mathbf{e}^i$; hence, the terms $\partial\phi/\partial u^i$ are the covariant components of $\nabla\phi$. In fact, under a change of coordinates, we see that

$$\frac{\partial\phi}{\partial u^{*i}} = \frac{\partial\phi}{\partial u^j}\frac{\partial u^j}{\partial u^{*i}}$$

which agrees with (6.70) for the covariant components of a vector. For this reason, $\nabla\phi$ is called a covariant vector field. Of course, $\nabla\phi$ has contravariant components as well, given by $g^{ij}(\partial\phi/\partial u^j)$.

It is worthwhile to note that the expression $\nabla\phi = (\partial\phi/\partial u^i)\mathbf{e}^i$ is invariant under any coordinate transformation. In fact, by (6.68), we see that

$$\frac{\partial\phi}{\partial u^i}\mathbf{e}^i = \frac{\partial\phi}{\partial u^{*j}}\frac{\partial u^{*j}}{\partial u^i}\mathbf{e}^i = \frac{\partial\phi}{\partial u^{*j}}\mathbf{e}^{*j}$$

Thus, in cylindrical coordinates r, θ, z, we have

$$\nabla\phi = \frac{\partial\phi}{\partial r}\mathbf{e}^1 + \frac{\partial\phi}{\partial\theta}\mathbf{e}^2 + \frac{\partial\phi}{\partial z}\mathbf{e}^3$$

where $\mathbf{e}^1 = \cos\theta\,\mathbf{i}_1 + \sin\theta\,\mathbf{i}_2$, $\mathbf{e}^2 = (-\sin\theta\,\mathbf{i}_1 + \cos\theta\,\mathbf{i}_2)/r$, $\mathbf{e}^3 = \mathbf{i}_3$, as derived in Example 1, Sec. 6.6.

Example 2. Find the gradient of $\phi = x_1 x_3 + x_2$ in the curvilinear coordinates v, w, z, where

$$x_1 = vw, \quad x_2 = (v^2 + w^2)/2, \quad x_3 = z$$

Solution: From the position vector $\mathbf{R} = uv\,\mathbf{i}_1 + [(v^2 + w^2)/2]\,\mathbf{i}_2 + z\,\mathbf{i}_3$ with $u^1 = v$, $u^2 = w$, and $u^3 = z$, we find

$$\mathbf{e}_1 = \frac{\partial \mathbf{R}}{\partial u^1} = w\,\mathbf{i}_1 + v\,\mathbf{i}_2, \quad \mathbf{e}_2 = \frac{\partial \mathbf{R}}{\partial u^2} = v\,\mathbf{i}_1 + w\,\mathbf{i}_2, \quad \mathbf{e}_3 = \frac{\partial \mathbf{R}}{\partial u^3} = \mathbf{i}_3$$

and $\mathbf{e}_1 \cdot \mathbf{e}_2 \times \mathbf{e}_3 = w^2 - v^2$. Hence, the reciprocal base vectors are given by

$$\mathbf{e}^1 = \frac{\mathbf{e}_2 \times \mathbf{e}_3}{w^2 - v^2} = \frac{w\,\mathbf{i}_1 - v\,\mathbf{i}_2}{w^2 - v^2}$$

$$\mathbf{e}^2 = \frac{\mathbf{e}_3 \times \mathbf{e}_1}{w^2 - v^2} = \frac{-v\,\mathbf{i}_1 + w\,\mathbf{i}_2}{w^2 - v^2}$$

$$\mathbf{e}^3 = \frac{\mathbf{e}_1 \times \mathbf{e}_2}{w^2 - v^2} = \mathbf{i}_3$$

In the new coordinates v, w, z, we have $\phi = vwz + (v^2 + w^2)/2$. Hence,

$$\nabla\phi = \frac{\partial\phi}{\partial v}\mathbf{e}^1 + \frac{\partial\phi}{\partial w}\mathbf{e}^2 + \frac{\partial\phi}{\partial z}\mathbf{e}^3$$

$$= (wz + v)\mathbf{e}^1 + (vz + w)\mathbf{e}^2 + vw\mathbf{e}^3$$

It is instructive to observe that if we substitute the expressions of the vectors $\mathbf{e}^1, \mathbf{e}^2, \mathbf{e}^3$ in $\nabla\phi$ and collect the coefficients of $\mathbf{i}_1, \mathbf{i}_2, \mathbf{i}_3$, we find

$$\nabla\phi = x_3\,\mathbf{i}_1 + \mathbf{i}_2 + x_1\,\mathbf{i}_3$$

which is precisely the gradient of ϕ in rectangular cartesian coordinates.

Example 3. Consider the coordinate transformation $u^{*i} = u^i(u^1, u^2, u^3)$ $(i = 1, 2, 3)$. Taking the differential, we find

$$du^{*i} = \frac{\partial u^{*i}}{\partial u^j} du^j$$

By (6.71) this shows that the quantities du^i transform as contravariant components of a vector. This vector is exactly the differential $d\mathbf{R}$ of the position vector of $\mathbf{R}$. Note that

$$d\mathbf{R} = \frac{\partial \mathbf{R}}{\partial u^i} du^i = \mathbf{e}_i du^i$$

Thus $d\mathbf{R}$ is sometimes called a contravariant vector. It follows that the inner product of $\nabla\phi$ and $d\mathbf{R}$ is a scalar:

$$\nabla\phi \cdot d\mathbf{R} = \frac{\partial \phi}{\partial u^i} \mathbf{e}^i \cdot \mathbf{e}_j du^j = \frac{\partial \phi}{\partial u^i} du^i = d\phi$$

Example 4. The Kronecker delta δ_i^j is a mixed tensor of second order (or, more precisely, forms the mixed components of a second order tensor). Indeed, under any transformations of coordinates u^i to u^{*i}, we know that

$$\delta_i^j = \delta^{*j}_{\ i}\,. \qquad \delta^{*j}_{\ i} = \frac{\partial u^p}{\partial u^{*i}} \frac{\partial u^{*j}}{\partial u^p}$$

Hence,

$$\delta^{*j}_{\ i} = \frac{\partial u^p}{\partial u^{*i}} \frac{\partial u^{*j}}{\partial u^q} \delta_p^q$$

Further, we see that the Kronecker delta has the same components in all coordinate systems. Such tensors are called *isotropic* tensors.

Notice that since the metric tensor is symmetric, that is, $g_{ij} = g_{ji}$ and $g^{ij} = g^{ji}$, it follows that

$$g_i^{\cdot j} = g_{ik} g^{kj} = g^{jk} g_{ki} = g^j_{\cdot i} = \delta_i^j$$

Example 5. The components of a second order cartesian tensor are given by

$$(a_{ij}) = \begin{bmatrix} 0 & 0 & x_1x_3 \\ 0 & (x_2)^2 & 0 \\ x_2x_3 & 0 & 0 \end{bmatrix}$$

Find the covariant, contravariant and mixed components of the tensor in cylindrical coordinates.

Solution: The transformation equations relating the cylindrical coordinates r, θ, z and the rectangular cartesian coordinates x_1, x_2, x_3 are given by

$$x_1 = r\cos\theta, \quad x_2 = r\sin\theta, \quad x_3 = z$$

and the inverse transformation are given by

$$r = \sqrt{(x_1)^2 + (x_2)^2}, \quad \theta = \arctan(x_2/x_1), \quad z = x_3$$

Let us set $x_i = u^i$ $(1 \le i \le 3)$ and $r = u^{*1}$, $\theta = u^{*2}$, $z = u^{*3}$. Then the different values of $\partial u^i/\partial u^{*j}$ and $\partial u^{*i}/\partial u^j$ are represented by the following matrices (the index i indicating row):

$$\left(\frac{\partial u^i}{\partial u^{*j}}\right) = \begin{bmatrix} \cos\theta & -r\sin\theta & 0 \\ \sin\theta & r\cos\theta & 0 \\ 0 & 0 & 1 \end{bmatrix}$$

$$\left(\frac{\partial u^{*i}}{\partial u^j}\right) = \begin{bmatrix} \cos\theta & \sin\theta & 0 \\ \dfrac{-\sin\theta}{r} & \dfrac{\cos\theta}{r} & 0 \\ 0 & 0 & 1 \end{bmatrix}$$

Hence, by [6.72(i)], and dropping all zero terms, we find

$$a^*_{11} = \frac{\partial u^2}{\partial u^{*1}}\frac{\partial u^2}{\partial u^{*1}} a_{22} = (\sin\theta)^2 (r\sin\theta)^2 = r^2 \sin^4\theta$$

$$a^*_{12} = \frac{\partial u^2}{\partial u^{*1}} \frac{\partial u^2}{\partial u^{*2}} a_{22} = \sin\theta(r\cos\theta)(r\sin\theta)^2$$

$$= r^3 \cos\theta \sin^3\theta$$

$$a^*_{13} = \frac{\partial u^1}{\partial u^{*1}} \frac{\partial u^3}{\partial u^{*3}} a_{13} = (\cos\theta)(r\cos\theta)z = rz\cos^2\theta$$

$$a^*_{21} = \frac{\partial u^2}{\partial u^{*2}} \frac{\partial u^2}{\partial u^{*1}} a_{22} = (r\cos\theta)(\sin\theta)(r\sin\theta)^2$$

$$= r^3 \cos\theta \sin^3\theta$$

$$a^*_{22} = \frac{\partial u^2}{\partial u^{*2}} \frac{\partial u^2}{\partial u^{*2}} a_{22} = (r\cos\theta)^2 (r\sin\theta)^2 = r^4 \sin^2\theta \cos^2\theta$$

$$a^*_{23} = \frac{\partial u^1}{\partial u^{*2}} \frac{\partial u^3}{\partial u^{*3}} a_{13} = (-r\sin\theta)(r\cos\theta)z = -r^2 z\cos\theta\sin\theta$$

$$a^*_{31} = \frac{\partial u^3}{\partial u^{*3}} \frac{\partial u^1}{\partial u^{*1}} a_{31} = (\cos\theta)(r\sin\theta)z = rz\cos\theta\sin\theta$$

$$a^*_{32} = \frac{\partial u^3}{\partial u^{*3}} \frac{\partial u^1}{\partial u^{*2}} a_{31} = (-r\sin\theta)(r\sin\theta)z = -r^2 z\sin^2\theta$$

$$a^*_{33} = \frac{\partial u^i}{\partial u^{*3}} \frac{\partial u^j}{\partial u^{*3}} a_{ij} = 0$$

On the other hand, by [(6.72(ii)], we find (note that $a_{ij} = a^{ij}$)

$$a^{*11} = \frac{\partial u^{*1}}{\partial u^2} \frac{\partial u^{*1}}{\partial u^2} a^{22} = (\sin\theta)^2 (r\sin\theta)^2 = r^2 \sin^4\theta$$

$$a^{*12} = \frac{\partial u^{*1}}{\partial u^2}\frac{\partial u^{*2}}{\partial u^2} a^{22} = (\sin\theta)\frac{\cos\theta}{r}(r\sin\theta)^2$$

$$= r\cos\theta\sin^3\theta$$

$$a^{*13} = \frac{\partial u^{*1}}{\partial u^1}\frac{\partial u^{*3}}{\partial u^3} a^{13} = \cos\theta\,(r\cos\theta)z = rz\cos^2\theta$$

$$a^{*21} = \frac{\partial u^{*2}}{\partial u^2}\frac{\partial u^{*1}}{\partial u^2} a^{22} = (\frac{\cos\theta}{r})(\sin\theta)\,(r\sin\theta)^2$$

$$= r\cos\theta\sin^3\theta$$

$$a^{*22} = \frac{\partial u^{*2}}{\partial u^2}\frac{\partial u^{*2}}{\partial u^2} a^{22} = \left(\frac{\cos\theta}{r}\right)^2 (r\sin\theta)^2 = \cos^2\theta\sin^2\theta$$

$$a^{*23} = \frac{\partial u^{*2}}{\partial u^1}\frac{\partial u^{*3}}{\partial u^3} a^{13} = (-\frac{\sin\theta}{r})\,(r\cos\theta)z = -z\cos\theta\sin\theta$$

$$a^{*31} = \frac{\partial u^{*3}}{\partial u^3}\frac{\partial u^{*1}}{\partial u^1} a^{31} = (\cos\theta)\,(r\sin\theta)z = rz\cos\theta\sin\theta$$

$$a^{*32} = \frac{\partial u^{*3}}{\partial u^3}\frac{\partial u^{*2}}{\partial u^1} a^{31} = (-\frac{\sin\theta}{r})\,(r\sin\theta)z = -z\sin^2\theta$$

$$a^{*33} = \frac{\partial u^{*3}}{\partial u^i}\frac{\partial u^{*3}}{\partial u^j} a^{ij} = 0$$

The components a^{*ij} may also be obtained from the formula (6.46).

Next, to find the components $a^{*}{}_{i}{}^{\cdot k}$ and $a^{*i}{}_{\cdot k}$, we use the relations

$$a^{*}{}_{i}{}^{\cdot k} = g^{*}_{ij}\, a^{*jk}, \quad a^{*i}{}_{\cdot k} = g^{*ij}\, a^{*}_{jk}$$

from (4.47) and (4.48). The quantities g^*_{ij} and g^{*ij} of the metric tensor in cylindrical coordinates are obtained in Sec. 6.6, Example 2. Thus we find

$$a^{*\ \cdot k}_{i} = g^*_{ij}\, a^{*jk} = g^*_{ii}\, a^{*ik}$$

so that

$$a^{*\ \cdot k}_{1} = a^{*1k}, \qquad a^{*\ \cdot k}_{2} = r^2\, a^{*2k}, \qquad a^{*\ \cdot k}_{3} = a^{*3k}$$

for k = 1, 2, 3. Similarly, we have

$$a^{*i}_{\ \ \cdot k} = g^{*ij}\, a^*_{jk} = g^{*ii}\, a^*_{ik}$$

so that

$$a^{*1}_{\ \ \cdot k} = a^*_{1k}, \qquad a^{*2}_{\ \ \cdot k} = a^*_{2k}/r^2, \qquad a^{*3}_{\ \ \cdot k} = a^*_{3k}$$

for k = 1, 2, 3.

6.5 EXERCISES

1. Show that the quantities

$$g^{ij} = \nabla u^i \cdot \nabla u^j = \sum_{k=1}^{3} \frac{\partial u^i}{\partial x_k}\frac{\partial u^j}{\partial x_k}$$

transform as contravariant components of the metric tensor.

2. From the definitions of g^{ij} and g_{ij}, verify that

$$g^{ij} g_{jk} = \delta^i_k$$

3. The components of a cartesian vector field are given by

$$a_1 = x_1 x_2, \quad a_2 = x_2 x_3, \quad a_3 = x_1 x_3 .$$

Find the covariant and contravariant components of the vector in (a) parabolic cylindrical coordinates, (b) spherical coordinates, (c) elliptic cylindrical coordinates. [Note: Use formula (6,42) for determining a^{*i} and the fact that if the matrix (g_{ij}) is diagonal, i.e., $g_{ij} = 0$ for $i \neq j$, then (g^{ij}) consists only of $1/g_{ii}$ along the diagonal.]

4. The components of a cartesian tensor are given by $a_1 = x_1x_2$, $a_2 = 2x_2 - (x_3)^2$, $a_3 = (x_1)^2$. Find the covariant and contravariant components of the vector in (a) parabolic coordinates, (b) spherical coordinates.
5. The components of a cartesian tensor field of order two are given by

$$(a_{ij}) = \begin{bmatrix} x_1 & 0 & 0 \\ 0 & x_2 & 0 \\ 0 & 0 & x_3 \end{bmatrix}$$

Find the covariant, contravariant, and mixed components of the tensor in (a) cylindrical coordinates, (b) spherical coordinates, (c) parabolic coordinates.
6. Let $\phi = x_1^2 + x_2^2 + x_3^2$. Find the gradient of ϕ in (a) parabolic coordinates, (b) spherical coordinates.
7. Find the gradient of $\phi = x_1x_2 + x_2x_3 + x_1x_3$ in (a) parabolic cylindrical coordinates, (b) spherical coordinates.
8. Find the covariant and contravariant components of the cartesian tensor

$$(a_{ij}) = \begin{bmatrix} 0 & -x_1x_3 & x_2^2 \\ x_1x_3 & 0 & -x_1x_2 \\ -x_2^2 & x_1x_2 & 0 \end{bmatrix}$$

in cylindrical coordinates and spherical coordinates.
9. Let a^i_{jk} denote the components of a third order tensor in a coordinate system u^i. Show that if

$$a^i_{jk} = a^i_{kj} \qquad (\text{or } a^i_{jk} = -a^i_{kj})$$

then

$$a^{*i}_{jk} = a^{*i}_{kj} \qquad (\text{or } a^{*i}_{jk} = -a^{*i}_{kj})$$

in any coordinate system u^{*i}.
10. Let b_i^{jk} denote the components of a third order tensor in a coordinate system u^i. Show that if

$$b_i^{\ jk} = b_i^{\ kj} \qquad (\text{or } b_i^{\ jk} = - b_i^{\ kj})$$

then

$$b^{*\ jk}_{\ i} = b^{*\ kj}_{\ i} \qquad (\text{or } b^{*\ jk}_{\ i} = - b^{*\ kj}_{\ i})$$

11. Show that $a^i_{\ jk} = a^j_{\ ik}$ does not imply $a^{*i}_{\ \ jk} = a^{*j}_{\ \ ik}$.
(The last three problems show that the symmetric and skew-symmetric property of a cartesian tensor is also true for general tensors provided that the indices to be interchanged are either both covariant or both contravariant.)

6.8 COVARIANT DERIVATIVE OF A VECTOR

Let **A** be a vector field with components a_i in a rectangular cartesian coordinate x_i with orthonormal basis $\mathbf{i}_i$ $(1 \le i \le 3)$. The partial derivative of **A** with respect to x_k $(1 \le k \le 3)$ is given by

$$\frac{\partial \mathbf{A}}{\partial x_k} = \frac{\partial a_i}{\partial x_k} \mathbf{i}_i \tag{6.73}$$

since the basis vectors $\mathbf{i}_i$ are constants. Thus, for each value of the index k, the derivative $\partial \mathbf{A}/\partial x_k$ is a vector field with components $\partial a_i/\partial x_k$. In fact, as we saw in Sec. 5.9, the nine quantities $\partial a_i/\partial x_k$ (i, k = 1, 2, 3) are the components of a second order tensor called the tensor gradient of **A**. It is clear from (6.73) that

$$\frac{\partial a_i}{\partial x_k} = \frac{\partial \mathbf{A}}{\partial x_k} \cdot \mathbf{i}_i \tag{6.74}$$

Likewise, if **A** is a vector field with covariant component a_i or contravariant components a^i in an oblique coordinate system x^i with the basis $\mathbf{e}_i$ and reciprocal basis $\mathbf{e}^i$, its derivative with respect to x^k (k = 1, 2, 3) is given by

$$\frac{\partial \mathbf{A}}{\partial x^k} = \frac{\partial a_i}{\partial x^k} \mathbf{e}^i = \frac{\partial a^i}{\partial x^k} \mathbf{e}_i \tag{6.75}$$

As in the case of a cartesian tensor, it can be shown that the quantities

$\partial a_i/\partial x^k$ and $\partial a^i/\partial x^k$ are the covariant and mixed components of a second order tensor (Problem 5). We deduce from (6.75) that

$$\frac{\partial a_i}{\partial x^k} = \frac{\partial \mathbf{A}}{\partial x^k}\cdot \mathbf{e}_i, \qquad \frac{\partial a^i}{\partial x^k} = \frac{\partial \mathbf{A}}{\partial x^k}\cdot \mathbf{e}^i \tag{6.76}$$

Now let us consider a vector field **A** in a general coordinate system u^i. At each point in space we can write

$$\mathbf{A} = a_i\mathbf{e}^i = a^i\mathbf{e}_i$$

where a_i and a^i are the covariant and contravariant components of **A**, and $\mathbf{e}_i$ and $\mathbf{e}^i$ are the local and reciprocal bases. Taking the partial derivative with respect to u^k, we find

$$\frac{\partial \mathbf{A}}{\partial u^k} = \frac{\partial a_i}{\partial u^k}\mathbf{e}^i + a_i\frac{\partial \mathbf{e}^i}{\partial u^k} \tag{6.77a}$$

$$= \frac{\partial a^i}{\partial u^k}\mathbf{e}_i + a^i\frac{\partial \mathbf{e}_i}{\partial u^k} \tag{6.77b}$$

The derivatives $\partial \mathbf{e}_i/\partial u^i$ and $\partial \mathbf{e}^i/\partial u^i$ are no longer zero since the base vectors now depend on the coordinates u^i of the point. Therefore, the rate of change of **A** involves not only the rate of change of its components ($\partial a_i/\partial u^i$, $\partial a^i/\partial u^i$), but also the rate of change of the local bases ($\partial \mathbf{e}_i/\partial u^i$, $\partial \mathbf{e}^i/\partial u^i$). In analogy to (6.76), the two kinds of components of the derivative $\partial \mathbf{A}/\partial u^i$, which we denote by $a_{i,\,k}$ and $a^i{}_{,\,k}$, are defined by

$$a_{i,\,k} = \frac{\partial \mathbf{A}}{\partial u^k}\cdot \mathbf{e}_i = \frac{\partial a_i}{\partial u^k} + a_j\frac{\partial \mathbf{e}^j}{\partial u^k}\cdot \mathbf{e}_i \tag{6.78}$$

and

$$a^i{}_{,\,k} = \frac{\partial \mathbf{A}}{\partial u^k}\cdot \mathbf{e}^i = \frac{\partial a^i}{\partial u^k} + a^j\frac{\partial \mathbf{e}_j}{\partial u^k}\cdot \mathbf{e}^i \tag{6.79}$$

respectively. The comma in the notation $a_{i,k}$ or $a^i_{,k}$ indicates differentiation with respect to the coordinate u^k. We will show that the quantities $a_{i,k}$ or $a^i_{,k}$ are components of a second order general tensor called the covariant derivative of the vector **A**. More precisely, we will show that $a_{i,k}$ are the covariant components of the covariant derivative of $\mathbf{A} = a_i\mathbf{e}^i$, while $a^i_{,k}$ are the mixed components of the covariant derivative of **A** when it is represented by its contravariant components, i.e., $\mathbf{A} = a^i\mathbf{e}_i$. It is clear that in rectangular cartesian coordinates the covariant derivative of a vector field is just the ordinary derivative (6.73).

Christoffel Symbols

To facilitate the calculation of the covariant derivative of vector fields and tensors in general, we now introduce the so-called Christoffel symbols. First, let [i, jk] denote the dot product of $\mathbf{e}_i$ and $\partial\mathbf{e}_j/\partial u^k$, that is

$$[i,jk] = \mathbf{e}_i \cdot \frac{\partial \mathbf{e}_j}{\partial u^k} \qquad (i,j,k = 1,2,3) \tag{6.80}$$

This notation is called the *Christoffel symbol* of the first kind. If we multiply both sides of (6.80) by g^{ri} and take the sum with respect to i, noting that $g^{ri}\,\mathbf{e}_i = \mathbf{e}^r$, we obtain

$$\begin{aligned} g^{ri}[i,jk] &= g^{ri}\mathbf{e}_i \cdot \frac{\partial \mathbf{e}_j}{\partial u^k} \\ &= \mathbf{e}^r \cdot \frac{\partial \mathbf{e}_j}{\partial u^k} = \begin{Bmatrix} r \\ j\ k \end{Bmatrix} \end{aligned} \tag{6.81}$$

The notation on the right-hand side of (6.81) is called the *Christoffel symbol* of the second kind. Since $g_{ij} = g^{jk} = \delta_i^{\ k}$, it follows that

$$[i,jk] = g_{ir}\begin{Bmatrix} r \\ j\ k \end{Bmatrix} \tag{6.82}$$

Thus the Christoffel symbols are related to each other by the metric tensor. Moreover, if we assume that the coordinate transformation equations are twice continuously differentiable, then we have

$$\frac{\partial \mathbf{e}_j}{\partial u^k} = \frac{\partial}{\partial u^k}\left[\frac{\partial \mathbf{R}}{\partial u^j}\right] = \frac{\partial}{\partial u^j}\left[\frac{\partial \mathbf{R}}{\partial u^k}\right] = \frac{\partial \mathbf{e}_k}{\partial u^j} \tag{6.83}$$

Hence the Christoffel symbols [i, jk] and $\{^i{}_{jk}\}$ are both symmetric with respect to the indices j, k, that is, [i, jk] = [i, kj] and $\{^i{}_{jk}\} = \{^i{}_{kj}\}$. However, the Christoffel symbols do not transform as components of a tensor, as we shall see later.

Now from the fact that $\mathbf{e}_i \cdot \mathbf{e}^j = \delta_i^j$, it follows that

$$\mathbf{e}_i \cdot \frac{\partial \mathbf{e}^j}{\partial u^k} + \mathbf{e}^j \cdot \frac{\partial \mathbf{e}_i}{\partial u^k} = 0$$

so that

$$\mathbf{e}_i \cdot \frac{\partial \mathbf{e}^j}{\partial u^k} = -\mathbf{e}^j \cdot \frac{\partial \mathbf{e}_i}{\partial u^k} = -\left\{\begin{matrix} j \\ i\,k \end{matrix}\right\} \tag{6.84}$$

If we multiply both sides of (6.84) by g^{ir} and sum with respect to i, noting that $g^{ir}\mathbf{e}_i = \mathbf{e}^r$, we find

$$\mathbf{e}^r \cdot \frac{\partial \mathbf{e}^j}{\partial u^k} = -g^{ri}\left\{\begin{matrix} j \\ i\,k \end{matrix}\right\} \quad \text{or} \quad \mathbf{e}^i \cdot \frac{\partial \mathbf{e}^j}{\partial u^k} = -g^{ir}\left\{\begin{matrix} j \\ r\,k \end{matrix}\right\} \tag{6.85}$$

Thus we obtain the various dot products of the bases $\mathbf{e}_i$ and $\mathbf{e}^j$ with their derivatives $\partial \mathbf{e}^i/\partial u^k$ and $\partial \mathbf{e}_i/\partial u^k$.

Using (6.84) and (6.81), we can therefore write (6.78) and (6.79) as

$$a_{i,k} = \frac{\partial a_i}{\partial u^k} - \left\{\begin{matrix} j \\ i\,k \end{matrix}\right\} a_j \tag{6.86}$$

and

$$a^i_{\,,k} = \frac{\partial a^i}{\partial u^k} + \left\{\begin{matrix} i \\ j\,k \end{matrix}\right\} a^j \tag{6.87}$$

From the definition (6.78) and (6.79), we observe that

$$g^{ri} a_{i,k} = \frac{\partial \mathbf{A}}{\partial u^k} \cdot (g^{ri} \mathbf{e}_i) = \frac{\partial \mathbf{A}}{\partial u^k} \cdot \mathbf{e}^r = a^r_{,k} \tag{6.88}$$

and, conversely,

$$g_{ri} a^i_{,k} = \frac{\partial \mathbf{A}}{\partial u^k} \cdot (g_{ri} \mathbf{e}^i) = \frac{\partial \mathbf{A}}{\partial u^k} \cdot \mathbf{e}_r = a_{r,k} \tag{6.89}$$

These relations are to be expected since $a_{i,k}$ and $a^i_{,k}$ are the (covariaiant and mixed) components of the covariant derivative of the same vector **A**.

The results of the various dot products of the bases $\mathbf{e}_i$ and $\mathbf{e}^j$ with their derivatives $\partial \mathbf{e}_i/\partial u^k$ and $\partial \mathbf{e}^j/\partial u^k$ are shown in the following schematic diagram:

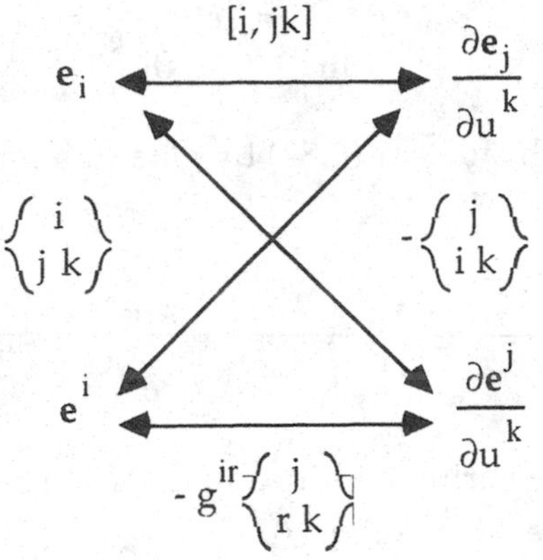

where the double-headed arrow points out the two vectors whose dot product is equal to the Christoffel symbol written beside the arrow.

Example 1. Find the values of the Christoffel symbols [i, jk] and $\{^i{}_{jk}\}$ in cylindrical coordinates.

Solution: Set $u^1 = r$, $u^2 = \theta$, and $u^3 = z$. From Example 1 of Sec. 6.6, we have

$$\mathbf{e}_1 = \cos\theta\, \mathbf{i}_1 + \sin\theta\, \mathbf{i}_2,\quad \mathbf{e}_2 = -r\sin\theta\, \mathbf{i}_1 + r\cos\theta\, \mathbf{i}_2,\quad \mathbf{e}_3 = \mathbf{i}_3$$

Hence we find

$$\frac{\partial \mathbf{e}_1}{\partial u^1} = 0,\qquad \frac{\partial \mathbf{e}_1}{\partial u^2} = -\sin\theta\, \mathbf{i}_1 + \cos\theta\, \mathbf{i}_2,\qquad \frac{\partial \mathbf{e}_3}{\partial u^3} = 0$$

$$\frac{\partial \mathbf{e}_2}{\partial u^1} = -\sin\theta\, \mathbf{i}_1 + \cos\theta\, \mathbf{i}_2,\qquad \frac{\partial \mathbf{e}_2}{\partial u^2} = -r\cos\theta\, \mathbf{i}_1 - r\sin\theta\, \mathbf{i}_2,\qquad \frac{\partial \mathbf{e}_2}{\partial u^3} = 0$$

$$\frac{\partial \mathbf{e}_3}{\partial u^1} = 0,\qquad \frac{\partial \mathbf{e}_3}{\partial u^2} = 0,\qquad \frac{\partial \mathbf{e}_3}{\partial u^3} = 0$$

Therefore, by (6.80), we see that among the 27 possible values of [i, jk], only three are nonzero and they are given by

$$\mathbf{e}_1 \cdot \frac{\partial \mathbf{e}_2}{\partial u^2} = [1, 22] = -r,\quad \mathbf{e}_2 \cdot \frac{\partial \mathbf{e}_1}{\partial u^2} = [2, 12] = r,\quad \mathbf{e}_2 \cdot \frac{\partial \mathbf{e}_2}{\partial u^1} = [2, 21] = r$$

From (6.81) it follows that

$$g^{11}[1, 22] = \left\{ {1 \atop 2\,2} \right\} = -r,\quad g^{22}[2, 12] = \left\{ {2 \atop 1\,2} \right\} = \frac{1}{r^2} r = \frac{1}{r}$$

$$g^{22}[2, 12] = \left\{ {2 \atop 2\,1} \right\} = \frac{1}{r}$$

Example 2. The covariant components of a vector field **A** in cylindrical coordinates $u^1 = r$, $u^2 = \theta$, and $u^3 = z$ are given by

$$a_1 = 2\sin\theta,\quad a_2 = r\cos\theta,\quad a_3 = z$$

Find the convariant and mixed components of the covariant derivative of **A**.

Solution: We calculate $a_{1,k}$, $a_{2,k}$, $a_{3,k}$ for $k = 1, 2, 3$. From Example 1, we recall that

$$\left\{ {1 \atop 2\,2} \right\} = -r, \quad \left\{ {2 \atop 1\,2} \right\} = \left\{ {2 \atop 2\,1} \right\} = \frac{1}{r}$$

and $\{^i{}_{jk}\} = 0$, otherwise. Now since $a_1 = 2 \sin\theta$, which depends only on θ, we find

$$a_{1,1} = 0, \quad a_{1,3} = 0$$

$$a_{1,2} = \frac{\partial a_1}{\partial \theta} - \left\{ {2 \atop 1\,2} \right\} a_2 = 2\cos\theta - \frac{1}{r}(r\cos\theta) = \cos\theta$$

Similarly, we find

$$a_{2,1} = \frac{\partial a_2}{\partial r} - \left\{ {2 \atop 2\,1} \right\} a_2 = \cos\theta - (\frac{1}{r})(r\cos\theta) = 0$$

$$a_{2,2} = \frac{\partial a_2}{\partial \theta} - \left\{ {1 \atop 2\,2} \right\} a_1 = -r\sin\theta - (-r)2\sin\theta = r\sin\theta$$

$$a_{2,3} = 0, \quad a_{3,1} = 0, \quad a_{3,2} = 0, \quad a_{3,3} = \partial a_3/\partial z = 1$$

These are the convariant components of the covariant derivative of the vector.

To find the mixed components, we use the relation (6.88). Since

$$g^{11} = 1, \; g^{22} = 1/r^2, \; g^{33} = 1, \text{ and } g^{ij} = 0 \text{ for } i \neq j$$

we obtain

$$a^1{}_{,2} = a_{1,2}, \quad a^2{}_{,1} = g^{22} a_{2,1} = 0$$

$$a^2{}_{,2} = (1/r^2)\, a_{2,2} = (\sin\theta)/r, \quad a^3{}_{,3} = a_{3,3}$$

while all the other components are zero.

Example 3. Determine the Christoffel symbols for the general coordinates given in Example 2, Sec. 6.7.

Solution: The general coordinates are defined by the equations

$$x_1 = vw, \quad x_2 = (v^2 + w^2)/2, \quad x_3 = z$$

Setting $u^1 = v$, $u^2 = w$, $u^3 = z$, we find from Example 2, Sec. 6.7

$$g_{11} = v^2 + w^2, \quad g_{12} = g_{21} = 2vw, \quad g_{22} = v^2 + w^2, \quad g_{33} = 1$$

and

$$g^{11} = \frac{v^2 + w^2}{(v^2 - w^2)^2}, \quad g^{12} = \frac{-2vw}{(v^2 - w^2)^2}$$

$$g^{22} = \frac{v^2 + w^2}{(v^2 - w^2)^2}, \quad g^{33} = 1$$

with $g_{ij} = 0$ and $g^{ij} = 0$ for the rest of the values of the indices i, j. Therefore, by (6.80), we have

$$[1, 1\,1] = v, \quad [1, 1\,2] = [1, 2\,1] = w, \quad [1, 2\,2] = v$$

$$[2, 1\,1] = w, \quad [2, 1\,2] = [2, 2\,1] = v, \quad [2, 2\,2] = w$$

and $[i, jk] = 0$ for all the other values of i, j, k.

It follows from (6.81) that

$$\left\{ {1 \atop 1\,1} \right\} = g^{1r}[r, 1\,1] = \frac{v}{v^2 - w^2}, \quad \left\{ {1 \atop 2\,2} \right\} = g^{1r}[r, 2\,2] = \frac{v}{v^2 - w^2}$$

$$\left\{ {1 \atop 1\,2} \right\} = \left\{ {1 \atop 2\,1} \right\} = g^{1r}[r, 1\,2] = \frac{-w}{v^2 - w^2}$$

$$\left\{ {2 \atop 1\,1} \right\} = g^{2r}[r, 1\,1] = \frac{-w}{v^2 - w^2}, \quad \left\{ {2 \atop 2\,2} \right\} = g^{2r}[r, 2\,2] = \frac{-w}{v^2 - w^2}$$

$$\left\{ {2 \atop 1\,2} \right\} = \left\{ {2 \atop 2\,1} \right\} = g^{2r}[r, 1\,2] = \frac{v}{v^2 - w^2}$$

and $\{{}^{i}{}_{jk}\} = 0$ for all the other values of i, j, k.

6.9 TRANSFORMATION OF CHRISTOFFEL SYMBOLS

We observe that the Christoffel symbols $[i, jk]$ and, $\{{}^{i}{}_{jk}\}$ are zero in all cartesian (rectangular or oblique) coordinate system since in any such system the base vectors are constant. However, as we saw in Example 1, Sec. 6.8, they do not all vanish in cylindrical coordinates. This implies that the Christoffel symbols cannot possibly be components of a tensor because a tensor that is zero in any particular coordinate system must be zero in all coordinate systems. By considering the transformation laws for [i, jk] and $\{^i{}_{jk}\}$, we can also see the non-tensorial character of these symbols. We illustrate this for the Christoffel symbol of the second kind, $\{^i{}_{jk}\}$.

Under a transformation of coordinates, say from u^i to u^{*i} defined by equations of the form (6.63), we have

$$\left\{ {j \atop i\,k} \right\}^* = \mathbf{e}^{*j} \cdot \frac{\partial \mathbf{e}^*_i}{\partial u^{*k}} = \left(\frac{\partial u^{*j}}{\partial u^p} \mathbf{e}^p \right) \cdot \frac{\partial}{\partial u^{*k}} \left(\frac{\partial u^q}{\partial u^{*i}} \mathbf{e}_q \right)$$

$$= \left(\frac{\partial u^{*j}}{\partial u^p} \mathbf{e}^p \right) \cdot \left(\frac{\partial u^q}{\partial u^{*i}} \frac{\partial u^r}{\partial u^{*k}} \frac{\partial \mathbf{e}_q}{\partial u^r} + \frac{\partial^2 u^q}{\partial u^{*i} \partial u^{*k}} \mathbf{e}_q \right) \tag{6.90}$$

$$= \frac{\partial u^{*j}}{\partial u^p} \frac{\partial u^q}{\partial u^{*i}} \frac{\partial u^r}{\partial u^{*k}} \left\{ {p \atop q\,r} \right\} + \frac{\partial^2 u^p}{\partial u^{*i} \partial u^{*k}} \frac{\partial u^{*j}}{\partial u^p}$$

Because of the presence of the second term in (6.90), the Christoffel symbol $\{^i{}_{jk}\}$ does not transform as components of a tensor.

Using the transformation law (6.90), we can now show that the quantities $a_{i,k}$ and $a^i{}_{,k}$ transform as covariant and mixed components of a second order general tensor, which is called the covariant derivative of the vector field **A**. Indeed, under a change of coordinates, we know that

$$a^*_i = \frac{\partial u^q}{\partial u^{*i}} a_q$$

and, hence, by (6.90) we have

$$a^*_{i,k} = \frac{\partial a^*_i}{\partial u^{*k}} - \left\{ {j \atop i\,k} \right\}^* a^*_j$$

$$= \frac{\partial u^q}{\partial u^{*i}}\frac{\partial u^p}{\partial u^{*k}}\frac{\partial a_q}{\partial u^p} + \frac{\partial^2 u^q}{\partial u^{*i}\partial u^{*k}} a_q$$

$$- \left(\frac{\partial u^{*j}}{\partial u^p}\frac{\partial u^q}{\partial u^{*i}}\frac{\partial u^r}{\partial u^{*k}}\left\{ {p \atop q\,r} \right\}\frac{\partial u^s}{\partial u^{*j}} a_s + \frac{\partial^2 u^p}{\partial u^{*i}\partial u^{*k}}\frac{\partial u^{*j}}{\partial u^p}\frac{\partial u^s}{\partial u^{*j}} a_s \right)$$

Since $(\partial u^{*j}/\partial u^p)(\partial u^s/\partial u^{*j}) = \delta_p{}^s$, this equation simplifies to

$$a^*_{i,\,k} = \frac{\partial u^q}{\partial u^{*i}}\frac{\partial u^r}{\partial u^{*k}}\left(\frac{\partial a_q}{\partial u^r} - \left\{ {p \atop q\,r} \right\} a_p \right)$$

$$= \frac{\partial u^q}{\partial u^{*i}}\frac{\partial u^r}{\partial u^{*k}} a_{q,\,r} \qquad (6.91)$$

Therefore, the quantities $a_{i,\,k}$ are components of a second order tensor.

It follows from (6.88) and the tensorial character of the inner product of two tensors that the quantities $a^i{}_{,k}$ are mixed components of the same second order tensor which is the covariant derivative of the vector field **A**. Of course, this fact can also be verified by exhibiting the transformation law for the quantities $a^i{}_{,k}$. This is requested in the exercises (see Prob. 13).

To obtain further insight into the significance of the components $a_{i,\,k}$ and $a^i{}_{,k}$ of $\partial \mathbf{A}/\partial u^k$, we derive formulas for $\partial \mathbf{A}/\partial u^k$ that resemble those given in (6.75). First, from (6.84), we deduce that

$$\frac{\partial \mathbf{e}^j}{\partial u^k} = -\left\{ {j \atop i\,k} \right\} \mathbf{e}^i$$

This says that the quantities $-\{{}^j{}_{ik}\}$ are the expansion coefficients of $\partial \mathbf{e}^j/\partial u^k$ with respect to the reciprocal basis $\mathbf{e}^i$. Subsitituting this in (6. 77a), we find

$$\frac{\partial \mathbf{A}}{\partial u^k} = \frac{\partial a_i}{\partial u^k}\mathbf{e}^i + \frac{\partial \mathbf{e}^j}{\partial u^k}a_j = \left[\frac{\partial a_i}{\partial u^k} - \begin{Bmatrix} j \\ i\,k \end{Bmatrix} a_j\right]\mathbf{e}^i$$

$$= a_{i,\,k}\,\mathbf{e}^i \qquad (6.92)$$

in view of (6.86). Similarly, from (6.81) we deduce

$$\frac{\partial \mathbf{e}_j}{\partial u^k} = \begin{Bmatrix} i \\ j\,k \end{Bmatrix}\mathbf{e}_i$$

Substituting this in (6.77b) leads to the formula

$$\frac{\partial \mathbf{A}}{\partial u^k} = a^i{}_{,k}\,\mathbf{e}_i \qquad (6.93)$$

Equations (6.92) and (6.93) imply that the quantities $a_{i,\,k}$ and $a^i{}_{,k}$ are precisely the components of $\partial\mathbf{A}/\partial u^k$ with respect to the bases $\mathbf{e}^i$ and $\mathbf{e}_i$, respectively.

We conclude this section with the derivation of the relation of the Christoffel symbols with the metric tensor. Using the fact that $\partial\mathbf{e}_i/\partial u^k = \partial\mathbf{e}_k/\partial u^i$, we see that

$$[i, jk] = \mathbf{e}_i\cdot\frac{\partial \mathbf{e}_j}{\partial u^k} = \frac{1}{2}\left[\mathbf{e}_i\cdot\frac{\partial \mathbf{e}_j}{\partial u^k} + \mathbf{e}_i\cdot\frac{\partial \mathbf{e}_k}{\partial u^j}\right]$$

$$= \frac{1}{2}\left[\frac{\partial}{\partial u^k}(\mathbf{e}_i\cdot\mathbf{e}_j) - \mathbf{e}_j\cdot\frac{\partial \mathbf{e}_i}{\partial u^k} + \frac{\partial}{\partial u^j}(\mathbf{e}_i\cdot\mathbf{e}_k) - \mathbf{e}_k\cdot\frac{\partial \mathbf{e}_i}{\partial u^j}\right]$$

But

$$\mathbf{e}_j\cdot\frac{\partial \mathbf{e}_i}{\partial u^k} + \mathbf{e}_k\cdot\frac{\partial \mathbf{e}_i}{\partial u^j} = \mathbf{e}_j\cdot\frac{\partial \mathbf{e}_k}{\partial u^i} + \mathbf{e}_k\cdot\frac{\partial \mathbf{e}_j}{\partial u^i}$$

$$= \frac{\partial}{\partial u^i}(\mathbf{e}_j\cdot\mathbf{e}_k)$$

therefore, we obtain

$$[i, jk] = \frac{1}{2}\left[\frac{\partial}{\partial u^k}(\mathbf{e}_i \cdot \mathbf{e}_j) + \frac{\partial}{\partial u^j}(\mathbf{e}_i \cdot \mathbf{e}_k) - \frac{\partial}{\partial u^i}(\mathbf{e}_j \cdot \mathbf{e}_k)\right]$$

$$= \frac{1}{2}\left[\frac{\partial g_{ij}}{\partial u^k} + \frac{\partial g_{ik}}{\partial u^j} - \frac{\partial g_{jk}}{\partial u^i}\right] \tag{6.94}$$

By (6.81) we further obtain

$$\left\{ {i \atop j\,k} \right\} = g^{ir}[r, jk] = \frac{1}{2}\, g^{ir}\left[\frac{\partial g_{rj}}{\partial u^k} + \frac{\partial g_{rk}}{\partial u^j} - \frac{\partial g_{jk}}{\partial u^i}\right] \tag{6.95}$$

Hence, if the g_{ij} are constants, the Christoffel symbols vanish identically, and thus the covariant derivative of a vector field reduces to the ordinary derivative. This is certainly the case when we are in a cartesian coordinate system.

Using (6.94) we readily obtain

$$\frac{\partial g_{ij}}{\partial u^k} = [i, jk] + [j, ik] \tag{6.96}$$

which, in view of (6.82), can be written as

$$\frac{\partial g_{ij}}{\partial u^k} = g_{ir}\left\{ {r \atop j\,k} \right\} + g_{jr}\left\{ {r \atop i\,k} \right\} \tag{6.97}$$

We can also derive a formula analogous to (6.97) for the partial derivatives of the contravariant components g^{ij} of the metric tensor. First, from the identity $g_{ij}\, g^{jk} = \delta_i^{\,j}$, we obtain by differentiation

$$g_{ij}\frac{\partial g^{jk}}{\partial u^m} = -\frac{\partial g_{ij}}{\partial u^m}\, g^{jk}$$

Now multiply both sides of this equation by g^{ir} and note that

$g_{ij}\, g^{ri} = \delta_j^{\ r}$. We obtain

$$\frac{\partial g^{rk}}{\partial u^m} = -g^{ri} g^{jk}([i, j\, m] + [j, i\, m])$$

where we have made use of (6.96). By (6.81) this can be written as

$$\frac{\partial g^{rk}}{\partial u^m} = -g^{jk}\left\{ {r \atop j\, m} \right\} - g^{ri}\left\{ {k \atop i\, m} \right\}$$

or, changing indices,

$$\frac{\partial g^{ij}}{\partial u^m} = -g^{rj}\left\{ {i \atop r\, k} \right\} - g^{ir}\left\{ {j \atop r\, k} \right\} \tag{6.98}$$

6.6 EXERCISES

1. If $\mathbf{A} = x_1 x_2\, \mathbf{i}_1 + x_3 x_2\, \mathbf{i}_2 + x_1 x_3\, \mathbf{i}_3$, find the derivatives of $\mathbf{A}$ with respect to x_1, x_2, x_3, and write down the matrix $(\partial a_i / \partial x_j)$.
2. Repeat Problem 1 when

$$\mathbf{A} = x_1 \cos x_2\, \mathbf{i}_1 + x_2 \sin x_3\, \mathbf{i}_2 + x_3 \exp(x_3)\, \mathbf{i}_3$$

3. Let $\mathbf{A} = x^2 x^3\, \mathbf{e}_1 + x^2 \cos x^1\, \mathbf{e}_2 - (1 + x^3)\, \mathbf{e}_3$, a contravariant vector in an oblique coordinate system x^i with basis $\mathbf{e}_i$. Find the derivatives of $\mathbf{A}$ with respect to the coordinates x^i $(1 \le i \le 3)$ and write down the matrix $(\partial a^i / \partial x^j)$.
4. Repeat Problem 3 for the covariant vector $\mathbf{A} = \sin x^2\, \mathbf{e}^1 + x^1 \exp(x^2)\, \mathbf{e}^2 + x^2 x^3\, \mathbf{e}^3$
5. Let a_i and a^i denote the covariant and the contravariant components of a vector field in an oblique cartesian coordinate system. Show that the quantities $\partial a_i / \partial x^k$ and $\partial a^i / \partial x^k$ are the covariant and mixed components of a second order tensor.

In each of Problems 6 through 9, determine the Christoffel symbols $[i, jk]$ and $\{^i{}_{jk}\}$ in the given coordinate system.

6. The spherical coordinates $u^1 = r$, $u^2 = \phi$, $u^3 = \theta$, where

$$x_1 = r \sin\phi \cos\theta, \quad x_2 = r \sin\phi \sin\theta, \quad x_3 = r \cos\phi.$$

7. The parabolic cylindrical coordinates $u^1 = v$, $u^2 = w$, $u^3 = z$, where

$$x_1 = (v^2 - w^2)/2, \quad x_2 = vw, \quad x_3 = z.$$

8. The parabolic coordinates $u^1 = v$, $u^2 = w$, $u^3 = \theta$, where

$$x_1 = vw \cos\theta, \quad x_2 = vw \sin\theta, \quad x_3 = (v^2 - w^2)/2.$$

9. The general coordinates $u^1 = u$, $u^2 = v$, $u^3 = w$, where

$$x_1 = uv + 1, \quad x_2 = vw - 1, \quad x_3 = (u^2 + w^2)/2.$$

10. The covariant components of a vector field **A** in spherical coordinates are given by

$$a_1 = r \sin^2\phi \cos^2\theta, \quad a_2 = r^2 \sin\phi \cos\phi \cos^2\theta,$$
$$a_3 = r \sin\phi \cos\phi \cos\theta.$$

Find the components $a_{i,\,k}$ and $a^i{}_{,\,k}$ of the covariant derivative of **A**.

11. The covariant components of a vector field **A** in parabolic cylindrical coordinates are given by

$$a_1 = vw^2, \quad a_2 = v^2 w, \quad a_3 = 0.$$

Find the components $a_{i,\,k}$ and $a^i{}_{,\,k}$ of the covariant derivative of **A**.

12. The covariant components of a vector field **A** in parabolic coordinates are given by

$$a_1 = w \cos\theta + v, \quad a_2 = v \cos\theta - w, \quad a_3 = -vw \sin\theta.$$

Find the covariant and mixed components of the covariant derivative of **A**.

13. Show that the mixed components $a^i{}_{,\,k}$ of the covariant derivative of a vector **A** transform under any changes of coordinates according to the law

$$a^{*i}{}_{,\,k} = \frac{\partial u^{*i}}{\partial u^p} \frac{\partial u^q}{\partial u^{*k}} a^{*p}{}_{,\,q}$$

14. Show that the Christoffel symbol of the first kind transforms under any changes of coordinates according to the law

$$[i, j\, k]^* = \frac{\partial u^p}{\partial u^{*i}} \frac{\partial u^q}{\partial u^{*j}} \frac{\partial u^r}{\partial u^{*k}} [p, q\, r] + \frac{\partial^2 u^q}{\partial u^{*j} \partial u^{*k}} \frac{\partial u^p}{\partial u^{*i}} g_{pq}$$

15. Deduce from (6.80) and (6.84) the formulas

$$\frac{\partial \mathbf{e}_j}{\partial u^k} = [i, j\, k]\, \mathbf{e}^i, \qquad \frac{\partial \mathbf{e}^j}{\partial u^k} = -g^{ri} \left\{ {j \atop i\,k} \right\} \mathbf{e}_r$$

16. Show that

$$\frac{\partial g_{ij}}{\partial u^k} - \frac{\partial g_{jk}}{\partial u^i} = [i, j\, k] - [k, i\, j]$$

17. Show that in any orthogonal coordinate system ($g_{ij} = 0$, if $i \neq j$), the Christoffel symbol $\{^i{}_{jk}\}$ vanishes whenever i, j, k have distinct values.

18. Noting that the covariant derivative of a scalar field is precisely the ordinary derivative, show that

$$(a_i b^i)_{,k} = \frac{\partial}{\partial u^k}(a_i b^i) = a_{i,k} b^i + a_i b^i{}_{,k}$$

where a_i and b^i are components of two vector fields. (Hint: Differentiate; then add and subtract the term $a_i\{^i{}_{jk}\}b^j$, observing that $a_i\{^i{}_{jk}\}b^j = a_j\{^j{}_{ik}\}b^i$.)

6.10 COVARIANT DERIVATIVE OF TENSORS

The concept of covariant derivative of a vector field can be extended to tensors of higher order. First, we observe that if ϕ is a scalar field, its covariant derivative is simply the ordinary derivative, that is,

$$\phi_{,k} = \frac{\partial \phi}{\partial u^k}$$

This is to be expected since the changes in the base vectors $\mathbf{e}_i$ and $\mathbf{e}^i$

have no effect on the field itself. Thus the covariant derivative of ϕ is precisely the covariant vector grad ϕ. Next, let us consider a second order tensor and suppose that a_{ij}, a^{ij}, and $a_i^{\cdot j}$ are its covariant, contravariant, and mixed components, respectively. Then the corresponding components of its covariant derivative are given by

$$a_{ij,k} = \frac{\partial a_{ij}}{\partial u^k} - \left\{ \begin{matrix} r \\ i\,k \end{matrix} \right\} a_{rj} - \left\{ \begin{matrix} r \\ j\,k \end{matrix} \right\} a_{ir} \tag{6.99}$$

$$a^{ij}_{\ ,k} = \frac{\partial a^{ij}}{\partial u^k} + \left\{ \begin{matrix} i \\ r\,k \end{matrix} \right\} a^{rj} + \left\{ \begin{matrix} j \\ r\,k \end{matrix} \right\} a^{ir} \tag{6.100}$$

$$a_{i,k}^{\cdot j} = \frac{\partial a_i^{\cdot j}}{\partial u^k} - \left\{ \begin{matrix} r \\ i\,k \end{matrix} \right\} a_r^{\cdot j} + \left\{ \begin{matrix} j \\ r\,k \end{matrix} \right\} a_i^{\cdot r} \tag{6.101}$$

It can be readily shown that these quantities transform as components of a third order mixed tensor under any transformations of coordinates. In particular, the quantities $a_{ij,k}$ transform as covariant components, the $a^{ij}_{\ ,k}$ as mixed components with two contravariant indices, and the $a_i^{\cdot j}{}_{,k}$ as mixed components with two covariant indices.

From the above definition it should be evident how the components of the covariant derivative of a tensor of order r may be defined. It consists of r+1 terms, the first term being the partial derivative of the components of the tensor with respect to the coordinate, say u^k. The remaining r terms are sums of quantities resembling components of inner products of the tensor and the Christoffel symbols of the second kind with the differentiation index k appearing as one of the subscripts in the Christoffel symbol. Each index in the components of the tensor that is replaced by a summation index gets transferred to the unoccupied spot of the character (subscript or superscript) in the Christoffel symbol. Finally, each of the remaining r terms has a plus sign if the summation is with respect to a contravariant index (superscript) and a minus sign if it is with respect to a covariant index (subscript) of the components of the tensor. It can be shown that the covariant derivative of a tensor of order r is a tensor of order r + 1, where the increase in the order occurs as a covariant index.

Example 1. If $a_i{}^{\cdot jm}$ are the mixed components of a third order tensor, then

$$a_{i,k}^{\cdot jm} = \frac{\partial a_i^{\cdot jm}}{\partial u^k} - \begin{Bmatrix} r \\ i\,k \end{Bmatrix} a_r^{\cdot jm} + \begin{Bmatrix} j \\ r\,k \end{Bmatrix} a_i^{\cdot rm} + \begin{Bmatrix} m \\ r\,i \end{Bmatrix} a_i^{\cdot jr}$$

are the mixed components of the covariant derivative of the tensor.

Example 2. Consider the mixed tensor defined by the Kronecker delta δ_i^j. By (101) we see that

$$\delta_{i,k}^{j} = \frac{\partial \delta_i^j}{\partial u^k} - \begin{Bmatrix} r \\ i\,k \end{Bmatrix} \delta_r^j + \begin{Bmatrix} j \\ r\,k \end{Bmatrix} \delta_i^r$$

$$= 0 - \begin{Bmatrix} j \\ i\,k \end{Bmatrix} \delta_r^j + \begin{Bmatrix} j \\ i\,k \end{Bmatrix} \delta_i^r = 0$$

Thus the mixed tensor defined by the Kronecker delta behaves like a constant with respect to covariant differentiation.

It is easy to show that the rules for covariant differentiation of sums and products of tensors are identical with those used in ordinary differentiation. The verification of the case involving sums is trivial. We illustrate the proof of the case involving products with two particular examples. First, consider the components $t_{ij} = a_i b_j$ of a second order tensor, which is the outer product of two vectors with components a_i and b_j. Then, in accordance with (6.99), we have

$$t_{ij,k} = \frac{\partial t_{ij}}{\partial u^k} - \begin{Bmatrix} r \\ i\,k \end{Bmatrix} t_{rj} - \begin{Bmatrix} r \\ j\,k \end{Bmatrix} t_{ir}$$

$$= \frac{\partial (a_i b_j)}{\partial u^k} - \begin{Bmatrix} r \\ i\,k \end{Bmatrix} a_r b_j - \begin{Bmatrix} r \\ j\,k \end{Bmatrix} a_i b_r$$

Expanding the first term and rearranging terms, we find

$$(a_i b_j)_{,k} = \left[\frac{\partial a_i}{\partial u^k} - \begin{Bmatrix} r \\ i\,k \end{Bmatrix} a_r \right] b_j + \left[\frac{\partial b_j}{\partial u^k} - \begin{Bmatrix} r \\ j\,k \end{Bmatrix} b_r \right] a_i$$

$$= a_{i,k} b_j + a_i b_{j,k}$$

which is the result we desired.

Next, let us consider the components $a_i = t_{ij} b^j$ of the inner product of a second order tensor and a vector. We find

$$a_{i,k} = \frac{\partial (t_{ij} b^j)}{\partial u^k} - \begin{Bmatrix} j \\ i\,k \end{Bmatrix} (t_{jm} b^m)$$

$$= \frac{\partial t_{ij}}{\partial u^k} b^j + t_{ij} \frac{\partial b^j}{\partial u^k} - \begin{Bmatrix} j \\ i\,k \end{Bmatrix} t_{jm} b^m$$

But from the expressions of $t_{ij,\,k}$ and $b^j{}_{,\,k}$, we have

$$\frac{\partial t_{ij}}{\partial u^k} = t_{ij,\,k} + \begin{Bmatrix} r \\ i\,k \end{Bmatrix} t_{rj} + \begin{Bmatrix} r \\ j\,k \end{Bmatrix} t_{ir}$$

and

$$\frac{\partial b^j}{\partial u^k} = b^j{}_{,k} - \begin{Bmatrix} j \\ r\,k \end{Bmatrix} b^r$$

Substituting these in the previous equation, we observe that all the terms involving the Christoffel symbol cancel out, yielding the desired result

$$(t_{ij} b^j)_{,\,k} = t_{ij,\,k} b^j + t_{ij} b^j{}_{,\,k}$$

Ricci's Theorem

We conclude this section with the verification of the fact that the metric tensor behaves like a constant with respect to covariant differentiation, that is, $g_{ij,\,k} = 0$. This is known as *Ricci's* theorem. Indeed, by definition, we find

$$g_{ij,k} = \frac{\partial g_{ij}}{\partial u^k} - \begin{Bmatrix} r \\ i\,k \end{Bmatrix} g_{rj} - \begin{Bmatrix} r \\ j\,k \end{Bmatrix} g_{ir}$$

In view of (6.97), the right-hand side of this equation vanishes identically. Thus, we have

$$g_{ij,k} = 0.$$

Similarly, by (6.100) and (6.98), we obtain

$$g^{ij}{}_{,k} = 0.$$

As a consequence of Ricci's theorem, we see that the components g_{ij} or g^{ij} of the metric tensor can be regarded as constant factors with respect to covariant differentiation. For example, we have

$$(g_{ir}\, a^{r}{}_{.jm})_{,k} = g_{ir}\, a^{r}{}_{.jm,k}$$

and

$$(g^{ir}\, a_{rjm})_{,k} = g^{ir}\, a_{rjm,k}$$

6.11 GRADIENT, DIVERGENCE, LAPLACIAN, AND CURL IN GENERAL COORDINATES

The results obtained in the preceding sections provide us with tools for expressing a number of physical laws in forms that are valid in any coordinate system. We illustrate this by expressing the familiar concepts of gradient, divergence, Laplacian, and curl in tensorial forms. First, we recall from Sec. 6.7 (Example 1) that the gradient of a scalar field ϕ in general coordinates u^i is given by

$$\text{grad}\,\phi = \nabla\phi = \mathbf{e}^i \frac{\partial\phi}{\partial u^i} \tag{102}$$

From this we deduce that the del operator ∇ has the form

$$\nabla = \mathbf{e}^i \frac{\partial}{\partial u^i}$$

Now let **A** be a vector field represented by its contravariant components a^i. Then, by definition, the divergence of **A** is given by

$$\text{div } \mathbf{A} = \nabla \cdot \mathbf{A} = \mathbf{e}^i \cdot \frac{\partial \mathbf{A}}{\partial u^i} \tag{6.103}$$

But from (6.93) we have

$$\frac{\partial \mathbf{A}}{\partial u^i} = a^j{}_{,i}\, \mathbf{e}_j$$

hence (6.103) becomes

$$\text{div } \mathbf{A} = \mathbf{e}^i \cdot a^j{}_{,i}\, \mathbf{e}_j = (\mathbf{e}^i \cdot \mathbf{e}_j) a^j{}_{,i} = a^i{}_{,i}$$

This says that the divergence of **A** is the contraction of the mixed components of the covariant derivative of **A**. In rectangular cartesian coordinates ($a^i = a_i$, $u^i = x_i$), this reduces to the familiar form div **A** = $\partial a_i / \partial x_i$, which is the contraction of the tensor gradient $\partial a_i / \partial x_j$.

Writing the term $a^i{}_{,i}$ explicitly, we find

$$\text{div } \mathbf{A} = \frac{\partial a^i}{\partial u^i} + \left\{ {i \atop i\,k} \right\} a^k \tag{6.104}$$

This can be written more concisely by calculating the Christoffel symbol $\{^i{}_{ik}\}$. In fact, since $g^{ir} = g^{ri}$, we have

$$\left\{ {i \atop i\,k} \right\} = g^{ir} [r, i\, k] = g^{ri} [i, r\, k] \tag{6.105}$$

By adding the two equations

$$\left\{ {i \atop i\,k} \right\} = g^{ir} [r, i\, k] = \frac{1}{2} g^{ir} \left[\frac{\partial g_{ir}}{\partial u^k} + \frac{\partial g_{rk}}{\partial u^i} - \frac{\partial g_{ik}}{\partial u^r} \right]$$

and

$$\left\{ {i \atop i\,k} \right\} = g^{ri} [i, r\, k] = \frac{1}{2} g^{ri} \left[\frac{\partial g_{ir}}{\partial u^k} + \frac{\partial g_{ik}}{\partial u^r} - \frac{\partial g_{rk}}{\partial u^i} \right]$$

we find

$$\left\{ {i \atop i\,k} \right\} = \frac{1}{2} g^{ir} \frac{\partial g_{ir}}{\partial u^k} \tag{6.106}$$

Now consider the determinant g of the metric tensor (g_{ij}). Expanding by the elements of the i-th row (i = 1, 2, 3), we find

$$g = g_{i1}G^{i1} + g_{i2}G^{i2} + g_{i3}G^{i3} \quad (i = 1, 2, 3)$$
$$= g_{ir}\, G^{ir} \quad \text{(sum on r for fixed i; no sum on i)}$$

where G^{ir} is the cofactor of the element g_{ir} of the matrix (g_{ij}). Differentiating this with respect to g_{ij}, noting that $\partial g_{ir} / \partial g_{ij} = \delta_{rj}$ and $G^{ir}/\partial g_{ij} = 0$, we find

$$\frac{\partial g}{\partial g_{ij}} = G^{ij}$$

Thus, by the chain rule, we have

$$\frac{\partial g}{\partial u^k} = \frac{\partial g}{\partial g_{ij}} \frac{\partial g_{ij}}{\partial u^k} = G^{ij} \frac{\partial g_{ij}}{\partial u^k}$$

Since $G^{ij} = g\, g^{ij}$, it follows that

$$\frac{1}{g} \frac{\partial g}{\partial u^k} = g^{ij} \frac{\partial g_{ij}}{\partial u^k}$$

and so (6.106) becomes

$$\left\{ \begin{matrix} i \\ i\,k \end{matrix} \right\} = \frac{1}{2g} \frac{\partial g}{\partial u^k} = \frac{1}{\sqrt{g}} \frac{\partial}{\partial u^k} (\sqrt{g}) \tag{6.107}$$

Substituting this in (6.104), we finally obtain

$$\operatorname{div} \mathbf{A} = \frac{\partial a^i}{\partial u^i} + \left\{ \begin{matrix} i \\ i\,k \end{matrix} \right\} a^k$$

$$= \frac{1}{\sqrt{g}} \left[\sqrt{g} \frac{\partial a^i}{\partial u^i} + \frac{\partial}{\partial u^k} (\sqrt{g}) a^k \right]$$

$$= \frac{1}{\sqrt{g}} \frac{\partial}{\partial u^i} \left[\sqrt{g}\, a^i \right] \tag{6.108}$$

This is the general expression of the divergence of a vector field **A** in any coordinate system u^i in terms of the contravariant components a^i. Because $a^i = g^{ij} a_j$, the expression (6.108) can also be written in terms of the covariant components a_i as

$$\text{div}\,\mathbf{A} = \frac{1}{\sqrt{g}}\frac{\partial}{\partial u^i}\left[\sqrt{g}\,g^{ij}a_j\right] \tag{6.109}$$

To derive the expression of the Laplacian in general coordinates u^i, we observe that grad $\phi = \mathbf{e}^i\,\partial\phi/\partial u^i$, so that $\partial\phi/\partial u^i$ are the covariant components of grad ϕ. Therefore, from (6.109) we obtain

$$\nabla^2\phi = \text{div}(\text{grad}\,\phi) = \frac{1}{\sqrt{g}}\frac{\partial}{\partial u^i}\left[\sqrt{g}\,g^{ij}\frac{\partial\phi}{\partial u^j}\right] \tag{6.110}$$

This is the expression of the Laplacian of a scalar field ϕ in general coordinate system u^i.

Finally, to obtain the formula for the curl of a vector field in a general coordinate system u^i, suppose $\mathbf{A} = a_i\mathbf{e}^i$ where we assume that the local basis forms a right-handed triple at each point. Then, by definition and by (6.92), we have

$$\begin{aligned}\text{curl}\,\mathbf{A} &= \nabla \times \mathbf{A} = \mathbf{e}^i \times \frac{\partial \mathbf{A}}{\partial u^i} \\ &= \mathbf{e}^i \times (a_{j,i}\mathbf{e}^j) = a_{j,i}(\mathbf{e}^i \times \mathbf{e}^j)\end{aligned} \tag{6.111}$$

But from (6.5) and (6.51), we note that

$$\mathbf{e}^i \times \mathbf{e}^j = \frac{\mathbf{e}_k}{V} = \frac{\mathbf{e}_k}{\sqrt{g}}$$

where i, j, k are cyclic permutations of 1, 2, 3. (Note that if the local basis is left-handed, a minus sign appears in this equation.) Therefore, (6.111) becomes

$$\text{curl}\,\mathbf{A} = \frac{1}{\sqrt{g}}(a_{j,i} - a_{i,j})\mathbf{e}_k \tag{6.112}$$

Since $\{^k{}_{ij}\} = \{^k{}_{ji}\}$, it follows that

$$a_{j,i} - a_{i,j} = \frac{\partial a_j}{\partial u^i} - \frac{\partial a_i}{\partial u^j}$$

and so (6.112) simplifies to the form

$$\text{curl } \mathbf{A} = \frac{1}{\sqrt{g}}\left[\frac{\partial a_i}{\partial u^j} - \frac{\partial a_j}{\partial u^i}\right]\mathbf{e}_k$$

This formula may be conveniently written in the determinant form

$$\text{curl } \mathbf{A} = \frac{1}{\sqrt{g}}\begin{vmatrix} \mathbf{e}_1 & \mathbf{e}_2 & \mathbf{e}_3 \\ \dfrac{\partial}{\partial u^1} & \dfrac{\partial}{\partial u^2} & \dfrac{\partial}{\partial u^3} \\ a_1 & a_2 & a_3 \end{vmatrix} \tag{6.113}$$

which resembles the formula in rectangular cartesian coordinates.

Orthogonal Coordinate Systems

It is instructive to note that when the general coordinate system is orthogonal, that is, the base vectors $\mathbf{e}_i$ $(1 \le i \le 3)$ are orthogonal at every point, the various formulas we have derived above reduce to the corresponding formulas we obtained in Sec. 3.9. To see this, we must express the vector field in terms of its so-called physical components. These are simply the components of the vector field when referred to a normalized basis. First, we observe that when the local basis $\mathbf{e}_i$ forms an orthogonal set, the reciprocal basis $\mathbf{e}^i$ is given by

$$\mathbf{e}^i = \frac{\mathbf{e}_i}{h_i^2} \qquad (h_i = |\mathbf{e}_i|, i = 1, 2, 3) \tag{6.114}$$

Thus if ϕ is a differentiable scalar field, then by (6.102) its gradient becomes

$$\nabla\phi = \mathbf{e}^i \frac{\partial\phi}{\partial u^i} = \sum_{i=1}^{3} \frac{\mathbf{e}_i}{h_i^2}\frac{\partial\phi}{\partial u^i}$$

$$= \sum_{i=1}^{3} \frac{1}{h_i} \frac{\partial \phi}{\partial u^i} \mathbf{u}_i \tag{6.115}$$

in terms of the unit base vectors $\mathbf{u}_i = \mathbf{e}_i / h_i$ (i = 1, 2, 3). The quantities

$$\frac{1}{h_i} \frac{\partial \phi}{\partial u^i}, \; i = 1, 2, 3 \quad \text{(no sum on i)}$$

are called the physical components of the gradient vector $\nabla\phi$. It is obvious that (6.115) coincides with (3.53).

Next, in orthogonal coordinates we note that $g_{ij} = h_i^2$ (no sum on i) and $g_{ij} = 0$ for $i \neq j$, so that $g = \det(g_{ij}) = (h_1h_2h_3)^2$ or $\sqrt{g} = h_1h_2h_3$. Now consider the vector field $\mathbf{A} = a^i \mathbf{e}_i$. In terms of the normalized basis $\mathbf{u}_i = \mathbf{e}_i/h_i$ $(1 \le i \le 3)$, we can write this vector field as

$$\mathbf{A} = \sum_{i=1}^{3} (a^i h_i) \frac{\mathbf{e}_i}{h_i} = a'^i \mathbf{u}_i$$

Thus the physical components of the vector field $\mathbf{A}$ are given by $a'_i = a'^i = a^i h_i$ (no sum on i) for i = 1, 2, 3, so that $a^i = a'^i/h_i$. Hence, in terms of the physical components, we obtain from (6.108) the formula

$$\text{div } \mathbf{A} = \frac{1}{h_1h_2h_3} \sum_{i=1}^{3} \frac{\partial}{\partial u^i} \left[h_1h_2h_3 \left(\frac{a'_i}{h_i} \right) \right]$$

$$= \frac{1}{h_1h_2h_3} \left[\frac{\partial}{\partial u^1}(h_2h_3a'_1) + \frac{\partial}{\partial u^2}(h_1h_3a'_2) + \frac{\partial}{\partial u^3}(h_1h_2a'_3) \right] \tag{6.116}$$

This coincides with (3.59) of Sec. 3.9. If $\mathbf{A} = \nabla\phi$, so that $a'_i = (1/h_i)\, \partial\phi/\partial u^i$, then (6.116) yields the Laplacian

$$\nabla^2 \phi = \frac{1}{h_1h_2h_3} \left[\frac{\partial}{\partial u^1}\left(\frac{h_2h_3}{h_1} \frac{\partial \phi}{\partial u^1} \right) + \frac{\partial}{\partial u^2}\left(\frac{h_3h_1}{h_2} \frac{\partial \phi}{\partial u^2} \right) + \frac{\partial}{\partial u^3}\left(\frac{h_1h_2}{h_3} \frac{\partial \phi}{\partial u^3} \right) \right]$$

in accord with (3.60) of Sec. 3.9.

Finally, we consider $\mathbf{A} = a_i \mathbf{e}^i$. By (6.114) we observe that

$$\mathbf{A} = \sum_{i=1}^{3} a_i \left(\frac{\mathbf{e}_i}{h_i^2} \right) = a'^i \mathbf{u}_i$$

where $a'_i = a'^i$ are the physical components defined by $a'_i = a_i / h_i$ $(i = 1, 2, 3)$. Thus, in terms of the physical components a'_i and the normalized basis $\mathbf{u}_i$, we have $a_i = a_i' h_i$ and $\mathbf{e}_i = h_i \mathbf{u}_i$. Substituting these in (6.113), we obtain

$$\text{curl } \mathbf{A} = \frac{1}{h_1 h_2 h_3} \begin{vmatrix} h_1 \mathbf{u}_1 & h_2 \mathbf{u}_2 & h_3 \mathbf{u}_3 \\ \dfrac{\partial}{\partial u^1} & \dfrac{\partial}{\partial u^2} & \dfrac{\partial}{\partial u^3} \\ a'_1 h_1 & a'_2 h_2 & a'_3 h_3 \end{vmatrix} \tag{6.117}$$

which coincides with (3.61) of Sec. 3.9.

6.7 EXERCISES

1. Let a_{ijm}, a^{ijm}, $a^i{}_{jm}$ denote the various kinds of components of a third order tensor. Write down the formulas for the corresponding components of the covariant derivative of the tensor.
2. Write down the formula for the components of the covariant derivative of a fourth order mixed tensor defined by the components $a^{ij}{}_{pq}$.
3. Verify that the components $a_{ij,k}$ and $a^{ij}{}_{,k}$ transform as components of a third order tensor.
4. Verify the following formulas:

   (a) $(a_{ij} + b_{ij})_{,k} = a_{ij,\,k} + b_{ij,\,k}$

   (b) $(\phi\, a_i)_{,k} = \phi\, a_{i,k} + a_i(\partial\phi / \partial u^k)$, where f is a differentiable scalar field.

   (c) $(a^{ij}\, b_j)_{,k} = a^{ij}{}_{,k}\, b_j + a^{ij}\, b_{j,k}$

   (d) $(g_{ir}\, a^r{}_{.j})_{,\,k} = g_{ir}\, a^r{}_{.j,\,k}$

5. Show that

$$a_{ij,k} = g_{ip} g_{jq} a^{pq}{}_{,k}$$

and

$$a^{ij}{}_{,k} = g^{ip} g^{jq} a_{pq,k}$$

6. Show that

$$(a_{ij} b^m)_{,k} = a_{ij,k} b^m + a_{ij} b^m{}_{,k}$$

and

$$(a_i b_j c_m)_{,k} = a_{i,k} b_j c_m + a_i b_{j,k} c_m + a_i b_j c_{m,k}$$

7. Show that (6.106) can be written as

$$\left\{ \begin{matrix} i \\ i\,k \end{matrix} \right\} = \frac{\partial}{\partial u^k} (\ln \sqrt{g})$$

8. Suppose that a coordinate system u^i is such that $g_{ij} = 0$ for $i \neq j$, show that

(a) $$\left\{ \begin{matrix} i \\ j\,j \end{matrix} \right\} = -\frac{1}{2g_{ii}} \frac{\partial g_{jj}}{\partial u^i}$$

(b) $$\left\{ \begin{matrix} i \\ j\,i \end{matrix} \right\} = \frac{\partial}{\partial u^j} \ln(\sqrt{g_{ii}})$$

(c) $$\left\{ \begin{matrix} i \\ j\,i \end{matrix} \right\} = \frac{\partial}{\partial u^i} \ln(\sqrt{g_{ii}})$$

where the summation convention is not applied and $i \neq j$.
Hint: Use (6.105).

9. Prove that

$$(g_{ij} a^i b^j)_{,k} = a_{i,k} b^i + a^i b_{i,k}$$

10. Prove that the covariant derivative of the magnitude of a vector field **A** is given by

$$|\mathbf{A}|_{,k} = \frac{1}{|\mathbf{A}|} a_{i,k} a^i$$

Hint: Consider $|\mathbf{A}| = (g_{ij} a^i a^j)^{1/2}$.

11. The contravariant components of a vector field **A** in spherical coordinates $u^1 = r$, $u^2 = \phi$, $u^3 = \theta$ are given by

$$a^1 = r^2 \sin^2 \phi \sin 2\theta, \quad a^2 = \sin 2\phi \cos \theta, \quad a^3 = r \cos 2\phi .$$

Find the physical component, the divergence, and the curl of the vector.

12. The covariant components of a vector field **A** in spherical coordinates $u^1 = r$, $u^2 = \phi$, $u^3 = \theta$ are given by

$$a_1 = r^2 \sin^2 2\phi \sin \theta, \quad a_2 = \sin^2 \phi, \quad a^3 = r \cos \phi .$$

Find the physical components, the divergence, and the curl of the vector.

6.12 EQUATIONS OF MOTION OF A PARTICLE

In this last section, we illustrate an application of covariant differentiation by analyzing the motion of a particle. The particle is assumed to be moving in the three dimensional Euclidean space and that its position at a given time t is referred to some curvilinear coordinates u^i.

So let the curvilinear coordinates of the particle at time t be given by $u^i = u^i(t)$ $(i = 1, 2, 3)$, where the functions $u^i(t)$ are twice continuously differentiable for $t > 0$. Let $\mathbf{R} = \mathbf{R}(u^1, u^2, u^3)$ denote the position vector of the particle at time t. Then the velocity of the particle is given by

$$\mathbf{v} = \frac{d\mathbf{R}}{dt} = \frac{\partial \mathbf{R}}{\partial u^j}\frac{du^j}{dt} = v^j \mathbf{e}_j \tag{6.118}$$

where

$$v^j(t) = \frac{du^j(t)}{dt} \tag{6.119}$$

and $\mathbf{e}_j = \partial \mathbf{R}/\partial u^j$ $(j = 1, 2, 3)$ is the local basis. The acceleration is given by

$$\mathbf{a} = \frac{d\mathbf{v}}{dt} = \frac{\partial \mathbf{v}}{\partial u^i}\frac{du^i}{dt} \tag{6.120}$$

But from (6.93) we note that

$$\frac{\partial \mathbf{v}}{\partial u^i} = v^j_{\ ,i}\mathbf{e}_j = \left(\frac{\partial v^j}{\partial u^i} + \left\{ {j \atop i\,k} \right\} v^k\right)\mathbf{e}_j$$

hence (6.120) becomes

$$\mathbf{a} = \left(\frac{\partial v^j}{\partial u^i} + \left\{ {j \atop i\,k} \right\} v^k\right)\frac{du^i}{dt}\mathbf{e}_j$$

$$= \left(\frac{dv^j}{du^i} + \left\{ {j \atop i\,k} \right\} v^k \frac{du^i}{dt}\right)\mathbf{e}_j = a^j\mathbf{e}_j \qquad (6.121)$$

where

$$a^j = \frac{dv^j}{du^i} + \left\{ {j \atop i\,k} \right\} v^k \frac{du^i}{dt} \qquad (j = 1, 2, 3) \qquad (6.122)$$

The Christoffel symbols appearing in (6.122) must be calculated from the metric tensor associated with the coordinate system u^i.

The expression on the right-hand side of (6.122) is commonly written as $\delta v^j/\delta t$. Thus we can write

$$\frac{d\mathbf{v}}{dt} = \frac{\delta v^j}{\delta t}\mathbf{e}_j$$

where $\delta v^j/\delta t = v^j_{\ ,k}\, du^k/dt$. This is known as the absolute or intrinsic derivative of the velocity vector $\mathbf{v} = v^j\mathbf{e}_j$. The process of intrinsic differentiation can be extended to tensors of higher order (see Problem 1).

Now let m be the mass of the particle and $\mathbf{F} = F^j\mathbf{e}_j$ the force acting on the particle. By Newton's second law of motion, we have $F^j\mathbf{e}_j = ma^j\mathbf{e}_j$, which implies that

$$F^j = ma^j = m\frac{\delta v^j}{\delta t} = mv^j_{\ ,k}\frac{du^k}{dt} \qquad (j = 1, 2, 3) \qquad (6.123)$$

In rectangular cartesian coordinates, the right-hand side reduces to

$m(d^2x/dt^2)$, and F^j denotes the physical components of the force. By definition, the amount of work done by the force in displacing the particle through a displacement $d\mathbf{R}$ is given by the scalar product

$$dW = \mathbf{F} \cdot d\mathbf{R} = (F^i \mathbf{e}_i) \cdot (\mathbf{e}_j \, du^j)$$

$$= g_{ij} F^i du^j, \qquad g_{ij} = \mathbf{e}_i \cdot \mathbf{e}_j$$

Thus the work done by the force in displacing the particle from a point P_1 to a point P_2 is given by the line integral

$$W = \int_{P_1}^{P_2} g_{ij} F^i du^j = \int_{P_1}^{P_2} m g_{ij} \frac{\delta v^i}{\delta t} du^j$$

$$= \int_{P_1}^{P_2} m g_{ij} \frac{\delta v^i}{\delta t} v^j dt \tag{6.124}$$

Since $\phi = g_{ij} v^i v^j$ is a scalar, it is clear that $d\phi/dt = \delta\phi/\delta t$. Moreover, it can be shown (Problem 1) that

$$\frac{\delta}{\delta t}(g_{ij} v^i v^j) = 2 g_{ij} v^i \frac{\delta v^j}{\delta t} \tag{6.125}$$

Hence (6.124) becomes

$$W = \frac{1}{2} \int_{t_1}^{t_2} m \frac{d}{dt} [g_{ij} v^i v^j] \, dt = \frac{1}{2} m g_{ij} v^i v^j \Big|_{t_1}^{t_2}$$

$$= T_2 - T_1 \tag{6.126}$$

where

$$T = \frac{1}{2} m g_{ij} v^i v^j = \frac{1}{2} m |\mathbf{v}|^2 \tag{6.127}$$

is the kinetic energy of the particle. Therefore, the work done by the force in displacing the particle from P_1 to P_2 is equal to the change in the kinetic energy of the particle as it moves from P_1 to P_2. (Cf. Example 5, Sec. 4.3.)

Example 1. Find the expression for the kinetic energy and the contravariant components of the acceleration vector in cylindrical coordinates.

Solution: We recall that in cylindrical coordinates $u^1 = r$, $u^2 = \theta$, $u^3 = z$, the covariant components of the metric tensor are

$$g_{11} = 1, \qquad g = r^2, \qquad g = 1, \qquad g_{ij} = 0 \text{ for } i \neq j$$

Hence, by (6.127), the kinetic energy is given by

$$T = \frac{1}{2} m(\dot{r}^2 + r^2\dot{\theta}^2 + \dot{z}^2)$$

where the dot indicates derivative with respect to time t.

Next, in cylindrical coordinates the only nonzero Christoffel symbols are

$$\left\{ {1 \atop 2\,2} \right\} = -r, \quad \left\{ {2 \atop 1\,2} \right\} = \left\{ {2 \atop 2\,1} \right\} = \frac{1}{r}$$

Therefore, by (6.122), the contravariant components of the acceleration vector are

$$a^1 = \ddot{r} - r\dot{\theta}^2, \quad a^2 = \ddot{\theta} + \frac{2}{r}\dot{r}\dot{\theta}, \quad a^3 = \ddot{z}$$

If the equations expressing r, θ, and z as functions of time t are known, then the kinetic energy and the components of the acceleration vector can be calculated explicitly from the formulas given above.

Example 2. Find the expression for the kinetic energy and the contravariant components of the acceleration vector in the general coordinates v, w, z defined by the equations

$$x_1 = vw, \quad x_2 = (v^2 + w^2)/2, \quad x_3 = z$$

(Cf. Example 3, Sec. 6.8).

Solution: From Example 3 of Sec. 6.8, we recall that

$$g_{11} = g_{22} = v^2 + w^2, \quad g_{12} = g_{21} = 2vw, \quad g_{33} = 1$$

and

$$\left\{ {1 \atop 1\,1} \right\} = \left\{ {1 \atop 2\,2} \right\} = \left\{ {2 \atop 1\,2} \right\} = \left\{ {2 \atop 2\,1} \right\} = \frac{v}{v^2 - w^2}$$

$$\left\{ {2 \atop 1\,1} \right\} = \left\{ {2 \atop 2\,2} \right\} = \left\{ {1 \atop 1\,2} \right\} = \left\{ {1 \atop 2\,1} \right\} = \frac{-w}{v^2 - w^2}$$

with $g_{ij} = 0$, $\{^i{}_{jk}\} = 0$ for all other values of i, j, k. Hence, by (6.127), we find

$$T = \frac{1}{2} m g_{ij} \dot{u}^i \dot{u}^j = \frac{1}{2} m[(v^2 + w^2)(\dot{v}^2 + \dot{w}^2) + 4vw\dot{v}\dot{w} + \dot{z}^2]$$

and, by (6.122), we have

$$a^1 = \ddot{v} + \frac{1}{v^2 - w^2}(v\dot{v}^2 - 2w\dot{v}\dot{w} + v\dot{w}^2)$$

$$a^2 = \ddot{w} - \frac{1}{v^2 - w^2}(w\dot{v}^2 - 2v\dot{v}\dot{w} + w\dot{w}^2)$$

$$a^3 = \ddot{z}$$

The equations of motion (6.123) can be given in another form phrased in terms of the kinetic energy of the particle. We consider the expression (6.127) for kinetic energy

$$T = \frac{1}{2} m g_{ij} v^i v^j = \frac{1}{2} m g_{ij} \dot{u}^i \dot{u}^j \qquad (\dot{u}^i = du^i/dt)$$

Differentiating this with respect to u^i and u^i, treating u^i and u^i as independent variables, we find

$$\frac{\partial T}{\partial u^i} = \frac{1}{2} m \frac{\partial g_{jk}}{\partial u^i} \dot{u}^j \dot{u}^k \tag{6.128}$$

and

$$\frac{\partial T}{\partial \dot{u}^i} = m g_{ij} \dot{u}^j \tag{6.129}$$

Differentiating (6.129) with respect to t by the chain rule, we obtain

$$\frac{d}{dt}\left(\frac{\partial T}{\partial \dot{u}^i}\right) = m\left(g_{ij}\ddot{u}^j + \frac{\partial g_{ij}}{\partial u^k}\dot{u}^j \dot{u}^k\right)$$

If we subtract from this the expression (6.128) and note that

$$\frac{\partial g_{ij}}{\partial u^k}\dot{u}^j \dot{u}^k = \frac{\partial g_{ik}}{\partial u^j}\dot{u}^j \dot{u}^k$$

we find

$$\frac{d}{dt}\left(\frac{\partial T}{\partial \dot{u}^i}\right) - \frac{\partial T}{\partial u^i} = m\left[g_{ij}\ddot{u}^j + \frac{1}{2}\left(\frac{\partial g_{ij}}{\partial u^k} + \frac{\partial g_{ik}}{\partial u^j} - \frac{\partial g_{jk}}{\partial u^i}\right)\left(\dot{u}^j \dot{u}^k\right)\right]$$

$$= m\left(g_{ij}\ddot{u}^j + [i, j\, k]\dot{u}^j \dot{u}^k\right)$$

$$= m g_{ir}\left(\ddot{u}^r + \left\{ {r \atop j\,k} \right\}\dot{u}^j \dot{u}^k\right)$$

In view of (6.122) this can be written as

$$\frac{d}{dt}\left(\frac{\partial T}{\partial \dot{u}^i}\right) - \frac{\partial T}{\partial u^i} = m g_{ir} a^r = m a_i = F_i \qquad (i = 1, 2, 3) \tag{6.130}$$

where F_i denote the covariant components of the force field. These equations are known as *Lagrange's* equations of motion.

Now if **F** is a conservative field, then we know there exists a scalar field ϕ such that $\mathbf{F} = \nabla\phi$, so that $F_i = \partial\phi/\partial u^i$ $(1 \le i \le 3)$. It is common among physicists to set $V = -\phi$ and call V the potential energy of the particle. Doing this, Lagrange's equations (6.130) become

$$\frac{d}{dt}\left(\frac{\partial T}{\partial \dot{u}^i}\right) - \frac{\partial T}{\partial u^i} = -\frac{\partial V}{\partial u^i} \qquad (i = 1, 2, 3)$$

Since the potential energy V depends only on the coordinates u^i, we can write the equations in the form

$$\frac{d}{dt}\left(\frac{\partial L}{\partial \dot{u}^i}\right) - \frac{\partial L}{\partial u^i} = 0 \qquad (i = 1, 2, 3) \qquad (6.131)$$

where L = T - V. The function L is known as the *Lagrangian* function.

Example 3. Find the equations of motion of a pendulum bob (mass m) that is suspended from an extensible string, neglecting the potential energy stored in the string.

Solution: Let us introduce spherical coordinates as shown in Fig. 6.6, and let $u^1 = r$, $u^2 = \phi$, $u^3 = \theta$. Then the components of the metric tensor are given by

$$g_{11} = 1, \qquad g_{22} = r^2, \qquad g_{33} = r^2 \sin^2 \phi, \qquad g_{ij} = 0 \text{ when } i \neq j$$

Hence, the expression for the kinetic energy is given by

$$T = \frac{1}{2} m g_{ij} \dot{u}^i \dot{u}^j = \frac{1}{2} m(\dot{r}^2 + r^2\dot{\phi}^2 + r^2 \sin^2 \phi \, \dot{\theta}^2)$$

Let us choose the origin as the point where the potential energy is zero. Then we have

$$V = - mgr \cos \phi$$

Thus (6.131) leads to the following equations:

$$\ddot{r} - r\dot{\phi}^2 - r\dot{\theta}^2 \sin^2 \phi = g \cos \phi$$

$$r\ddot{\phi} + 2\dot{r}\dot{\phi} - r\dot{\theta}^2 \sin \phi \cos \phi = - g \sin \phi$$

$$\frac{d}{dt}(r^2\dot{\theta} \sin^2 \phi) = 0$$

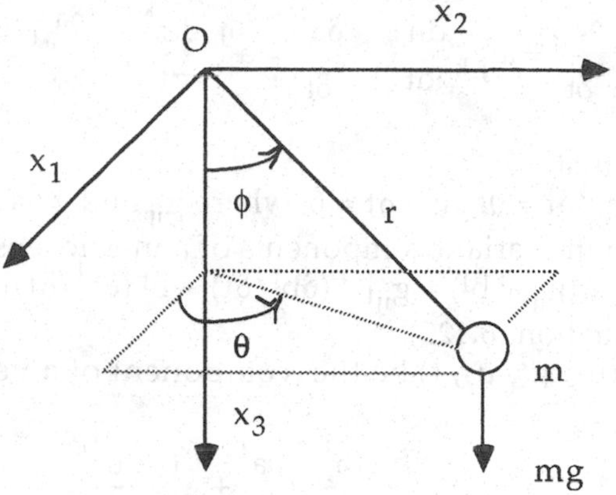

Fig. 6.6 Motion of a pendulum bob.

Now if the motion is confined to a plane, then $d\theta/dt = 0$ and the equations of motion simplify to

$$\ddot{r} - r\dot{\phi}^2 = g\cos\phi, \qquad r\ddot{\phi} + 2\dot{r}\dot{\phi} = -g\sin\phi$$

If, in addition, the string is unstretchable so that $dr/dt = 0$, then we obtain

$$\ddot{\phi} + \frac{g}{r}\sin\phi = 0$$

which is the familiar equation of the simple pendulum. For small angles of oscillation, the equation is linearized by replacing $\sin\phi$ by ϕ, in which case the motion becomes simple harmonic. For large oscillation, the solution of the equation is given in terms of elliptic functions.

6.8 EXERCISES

1. Let a_{ij}, a^{ij}, $a^i_{.j}$ denote the covariant, contravariant, and mixed components of a second order tensor, respectively. The corresponding components of the intrinsic derivative of the tensor are defined as follows:

$$\frac{\delta a_{ij}}{\delta t} = a_{ij,k}\frac{du^k}{dt}, \quad \frac{\delta a^{ij}}{\delta t} = a^{ij}_{,k}\frac{du^k}{dt}, \quad \frac{\delta a^i_{.j}}{\delta t} = a^i_{.j,k}\frac{du^k}{dt}$$

Show that
(a) $\delta g_{ij}/\delta t = 0$, $\delta g^{ij}/\delta t = 0$, where g_{ij} and g^{ij} are the covariant and contravariant components of a metric tensor;
(b) $\delta/\delta t(g_{ij}\, a^i b^j) = g_{ij}[a^i (\delta b^j/\delta t) + b^j (\delta a^i/\delta t)]$, and thus deduce the equation (6.125).

2. Let $a^i(u^1, u^2, u^3, t)$ be the component of a vector field. Show that

$$\frac{\delta a^i}{\delta t} = \frac{\partial a^i}{\partial t} + a^i_{,j}\frac{du^j}{dt}$$

and that $\delta a^i/\delta t$ $(1 \leq i \leq 3)$ transform as contravariant components of a vector field.

3. Find the expression of the kinetic and the covariant and contravariant components of the acceleration in a (a) parabolic coordinate system, (b) parabolic cylindrical coordinate system.

4. Show from Lagrange's equations that if a particle is not subjected to the influence of forces, then its trajectory is a straight line given by $x_i = a_i t + b_i$ $(1 \leq i \leq 3)$, where the a_i and b_i are constants and the x_i are rectangular cartesian coordinates.

5. The kinetic and potential energy of a particle moving in a constant gravitational field are given by

$$T = \frac{1}{2} m \dot{x}_i \dot{x}_i \qquad \text{and} \qquad V = mgx_3$$

where x_1, x_2, x_3 are rectangular cartesian coordinates with the positive x_3-axis directed upward. Find the equations of motion of the particle.

6. Prove that if a particle moves so that its velocity is constant in magnitude, then its acceleration vector is either orthogonal to the velocity vector or it is zero. (Hint: Compute the intrinsic derivative of $|\mathbf{v}|^2 = g_{ij}v^i v^j$.)

7. A particle slides in a frictionless tube which rotates in a horizontal plane at a constant angular speed ω. If the only horizontal force is the reaction F of the tube on the particle, find the equations of motion of the particle. Hint: Introduce polar

coordinates and note that

$$T = \frac{1}{2}m(\dot{r}^2 + r^2\dot{\theta}^2), \quad V = 0$$

SOLUTIONS TO SELECTED PROBLEMS

EXERCISES 1.1

1. $[3, -3]$, $|PQ| = 3\sqrt{2}$
2. $[4, 2]$, $|PQ| = 2\sqrt{5}$
3. $[-2, -4]$, $|PQ| = 2\sqrt{5}$
4. $[3, -2, 1]$, $|PQ| = \sqrt{14}$
5. $[-1, -6, -3]$, $|PQ| = \sqrt{46}$
6. $[2, 1, 2]$, $|PQ| = 3$
7. Q: $(-1, 1)$
8. P: $(-5, 6)$
9. Q: $(1, 2/3)$
10. Q: $(3, 0, 5)$
11. Q: $(1, 2, -5)$
12. P: $(1, 2, 8)$
13. P: $(-5, 0, 5)$
14. $\cos\alpha = -1/3$, $\cos\beta = -2/3$, $\cos\gamma = 2/3$
15. $\cos\alpha = 1/\sqrt{14}$, $\cos\beta = -3/\sqrt{14}$, $\cos\gamma = 2/\sqrt{14}$
16. $[-5\sqrt{3}/2, 5/2]$
17. $[-15\sqrt{2}/2, -15\sqrt{2}/2]$
19. $\mathbf{u} = [2/3, -1/3, -2/3]$ the same direction as **A**
20. 4.31 miles high and 16 miles from the airport

EXERCISES 1.2

3. (a) $7\mathbf{i} + 2\mathbf{k}$, (b) $7\mathbf{j} - 17\mathbf{k}$, (c) $|\mathbf{A} + \mathbf{B}| = \sqrt{27}$
   (d) $|\mathbf{A} - \mathbf{B}| = \sqrt{59}$
4. (a) $4\mathbf{A} - 2\mathbf{B} = -6\mathbf{k}$, (b) $2\mathbf{A} - 3\mathbf{B} = -4\mathbf{i} + 4\mathbf{j} - 9\mathbf{k}$
   (c) $|2(\mathbf{A} + \mathbf{B})| = 6\sqrt{3}$, (d) $|\mathbf{A}| + |\mathbf{B}| = \sqrt{2} + \sqrt{17}$
5. $\mathbf{X} = [4/3, 7/3, -8/3]$
6. $\mathbf{X} = [4, 3/2, -2]$
7. $m = 2$, $n = -1$
8. $m = 3$, $n = -2$
9. $a = -1$, $b = 1$, $c = 2$
10. $a = 1$, $b = -1$, $c = 1$
11. $|\mathbf{A} - \mathbf{B}| = |-(\mathbf{B} - \mathbf{A})| = |\mathbf{B} - \mathbf{A}|$, $|\mathbf{A} + \mathbf{B}| = |\mathbf{A}| + |\mathbf{B}|$ when the vectors are in the same direction.
12. $|\mathbf{A} - \mathbf{B}| = ||\mathbf{A}| - |\mathbf{B}||$ when the vectors are in the same direction.
13. Max $|\mathbf{A} - \mathbf{B}| = |\mathbf{A}| + |\mathbf{B}|$, min $|\mathbf{A} - \mathbf{B}| = ||\mathbf{A}| - |\mathbf{B}||$
14. $\mathbf{v} = (250 + 25\sqrt{2})\mathbf{i} + 25\sqrt{2}\mathbf{j}$, 83 deg. east of north or N 83° E
15. N 12.25° E
16. $75\sqrt{3}\mathbf{i} + 125\mathbf{j}$, N 46.1° E
17. 2:30 PM, 16.7 miles from the lighthouse.
18. 28.96 miles apart

19. Let $\mathbf{F}_1 = F_1\mathbf{i}$. Then $\mathbf{F}_2 = F_1(-\mathbf{i} + \sqrt{3}\mathbf{j})$ and $\mathbf{F}_3 = 3F_1(-\mathbf{i} - \sqrt{3}\mathbf{j})/2$ so that $\mathbf{F}_1 + \mathbf{F}_2 + \mathbf{F}_3 = (-3\mathbf{i} - \sqrt{3}\mathbf{j})F_1/2$.
20. Tension in shorter wire $50\sqrt{3}$ kg, in longer wire 50 kg
22. $\mathbf{v} = 30\mathbf{i} + 150\mathbf{j}$, $t = 9.8$ min, 294 ft downstream

EXERCISES 1.3

2. Let the sides AB, BC, and CA be represented by the vectors **A**, **B**, and **C**, respectively. Then $\boldsymbol{PQ} = \boldsymbol{PB} + \boldsymbol{BQ} = (\mathbf{A} + \mathbf{B})/2 = -(1/2)\mathbf{C} = \boldsymbol{AR}$ which shows that PQ and AR are parallel and equal in length.
4. Consider a trapezoid ABCD, and let P and Q denote the respective midpoints of the nonparallel sides BC and DA. Since $\boldsymbol{AB} + \boldsymbol{BP} + \boldsymbol{PQ} + \boldsymbol{QA} = \mathbf{0}$, it follows that $\boldsymbol{QP} = \boldsymbol{AB} + (1/2)\,\boldsymbol{BC} + (1/2)\boldsymbol{DA} = \boldsymbol{AB} + (1/2)(\boldsymbol{BC} + \boldsymbol{DA}) = \boldsymbol{AB} + (1/2)(-\boldsymbol{AB} - \boldsymbol{CD}) = (1/2)(\boldsymbol{AB} + \boldsymbol{DC})$.
6. The hypothesis gives $(\mathbf{B} + 2\mathbf{C})/3 = (\mathbf{A} + 2\mathbf{D})/3$, which says that the line segments AD and BC intersect. The point of intersection divides the line segments in the ratio 2 : 1.
9. Let **A**, **B**, **C**, and **D** denote the position vectors of the respective vertices A, B, C, and D of a quadrilateral ABCD. Then the position vectors of the midpoints of the sides AB, BC, CD and DA are given by $\mathbf{P} = (\mathbf{A} + \mathbf{B})/2$, $\mathbf{Q} = (\mathbf{B} + \mathbf{C})/2$, $\mathbf{R} = (\mathbf{C} + \mathbf{D})/2$, $\mathbf{S} = (\mathbf{D} + \mathbf{A})/2$, respectively. Thus it follows that $(\mathbf{P} + \mathbf{R})/2 = (\mathbf{Q} + \mathbf{S})/2$, which is what needs to be proved.
10. The point divides AE in the ratio 6 : 1 and divides CD in the ratio 3 : 4.
12. The point divides BC in the ratio 2 : 1.

EXERCISES 1.4

1. $4/3$, $\cos\theta = 4/9$ 2. $-5/\sqrt{14}$, $\cos\theta = -5/14$
3. $-11/\sqrt{29}$, $\cos\theta = -11/5\sqrt{58}$ 4. $-28/\sqrt{21}$, $\cos\theta = -4/\sqrt{21}$
5. $\mathbf{A} = (4/9)\mathbf{B} + \mathbf{C}$, where $\mathbf{C} = (1/9)[22, -10, 1]$
6. $\mathbf{B} = (-4/7)\mathbf{A} + \mathbf{C}$, where $\mathbf{C} = [15/7, -2/7, 4/7]$
8. $\mathbf{i} - \mathbf{j} + \mathbf{k}$ 9. (a) $m = 26/27$, (b) $m = 27/38$
10. (a) $m = 1/7$, (b) $m = 29/5$, 11. $m = 2/7$, $n = -8/7$
13. $\mathbf{u} = \pm(2\mathbf{i} + \mathbf{j} - 3\mathbf{k})/\sqrt{14}$, 14. work = - 6, 15. work = 24
23. $|\mathbf{A} + \mathbf{B}|^2 = (\mathbf{A} + \mathbf{B})\cdot(\mathbf{A} + \mathbf{B}) = \mathbf{A}\cdot\mathbf{A} + 2\mathbf{A}\cdot\mathbf{B} + \mathbf{B}\cdot\mathbf{B} = \mathbf{A}\cdot\mathbf{A} + \mathbf{B}\cdot\mathbf{B} = (\mathbf{A} - \mathbf{B})\cdot(\mathbf{A} - \mathbf{B}) = |\mathbf{A} - \mathbf{B}|^2$ if and only if $\mathbf{A}\cdot\mathbf{B} = 0$.
24. Show that $\mathbf{A}\cdot\mathbf{C} = \mathbf{B}\cdot\mathbf{C}$. 25. Write $\boldsymbol{AD} = (\boldsymbol{AB} + \boldsymbol{BC})/2$ and $\boldsymbol{BC} = \boldsymbol{AC} - \boldsymbol{AB}$, then consider $\boldsymbol{AD}\cdot\boldsymbol{AD}$ and $\boldsymbol{BC}\cdot\boldsymbol{BC}$.

EXERCISES 1.5

1. (a) -3, (b) -10, (c) 2,2. (a) 10, (b) -2, (c) abc, (d) 0
9. (a) $(-\mathbf{i} + 7\mathbf{j} + 5\mathbf{k})/\sqrt{75}$, (b) $(-25\mathbf{i} + 3\mathbf{j} + 13\mathbf{k})/\sqrt{803}$
10. (a) $7\mathbf{i} - 7\mathbf{j} + 7\mathbf{k}$, (b) $27\mathbf{i} + 9\mathbf{j}$, (c) 63, (d) 63, (e) $9\mathbf{i} - 27\mathbf{j} - 9\mathbf{k}$ (f) $-14\mathbf{i} - 28\mathbf{j} - 14\mathbf{k}$
11. (a) $-19\mathbf{i} - 2\mathbf{j} + 9\mathbf{k}$, (b) $-3\mathbf{i} + 6\mathbf{j} + 3\mathbf{k}$, (c) $38\mathbf{i} + 4\mathbf{j} - 18\mathbf{k}$ (d) $-60\mathbf{i} + 30\mathbf{j} - 108\mathbf{k}$, (e) $-48\mathbf{i} - 24\mathbf{j}$
12 (a) 7, (b) $\sqrt{19}$, (c) $2\sqrt{13}$, (d) $(\sqrt{819})/2$
13. (a) $-6\mathbf{i} - 14\mathbf{j} - 2\mathbf{k}$, (b) $-18\mathbf{i} - 12\mathbf{j} + 6\mathbf{k}$
14. (a) $-15\mathbf{i} - 9\mathbf{j} - 17\mathbf{k}$, (b) $-29\mathbf{i} + 2\mathbf{j} - 7\mathbf{k}$
15. (a) $19\mathbf{k}$, (b) $14\mathbf{k}$
16. $\mathbf{v} = (42\mathbf{i} - 18\mathbf{j} - 4\mathbf{k})/\sqrt{19}$, speed = $|\mathbf{v}|$
17. $\mathbf{v} = (4\mathbf{i} - 16\mathbf{j} + 12\mathbf{k})/\sqrt{3}$, speed = $|\mathbf{v}|$, 18. (3, 1, 3), (1, -1, -1), (-3, 3, 3), 19. (a) $\mathbf{A}\cdot\mathbf{C} < 0$, (b) $\mathbf{B}\cdot\mathbf{D} > 0$
21. (a) $\mathbf{B} = (1 + t)\mathbf{i} - (1 + 2t)\mathbf{j} + t\mathbf{k}$ for arbitrary value of t.
(b) $\mathbf{B} = (2\mathbf{i} - \mathbf{j} - \mathbf{k})/3$

EXERCISES 1.6

1. $x = -1 + 2t,\ y = 3 - 3t,\ z = 2 + 5t\ (-\infty < t < \infty)$
2. $(x - 3)/5 = (y + 1)/(-4) = (z - 4)/(-1)$ or $(x + 2)/5 = (y - 3)/(-4) = (z - 5)/(-1)$
3. $x = 1 + 2t,\ y = -2 - 3t,\ z = 1 + 4t\ (-\infty < t < \infty)$
4. $x - 3 = -(y - 4) = -(z - 6)$, (z is arbitrary).
5. $x = 17/7 + z,\ y = -2/7 + z$, 6. (3, -2, 3), $\cos\theta = 8/\sqrt{66}$
7 (8, 8, 9), $\cos\theta = \sqrt{6}/3$, 8. (5, -4, 1)
9. $x + 3y - 7 = 0$, 10. $3x - 12y + 19z - 34 = 0$
11. $x = z$, 12. $x - z + 1 = 0$,
13. $2x - 7y - 4z + 28 = 0$, 14. $\sqrt{(3/2)}$
15. $\sqrt{(35/26)}$, 16. $2/\sqrt{38}$,
17. $5/\sqrt{38}$, 18. $21/\sqrt{494}$
19. $10/\sqrt{65}$, 20. $23/\sqrt{381}$
21. center at (18, 8, 5), radius = $\sqrt{(3/2)}$

EXERCISES 1.7

1. (a) 4 (b) -22 (c) 29 (d) -67, 2. 2, 3. 14, 4. 30
6. $\mathbf{D} = 2\mathbf{A} - 3\mathbf{B} + 2\mathbf{C}$, 7. $\mathbf{D} = \mathbf{A} + 3\mathbf{B} - 2\mathbf{C}$, 8. $\mathbf{D} = 2\mathbf{A} + \mathbf{B}$
9. $\mathbf{D} = 2\mathbf{A} + 3\mathbf{B}$, 10. $\mathbf{A} = 3\mathbf{B} + 2\mathbf{C}$, 11. $\mathbf{A} = -3\mathbf{B} - 2\mathbf{C}$
15. $\mathbf{A} \times (\mathbf{B} \times \mathbf{C}) = -15\mathbf{i} + 5\mathbf{j} - 25\mathbf{k}$, 16. Consider $(\mathbf{A} \times \mathbf{B}) \times (\mathbf{C} \times \mathbf{D})$ and expand with respect to $\mathbf{C}$ and $\mathbf{D}$.
17. Set $\mathbf{U} = \mathbf{A} \times \mathbf{B}$ and apply (1.45) and (1.46).
19. Apply (1.45).

EXERCISES 2.1

1. (a) $[2t - \ln(1 + t)]\mathbf{i} + [2(1 + t^2) + e^t]\mathbf{j} - (2 \sin t + 1)\mathbf{k}$
(b) $t \ln (1 + t) - e^t(1 + t^2) - \sin t$, (c) $[(1 + t^2) - e^t \sin t]\mathbf{i} - [t + (\sin t) \ln(1 + t)]\mathbf{j} - [te^t + (1 + t^2)\ln(1 + t)]\mathbf{k}$, (d) $|\mathbf{F}(t)| = [1 + 3t^2 + t^4 + \sin^2 t]^{1/2}$, $|\mathbf{G}(t)| = [\ln^2(1 + t) + e^{2t} + 1]^{1/2}$
2. (b) $2e^t(3t - 1) - (t^2 - 2)\cos t + t^3$, (c) $(t^4 - 2t^2 + t \cos t)\mathbf{i} + (2te^t - 3t^3 + t^2)\mathbf{j} - [(3t - 1) \cos t + 2t^2 e^t - 4e^t]\mathbf{k}$, (d) $[(2e^t + 3t - 1)^2 + (t^2 - 2 - \cos t)^2 + t^2(t + 1)^2]^{1/2}$, 3. $h(t) = [t^2 + e^{2t}]^{-1/2}$
4. $h(t) = [e^{-2t} + t^2]^{-1/2}$, 5. $(-b \cos t\, \mathbf{i} - b \sin t\, \mathbf{j} + a\mathbf{k})/(a^2 + b^2)^{-1/2}$
6. $[(\sin t - \cos t)\mathbf{i} - (\cos t + \sin t)\mathbf{j} + 2\mathbf{k}]/\sqrt{6}$,
8. $(\mathbf{F} \times \mathbf{G}) \times \mathbf{H} = t^3\mathbf{G} - te^t\mathbf{F}$

EXERCISES 2.2

1. $[(1 + \sqrt{3})/2]\mathbf{i} + (1 - \sqrt{3})\mathbf{j}$, 2. $\sqrt{2}$, 3. $\sqrt{3}$,
4. $(3 + \sqrt{3})/2\mathbf{i} + (\sqrt{3} - 3)/2\mathbf{j}$, 5. 5, 6. 4, 7. $4\mathbf{i} - \mathbf{j} - 3\mathbf{k}$,
8. $\sqrt{26}$, 9. (a) $t \cos t - 6t^2$, $(1 + 4t \cos t - 2t^2 \sin t)\mathbf{i} - (2t + 4t \sin t + 2t^2 \cos t)\mathbf{j} - t \sin t\mathbf{k}$, (b) $2 - 2t$, $-(4t^3 + 4t + 1)\mathbf{i} + (1 + 3t^2)\mathbf{k}$
11. 0, 12. $\mathbf{F}(t)\cdot \mathbf{F}'(t) = 0$ $(\theta = 90^\circ)$, 13. $\mathbf{F}''(t) = a^2\mathbf{F}(t)$
15. Differentiate $[\mathbf{F}(t)]^2 = \mathbf{F}(t)\cdot \mathbf{F}(t)$, 20. Note that if $\mathbf{r} = \mathbf{R}(t)/|\mathbf{R}(t)|$, then $d\mathbf{r} = d\mathbf{R}/|\mathbf{R}| - \mathbf{R}\, d|\mathbf{R}|/|\mathbf{R}^2$

EXERCISES 2.3

1. $\mathbf{R}(\tau) = \tau \cos (\ln \tau)\mathbf{i} + \tau \sin (\ln \tau)\mathbf{j}$ $(\tau \geq 1)$, $\mathbf{R}'(t) = \tau\, \mathbf{R}'(\tau)$
2. $\mathbf{R}(\tau) = \sqrt{(1 - \tau)}\mathbf{i} + (1 - \tau)\mathbf{j} + (1 - \tau)^{3/2}\mathbf{k}$ $(\tau \leq 1)$, $\mathbf{R}'(t) = -2t\mathbf{R}'(\tau)$
3. $x^2 + y^2 = a^2$, $z = a + y$, 4. $x^2 + y^2 = z^2$
5. $\mathbf{t} = [(3\sqrt{3} - \pi/2)\mathbf{i} + (3 + \sqrt{3}\pi/2)\mathbf{j}]/\sqrt{(36 + \pi^2)}$
6. $\mathbf{t} = (-a\mathbf{i} + b\mathbf{j})/\sqrt{(a^2 + b^2)}$, 7. $\mathbf{t} = (4\mathbf{i} + \mathbf{j})/\sqrt{17}$
8. $\mathbf{t} = (-\mathbf{i} + \sqrt{3}\mathbf{j} + \sqrt{3}\mathbf{k})/\sqrt{7}$, 9. $\mathbf{t} = (\mathbf{i} + \sqrt{3}\mathbf{j})/2$
10. $\mathbf{t} = (e^2\, \mathbf{i} - e^{-2}\, \mathbf{j} + \mathbf{k})/\sqrt{(e^4 + e^{-4} + 1)}$
11. $s = \sqrt{3}(e^t - 1)$, $t = \ln (1 + s/\sqrt{3})$
12. $s = t[\sqrt{(t^2 + 2)}]/2 + \ln [t/\sqrt{2} + (t^2 + 2)^{1/2}/2]$
13. $s = \sqrt{2}\, t^2/2$, 14. $s = 2\sqrt{2}\pi$, 15. $s = 2\pi\sqrt{(a^2 + b^2)}$
16. (a) $ds = a\sqrt{(1 + \cos^2 t)}\, dt$, (b) $ds = \sqrt{(t^2 + 2)}dt$

EXERCISES 2.4

1. $\kappa = (t^2 + 1)^{-3/2}/2$, $\mathbf{n} = [\mathbf{i} - (t/2)\mathbf{j} - (\sqrt{3}t/2)\mathbf{k}]/\sqrt{(1 + t^2)}$, $\mathbf{b} = (\sqrt{3}/2)\mathbf{j} - (1/2)\mathbf{k}$
2. $\kappa = \sqrt{2}e^{-t}/3$, $\mathbf{n} = [(-\sin t - \cos t)\mathbf{i} + (\cos t - \sin t)\mathbf{j}]/\sqrt{2}$, $\mathbf{b} = [(\sin t - \cos t)\mathbf{i} - (\sin t + \cos t)\mathbf{j} + 2\mathbf{k}]/\sqrt{6}$

4. $\kappa = (1/2)\,\text{sech}^2 t$, $\mathbf{n} = \text{sech}\, t\,\mathbf{i} - \tanh t\,\mathbf{k}$
   $\mathbf{b} = (-\tanh\, t\,\mathbf{i} + \mathbf{j} - \text{sech}\, t\,\mathbf{k})/\sqrt{2}$
5. $\kappa = 1/2t$, $\mathbf{n} = \cos t\,\mathbf{i} - \sin t\,\mathbf{j}$, $\mathbf{b} = (\sin t\,\mathbf{i} + \cos t\,\mathbf{j} - \mathbf{k})/\sqrt{2}$
6. $\kappa = \sqrt{2}/(1 + \sin^2 t)^{3/2}$, $\rho = 1/\kappa$. At $t = \pi/4$, the center of the circle of curvature is given by $(\sqrt{2}/8, -\sqrt{2}/4, \sqrt{2}/8)$
7. Center at $(\pi/2 - 3, 5, 2\sqrt{2})$ with radius $\rho = 4\sqrt{2}$
8. $\kappa = \sqrt{\text{sech}\, t}$, 9. (a) $\sqrt{2}y - 2z + e^{\pi/4} = 0$

EXERCISES 2.5

1. $\tau = 0$, 2. $\tau = e^t/3$, 3. $\tau = (t^2 + 6)/(t^4 + 5t^2 + 8)$
4. $\tau = 1/(\sinh^2 t + \cosh^2 t + 1) = 1/(2\cosh^2 t)$
5. $\tau = -1/(2t)$, 6. $\tau = 2a/(2a^2 t^2 + 1)^2$, 10. $\kappa = \tau = 1/[3(1 + t^2)^2]$
11. $d\mathbf{t}/ds = 1/(2t)\mathbf{n}$, $\mathbf{n} = \cos t\,\mathbf{i} - \sin t\,\mathbf{j}$, $d\mathbf{n}/ds = -(1/2t)\mathbf{t} - (1/2t)\mathbf{b}$,
    $\mathbf{t} = (\sin t\,\mathbf{i} + \cos t\mathbf{j} + \mathbf{k})/\sqrt{2}$, $\mathbf{b} = (\sin t\,\mathbf{i} + \cos t\,\mathbf{j} - \mathbf{k})/\sqrt{2}$,
    $d\mathbf{b}/ds = 1/(2t)\mathbf{n}$
12. $\mathbf{t} = (\cos t\,\mathbf{i} + \mathbf{j} + \sin t\,\mathbf{k})/\sqrt{2}$, $d\mathbf{t}/ds = (1/2)\mathbf{n}$, $\mathbf{n} = -\sin t\,\mathbf{i} + \cos t\,\mathbf{k}$
    $d\mathbf{n}/ds = (-1/2)[\mathbf{t} + \mathbf{b}]$, $\mathbf{b} = (\cos t\,\mathbf{i} - \mathbf{j} + \sin t\,\mathbf{k})/\sqrt{2}$
    $d\mathbf{b}/ds = (1/2)\mathbf{n}$

EXERCISES 2.6

1. $\mathbf{v} = 3\mathbf{i} + 6\mathbf{j} + 6\mathbf{k}$, $v = 9$, $\mathbf{a} = 6\mathbf{j} + 12\mathbf{k}$
2. $\mathbf{v} = (a/4)(\mathbf{i} + 3\mathbf{j})$, $v = (a/4)\sqrt{10}$, $\mathbf{a} = (a/4)(3\mathbf{i} + 5\mathbf{j})$
3. $\mathbf{v} = -\mathbf{i} - \pi\mathbf{j} + \pi\mathbf{k}$, $v = \sqrt{(2\pi^2 + 1)}$, $\mathbf{a} = \pi\mathbf{i} - 2\mathbf{j} + \mathbf{k}$
4. $\mathbf{v} = -(a/2)\sqrt{3}\mathbf{i} + (b/2)\mathbf{j} - (a/2)\sqrt{3}\mathbf{k}$, $\mathbf{a} = -(a/2)\mathbf{i} - (b/2)\sqrt{3}\mathbf{j} - (a/2)\mathbf{k}$
5. $\mathbf{v} = -\sinh \pi\,\mathbf{i} - \cosh \pi\,\mathbf{j} + \cosh \pi\,\mathbf{k}$, $\mathbf{a} = (\sinh \pi)(-2\mathbf{j} + \mathbf{k})$
6. $\mathbf{a} = a\omega^2\mathbf{n}$, $\mathbf{n} = -\cos \omega t\,\mathbf{i} - \sin \omega t\,\mathbf{j}$
7. $\mathbf{a} = \mathbf{n}$, $\mathbf{n} = -\sin t\,\mathbf{i} + \cos t\,\mathbf{j}$
8. $\mathbf{a} = [1 + \cos^2 (t/2)]^{1/2}\mathbf{n}$, $\mathbf{n} = [-\sin t\,\mathbf{i} + \cos t\,\mathbf{j} - \cos (t/2)\,\mathbf{k}]/|\mathbf{n}|$
9. $\mathbf{a} = a\mathbf{n}$, $\mathbf{n} = -(\cos t\,\mathbf{i} + \sin t\,\mathbf{j})$
10. $a_t = a \sin t \cos t/\sqrt{(1 + \sin^2 t)}$, $a_n = \sqrt{2}a/\sqrt{(1 + \sin^2 t)}$
11. $a_t = 2$, $a_n = 8t^2$, $\rho = 1/2$
13. If $\mathbf{v} = (ds/dt)\mathbf{t} =$ const, then $\mathbf{a} = (ds/dt)^2\, d\mathbf{t}/ds = (ds/dt)^2\, \kappa\mathbf{n}$.
    Thus $\mathbf{v}\cdot \mathbf{a} = 0$.
14. If $|\mathbf{v}| = c$ and $|\mathbf{a}| = a$ are constants, then $\mathbf{v} = c\mathbf{t}$ so that $\mathbf{a} = c^2\kappa\mathbf{n}$.
    Since $|\mathbf{a}| = a$, it follows that $c^2\kappa = a$ or $\kappa = a/c^2 =$ constant.
15. The force is given by $\mathbf{F} = m\mathbf{a} = m\mathbf{R}''(t)$ and the direction of motion is indicated by the velocity vector $\mathbf{R}'(t)$. Then $\mathbf{F}\cdot \mathbf{R}' = m\mathbf{R}''(t)\cdot \mathbf{R}'(t) = 0$ implies $(d/dt)[\mathbf{R}'(t)\cdot \mathbf{R}'(t)] = (d/dt)v^2 = 0$, which means $v =$ const.
16. $\mathbf{R} \times \mathbf{R}'' = 0$ if and only if $(\mathbf{R} \times \mathbf{R}')' = (\mathbf{R} \times \mathbf{v})' = 0$.

EXERCISES 2.7

1. $\mathbf{v} = \omega \sinh \omega r\, \mathbf{u}_r + \cosh \omega t\, \mathbf{u}_\theta$, $\mathbf{a} = (\omega^2 - 1)\cosh \omega t\, \mathbf{u}_r + 2\omega \sinh \omega t\, \mathbf{u}_\theta$
2. $\mathbf{v} = 2t\, \mathbf{u}_r + \omega t^2\, \mathbf{u}_\theta$, $\mathbf{a} = (2 - \omega^2 t^2)\mathbf{u}_r + 4\omega t \mathbf{u}_\theta$
3. $\mathbf{v} = r'\mathbf{u}_r + 3r\, \mathbf{u}_\theta$, $\mathbf{a} = (r'' - 9r)\mathbf{u}_r + 6r'\mathbf{u}_\theta$, where $r' = 3a \sin 3t$, $r'' = 9a \cos 3t$
4. $\mathbf{v} = r'\mathbf{u}_r + e^{-t} r\, \mathbf{u}_\theta$, $\mathbf{a} = (r'' - re^{-2t})\mathbf{u}_r + (2e^{-t} r' - e^{-t} r)\mathbf{u}_\theta$ where $r' = a \cos t$, $r'' = -a \sin t$
5. $\mathbf{v} = r'\mathbf{u}_r + r\omega\, \mathbf{u}_\theta$, $\mathbf{a} = (r'' - r\omega^2)\mathbf{u}_r + 2r'\omega\, \mathbf{u}_\theta$, where $r' = a \sin t/(1 + \cos t)^2$, $r'' = a(2 - \cos t)/(1 + \cos t)^2$
7. $d^3\mathbf{R}/dt^3 = (r''' - 3r''\theta'^2 - 3r\theta'\theta'')\mathbf{u}_r + (3r''\theta' + 3r'\theta'' - r\theta'^3 + r\theta''')\mathbf{u}_\theta$
8. Let $\mathbf{R}(t) = r(t)\, \mathbf{u}_r(\theta)$. Then $\mathbf{v} = r'\mathbf{u}_r + r\omega\mathbf{u}_\theta$, $\mathbf{a} = (r'' - r\omega^2)\, \mathbf{u}_r + 2\omega r'\mathbf{u}_\theta$. Since $(d/dt)\mathbf{a} = k\mathbf{u}_r$, it follows that the transversal component of $(d/dt)\mathbf{a}$ must be zero, that is, $(r'' - r\omega^2)\,\omega\mathbf{u}_\theta + 2\omega r''\mathbf{u}_\theta = (3\omega r'' - r\omega^3)\mathbf{u}_\theta = 0$. Thus $r'' = r\omega^2/3$.
9. Centrifugal force = π^2, Coriolis force = 4π

EXERCISES 2.8

1. $\mathbf{v} = 6\, \mathbf{u}_\theta + 2\mathbf{k}$, $\mathbf{a} = -18\mathbf{u}_r$
2. $\mathbf{v} = \mathbf{u}_r + t\mathbf{u}_\theta + 2\mathbf{k}$, $\mathbf{a} = -t\mathbf{u}_r + 2\mathbf{u}_\theta$
3. $\cos \theta = [5/(5 + t^2)]^{1/2}$
4. (b) $\mathbf{v} = \cos t\mathbf{u}_r + \sin t\mathbf{u}_\theta + \mathbf{k}$, $\mathbf{a} = -2 \sin t\mathbf{u}_r + 2 \cos t\mathbf{u}_\theta$
   (c) $\theta = 45^\circ$
5. (a) $\mathbf{v} = \sqrt{3}\omega\mathbf{u}_\theta$, $\mathbf{a} = -\sqrt{3}\omega^2\, \mathbf{u}_r$
   (b) $\mathbf{v} = \sqrt{3}(\omega + \gamma)\mathbf{u}_\theta$, $\mathbf{a} = -\sqrt{3}(\omega + \gamma)^2\mathbf{u}$
   (c) $\mathbf{v} = \pm\sqrt{3}(\omega + \gamma)\mathbf{u}_\theta$. $\mathbf{a} = \sqrt{3}(\omega + \gamma)^2\mathbf{u}_r$
6. $a_r = -4at^2 \sin t^2 + 2a \cos t^2$, $a_t = -4at^2 \cos t^2 + 2a \sin t^2$
7. $\mathbf{v} = a\gamma \cos \gamma t\, \mathbf{u}_r + 2at \sin \gamma t\, \mathbf{u}_\theta - a\gamma \sin \gamma t\, \mathbf{k}$
   $\mathbf{a} = -\gamma^2 \mathbf{R} - 4at^2 \sin \gamma t\, \mathbf{u}_r + (2a \sin \gamma t + 4at\gamma \cos \gamma t)\mathbf{u}_\theta$
   Coriolis acceleration = $2a \sin \gamma t + 4at\gamma \cos \gamma t$
8. $\mathbf{v} = -a\alpha \sin \alpha t\, \mathbf{i} + a\alpha \cos \alpha t\, \mathbf{j}^*$, $\mathbf{a} = -\alpha^2 \mathbf{R}$
9. $\mathbf{v} = -a\alpha \sin \alpha t\, \mathbf{i} + (a\alpha \cos \alpha t \sin \gamma t + a\gamma \sin \alpha t \cos \gamma t)\mathbf{j} + (a\alpha \cos \alpha t \cos \gamma t - a\gamma \sin \alpha t \sin \gamma t)\mathbf{k}$,
   $\mathbf{a} = -\alpha^2 \mathbf{R} + 2a\alpha\gamma \cos \alpha t(\cos \gamma t\, \mathbf{j} - \sin \gamma t\, \mathbf{k}) - a\gamma^2 \sin \alpha t\, \mathbf{j}^*$
12. $\mathbf{v} = \pi(a\mathbf{i} + b\mathbf{k})$, $s = 2\pi(a^2 + b^2)^{1/2}$
13. $\theta = gbt^2/[2(a^2 + b^2)]$, $s = 2gb/(a^2 + b^2)^{1/2}$

EXERCISES 3.1

1. Planes 2. Paraboloids 3. An ellipsoid 4. Hyperboloid of two sheets 5. Hyperbolas with axes $x = y$ and $x = -y$
6. Hyperbolas with axes $x = 0$ and $y = 0$
12. (a) $(\sqrt{2}/2 + 1/2)\mathbf{i} + (1 - \sqrt{2}/2)\mathbf{j}$ (b) $\sqrt{2}$ (c) $(\sqrt{2}\pi/2)\mathbf{i}$
13. (a) $\mathbf{j} + 2\mathbf{k}$ (b) $|\mathbf{F}| = a$
14. (a) $(\sin x + e^x\cos y)\mathbf{i} + (\cos y + e^x\sin y)\mathbf{j} + \ln z\,\mathbf{k}$
    (b) $e^x(\sin x + \sin y)\cos y$
    (c) $-e^x\cos y \ln z\,\mathbf{i} + e^x\cos y \ln z\,\mathbf{j} + e^x(\sin x \sin y - \cos^2 y)\mathbf{k}$

EXERCISES 3.2

1. $3\sqrt{2}$, 2. $34\sqrt{5}/5$, 3. $(1 + \sqrt{3})e/(2\sqrt{2})$, 4. $-38/\sqrt{5}$, 5. $-4/3$
6. $-8/3$, 7. $-9/\sqrt{34}$, 8. $\sqrt{3}/3$, 10. $1/2, \theta = -90$ deg.
11. $\sqrt{2}, \theta = 45$ deg. 12. $\partial f/\partial x = -1, \partial f/\partial y = 3$; $11/\sqrt{17}$
13. $20 + e^{\sqrt{3}}(\ln 2 - \sqrt{3})$, 14. $4t^3 - 3t^2 + 10t - 2$
16. $2\sqrt{3} + 4 + 2\pi/3$, 17. 12

EXERCISES 3.3

1. (a) $2\mathbf{i} + \mathbf{j}$ (b) $(1/\sqrt{2})[(1 - \pi/4)\mathbf{i} + (1 + \pi/4)\mathbf{j}]$ (c) $(\mathbf{i} + \mathbf{j})/2$
   (d) $2\mathbf{i} + \mathbf{j} + \mathbf{k}$ (e) $(2/3)(\mathbf{i} - \mathbf{j} - \mathbf{k})$ (f) $(-1/e)\mathbf{j} - \mathbf{k}$
2. (a) $\sqrt{5}$ (b) $2\sqrt{5}/3$ (c) $2\sqrt{17}$ (d) $(32 + 9\pi^2)^{1/2}/4$
3. $(-1/\sqrt{2}, 1/\sqrt{2})$ and $(1/\sqrt{2}, -1/\sqrt{2})$; 1
4. (a) $14/3$ (b) $(2/e - 3)/\sqrt{17}$
5. $2(1 + x^2 + y^2)/[4(x^2 + y^2) + 1]^{1/2}$
6. $df/dn = y + z$ on $x = 1$ and $-(y + z)$ on $x = -1$
   $df/dn = z + x$ on $y = 1$ and $-(z + x)$ on $y = -1$
   $df/dn = x + y$ on $z = 1$ and $-(x + y)$ on $z = -1$
7. $2x - 2y + z + 1 = 0$ 8. $\mathbf{t} = \mathbf{i}$
9. (a) $2\mathbf{j} + \mathbf{k}$, $\cos\theta = -8\sqrt{21}/63$ (b) $\mathbf{i} - \mathbf{j} - \mathbf{k}$, $\cos\theta = 1/2$
10. $b = \pm 2$ 11. $a = 3/4$ 15. (a) $\nabla f(r) = f'(r)\mathbf{R}/r$ (b) $\nabla r^n = nr^{n-2}\mathbf{R}$

EXERCISES 3.4

1. $2e^x \cos y$ 2. $2(x^2 - y^2)/(x^2 + y^2)^2$ 3. $x(2 + ze^{xy})$
4. 0 5. $e^x + e^y + e^z$ 6. $\sin y + \sin(xz) - e^z \sin e^z$
7. $2/(x^2 + y^2 + z^2)^{1/2}$ 10. $\operatorname{div}\mathbf{F} = (3 + n)r^n$
11. $\operatorname{grad}\operatorname{div}\mathbf{F} = \mathbf{i} + \mathbf{j} + \mathbf{k}$ 14. $\phi(x, y) = (1/2)\ln(x^2 + y^2) + C$
15. $\phi(x, y) = \arctan(y/x) + C$ 16. $\phi(x, y, z) = x^2y + xz^2 + zy^2 + C$
17. $\phi(x, y, z) = e^x\cos z + y^2 + C$
18. $\phi(x, y, z) = xe^y + zy^2 \ln z + z^2 \sin x$

EXERCISES 3.5

1. 0 2. $1/\sqrt{(x^2 + y^2)}$
3. $(x^2z^3 - 3x^2yz^2)\,\mathbf{i} + (3x^2yz^2 - 2xyz^3)\,\mathbf{j} + (2xyz^3 - x^2z^3)\,\mathbf{k}$
4. $2x(1 - z)\,\mathbf{i} + 2(yz - x - y)\mathbf{j}$ 5. $-x^2z\,\mathbf{i} + 4xyz\,\mathbf{j} + (2y^3 - xz^2)\,\mathbf{k}$
6. 0 8. $-2xy^2 - 2yz^2 - 2x^2z$
9. $(3xy^2 - x^3 + 2y^2z)\,\mathbf{i} - (y^3 - 2xz^2 - 3yz^2)\,\mathbf{j} + (3x^2z - z^3 + 2x^2y)\,\mathbf{k}$
10. $(y^2z - x^3)\,\mathbf{i} + (xz^2 - y^3)\,\mathbf{j} + (x^2y - z^3)\,\mathbf{k}$
11. $(x^2 - y^2)\,\mathbf{i} + (y^2 - z^2)\,\mathbf{j} + (z^2 - x^2)\,\mathbf{k}$ 14. 0
15. $\phi(x, y, z) = x^2yz + C$

EXERCISES 3.6

1. $2yz\,\mathbf{i} - xz\,\mathbf{j} + \text{grad}\,\phi$, ϕ any differentiable function of x, y, z.
2. $(xz^2/2)\,\mathbf{i} + (x^2y/2 - yz^2/2)\,\mathbf{j}$

12. $\mathbf{F} \times (\mathbf{H}\cdot\nabla)\mathbf{G} - \mathbf{F} \times (\mathbf{G}\cdot\nabla)\mathbf{H} - \nabla\cdot\mathbf{G}(\mathbf{F} \times \mathbf{H}) + \nabla\cdot\mathbf{H}(\mathbf{F} \times \mathbf{G})$
$+ (\mathbf{F}\cdot\nabla)(\mathbf{G} \times \mathbf{H}) + (\mathbf{G} \times \mathbf{H}) \times (\nabla \times \mathbf{F}) + [(\mathbf{G} \times \mathbf{H})\cdot\nabla]\mathbf{F}$
13. $\nabla(\mathbf{F}\cdot\mathbf{H})\cdot\mathbf{G} + (\mathbf{F}\cdot\mathbf{H})\nabla\cdot\mathbf{G} - \nabla(\mathbf{F}\cdot\mathbf{G})\cdot\mathbf{H} - (\mathbf{F}\cdot\mathbf{G})\nabla\cdot\mathbf{H}$
14. $\nabla(\mathbf{F}\cdot\mathbf{H}) \times \mathbf{G} + (\mathbf{F}\cdot\mathbf{H})\nabla \times \mathbf{G} - \nabla(\mathbf{F}\cdot\mathbf{G}) \times \mathbf{H} - (\mathbf{F}\cdot\mathbf{G})\nabla \times \mathbf{H}$

EXERCISES 3.7

6. (a) u = const. defines a parabola opening to the left
v = const. defines a parabola opening to the right
u = 0 corresponds to the nonpositive x-axis
u = v corresponds to the nonnegative y-axis
u = -v corresponds to the nonpositive y-axis
(b) $\mathbf{R}_u = u\mathbf{i} + v\mathbf{j}$, $\mathbf{R}_v = -v\mathbf{i} + u\mathbf{j}$, $\mathbf{R}_z = \mathbf{k}$
(c) $ds^2 = (u^2 + v^2)(du^2 + dv^2) + dz^2$
8. $ds^2 = (u^2 + v^2)(du^2 + dv^2) + u^2 v^2\, d\theta^2$
9. $\partial(u, v, w)/\partial(x, y, z) = 1/(u^2 - v^2)$
10. (a) $x = (u + v + w)/2$, $y = (-u + v + w)/2$, $z = (u + v - w)/2$
(b) $\partial\mathbf{R}/\partial u\cdot\partial\mathbf{R}/\partial v \times \partial\mathbf{R}/\partial w = -1/2$
(e) $ds^2 = 3(du^2 + dv^2 + dw^2)/4 + (du\,dv + dv\,dw - dw\,du)/2$

EXERCISES 3.8

1. $$\nabla^2 f = \frac{1}{r}\left[\frac{\partial}{\partial r}\left(r\frac{\partial f}{\partial r}\right) + \frac{\partial}{\partial\theta}\left(\frac{1}{r}\frac{\partial f}{\partial\theta}\right) + \frac{\partial}{\partial z}\left(r\frac{\partial f}{\partial z}\right)\right]$$

$$\text{curl } \mathbf{F} = \frac{1}{r}\begin{vmatrix} \mathbf{u}_1 & \mathbf{u}_2 & \mathbf{u}_3 \\ \partial/\partial r & \partial/\partial\theta & \partial/\partial z \\ F_1 & rF_2 & F_3 \end{vmatrix}$$

$\mathbf{F} = F_1\,\mathbf{u}_1 + F_2\,\mathbf{u}_2 + F_3\,\mathbf{u}_3$, $\mathbf{u}_1 = \cos\theta\,\mathbf{i} + \sin\theta\,\mathbf{j}$,
$\mathbf{u}_2 = -\sin\theta\,\mathbf{i} + \cos\theta\,\mathbf{j}$, $\mathbf{u}_3 = \mathbf{k}$

2. $$\nabla^2 f = \frac{1}{r^2\sin\phi}\left[\frac{\partial}{\partial r}\left(r^2\sin\phi\,\frac{\partial f}{\partial r}\right) + \frac{\partial}{\partial\phi}\left(\sin\phi\,\frac{\partial f}{\partial\phi}\right) + \frac{\partial}{\partial\theta}\left(\frac{1}{\sin\phi}\frac{\partial f}{\partial\theta}\right)\right]$$

$$\nabla\cdot\mathbf{F} = \frac{1}{r^2\sin\phi}\left[\frac{\partial}{\partial r}\left(r^2\sin\phi\,F_1\right) + \frac{\partial}{\partial\phi}\left(r\sin\phi\,F_2\right) + \frac{\partial}{\partial\theta}\left(r\,F_{?3}\right)\right]$$

$$\text{curl } \mathbf{F} = \frac{1}{r^2\sin\phi}\begin{vmatrix} \mathbf{u}_1 & r\,\mathbf{u}_2 & r\sin\phi\,\mathbf{u}_3 \\ \partial/\partial r & \partial/\partial\phi & \partial/\partial\theta \\ F_1 & rF_2 & r\sin\phi\,F_3 \end{vmatrix}$$

$\mathbf{F} = F_1\mathbf{u}_1 + F_2\mathbf{u}_2 + F_3\mathbf{u}_3$, $\mathbf{u}_1 = \sin\phi\,\cos\theta\,\mathbf{i} + \sin\phi\,\sin\theta\,\mathbf{j} + \cos\phi\,\mathbf{k}$,
$\mathbf{u}_2 = \cos\phi\,\cos\theta\,\mathbf{i} + \cos\phi\,\sin\theta\,\mathbf{j} - \sin\phi\,\mathbf{k}$, $\mathbf{u}_3 = -\sin\theta\,\mathbf{i} + \cos\theta\,\mathbf{j}$

3. (a) In cylindrical coordinates, $\nabla f = (2r\sin\theta\cos\phi + z\sin\theta + z\cos\theta)\mathbf{u}_1 + [r(\cos^2\theta - \sin^2\theta) + z(\cos\theta - \sin\theta)]\mathbf{u}_2 + r(\sin\theta + \cos\theta)\mathbf{u}_3$, $\nabla^2 f = 0$
   (b) In spherical coordinates, $\nabla f = 2r\,\mathbf{e}_1$, $\nabla^2 f = 6$
4. $\mathbf{i} = \cos\theta\,\mathbf{u}_1 - \sin\theta\,\mathbf{u}_2$, $\mathbf{j} = \sin\theta\,\mathbf{u}_1 + \cos\theta\,\mathbf{u}_2$, $\mathbf{k} = \mathbf{u}_3$
   $\mathbf{F} = (z\cos\theta - 2r\sin\theta\cos\theta)\mathbf{u}_1 - (z\sin\theta + 2r\cos^2\theta)\mathbf{u}_2 + 2r\sin\theta\,\mathbf{u}_3$
5. div $\mathbf{F} = 0$, curl $\mathbf{F} = (2\cos\theta + \sin\theta)\mathbf{u}_1 + (\cos\theta - 2\sin\theta)\mathbf{u}_2 - 2\mathbf{u}_3$
6. div $\mathbf{F} = 0$, curl $\mathbf{F} = -2z\cos\theta\,\mathbf{u}_1 + 2(z\sin\theta - r)\,\mathbf{u}_2 - \mathbf{u}_3$
7. $\mathbf{i} = \sin\phi\,\cos\theta\,\mathbf{u}_r + \cos\phi\,\cos\theta\,\mathbf{u}_\phi - \sin\theta\,\mathbf{u}_\theta$,
   $\mathbf{j} = \sin\phi\,\sin\theta\,\mathbf{u}_r + \cos\phi\,\sin\theta\,\mathbf{u}_\phi - \cos\theta\,\mathbf{u}_\theta$,
   $\mathbf{k} = \cos\phi\,\mathbf{u}_r - \sin\phi\,\mathbf{u}_\phi$
8. $\mathbf{F} = (r\sin^2\phi\,\sin\theta\cos\theta - 2r\sin\phi\,\cos\phi\,\sin\theta + r\sin\phi\,\cos\phi\,\cos\theta)\mathbf{u}_r +$

$(r\sin\phi\cos\phi\sin\theta\cos\theta - 2r\cos^2\phi\sin\theta - r\sin^2\phi\cos\theta)\mathbf{u}_\phi -$
$(r\sin\phi\sin^2\theta + 2r\cos\phi\cos\theta)\mathbf{u}_\theta$, $\nabla\cdot\mathbf{F} = 0$

$\nabla\times\mathbf{F} = (2\sin\phi\cos\theta - \sin\phi\sin\theta - \cos\phi)\mathbf{u}_r + (2\cos\phi\cos\theta -$
$\cos\phi\sin\theta + \sin\phi)\mathbf{u}_\phi - (2\sin\theta + \cos\theta)\mathbf{u}_\theta$

9. $\mathbf{F} = (r\sin\phi\cos\phi\cos\theta + 2r\sin^2\phi\sin^2\theta + 3r\cos^2\phi)\mathbf{u}_r +$
$(r\cos^2\phi\cos\theta + 2r\sin\phi\cos\phi\sin^2\theta - 3r\sin\phi\cos\phi)\,\mathbf{u} +$
$(2r\sin\phi\sin\theta\cos\theta - r\cos\phi\sin\theta)\mathbf{u}_\theta$

$\nabla\cdot\mathbf{F} = 5$, $\quad \nabla\times\mathbf{F} = \sin\phi\sin\theta\,\mathbf{u}_r + \cos\phi\sin\theta\,\mathbf{u}_\phi + \cos\theta\,\mathbf{u}_\theta$

10. $$\nabla f = \frac{1}{\sqrt{u^2+v^2}}\left(\frac{\partial f}{\partial u}\mathbf{u}_1 + \frac{\partial f}{\partial v}\mathbf{u}_2\right) + \frac{\partial f}{\partial z}\mathbf{u}_3$$

$$\nabla\cdot\mathbf{F} = \frac{1}{u^2+v^2}\left(\frac{\partial}{\partial u}\left(\sqrt{u^2+v^2}F_1\right) + \frac{\partial}{\partial v}\left(\sqrt{u^2+v^2}F_2\right)\right)$$
$$+ \frac{1}{u^2+v^2}\frac{\partial}{\partial z}[(u^2+v^2)F_3]$$

$$\nabla\times\mathbf{F} = \frac{1}{u^2+v^2}\begin{vmatrix} \sqrt{u^2+v^2}\,\mathbf{u}_1 & \sqrt{u^2+v^2}\,\mathbf{u}_2 & \mathbf{u}_3 \\ \partial/\partial u & \partial/\partial v & \partial/\partial z \\ \sqrt{u^2+v^2}\,F_1 & \sqrt{u^2+v^2}\,F_2 & F_3 \end{vmatrix}$$

$\mathbf{F} = F_1\,\mathbf{u}_1 + F_2\,\mathbf{u}_2 + F_3\,\mathbf{u}_3$

$\mathbf{u}_1 = (u\,\mathbf{i} + v\,\mathbf{j})/\sqrt{u^2+v^2}$, $\mathbf{u}_2 = (-v\,\mathbf{i} + u\,\mathbf{j})/\sqrt{u^2+v^2}$, $\mathbf{u}_3 = \mathbf{k}$

11. $$\nabla f = \frac{1}{\sqrt{u^2+v^2}}\left(\frac{\partial f}{\partial u}\mathbf{u}_1 + \frac{\partial f}{\partial v}\mathbf{u}_2\right) + \frac{1}{uv}\frac{\partial f}{\partial\theta}\mathbf{u}_3$$

$$\nabla \cdot \mathbf{F} = \left[\frac{\partial}{\partial u}\left(uv\sqrt{u^2+v^2}\,F_1\right) + \frac{\partial}{\partial v}\left(uv\sqrt{u^2+v^2}\,F_2\right) + \frac{\partial}{\partial \theta}\left((u^2+v^2)F_3\right)\right]$$

$$\cdot \frac{1}{uv(u^2+v^2)}$$

$$\nabla \times \mathbf{F} = \frac{1}{uv(u^2+v^2)}\begin{vmatrix} \sqrt{u^2+v^2}\,\mathbf{u}_1 & \sqrt{u^2+v^2}\,\mathbf{u}_2 & uv\mathbf{u}_3 \\ \partial/\partial u & \partial/\partial v & \partial/\partial \theta \\ \sqrt{u^2+v^2}\,F_1 & \sqrt{u^2+v^2}\,F_2 & uvF_3 \end{vmatrix}$$

$\mathbf{F} = F_1\mathbf{u}_1 + F_2\mathbf{u}_2 + F_3\mathbf{u}_3$, $\mathbf{u}_1 = (v\cos\theta\,\mathbf{i} + v\sin\theta\,\mathbf{j} + u\mathbf{k})/\sqrt{(u^2+v^2)}$
$\mathbf{u}_2 = (u\cos\theta\,\mathbf{i} + u\sin\theta\,\mathbf{j} - v\mathbf{k})/\sqrt{(u^2+v^2)}$, $\mathbf{u}_3 = -\sin\theta\,\mathbf{i} + \cos\theta\,\mathbf{j}$

EXERCISES 4.1

1. 0, 2. $2\sqrt{2}$,
3. $\sqrt{(2+\pi^2)}[(2+\pi^2)/3 + \pi/2] + \ln[\sqrt{(2+\pi^2)} + \pi] - (2/3)\sqrt{2} - (\ln 2)2$
4. $\sqrt{(a^2+b^2)}(a^2/4 - 9b\pi^2/32)$, 5. $8(2\sqrt{2}-1)/3$
6. (a) $3\sqrt{2}$, (b) $(1/3, 1/3, 1/3)$
7. (a) $6a^2$, (b) $(0, 0)$, (c) $7a^4/3$, (d) $11a^4/3$
8. $(4/3)(\sqrt{2}+1)/[\sqrt{10} + \sqrt{5}\ln(1+\sqrt{2})]$
9. $(2/3)[(\pi^2+5)^{3/2} - 5\sqrt{5}]$, 10. $88/15$, 11. $-8/3$
12. 0 13. 0, 14. $(2 + 9\pi^3/8)\sqrt{2}$, 15. π

EXERCISES 4.2

1. 1, 2. $18/5$, 3. $-8/3$, 4. 0, 5. $137/6$,
6. $-1/60$, 7. $1/3$, 8. $-2a + \pi^2$
9. (a) 2π, (b) 2π, 10. -2π, 11. -2π, 12. $-\pi$
13. (a) 0, (b) 0, 14. -2π, 15. $\pi/2 - 3\pi^2/8$,
16. $2 - \sqrt{2}/6$ 17. $-1 + \sqrt{2}/6$

EXERCISES 4.3

3. $(1/2)\ln(x^2+y^2) + C$, 4. $y^2\sin x + x^2e^y + C$
5. $z^2e^{xy} + y\cos x + z + C$, 6. $xye^z - xz + y^2 + C$
7. $x^2z + ye^{-x} + y\cos(z-1) + C$, 8. $(1/2)\ln(x^2+y^2+z^2)$
9. $3xe^z + x^2y - z\cos y + C$, 10. **F** is not a conservative field
12. 32, 13. $\pi^2 + e + 4$, 14. $-2 - \pi/2$, 15. $(9 + \ln 5)/2$

EXERCISES 4.4

1\. 16, 2. 0, 3. 0, 4. $\pi a^4/4$, 5. 20π
6\. $3\pi/2$, 10. 4π, 11. 0, 12. 6π, 13. 0
14\. 4π, 15. (a) Let $F = v\,\nabla u$ in (4.21)

EXERCISES 4.5

1. $x^2 + y^2 = a^2$ $(z \geq 0)$, a cylinder, 2. $x^2 + y^2 = z$, a paraboloid
3. $a^2(x^2 + y^2) = z^2$, an upright circular cone
4. $(x/a)^2 + (y/b)^2 + (z/c)^2 = 1$, an ellipsoid
5. $(x/a)^2 - (y/b)^2 = 4z$, a hyperbolic paraboloid
6. $(x/a)^2 + (y/b)^2 - (z/c)^2 = 1$, an elliptic hyperboloid of one sheet
7. $(x/a)^2 - (y/b)^2 = z^2$, hyperboloid of two sheets
8. $x = u,\ y = v,\ z = (d - au - bv)/c$, $(|u| < \infty,\ |v| < \infty)$
9. $x = u,\ y = v,\ z = u^2$, $(|u| < \infty,\ |v| < \infty)$
10. $x = au\cos v,\ y = bu\sin v,\ z = 1 - u^2$ $(u \geq 0,\ 0 \leq v \leq 2\pi)$
11. $x = a\sin u\cos v,\ y = b\sin u\sin v,\ z = c\cos u$ $(0 \leq u \leq \pi,\ 0 \leq v < 2\pi)$
12. $x = a\cos u\cosh v,\ y = b\sin u\cosh v,\ z = c\sinh v$ $(|v| < \infty,\ 0 \leq u \leq 2\pi)$
13. $x = \pm a\cosh u,\ y = b\cos v\sinh u,\ z = c\sin v\sinh u$ $(|u| < \infty,\ 0 \leq v \leq 2\pi)$
14. $x = a(u + v)/2,\ y = b(u - v)/2,\ z = cuv$ $(|u| < \infty,\ |v| < \infty)$
15. $2x - 2y - z = 2$, 16. $\sqrt{3}x + y - 2z = 0$
17. $2x + y + 4z = 8$, 18. $x + 6y - 4z + 8 = 0$
19. $3x + \sqrt{3}y - 12 = 0$

EXERCISES 4.6

1\. $\sqrt{3}/2$, 2. $\sqrt{3}\pi a^2$, 3. $\sqrt{2}\pi a^2$, 4. $(5\sqrt{5} - 1)\pi/6$
5\. $2\sqrt{(2ab)}(a + b)/3$, 6. $2a^2(\pi - 2)$, 7. $8a^2$
8\. $\pi[a\sqrt{(1 + a^2)} + \ln(a + \sqrt{(1 + a^2)})]$, 9. $4\pi a^2(1 - 1/\sqrt{2})$
10\. $(2/3)\sqrt{(2/3)}\,\pi$, 11. $2\sqrt{2}\pi a^2$
12\. $2\pi[(a^2 + 1)^{3/2} - 1]/3$, 13. a^2
14\. (a) $4\pi a^2[2 - ((7/2) - \sqrt{13/2})^{1/2}]$, (b) $\pi a^2[(2\sqrt{13} + 3)^{3/2} - 1]/6$
15\. $4\sqrt{5}$, 16. $2\pi(2\sqrt{2} - 1)/3$
17\. $(3 + 25\sqrt{5} - 16\sqrt{2})/15$

EXERCISES 4.7

1\. $2\sqrt{3}/15$, 2. $a^3b\pi + 2ab^3\pi/3$, 3. $2\pi a^4/3$
4\. 0, 5. 0, 8. $2\pi a^3h$, 9. $5\sqrt{5}\pi/4 - 3\pi/20$
10\. (a) $\pi a^5/2$, (b) $\pi a^5/4$, 11. 64
12\. $(25\sqrt{5} + 1)/(75\sqrt{5} - 9)$, 13. 1/2, 14. 2

15. $3\pi/2$, 16. $\pi^2 a^2/2$, 17. $28/3$, 18. 0
19. $\pi/8 - 1/21$, 20. 0, 21. 24, 22. 16π
23. 0, 24. $8\pi a^3/3$, 25. π, 26. $4\pi R^3/3$
27. 16π, 28. 0, 29. -2π

EXERCISES 4.8

1. $3/2$, 2. $1/24$, 3. 6π, 4. 7π, 5. 0
6. 5π, 8. $n = 1$, $4\pi a^3$; $n = 2$, $16\pi a^5/5$
10. (a) $4\pi a^3$, (b) $16\pi a^5/5$, 11. $16\pi a^3/3$, 12. 16, 16. 1

EXERCISES 4.9

1. $1/6$, 2. -8π, 3. -2π, 4. 0, 5. 0
6. 0, 7. 0, 8 $-\pi a^4/2$, 9. $2\pi R^2(a^2 + b^2 + c^2)/c$
10. (c) 2π

EXERCISES 5.1

1. (a) $\delta_{ij}a_j = \delta_{i1}a_1 + \delta_{i2}a_2 + \delta_{i3}a_3$ (1 = 1, 2, 3) or (a_1, a_2, a_3)
   (b) $a_1b_1 + a_2b_2 + a_3b_3$
   (c) $a_{11}x_1^2 + (a_{12} + a_{21})x_1x_2 + a_{22}x_2^2 + (a_{13} + a_{31})x_1x_3 + (a_{23} + a_{32})x_2x_3 + a_{33}x_3^2$
   (d) $a_{i1}b_{1j} + a_{i2}b_{2j} + a_{i3}b_{3j} = \delta_{ij}$ $(i, j = 1, 2, 3)$
   (e) $s = a_{11}b_{11} + a_{12}b_{12} + a_{13}b_{13} + a_{21}b_{21} + a_{22}b_{22} + a_{23}b_{23} + a_{31}b_{31} + a_{32}b_{32} + a_{33}b_{33}$
   (f) $$g_{ij} = \frac{\partial x_i}{\partial u_1}\frac{\partial x_j}{\partial u_1} + \frac{\partial x_i}{\partial u_2}\frac{\partial x_j}{\partial u_2} + \frac{\partial x_i}{\partial u_3}\frac{\partial x_j}{\partial u_3} \quad (i, j = 1, 2, 3)$$

2. $$\frac{d\phi}{dt} = \frac{\partial\phi}{\partial x_i} f_i'(t)$$

3. $$\frac{\partial\phi}{\partial u_i} = \frac{\partial\phi}{\partial x_k}\frac{\partial g_k}{\partial u_i} \quad (i = 1, 2, 3)$$
   $$\frac{\partial^2\phi}{\partial u_j \partial u_i} = \frac{\partial^2\phi}{\partial x_m \partial x_k}\frac{\partial g_k}{\partial u_i}\frac{\partial g_m}{\partial u_j} + \frac{\partial\phi}{\partial x_k}\frac{\partial^2 g_k}{\partial u_j \partial u_i}$$

4. $$\frac{\partial\phi}{\partial x^*_i} = \frac{\partial\phi}{\partial x_j}\frac{\partial x_j}{\partial x^*_i} = \alpha_{ji}\frac{\partial\phi}{\partial x_j}$$

EXERCISES 5.2

1. $x^*_1 = x_3$, $x^*_2 = -x_1$, $x^*_3 = x_2$, left-handed
2. $x^*_1 = x_1 \cos\theta + x_2 \sin\theta$, $x^*_2 = -x_1 \sin\theta + x_2 \cos\theta$, $x^*_3 = x_3$
3. $x^*_1 = x_2$, $x^*_2 = x_3$, $x^*_3 = x_1$, right-handed
4. $x^*_1 = (x_1 + x_2 + x_3)/\sqrt{3}$, $x^*_2 = (2x_1 - x_2 - x_3)/\sqrt{6}$, $x^*_1 = (x_2 - x_3)/\sqrt{2}$,

EXERCISES 5.3

2. (a) $a^*_1 = 1 - \sqrt{3}/2$, $a^*_2 = 1/2 + \sqrt{3}/4$, $a^*_3 = -1/4 - 3\sqrt{3}/2$
   (b) $b_1 = -3/4 + \sqrt{3}/2$, $b_1 = 5/2 + 3\sqrt{3}/4$, $b_1 = 1 - 3\sqrt{3}/2$
   (c) $4\sqrt{3} + 15/4$
3. (b) $[4\sqrt{3}, -3\sqrt{2}, 5\sqrt{6}]$, (c) $[3, 1, 1]$

EXERCISES 5.4

3. $I_{11} = 13M$, $I_{22} = 13M$, $I_{33} = 8M$, $I_{12} = I_{21} = 4M$, $I_{23} = I_{32} = 6M$,
   $I_{13} = I_{31} = -6M$,
4. 17M, 5. $(a + b + c)/3$, 6. 5M
8. $I^*_{11} = 11/4$, $I^*_{22} = (75 + 4\sqrt{3})/16$, $I^*_{33} = -(39 + 4\sqrt{3})/16$
11. $(5\sqrt{3}, -4\sqrt{3}, 1\sqrt{3})$, $s_n = 10/3$, $s_t = \sqrt{(26)}/3$
12. $(4/3, -1/3, -4/3)$, $s_n = 5/3$, $s_t = \sqrt{8}/3$
13. $(8/\sqrt{5}, -1/\sqrt{5})$, $s_n = 6/5$, $s_t = \sqrt{(289)}/5$
14. $s^*_{11} = (s_1 + 3s_2)/4$, $s^*_{12} = 0$, $s^*_{13} = \sqrt{3}(s_1 - s_2)/4$, $s^*_{21} = 0$,
    $s^*_{22} = s_3$, $s^*_{23} = 0$, $s^*_{31} = s^*_{13}$, $s^*_{32} = 0$, $s^*_{33} = (3s_1 + s_2)/4$

EXERCISES 5.5

1. (a) $[-1, 8, 15]$, (b) $[-5, 1, 0]$, (c) 3
2. (a) $(c_{ij}) = \begin{bmatrix} 1 & 4 & 0 \\ 7 & -18 & 3 \\ 7 & 8 & 8 \end{bmatrix}$, where $c_{ij} = a_{ik}b_{kj}$

   (b) $(c_{ij}) = \begin{bmatrix} 4 & 10 & 1 \\ 0 & -13 & 7 \\ 0 & -2 & 7 \end{bmatrix}$, where $c_{ij} = a_{ik}b_{jk}$

   (c) $(c_{ij}) = \begin{bmatrix} 3 & -10 & 0 \\ 9 & -9 & 7 \\ 2 & 20 & 4 \end{bmatrix}$, where $c_{ij} = a_{ki}b_{kj}$, (d) -2

5. $(s_{ij}) = \begin{bmatrix} 3 & 2 & 1 \\ 2 & 3 & 2 \\ 1 & 2 & 1 \end{bmatrix}$, $(t_{ij}) = \begin{bmatrix} 0 & 2 & -3 \\ -2 & 0 & 0 \\ 3 & 0 & 0 \end{bmatrix}$

EXERCISES 5.6

1. (a) $\lambda = 0$, $[2, -1, -1]/\sqrt{6}$; $\lambda = 2$, $[0, 1, -1]/\sqrt{2}$; $\lambda = 3$, $[1, 1, 1]/\sqrt{3}$

   (b) $(I^*_{ij}) = \begin{bmatrix} 0 & 0 & 0 \\ 0 & 2 & 0 \\ 0 & 0 & 3 \end{bmatrix}$

   (c) $x^*_i = \alpha_{ij} x_j$, where

$$(\alpha_{ij}) = \begin{bmatrix} 2/\sqrt{6} & -1/\sqrt{6} & -1/\sqrt{6} \\ 0 & 1/\sqrt{2} & -1/\sqrt{2} \\ 1/\sqrt{3} & 1/\sqrt{3} & 1/\sqrt{33} \end{bmatrix}$$

2. (a) $\lambda = 1$, $[0, -1, 1]/\sqrt{2}$; $\lambda = 2$, $[1, 0, 0]$; $\lambda = 5$, $[0, 1, 1]/\sqrt{2}$

   (b) $(I^*_{ij}) = \begin{bmatrix} 1 & 0 & 0 \\ 0 & 2 & 0 \\ 0 & 0 & 5 \end{bmatrix}$

   (c) $(\alpha_{ij}) = \begin{bmatrix} 0 & -1/\sqrt{2} & 1/\sqrt{2} \\ 1 & 0 & 0 \\ 0 & 1/\sqrt{2} & 1/\sqrt{2} \end{bmatrix}$

3. $\lambda = 1$, $[1, 1, -1]/\sqrt{3}$; $\lambda = -2$, $[1, -1, 0]/\sqrt{2}$; $\lambda = 5$, $[0, 1, 1]/\sqrt{2}$
4. $\lambda = -3, -3$, $[a_1, a_2, a_3]$ and $[b_1, b_2, b_3]$ where $2a_1 - 2a_2 + a_3 = 0$, $2b_1 - 2b_2 + b_3 = 0$, and the vectors are chosen so that they are orthogonal; for example, $[1, 1, 0]/\sqrt{2}$ and $[1, -1, 4]/\sqrt{18}$; $\lambda = 6$, $[2, -2, 1]/3$
5. $s^*_{11} = 0$, $s^*_{22} = [1 + \sqrt{(1 + 4(a^2 + b^2))}]/2$, $s^*_{33} = [1 - \sqrt{(1 + 4(a^2 + b^2))}]/2$
7. $x^{*2}_1 + x^{*2}_2 + 4x^{*2}_3 = 1$, ellipsoid (the variables may be interchanged)
8. $3x^{*2}_1 + 3x^{*2}_2 + 9x^{*2}_3 = 1$, the variables may be interchanged
9. $x^{*2}_1 + 6x^{*2}_2 = 1$, the variables may be interchanged

EXERCISES 5.7

4. (a) $$\left(\frac{\partial a_i}{\partial x_j}\right) (c_{ij}) = \begin{bmatrix} \sin x_3 & 0 & x_1 \cos x_3 \\ -x_2 e^{x_1} & -e^{x_1} & 0 \\ 0 & x_3 & x_3 \end{bmatrix}$$

(b) $$\begin{bmatrix} 1 & -2x_2 & 0 \\ \dfrac{2x_1 x_3}{1 + x_1^2} & 0 & \ln(1 + x_1^2) \\ -2x_1 \cos x_1 & 0 & x_1^2 \sin x_3 \end{bmatrix}$$

5. (a) $c_1 = 4x_1$, $c_2 = 3x_2$, $c_3 = 4x_3$, where $c_i = \partial s_{ji}/\partial x_j$ $(i = 1, 2, 3)$;
$c_1 = 3x_1 + x_3$, $c_2 = 4x_2$, $c_3 = 4x_3$, where $c_i = \partial s_{ij}/\partial x_j$ $(i = 1, 2, 3)$
(b) $c_1 = -1$, $c_2 = 0$, $c_3 = 1$, where $c_i = \partial t_{ij}/\partial x_j$ $(i = 1, 2, 3)$;
$c_1 = 1$, $c_2 = 0$, $c_3 = x_2$, where $c_i = \partial t_{ji}/\partial x_j$ $(i = 1, 2, 3)$

8. $$(s_{ij}) = \begin{bmatrix} x_2 & x_1/2 & [-2x_3 + x_3 \cos(x_1 x_3)]/2 \\ x_1/2 & e^{x_3} & x_2 e^{x_3}/2 \\ [-2x_3 + x_3 \cos(x_1 x_3)]/2 & x_2 e^{x_3}/2 & x_1 \cos(x_1 x_3) \end{bmatrix}$$

$$(t_{ij}) = \begin{bmatrix} 0 & x_1/2 & [-2x_3 - x_3 \cos(x_1 x_3)]/2 \\ -x_1/2 & 0 & x_2 e^{x_3}/2 \\ [2x_3 + x_3 \cos(x_1 x_3)]/2 & -x_2 e^{x_3}/2 & 0 \end{bmatrix}$$

EXERCISES 6.1

1. $(a_i b^j) = \begin{bmatrix} -2 & -4 & -6 \\ 1 & 2 & 3 \\ 3 & 6 & 9 \end{bmatrix}$, $a_i b^i = 9$,   2. Trace of the matrix.

3. $a^i = \frac{\partial u^i}{\partial x^1} b^1 + \frac{\partial u^i}{\partial x^2} b^2 + \frac{\partial u^i}{\partial x^3} b^3 \quad (i = 1, 2, 3)$

$$a_i = \frac{\partial u^1}{\partial x^i} b_1 + \frac{\partial u^2}{\partial x^i} b_2 + \frac{\partial u^3}{\partial x^i} b_3 \quad (i = 1, 2, 3)$$

5. $\frac{\partial a^i}{\partial x^k} = \frac{\partial^2 u^i}{\partial x^k \partial x^j} b^j + \frac{\partial u^i}{\partial x^j} \frac{\partial b^j}{\partial x^k} \quad (i, k = 1, 2, 3)$

7. (-1, 1), 8. (-2/3, 4/3), 9. (-1/2, -1/2, 5/2)
10. (4, -3, 2), 11. (17/2, -7/2, -3/2),
14. $\mathbf{e}^1 = [1, -1, 0]$, $\mathbf{e}^2 = [0, 1, -1]$, $\mathbf{e}^3 = [0, 0, 1]$
15. $\mathbf{e}_1 = [-1/2, 1/2, 1/2]$, $\mathbf{e}_2 = [1/2, -1/2, 1/2]$, $\mathbf{e}_3 = [1/2, 1/2, -1/2]$
16. $\mathbf{e}^1 = [-1, 3, -5]/2$, $\mathbf{e}^2 = [1, -1, 3]/2$, $\mathbf{e}^3 = [1, -1, 1]/2$
17. $x^i = \alpha^i{}_{*j} x^{*j}$, where

$$(\alpha^i_{*j}) = \begin{bmatrix} 1/2 & 1/2 & 1 \\ -1/2 & -1/2 & 1 \\ -1/2 & 1/2 & 0 \end{bmatrix}, \ \det(\alpha^i_{*j}) = -1$$

The transformation reverses the orientation of the coordinate system.

18. $x^i = \alpha_*{}^i{}_j x^{*j}$, where

$$(\alpha^i_{*j}) = \begin{vmatrix} 1 & -1/3 \\ 1 & 4/3 \end{vmatrix}, \ \det(\alpha^i_{*j}) = 5/3$$

19. $x^{*i} = \alpha^{i*}{}_j x^j$, where

$$(\alpha^{*i}{}_j) = \begin{bmatrix} 0 & -3/2 & 0 \\ -1/2 & 0 & 1/2 \\ 1/2 & 1/2 & 1/2 \end{bmatrix}, \ \det(\alpha^{*i}_j) = -3/4$$

20. $x^{*i} = \alpha^{i*}{}_j x^j$, where

$$(\alpha^{*i}{}_j) = \begin{vmatrix} 1 & 1/2 \\ -1 & 3/2 \end{vmatrix}, \ \det(\alpha^{*i}_j) = 2$$

21. $x^{*i} = \alpha^{i*}{}_j x^j$, where

$$(\alpha^{*i}{}_j) = \begin{bmatrix} -1 & 0 & -1/2 \\ 9/5 & 7/5 & 3/2 \\ -2/5 & 0 & 1/2 \end{bmatrix}, \ x^{*i} \text{ coordinate system is left-handed}$$

22. $x^{*i} = \alpha^{i*}{}_j x^j$, where

$$(\alpha^{*i}{}_j) = \begin{bmatrix} 3/2 & 5/2 & 0 \\ 11/2 & 13/2 & -5 \\ 4 & 5 & -3 \end{bmatrix}$$

EXERCISES 6.2

2. (a) $a^*{}_1 = 3,\ a^*{}_2 = 3;\ a^{*1} = 1/3,\ a^{*2} = 4/3$
   (b) $a^*{}_1 = 3,\ a^*{}_2 = 5;\ a^{*1} = 1/3,\ a^{*2} = 7/3$
   (c) $a^*{}_1 = 3,\ a^*{}_2 = -1,\ a^*{}_3 = -2;\ a^{*1} = 3,\ a^{*2} = -1,\ a^{*3} = -2$
   (d) $a^*{}_1 = 4,\ a^*{}_2 = -3,\ a^*{}_3 = 5;\ a^{*1} = 1,\ a^{*2} = 0,\ a^{*3} = 2$
3. $a^{*1} = (x_1x_2 + x_3x_2 - x_1x_3)/2,\ a^{*2} = (-x_1x_2 + x_3x_2 + x_1x_3)/2,$
   $a^{*3} = (x_1x_2 - x_3x_2 + x_1x_3)/2$, where $x_1 = x^{*1} + x^{*3},\ x_2 = x^{*1} + x^{*2},$
   $x_3 = x^{*2} + x^{*3}$
4. $a^{*1} = (x_1^2 - x_2^2 - x_3^2)/2,\ a^{*2} = (x_1^2 + x_2^2 - x_3^2)/2,$
   $a^{*3} = (x_1^2 + x_2^2 + x_3^2)/2$, where $x_1 = x^{*1} + x^{*3},\ x_2 = -x^{*1} + x^{*2},$
   $x_3 = -x^{*2} + x^{*3}$
5. $$(a^*{}_{ij}) = \begin{bmatrix} 6 & 10 & 4 \\ 6 & 6 & 2 \\ -4 & -2 & 1 \end{bmatrix},\ (a^{*ij}) = \begin{bmatrix} 1/4 & 3/4 & 1/2 \\ 1 & -1/2 & 0 \\ -3/2 & 1/2 & 0 \end{bmatrix}$$
6. $$(a^*{}_{ij}) = \begin{bmatrix} x_1 & x_2 + x_3 & -x_1 + x_2 - x_3 \\ -x_2 - x_3 & x_1 & -x_1 + x_2 - x_3 \\ -x_1 - x_2 + x_3 & -x_1 - x_2 + x_3 & 2x_1 \end{bmatrix},$$

   $$(a^{*ij}) = \frac{1}{2}\begin{bmatrix} x_1 & -x_1 + x_2 + x_3 & x_2 \\ -x_1 - x_2 - x_3 & x_1 & -x_3 \\ -x_2 & x_3 & x_1 \end{bmatrix},$$

   where $x_1 = -x^{*1} + x^{*3},\ x_2 = x^{*1} + x^{*2},\ x_3 = -x^{*2} - x^{*3}$
7. $$a^*{}_{ijk} = \frac{\partial x^p}{\partial x^{*i}} \frac{\partial x^q}{\partial x^{*j}} \frac{\partial x^r}{\partial x^{*k}} a_{pqr},\quad a^{*ijk} = \frac{\partial x^{*i}}{\partial x^p} \frac{\partial x^{*j}}{\partial x^q} \frac{\partial x^{*k}}{\partial x^r} a^{pqr}$$

   $$a^{*k}{}_{ij} = \frac{\partial x^p}{\partial x^{*i}} \frac{\partial x^q}{\partial x^{*j}} \frac{\partial x^{*k}}{\partial x^r} a^r{}_{pq}$$

8. $$a^{*kmn}_{\ \ ij} = \frac{\partial x^p}{\partial x^{*i}} \frac{\partial x^q}{\partial x^{*j}} \frac{\partial x^{*k}}{\partial x^r} \frac{\partial x^{*m}}{\partial x^s} \frac{\partial x^{*n}}{\partial x^t} a^{rst}_{pq}$$

10. $A^* = \Lambda^* A \Lambda^{*T}$, where $A^* = (a^{*ij})$ and $\Lambda^* = (\alpha^{*}{}_j{}^i)$

EXERCISES 6.3

1. $$(a^{ij}) = \begin{bmatrix} -16 & -18 & 25 \\ -18 & -25 & 32 \\ 25 & 32 & -42 \end{bmatrix}, \quad (a^i_{.k}) = (a_k^{.i}) = \begin{bmatrix} 0 & -2 & 7 \\ 3 & -4 & 10 \\ -2 & 5 & -12 \end{bmatrix}$$

2. $$(a^{ij}) = \begin{bmatrix} 4x^1 + x^2 & -2x^2 - 4x^3 & -3x^1 + 2x^2 + 6x^3 \\ -2(x^1 + 3x^2 - x^3) & -3x^1 + 6x^2 - 2x^3 & 6x^1 - 4x^2 + x^3 \\ 8x^2 - 3x^3 & 4x^1 - 7x^2 + 6x^3 & -6x^1 + 4x^2 - 6x^3 \end{bmatrix}$$

$$(a^i_{.k}) = \begin{bmatrix} 2x^1 & -x^2 & -x^1 + 2x^3 \\ -x^1 - 2x^2 & 2x^2 - 2x^3 & 2x^1 - x^3 \\ 3x^2 & -2x^2 + 3x^3 & -2x^1 \end{bmatrix}$$

$$(a_k^{.i}) = \begin{bmatrix} 2x^1 & -x^2 & 2x^2 - x^3 \\ -x^1 - 2x^3 & 2x^2 - 2x^1 & -x^2 + 2x^3 \\ 3x^3 & 3x^1 - 2x^2 & -2x^3 \end{bmatrix}$$

3. $$(a_{ij}) = \begin{bmatrix} 0 & -2x^1 + 3x^2 - 2x^3 & -x^1 + 2x^2 - 2x^3 \\ 2x^1 - 3x^2 + 2x^3 & 0 & 2x^1 - 2x^2 + x^3 \\ x^1 - 2x^2 + 2x^3 & -2x^1 + 2x^2 - x^3 & 0 \end{bmatrix}$$

$$(a_i^{.k}) = -(a^k_{.i}) = \begin{bmatrix} -x^2 + 2x^3 & -2x^1 + 2x^2 & x^1 - 2x^3 \\ -2x^2 + 2x^3 & -2x^1 + x^2 & 2x^1 - x^3 \\ -2x^2 + 3x^3 & -3x^1 + 2x^2 & 2x^1 - 2x^3 \end{bmatrix}$$

The index i denotes rows.

4. (a) $\sqrt{30}$, (b) -17, (c) $16\mathbf{e}^1 - 4\mathbf{e}^2 + 14\mathbf{e}^3$

5. $\mathbf{A}\cdot\mathbf{B} = 23$, $\mathbf{A}\times\mathbf{B} = -10\mathbf{e}^1 - 8\mathbf{e}^2 + 11\mathbf{e}^3$

EXERCISES 6.4

1. $\mathbf{e}_1 = [\sin\phi\cos\theta, \sin\phi\sin\theta, \cos\phi]$
$\mathbf{e}_2 = [r\cos\phi\cos\theta, r\cos\phi\sin\theta, -r\sin\phi]$
$\mathbf{e}_3 = [-r\sin\phi\sin\theta, r\sin\phi\cos\theta, 0]$
$\mathbf{e}^1 = \mathbf{e}_1$, $\mathbf{e}^2 = \mathbf{e}_2/r^2$, $\mathbf{e}^3 = \mathbf{e}_3/(r\sin\phi)^2$

$$(g_{ij}) = \begin{bmatrix} 1 & 0 & 0 \\ 0 & r^2 & 0 \\ 0 & 0 & r^2\sin^2\phi \end{bmatrix}$$

$$(g^{ij}) = \begin{bmatrix} 1 & 0 & 0 \\ 0 & 1/r^2 & 0 \\ 0 & 0 & 1/(r^2\sin^2\phi) \end{bmatrix}$$

2. $\mathbf{e}_1 = [\mathbf{v}, \mathbf{w}, 0]$, $\mathbf{e}_2 = [-\mathbf{w}, \mathbf{v}, 0]$, $\mathbf{e}_1 = [0, 0, 1]$,
$\mathbf{e}^1 = \mathbf{e}_1/(u^2 + v^2)$, $\mathbf{e}^2 = \mathbf{e}_2/(u^2 + v^2)$, $\mathbf{e}_3 = \mathbf{e}^3$

$$(g_{ij}) = \begin{bmatrix} w^2 + v^2 & 0 & 0 \\ 0 & w^2 + v^2 & 0 \\ 0 & 0 & w^2 + v^2 \end{bmatrix}$$

$$(g^{ij}) = \begin{bmatrix} (w^2 + v^2)^{-1} & 0 & 0 \\ 0 & (w^2 + v^2)^{-1} & 0 \\ 0 & 0 & (w^2 + v^2)^{-1} \end{bmatrix}$$

$\mathbf{e}_1 = [w\cos\theta, w\sin\theta, v]$, $\mathbf{e}_2 = [v\cos\theta, v\sin\theta, -w]$
$\mathbf{e}_3 = [-vw\sin\theta, vw\cos\theta, 0]$,

$$(g_{ij}) = \begin{bmatrix} w^2 + v^2 & 0 & 0 \\ 0 & w^2 + v^2 & 0 \\ 0 & 0 & w^2 v^2 \end{bmatrix}$$

4. $\mathbf{e}_1 = [a \sinh v \cos w, a \cosh v \sin w, 0]$
   $\mathbf{e}_2 = [-a \cosh v \sin w, a \sinh v \cos w, 0]$, $\mathbf{e}_3 = [1, 0, 0]$
   $\mathbf{e}^1 = \mathbf{e}_1/[a^2(\sinh^2 v + \sin^2 w)]$, $\mathbf{e}^2 = \mathbf{e}_2/[a^2(\sinh^2 v + \sin^2 w)]$
   $\mathbf{e}^3 = \mathbf{e}_3$

$$(g_{ij}) = \begin{bmatrix} a^2(\sinh^2 v + \sin^2 w) & 0 & 0 \\ 0 & a^2(\sinh^2 v + \sin^2 w) & \\ 0 & 0 & 1 \end{bmatrix}$$

5. $\mathbf{e}_1 = [u^2, u^1, 0]$, $\mathbf{e}_2 = [u^1, u^2, 0]$, $\mathbf{e}_3 = [0, 0, 1]$
   $\mathbf{e}^1 = [u^2, -u^1, 0]/v$, $\mathbf{e}^2 = [-u^1, u^2, 0]/v$, $\mathbf{e}^3 = \mathbf{e}_3$, where
   $v = (u^2)^2 - (u^1)^2$
6. $r = (v^2 + w^2)/2$, $\theta = \arctan[2vw/(v^2 - w^2)]$, $z = z$
   $$\frac{\partial(r, \theta, z)}{\partial(v, w, z)} = 2$$
7. $r = vw$, $\theta = \theta$, $z = (v^2 - w^2)/2$; $\dfrac{\partial(r, \theta, z)}{\partial(v, w, \theta)} = v^2 + w^2$
8. $R^2 = r^2 + z^2$, $\theta = \theta$, $\phi = \arctan(r/z)$; $r = R \sin\phi$, $\theta = \theta$, $z = R\cos\phi$

   (b) $$\left(\frac{\partial u^i}{\partial u^{*j}}\right) = \begin{bmatrix} \sin\phi & R\cos\phi & 0 \\ 0 & 0 & 1 \\ \cos\phi & -R\cos\phi & 0 \end{bmatrix}$$

   $$\left(\frac{\partial u^{*i}}{\partial u^j}\right) = \begin{bmatrix} r/R & 0 & z/R \\ z/R^2 & 0 & -r/R^2 \\ 0 & 1 & 0 \end{bmatrix}$$

EXERCISES 6.5

3. (a) $a^*_1 = (v^2 - w^2)v^2 w/2 + vw^2 z$, $a^*_2 = -(v^2 - w^2)w^2 v/2 + v^2 wz$,
   $a^*_3 = (v^2 - w^2)z/2$; $a^{*1} = a^*_1/(v^2 + w^2)$, $a^{*2} = a^*_2/(v^2 + w^2)$,
   $a^{*3} = a^*_3$

(b) $a^*_1 = (x_1^2x_2 + x_2^2x_3 + x_1x_3^2)/r$

$a^*_2 = (x_1x_2x_3 \cos\theta + x_2x_3^2 \sin\theta - x_1x_3\, r \sin\phi$

$a^*_3 = -x_1x_2^2 + x_1x_2x_3$, where $x_1 = r\sin\phi\cos\theta$

$x_2 = r\sin\phi\sin\theta,\ x_3 = r\cos\phi$

$a^{*1} = a^*_1,\ a^{*2} = a^*_1/r^2,\ a^{*3} = a^*_3{}^2/(r^2 \sin^2\phi)$

(c) $a^*_1 = ax_1x_2 \sinh v \cos w + ax_3x_2 \cosh v \sin w$

$a^*_2 = -ax_1x_2 \cosh v \sin w + ax_3x_2 \sinh v \cos w$

$a^*_3 = -x_1x_2$, where $x_1 = a\cosh v \cos w,\ x_2 = a\sinh v \sin w$,

$x_3 = z$

4. $a^*_1 = x_1x_2\, w \cos\theta + (2x_2 - x_3^2)w \sin\theta + x_1^2 v$

$a^*_2 = x_1x_2\, v \cos\theta + (2x_2 - x_3^2)v \sin\theta - x_1^2 w$

$a^*_3 = -x_1x_2\, vw \sin\theta + (2x_2 - x_3^2)vw \cos\theta$, where $x_1 = vw\cos\theta$

$x_2 = vw\sin\theta,\ x_3 = (v^2 - w^2)/2$

$a^{*1} = a^*_1/(v^2 + w^2),\ a^{*2} = a^*_2/(v^2 + w^2),\ a^{*3} = a^*_3/(v^2w^2)$

5. (a) $$(a^*_{ij}) = \begin{bmatrix} x_1\cos^2\theta + x_2\sin^2\theta & r(x_2 - x_1)\sin\theta\cos\theta & 0 \\ r(x_2 - x_1)\sin\theta\cos\theta & r^2(x_1\cos^2\theta + x_2\sin^2\theta) & 0 \\ 0 & 0 & x_3 \end{bmatrix}$$

$$(a^{*ij}) = \begin{bmatrix} a^*_{11} & a^*_{12}/r^2 & 0 \\ a^*_{21}/r^2 & a^*_{22}/r^4 & 0 \\ 0 & 0 & x_3 \end{bmatrix}$$

$$(a^{*i}{}_{.j}) = (a^*{}_j{}^{.i}) = \begin{bmatrix} a^*_{11} & a^*_{12} & 0 \\ a^*_{21}/r^2 & a^*_{22}/r^2 & 0 \\ 0 & 0 & x_3 \end{bmatrix}$$

(b) $a^*_{11} = (x_1^3 + x_2^3 + x_3^3)/r^2$

$a^*_{12} = a^*_{21} = x_1^2 x_3 \cos\theta/r + x_2^2 x_3 \sin\theta/r - x_2 x_3^2/(r\sin\theta)$

$a^*_{13} = a^*_{31} = x_1 x_2 (x_2 - x_1)/r$

$a^*_{22} = x_1x_3^2 \cos^2\theta + x_2 x_3^2 \sin^2\theta + x_2^2 x_3/\sin\theta$

$$a^*_{23} = a^*_{32} = x_1\, x_2\, x_3(\sin\theta - \text{cso}\ \theta),\ a^*_{33} = x_1x_2^2 - x_2x_1^2$$

$$(a^{*ij}) = \begin{bmatrix} a^*_{11} & a^*_{12}/r^2 & a^*_{13}/(r^2\sin^2\theta) \\ a^*_{21}/r^2 & a^*_{22}/r^4 & a^*_{23}/(r^4\sin^2\theta) \\ a^*_{31}/(r^2\sin^2\theta) & a^*_{32}/(r^4\sin^2\theta) & a^*_{33}/(r^4\sin^4\theta) \end{bmatrix}$$

$$(a^{*i}_{.j}) = \begin{bmatrix} a^*_{11} & a^*_{12} & a^*_{13} \\ a^*_{21}/r^2 & a^*_{22}/r^2 & a^*_{23}/r^2 \\ a^*_{31}/(r^2\sin^2\theta) & a^*_{32}/(r^2\sin^2\theta) & a^*_{33}/(r^4\sin^4\theta) \end{bmatrix}$$

$$= a^{*.i}_{j}$$

(c) $a^*_{11} = w^2(x_1\cos^2\theta + x_2\sin^2\theta) + x_3v^2$
$a^*_{12} = a^*_{21} = vw(x_1\cos^2\theta + x_2\sin^2\theta - x_3)$
$a^*_{13} = a^*_{31} = vw^2\,(x_2 - x_1)\sin\theta\cos\theta$
$a^*_{22} = v^2(x_1\cos^2\theta + x_2\sin^2\theta) + x_3w^2$
$a^*_{23} = a^*_{32} = wv^2\,(x_2 - x_1)\sin\theta\cos\theta$
$a^*_{33} = v^2w^2(x_1\sin^2\theta + x_2\cos^2\theta)$, where $x_1 = vw\cos\theta$,
$x_2 = vw\sin\theta$, $x_3 = (v^2 - w^2)/2$
$a^{*11} = a^*_{11}/(v^2 + w^2)^2$, $a^{*21} = a^*_{21}/(v^2 + w^2)^2$
$a^{*12} = a^*_{12}/(v^2 + w^2)^2$, $a^{*22} = a^*_{22}/(v^2 + w^2)^2$
$a^{*31} = a^*_{13}/[v^2w^2(v^2 + w^2)]$, $a^{*32} = a^*_{23}/[v^2w^2(v^2 + w^2)]$
$a^{*33} = a^*_{33}/(v^4w^4)$

6. (a) $\nabla\phi = v\,(v^2 + w^2)\mathbf{e}^1 + w\,(v^2 + w^2)\mathbf{e}^2 +$, where $\mathbf{e}^1$, $\mathbf{e}^2$ are given in answer to Prob. 3,, Exer. 6.4.
(b) $\nabla\phi = 2r\,\mathbf{e}^1$, where $\mathbf{e}^1 = [\sin\phi\cos\theta,\ \sin\phi\ \sin\theta,\ \cos\phi]$
7. $\nabla\phi = [(3v^2w - w^3)/2 + (w + v)z]\mathbf{e}^1 + [(v^3 - 3vw^2v)/2 + (v - w)z]\mathbf{e}^2 + [(v^2 - w^2)/2 + wv]\mathbf{e}^3$, where $\mathbf{e}^1$, $\mathbf{e}^2$, $\mathbf{e}^3$ are given in answer to Prob. 2, Exer. 6.4.

EXERCISES 6.6

1. $\partial\mathbf{A}/\partial x_1 = [x_2, 0, x_3]$, $\partial\mathbf{A}/\partial x_2 = [x_1, x_3, 0]$, $\partial\mathbf{A}/\partial x_3 = [0, x_2, x_1]$

2. $$\left(\frac{\partial a_i}{\partial x_j}\right) = \begin{bmatrix} \cos x_2 & -x_1 \sin x_2 & 0 \\ 0 & \sin x_3 & x_2 \cos x_3 \\ x_3 e^{x_1} & 0 & e^{x_1} \end{bmatrix}$$

3. $$\begin{bmatrix} 0 & x^3 & x^2 \\ -x^2 \sin x^1 & \cos x^1 & 0 \\ 0 & 0 & -1 \end{bmatrix}$$

4. $$\begin{bmatrix} 0 & \cos x^2 & 0 \\ e^{x^2} & x^1 e^{x^2} & 0 \\ 0 & x^3 & x^2 \end{bmatrix}$$

6. $[1, 22] = -r$, $[2, 12] = [2, 21] = r$, $[1, 33] = - r \sin^2 \theta$
$[2. 33] = - r^2 \sin \phi \cos \phi$, $[3, 13] = [3, 31] = r \sin^2 \phi$
$[3, 23] = [3, 32] = r^2 \sin \phi \cos \phi$
$\{^1{}_{22}\} = - r$, $\{^2{}_{12}\} = \{^2{}_{21}\} = \{^3{}_{13}\} = \{^3{}_{31}\} = 1/r$
$\{^1{}_{33}\} = - r \sin^2 \phi$, $\{^3{}_{23}\} = \{^3{}_{32}\} = \cos \phi / \sin \phi$
$\{^2{}_{33}\} = - \cos \phi \sin \phi$; otherwise, zero.

7. $[1, 11] = [2, 12] = [2, 21] = -[1, 22] = v$
$[1, 12] = [1, 21] = -[2, 11] = [2, 22] = w$
$\{^1{}_{11}\} = \{^2{}_{12}\} = \{^2{}_{21}\} = - \{^1{}_{22}\} = v/(v^2 + w^2)$
$\{^1{}_{12}\} = \{^1{}_{21}\} = - \{^2{}_{11}\} = w/(v^2 + w^2) = \{^2{}_{22}\}$; otherwise, zero.

8. $[1, 11] = [2, 12] = [2,21] = - [1, 22] = v$
$[1, 12] = [1, 21] = [2,22] = - [2, 11] = w$
$[1, 33] = - vw^2$, $[2, 33] = - v^2 w$
$[3, 13] = [3, 31] = vw^2$, $[3, 23] = [3,32] = v^2 w$
$\{^1{}_{11}\} = \{^2{}_{12}\} = \{^2{}_{21}\} = - \{^1{}_{22}\} = v/(v^2 + w^2)$
$\{^1{}_{12}\} = \{^1{}_{21}\} = \{^2{}_{22}\} = - \{^2{}_{11}\} = w/(v^2 + w^2)$
$\{^3{}_{13}\} = \{^3{}_{31}\} = - \{^1{}_{33}\} = 1/v$
$\{^3{}_{23}\} = \{^3{}_{32}\} = - \{^2{}_{33}\} = 1/w$; otherwise, zero.

9. $[1, 11] = [1, 33] = [2, 21] = [2, 12] = u$
$[1, 12] = [1, 21] = [3, 23] = [3, 32] = v$
$[2, 23] = [2, 32] = [3, 11] = [3, 33] = w$

$\{^1{}_{11}\} = \{^1{}_{33}\} = \{^2{}_{21}\} = \{^2{}_{12}\} = u/(u^2 + w^2)$
$\{^1{}_{12}\} = \{^1{}_{21}\} = w^2/[v(u^2 + w^2)]$
$\{^2{}_{11}\} = \{^2{}_{33}\} = -v/(u^2 + w^2)$
$\{^1{}_{23}\} = \{^3{}_{12}\} = -uw/[v(u^2 + w^2)]$
$\{^2{}_{32}\} = \{^2{}_{23}\} = \{^3{}_{11}\} = \{^3{}_{33}\} = w/(u^2 + w^2)$
$\{^3{}_{23}\} = \{^3{}_{32}\} = u^2/[v(u^2 + w^2)]$; otherwise, zero.

10. $a_{1,1} = \sin^2\phi\cos^2\theta,\ a_{1,2} = r\sin\phi\cos\phi\cos^2\theta = a_{2,1}$
$a_{1,3} = -2r\sin^2\phi\cos\theta\sin\theta - \sin\phi\cos\phi\cos\theta$
$a_{2,1} = r\sin\phi\cos\phi\cos^2\theta,\ a_{2,2} = r^2\cos^2\phi\cos^2\theta$
$a_{2,3} = -(1/2)r^2\sin 2\phi\sin 2\theta - r\cos^2\phi\cos\theta,\ a_{3,1} = 0$
$a_{3,2} = r\sin^2\phi\cos\theta,\ a_{3,3} = -(1/2)r\sin 2\phi\sin\theta + r^2\sin^2\phi\cos^2\theta$
$a^1{}_{,1} = a_{1,1},\ a^1{}_{,2} = a_{1,2},\ a^1{}_{,3} = a_{1,3},\ a^2{}_{,1} = r^{-2}a_{2,1}$
$a^2{}_{,2} = r^{-2}a_{2,2},\ a^2{}_{,3} = r^{-2}a_{2,3},\ a^3{}_{,1} = a_{3,1}/(r^2\sin^2\phi)$
$a^3{}_{,2} = a_{3,2}/(r^2\sin^2\phi),\ a^3{}_{,3} = a_{3,3}/(r^2\sin^2\phi)$

11. $a_{1,1} = w^2,\quad a_{1,2} = vw^2,\quad a_{1,3} = 0$
$a_{2,1} = vw,\quad a_{2,2} = v^2,\quad a_{2,3} = 0$
$a_{3,1} = 0,\quad a_{3,2} = 0,\quad a_{3,3} = 0$
$a^j{}_{,i} = a_{j,i}/(v^2 + w^2)$ for $i = 1, 2, 3; j = 1, 2$
$a^j{}_{,i} = a_{j,i} = 0$, otherwise.

12. $a_{i,j} = 0,\ a^i{}_{,j} = 0$ for all $i, j = 1, 2, 3$

EXERCISES 6.7

11. (a) $[r^2\sin^2\phi\sin 2\theta,\ r^3\sin 2\phi\cos\theta,\ r^2\sin\phi\cos 2\phi]$
(b) $4r\sin^2\phi\sin 2\theta + 2r^2(2\cos^2\phi - \sin^2\phi)\cos\theta$

(c) $$\frac{1}{r^2\sin\phi}\begin{bmatrix} \mathbf{e}_1 & \mathbf{e}_2 & \mathbf{e}_1 \\ \partial/\partial r & \partial/\partial\phi & \partial/\partial\theta \\ r^2\sin^2\phi\sin 2\theta & r^4\sin^2 2\phi\cos\theta & r^3\sin^2\phi\ \mathrm{coss}\ 2\phi \end{bmatrix}$$

12. (a) $[r^2\sin 2\phi\sin\theta,\ r\sin^2\phi,\ \cos\phi/\sin\phi]$
(b) $4r\sin 2\phi\sin\theta + 3\sin\phi\cos\phi$

(c) $$\frac{1}{r^2 \sin\phi}\begin{bmatrix} \mathbf{e}_1 & \mathbf{e}_2 & \mathbf{e}_1 \\ \partial/\partial r & \partial/\partial\phi & \partial/\partial\theta \\ r^2 \sin 2\phi \sin\theta & r^2 \sin^2\phi & r \cos\phi \end{bmatrix}$$

EXERCISES 6.8

3. (a) $\frac{1}{2} m(v^2 + w^2)(\dot{v}^2 + \dot{w}^2) + \frac{1}{2} m v^2 w^2 \dot{\theta}^2$

   (b) $\frac{1}{2} m(v^2 + w^2)(\dot{v}^2 + \dot{w}^2) + \frac{1}{2} m \dot{z}^2$

5. $\ddot{x}_1 = 0,\ \ddot{x}_2 = 0,\ \ddot{x}_3 = -g$

7. $\ddot{r} - r\dot{\theta}^2 = 0,\ r\ddot{\theta} + 2\dot{r}\dot{\theta} = 0$

INDEX

Printed in the United States
by Baker & Taylor Publisher Services